CORE AND METRO NETWORKS

WILEY SERIES IN COMMUNICATIONS NETWORKING & DISTRIBUTED SYSTEMS

The 'Wiley Series in Communications Networking & Distributed Systems' is a series of expert-level, technically detailed books covering cutting-edge research, and brand new developments as well as tutorial-style treatments in networking, middleware and software technologies for communications and distributed systems. The books will provide timely and reliable information about the state-of-the-art to researchers, advanced students and development engineers in the Telecommunications and the Computing sectors.

Other titles in the series:

Wright: *Voice over Packet Networks* 0-471-49516-6 (February 2001)
Jepsen: *Java for Telecommunications* 0-471-49826-2 (July 2001)
Sutton: *Secure Communications* 0-471-49904-8 (December 2001)
Stajano: *Security for Ubiquitous Computing* 0-470-84493-0 (February 2002)
Martin-Flatin: *Web-Based Management of IP Networks and Systems* 0-471-48702-3 (September 2002)
Berman, Fox, Hey: *Grid Computing. Making the Global Infrastructure a Reality* 0-470-85319-0 (March 2003)
Turner, Magill, Marples: *Service Provision. Technologies for Next Generation Communications* 0-470-85066-3 (April 2004)
Welzl: *Network Congestion Control: Managing Internet Traffic* 0-470-02528-X (July 2005)
Raz, Juhola, Serrat-Fernandez, Galis: *Fast and Efficient Context-Aware Services* 0-470-01668-X (April 2006)
Heckmann: *The Competitive Internet Service Provider* 0-470-01293-5 (April 2006)
Dressler: *Self-Organization in Sensor and Actor Networks* 0-470-02820-3 (November 2007)
Berndt: *Towards 4G Technologies: Services with Initiative* 0-470-01031-2 (March 2008)
Jacquenet, Bourdon, Boucadair: *Service Automation and Dynamic Provisioning Techniques in IP/MPLS Environments* 0-470-01829-1 (March 2008)
Minei/Lucek: *MPLS-Enabled Applications: Emerging Developments and New Technologies, Second Edition* 0-470-98644-1 (April 2008)
Gurtov: *Host Identity Protocol (HIP): Towards the Secure Mobile Internet* 0-470-99790-7 (June 2008)
Boucadair: *Inter-Asterisk Exchange (IAX): Deployment Scenarios in SIP-enabled Networks* 0-470-77072-4 (January 2009)
Fitzek: *Mobile Peer to Peer (P2P): A Tutorial Guide* 0-470-69992-2 (June 2009)
Shelby: *6LoWPAN: The Wireless Embedded Internet* 0-470-74799-4 (November 2009)

CORE AND METRO NETWORKS

Editor

Alexandros Stavdas
University of Peloponnese, Greece

A John Wiley and Sons, Ltd, Publication

This edition first published 2010

Registered office
John Wiley & Sons Ltd, The Atrium, Southern Gate, Chichester, West Sussex, PO19 8SQ, United Kingdom

For details of our global editorial offices, for customer services and for information about how to apply for permission to reuse the copyright material in this book please see our website at www.wiley.com.

Library of Congress Cataloging-in-Publication Data

Core and metro networks / edited by Alexandros Stavdas.
p. cm.
Includes bibliographical references and index.
ISBN 978-0-470-51274-6 (cloth)
1. Metropolitan area networks (Computer networks) I. Stavdas, Alexandros A.
TK5105.85.C678 2010
004.67–dc22

2009044665

A catalogue record for this book is available from the British Library.

ISBN 9780470512746 (H/B)

Set in 10/12 pt Times Roman by Thomson Digital, Noida, India
Printed and Bound in Singapore by Markono Print Media Pte Ltd

Contents

Preface

It is commonly accepted today that optical fiber communications have revolutionized telecommunications. Indeed, dramatic changes have been induced in the way we interact with our relatives, friends, and colleagues: we retrieve information, we entertain and educate ourselves, we buy and sell, we organize our activities, and so on, in a long list of activities. Optical fiber systems initially allowed for a significant curb in the cost of transmission and later on they sparked the process of a major rethinking regarding some, generation-old, telecommunication concepts like the (OSI)-layer definition, the lack of cross-layer dependency, the oversegmentation and overfragmentation of telecommunications networks, and so on.

Traditionally, telecommunications are classified based on the physical properties of the channel; that is, fixed-line/wired-communications and wireless/radio communications. Following this classification, it can be safely argued that today's core networks and metropolitan area networks (metro networks for simplicity) are almost entirely based on optical fiber systems. Moreover, the penetration of optical fiber communications in the access segment is progressing at an astonishing rate, although, quite often, it is the competition between providers, the quest for higher profits based on the established technological framework, and the legislative gridlocks that prevent an even faster adoption of this technology. Thus, a full-scale deployment of optical fiber systems in the access networks, through fixed/wireless convergence, could further reduce the role of wireless technology in transporting bandwidth over a reasonably long distance. Evidently, optical-fiber-based networks are the dominant technology, literally the backbone, of the future Internet. The fields of this technology are diverse and its engineering requires knowledge that extends from layer 1 to layer 3.

Many excellent basic text and specialized books are available today aiming to educate and/or inform scientists, engineers and technicians on the essentials in the field of optical technology. However, there is a pressing need for books presenting both comprehensive guidelines for designing fiber-optic systems and core/metro network architectures and, simultaneously, illustrating the advances in the state of the art in the respective fields. IST-NOBEL (I and II) was a large-scale research project funded from the Framework Programme 6 of the European Commission, incorporating major operators, system vendors and leading European universities. Employing a large number of experts in several fields, the project decided to collectively produce such a book as part of the disseminating activities. Thus, a considerable part of this book is based on the deliverables of IST-NOBEL with significant effort made to provide the necessary introduction of concepts and notions. The objective was to make it readable for a non-highly specialized audience, as well as to demystify the necessity behind the introduction of this or that novelty by clearly stating the underlying "need." It is left to the readers to decide whether we have succeeded in our goals.

The contributors to this book would like to acknowledge the immense help and support of their colleagues in the IST-NOBEL project that contributed to the preparation of the respective deliverables. A separate, special, acknowledgment is for the IST-NOBEL I and II project leaders and colleagues from Telecom Italia, Antonio Manzalini, Marco Schiano, and Giuseppe Ferraris. Also, the editor is extremely grateful to Andreas Drakos and Penny Papageorgopoulou, PhD candidates in the University of Peloponnese, for their help in preparing the final manuscript.

Alexandros Stavdas
Department of Telecommunications Science and Technology
University of Peloponnese, Greece

1

The Emerging Core and Metropolitan Networks

Andrea Di Giglio, Angel Ferreiro and Marco Schiano

1.1 Introduction

1.1.1 Chapter's Scope and Objectives

The study of transport networks is a vast and highly multidisciplinary field in the modern telecommunication world. The beginner who starts studying this technical subject may remain astonished by the variety and complexity of network architectures and technologies that have proliferated in the last decade. Even an expert in the field may get disoriented in the huge variety of networks' functions and characteristics.

This introductory chapter is devoted to the definition of transport networks' fundamentals representing the very basic "toolbox" of any expert in the field. Furthermore, it investigates transport network architectural evolution in terms of new network services supporting emerging users' applications.

The chapter is structured as follows. Section 1.2 contains the definitions of the basic network concepts used throughout the book. Sections 1.3 and 1.4 describe the requirements and the architectural evolution roadmap of transport networks based on emerging users' applications. Finally, Section 1.5 shows the economic models and analysis techniques that enable the design and realization of economically sustainable transport services.

1.2 General Characteristics of Transport Networks

For more than a century, the traditional vision of telecommunication networks has been a smart combination of transmission and switching technologies. Even if transmission and switching are still the basic building blocks of any network, telecommunication networks fundamentals cover a much broader scope nowadays. This new vision is primarily due to the introduction of digital technologies paving the way to packet-based networks. In contrast to old analog networks, packet-based digital networks can be either connectionless or connection oriented, can have a control plane for the automation of some functions, can implement various resilience schemes, can perform a number of network services supporting users' applications, and so on.

The essential ideas are explained in this section as a background for the entire chapter.

1.2.1 Circuit- and Packet-Based Network Paradigms

Digital networks can transfer information between nodes by means of two fundamental paradigms: circuit switching or packet switching.

- In circuit-switched networks, data are organized in continuous, uninterrupted bit streams. In this mode of operation, a dedicated physical link between a couple of nodes is established. Before starting the data transfer on a specific connection, the connection itself must be "provisioned"; that is, the network switching nodes must be configured to provide the required physical link. This implies an exclusive allocation of network resources for the whole duration of the connection. Such a task is usually performed by dedicated elements belonging to the network control system; network resources are released when the connection ends.

  This is the way that the plain old telephony service (POTS) has been working so far. The private reservation of network resources prevents other connections from using them while the first one is working, and may lead to inefficient network use.
- In packet-switched networks, data are organized in packets of finite length that are processed one by one in network nodes and forwarded based on the packet header information. In this network scenario, each packet exploits switching and transmission devices just for the time of its duration, and these network resources are shared by all packets. This process of packet forwarding and aggregation is called statistical multiplexing and represents the major benefit of packet-switched networks with respect to the circuit-switched networks in terms of network exploitation efficiency.

Typical examples of circuit-switching and packet-switching technologies are synchronous digital hierarchy (SDH) and Ethernet respectively.

Packet-switched networks can, in turn, work in connectionless or connection-oriented network modes.

- In the connectionless network mode, packets are forwarded hop by hop from source node to destination node according to packet header information only, and no transfer negotiation is performed in advance between the network nodes involved in the connection; that is, the source node, optionally the intermediate node(s) and the destination node.
- In the connection-oriented network mode, packet transfer from source node to destination node is performed through defined resource negotiation and reservation schemes between the network nodes; that is, it is preceded by a connection set-up phase and a connection usage phase, followed by a connection tear-down phase.

Typical examples of packet-switched connectionless and connection-oriented network protocols are Internet protocol (IP) and asynchronous transfer mode (ATM) respectively.

The main characteristic of the connectionless network mode is that packets are routed throughout the network solely on the base of the forwarding algorithms working in each node; hence, packet routes may vary due to the network status. For instance, cable faults or traffic overloads are possible causes of traffic reroute: in the connectionless network mode, the new route of a packet connection is not planned in advance and, in general, is unpredictable.

On the contrary, in the connection-oriented network mode, the route of any connection is planned in advance and, in the case of faults, traffic is rerouted on a new path that can be determined in advance.

Since route and rerouting have strong impacts on the quality of a packet connection, the two network modes are used for different network services depending on the required quality and the related cost.

1.2.2 Network Layering

The functions of a telecommunication network have become increasingly complex. They include information transfer, traffic integrity and survivability aspects, and network management and performance monitoring, just to mention the main ones. To keep this growing complexity under control and to maintain a clear vision of the network structure, *layered network models* have been developed. According to these models, network functions are subdivided into a hierarchical structure of layers. Each layer encompasses a set of homogeneous network functions duly organized for providing defined services to the upper layer, while using the services provided by the lower layer. For example, in an Ethernet network, the physical layer provides data transmission services to the data link layer.

To define transport network architectures, it is essential to start from the description of the lowest three layers [1]: network, data link, and physical layers:

- **Network layer.** The main task of the network layer is to provide routing functions. It also provides fragmentation and reassembly of data at the endpoints. The most common layer 3 technology is the IP. It manages the connectionless transfer of data across a router-based network.
- **Data-link layer.** This provides frames, synchronization, and flow control. The data link layer also performs transfer of data coming from the network layer. Typical examples of data-link layers are point-to-point protocol and Ethernet MAC (medium/media access control) (IEEE 802.1xx).
- **Physical layer.** The physical layer defines the transmission media used to connect devices operating at the upper layer (e.g., data link). Physical media can be, for example, copper-wire pairs, coaxial cables or, more frequently, single-mode or multimode optical fibers. The physical layer also defines modulation encoding (e.g., Manchester, 8B/10B) or topology (e.g., ring, mesh) [2]. Most common technologies implementing layer 1 functionalities are Ethernet (physical layer, IEEE 802.3xx), SDH and optical transport network (OTN).

It is commonly agreed that the Open System Interconnection (OSI) model is an excellent place to begin the study of network architecture. Nevertheless, the network technologies commercially available do not map exactly with the levels described in the OSI basic model.

1.2.3 Data Plane, Control Plane, Management Plane

The layered network models encompass all network functions related to data transfer. However, modern transport networks are often provided with additional functions devoted to network management and automatic network control. Hence, the totality of network functions can be classified into three groups named planes: the data plane, the management plane and the control plane.

The functions that characterize each plane are summarized below.

- **Data plane.** The data plane aims at framing and carrying out the physical transportation of data blocks to the final destination. This operation includes all transmission and switching functions.
- **Control plane.** The control plane performs the basic functions of signaling, routing and resource discovery. These are essential operations to introduce automation on high level network functions such as: connection establishment (i.e., path computation, resource availability verification and connection signaling set-up and tear-down), reconfiguration of signaled connections and connection restoration in case of network faults.
- **Management plane.** The management plane performs management functions like alarm reporting, systems configuration and connection provisioning for data and control planes. The complexity of the management plane depends strongly on the availability of a control plane. For example, the management plane of traditional circuit-switched public switched telephone networks is more cumbersome than transport networks with a control plane, since, in the latter case, certain tasks (e.g., connection provisioning and restoration) are carried out by the control plane itself.

1.2.4 Users' Applications and Network Services

The current challenge of evolving telephony-dedicated transport networks towards enhanced communication architectures is set by two fundamental trends.

First, services offered today to final users are much richer than simple telephony. User services like video telephony, video on demand, and Web browsing require an advanced terminal, typically a personal computer with dedicated software; for this reason, they will be called "user applications" or simply "applications" from now on.

Second, to convey these end-user applications, transport networks are relying on "network services," which effectively refer to a number of transfer modes.

As an example, a point-to-point unprotected circuit connection at 2 Mbit/s represents a specific transfer mode. Other examples of network services are connections based on packet paradigms; for example, IP/multi-protocol label switching (MPLS), ATM or Ethernet. Today, all modern applications make reference to packet-based network services.

The idea of a transport network able to provide many different services is one of the most challenging of recent years and it will be analyzed in detail in the following chapters.

Network services and user applications can be provided by different actors. Network operators that own and manage the networks are typical providers of network services. Service providers sell and support user applications by means of network services supplied by network operators.

Important *user application* categories are:

- multimedia triple play – voice, video and high-speed Internet;
- data storage for disaster recovery and business continuity;
- grid computing; that is, computing services delivered from distributed computer networks.

The last two categories, storage and grid computing, are dedicated to business company customers and research institutions. On the contrary, multimedia applications address residential customers and the small office, home office.

Examples of *network services* are:

- time-division multiplexing (TDM) connections and wavelength connections (e.g., leased lines);
- Ethernet point-to-point, point-to-multipoint (p2mp) or rooted multipoint connections;
- virtual private networks (Section 1.2.8).

Each user application is enabled by a network service characterized by specific attributes. A list of the most important ones is shown below.

- **Protocols:** Ethernet and IP are the most common.
- **Bandwidth:** committed peak, committed average bit-rate, excess peak and excess bit-rate [3].
- **Quality of service (QoS):** regarding transport networks, this is defined by means of the maximum allowed packet loss rate (PLR), the packet latency (i.e., the packet transmission delay), and jitter (latency variation); see Section 1.2.6.
- **Resilience:** required connection availability (Section 1.2.5).

These service attributes are the main inputs for a network provider to design a multi-service network, in support of a number of defined applications.

1.2.5 Resilience

One of the most important features of transport networks is their ability to preserve live traffic even when faults occur. This feature is generally referred to as "resilience."

In transport networks, resilience is usually achieved by duplication of network resources. For example, a fiber-optic link between a couple of nodes can be duplicated to assure survivability to cable breaks. Similarly, the switching matrix of an exchange node can be duplicated to guarantee service continuity in the case of electronics faults.

The way these extra resources are used depends strongly on network topology (rings or meshed network configurations), equipment technology (packet or circuit switching, network mode, optical transmission), and traffic protection requirements. However, the following general definitions help understanding the fundamental resilience schemes.

1. If the connections for traffic protection are organized in advance, the resilience mechanism is called "protection."
   a. **1 + 1 protection (also called dedicated protection).** The whole traffic of a connection is duplicated and transmitted through two disjoint paths: the working and the protection path simultaneously. The receiving node switches between the two signals in the case of failure. The trigger of 1 + 1 protection is the received signal quality; for example, the received power level or the bit error rate (BER). Since no complex network protocols are needed, 1 + 1 protection works very quickly, typically within 50 ms. The drawback of this protection scheme is duplication of network resources.
   b. **1: 1 protection (also called protection with extra traffic).** The working connection is protected with one backup connection using a disjoint path. The working traffic is sent over only one of the connections at a time; this is in contrast to dedicated protection, where traffic is always bridged onto two connections simultaneously. Under normal conditions (no network failure) the protecting connection is either idle or is carrying some extra traffic (typically best-effort traffic). Configuring 1: 1 protection depends on the control plane's ability to handle extra traffic, that is, whether it supports the preemption of network resources for allocating them to the working traffic once it has been affected by the failure. The ingress node then feeds the working traffic on the protecting connection in the case of failure. The trigger of 1: 1 protection is the reception of network failure notification messages. Protection with extra traffic has two main drawbacks: the need to duplicate working traffic resources onto the protection path and, in the case of resource contention, the possibility that extra traffic may be interrupted without effective need.
   c. ***M*****: *****N*** **protection (also called shared protection).** *M* working connections are protected by *N* backup connections on a disjoint path ($N \leq M$). The traffic is no longer duplicated because backup connections can carry traffic initially transported by any one of the working connections in the case of fault. Thus, switching to backup connections requires first knowing their availability and then performing traffic switching. Signaling is needed for failure notification and backup connection activation. Once failure has been repaired, traffic is reassigned to the working connection and the resources of the backup connection are available again for protection. In any case, this protection mechanism allows resource savings with respect to 1 + 1 protection.

Both protection mechanisms, dedicated and shared, are used in rings and meshed network configurations. The main advantage of protection is its quick operation, since the backup path is predefined and network resources are pre-allocated.

2. Alternatively to protection, restoration is the resilience mechanism that sets up new backup connections after failure events by discovering, routing, and setting up new links "on the fly" among the network resources still available after the failure. This is achieved by the extension of signaling, routing, and discovery paradigms typical of IP networks. In fact, to restore a connection, switching nodes need to discover the network topology not affected by the failure, thus allowing one to compute a set of candidate routes, then to select a new route, and to set up the backup connections. Discovery, routing algorithms, and signaling functions embedded in commercial IP/MPLS routers can quite easily implement restoration. On the other hand, transport network equipment needs a dedicated control plane to perform such functions.

Table 1.1 Indicative figures for network availability

Availability (%)	*N*-Nines	Downtime time (minutes/year)
99	2-Nines	5000
99.9	3-Nines	500
99.99	4-Nines	50
99.999	5-Nines	5
99.9999	6-Nines	0.5

Usually, the resilience level of a network service (e.g., a leased line or an Ethernet connection, as defined in Section 1.2.4) is made precise through a number of parameters; the most important are:

- **Mean time to failure (MTTF):** the reciprocal of the failure rate, for systems being replaced after a failure.
- **Mean time to repair (MTTR):** this depends on the repair time of a network fault.
- **Mean time between failures (MTBF):** this is the sum of MTTF and MTTR and defines the mean time interval between successive failures of a repairable system; it is a measure of network component reliability.
- **Maximum recovery time:** this is the maximum delay between a failure injuring a network service and the restoration of the service over another path; in other words, the maximum time during which the network service is not available. It accounts for MTTR and all other possible delays affecting complete system recovery (signaling, rerouting).

  The same concept can be given a different flavor, insisting on network status instead of duration:
- **Unavailability:** the probability that the network service is not working at a given time and under specified conditions; it is the ratio MTTR/MTBF. Some indicative numbers for network availability are illustrated in Table 1.1.

1.2.6 Quality of Service

Network services are characterized by a set of parameters that define their quality (QoS).

- **BER:** this is a physical-layer parameter, manifesting the fraction of erroneous bits over the total number of transmitted bits. It is closely related to design rules applied to the physical layer transport network. It is studied in detail in Chapter 3.
- **PLR:** in packet-switched services, this is the fraction of data packets lost out of the total number of transmitted packets. Packets can be dropped due to congestion, or due to transmission errors or faults.
- **Latency:** the time needed for carrying data from the source node to the destination node. Latency is caused by the combination of signal propagation delay, data processing delays, and queuing delays at the intermediate nodes on the connection [3].
- **Latency variation:** the range of variation of the latency mainly due to variable queuing delays in network nodes or due to data segmentation and routing of data blocks, via different physical paths (a feature readily available in next-generation (NG)-synchronous optical network (SONET)/SDH). Also, queuing delay variations may occur in the case of traffic

overload in nodes or links. An excess of latency variation can cause quality degradation in some real-time or interactive applications such as voice over IP (VoIP) and video over IP (IP television (IPTV)).
- **Service unavailability:** this has already been defined in Section 1.2.5.

For connection-oriented network services, the definition of QoS also includes:

- **Blocking probability:** the ratio between blocking events (failure of a network to establish a connection requested by the user, because of lack of resources) and the number of attempts.
- **Set-up time:** delay between the user application request time and the network service actual delivery time.

Current packet-based networks are designed to satisfy the appropriate level of QoS for different network services. Table 1.2 shows suitable values of QoS parameters for the main users' applications. As an example, applications like voice or videoconference need tight values of latency and latency variation. Video distribution is more tolerant to latency variation, but it needs low packet loss, since lost packets are not retransmitted. File transfer (e.g., backup) does not have strong requirements about any QoS parameters, since the only requirement is to transfer a pre-established amount of data in a fixed time interval.

1.2.7 Traffic Engineering

In complex meshed networks, careful traffic engineering (TE) and resource optimization is a mandatory requirement providing network management and operation functions at reasonable capital expenditure (CAPEX) and operational expenditure (OPEX). Towards this end, the use of conventional algorithms to set up the working and protection (backup) paths and for traffic routing within the network is insufficient. To address this problem, use is made of TE, which is a network engineering mechanism allowing for network performance optimization by means of leveraging traffic allocation in conjunction with the available network resources.

The purpose of TE is to optimize the use of network resources and facilitate reliable network operations. The latter aspect is pursued with mechanisms enhancing network integrity and by embracing policies supporting network survivability. The overall operation leads to the minimizations of network vulnerability, service outages due to errors, and congestions and failures occurring during daily network operations. TE makes it possible to transport traffic via reliable network resources, minimizing the risk of losing any fraction of this traffic.

TE leverages on some instruments that are independent of the network layer and technology:

- A set of policies, objectives, and requirements (which may be context dependent) for network performance evaluation and performance optimization.
- A collection of mechanisms and tools for measuring, characterizing, modeling, and efficiently handling the traffic. These tools allow the allocation and control of network resources where these are needed and/or the allocation of traffic chunks to the appropriate resources.
- A set of administrative control parameters, necessary to manage the connections for reactive reconfigurations.

Table 1.2 QoS characterization of users' applications

User application	QoS				
	Max. latency (ms)	Max. latency variation (ms)	Packet loss (layer 3) (%)	Max. set-up time	Min. availability (%)
Storage					
Backup/restore	N.A.	N.A.	0.1	min	99.990
Storage on demand	10	1	0.1	s	99.999
Asyncrhonous mirroring	100	10	0.1	s	99.999
Synchronous mirroring	3	1		min	99.999
Grid computing					
Compute grid	100	20	0.0	s	99.990
Data grid	500	100	0.1	s	99.990
Utility grid	200	50	0.0	s	99.999
Multimedia					
Video on demand (entertainment quality, similar to DVD)	2–20 s	50	0.5	s	99.500%
Video broadcast (IP-TV), entertainment quality similar to DVD	2–20 s	50	0.5	s	99.500
Video download	2–20 s	1000	1.0	s	99.990
Video chat (SIF quality, no real-time coding penalty)	400	10	5.0	s	99.500
Narrowband voice, data (VoIP, ...)	100–400	10	0.5	ms	99.999
Telemedicine (diagnostic)	40–250	5-40	0.5	ms	99.999
Gaming	50–75	10	5.0	s	99.500
Digital distribution, digital cinema	120	80	0.5	s	99.990
Video conference (PAL broadcast quality 2.0 real-time coding penalty)	100	10	0.5		99.990

Note: latency is expressed in milliseconds with the exception of video on demand, video broadcast, and video download, where seconds are the unit.

The process of TE can be divided into four phases that may be applied both in core and in metropolitan area networks, as described by the Internet Engineering Task Force (IETF) in RFC 2702 [4]:

- Definition of a relevant control policy that governs network operations (depending on many factors like business model, network cost structure, operating constraints, etc.).
- Monitoring mechanism, involving the acquisition of measurement data from the actual network.
- Evaluation and classification of network status and traffic load. The performance analysis may be either *proactive* (i.e., based on estimates and predictions for the traffic load, scenarios

for the scheduling of network resources in order to prevent network disruptions like congestion) or *reactive* (a set of measures to be taken to handle unforeseen circumstances; e.g., in-progress congestion).
- Performance optimization of the network. The performance optimization phase involves a decision process, which selects and implements a set of actions from a set of alternatives.

1.2.8 Virtual Private Networks

A virtual private network (VPN) is a logical representation of the connections that makes use of a physical telecommunication infrastructure shared with other VPNs or services, but maintaining privacy through the use of tunneling protocols (Section 1.2.9) and security procedures. The idea of the VPN is to give a user the same services accessible in a totally independent network, but at much lower cost, thanks to the use of a shared infrastructure, rather than a dedicated one [5].

In fact, a common VPN application is to segregate the traffic from different user communities over the public Internet, or to separate the traffic of different service providers sharing the same physical infrastructure of a unique network provider.

VPNs are a hot topic also in the discussion within standardization bodies: different views exist on what a VPN *truly* is.

According to ITU-T recommendation Y.1311 [6] a VPN "provides connectivity amongst a limited and specific subset of the total set of users served by the network provider. A VPN has the appearance of a network that is dedicated specifically to the users within the subset." The restricted group of network users that can exploit the VPN services is called a closed user group.

The other standardization approach, used by the IETF, is to define a VPN's components and related functions (RFC 4026, [7]):

- **Customer edge (CE) device:** this is the node that provides access to the VPN service, physically located at the customer's premises.
- **Provider edge (PE) device:** a device (or set of devices) at the edge of the provider network that makes available the provider's view of the customer site. PEs are usually aware of the VPNs, and do maintain a VPN state.
- **Provider (P) device:** a device inside the provider's core network; it does not directly interface to any customer endpoint, but it can be used to provide routing for many provider-operated tunnels belonging to different customers' VPNs.

Standardization bodies specified VPNs for different network layers. For example, a transport layer based on SDH can be used to provide a layer 1 VPN [8, 9]. Layer 2, (e.g., Ethernet) allows the possibility to implement L2-VPN, also called virtual LAN (VLAN). Layer 3 VPNs are very often based on IP, and this is the first and the most common VPN concept.

In some situations, adaptation functions between the bit-stream that is provided from the "source" (of the applications) and the VPN are required. An example of an adaptation data protocol function is the mapping of Ethernet frames in NG-SDH containers.

1.2.9 Packet Transport Technologies

Packet technologies have been dominating the local area network (LAN) scenario for more than 25 years, and nowadays they are widely used also in transport networks, where many network services are based on packet paradigms. The main reason for this success is twofold: first, the superior efficiency of packet networks in traffic grooming due to the statistical aggregation of packet-based traffic; second, the inherent flexibility of packet networks that can support an unlimited variety of users' applications with a few fundamental network services, as shown in Section 1.2.4.

However, until now, the transport of packet traffic has been based on the underlying circuit-switched technology already available for telephony. A typical example is represented by Ethernet transport over NG-SDH networks. This solution is justified by the widespread availability of SDH equipment in already-installed transport networks, and by the excellent operation, administration, and maintenance (OAM) features of such technology. These features are fundamental for provisioning packet network services with the quality required for most users' applications, but they are not supported by the LAN packet technologies.

This situation is changing rapidly, because a new generation of packet-based network technologies is emerging. These new scenarios combine the efficiency and flexibility of packet networks with the effective network control and management features of circuit-based networks. These new technologies are referred to as packet transport technologies packet transport technology (PTTs).

There are proposals for introducing tunnels[1] facilitating to allow Ethernet attaining traffic engineering features rendering it into a connection-oriented platform. These developments are currently under standardization at IEEE and ITU-T where is known as Provider Backbone Bridge with Traffic Engineering (or simply PBB-TE).

An alternative approach under standardization at the ITU-T and IETF is to evolve the IP/MPLS protocol suites to integrate OAM functions for carrier-grade packet transport networks.

This PTT, known as MPLS-TP (MPLS transport profile) includes features traditionally associated with transport networks, such as protection switching and operation and maintenance (OAM) functions, in order to provide a common operation, control and management paradigm with other transport technologies (e.g., SDH, optical transport hierarchy (OTH), wavelength-division multiplexing (WDM)).

The trend imposed by the dramatic increase of packet traffic and the obvious advantages in evolving existing circuit-switched networks into advanced packet-switched networks is going to make PTTs a viable solution to building a unified transport infrastructure, as depicted in Figure 1.1. Incumbent network operators that have already deployed a versatile NG-SDH network for aggregated traffic may follow conservative migration guidelines for their core networks and keep circuit solutions based on optical technologies. These plausible solutions are discussed in Section 1.4.

[1] A tunnel is a method of communication between a couple of network nodes via a channel passing through intermediate nodes with no changes in its information content.

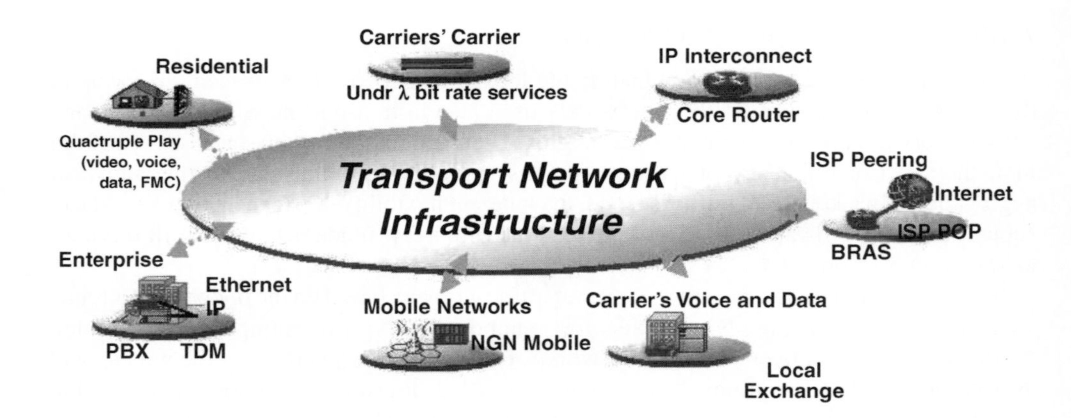

Figure 1.1 Unified transport network

1.3 Future Networks Challenges

1.3.1 Network Evolution Drivers

In the past decade, the proliferation of electronic and fiber-optic technologies has allowed network services to evolve from the exclusive support of plain telephony to an abundance of services which are transported based on the IP. These advances have had a major impact on the drivers for network evolution.

Nowadays, network design and planning is the outcome of the interplay between different technological, legal, and economic drivers:

- **Introduction of new services.** A network operator or a service provider can decide to offer new services based on customers' requests or market trends.
- **Traffic growth.** The growing penetration and the intensive use of new services increase the network load.
- **Availability of new technologies.** Electronic, optical, and software technologies keep on offering new advances in transmission, switching, and control of information flows based on circuits and packets.
- **Degree of standardization and interoperability of new network equipment.** Modern networks are very complex systems, requiring interaction of various kinds of equipment by means of dedicated protocols. Standardization and interoperability are key requirements for a proper integration of many different network elements.
- **Laws and regulations.** National laws and government regulations may set limitations and opportunities defining new business actors for network deployment and usage.
- **Market potential and amount of investments.** The financial resource availability and the potential of the telecommunication market are the key economic drivers for network development.

1.3.2 Characteristics of Applications and Related Traffic

In this section, the association between applications and network services is presented. The starting point of the analysis is the bandwidth requirement (traffic) of the various applications and the subsequent classification of this traffic into classes. Figure 1.2, illustrates a classification of user applications based on the following traffic characteristics:

- elasticity
- interactivity
- degree of resilience (availability)
- symmetry
- bandwidth.

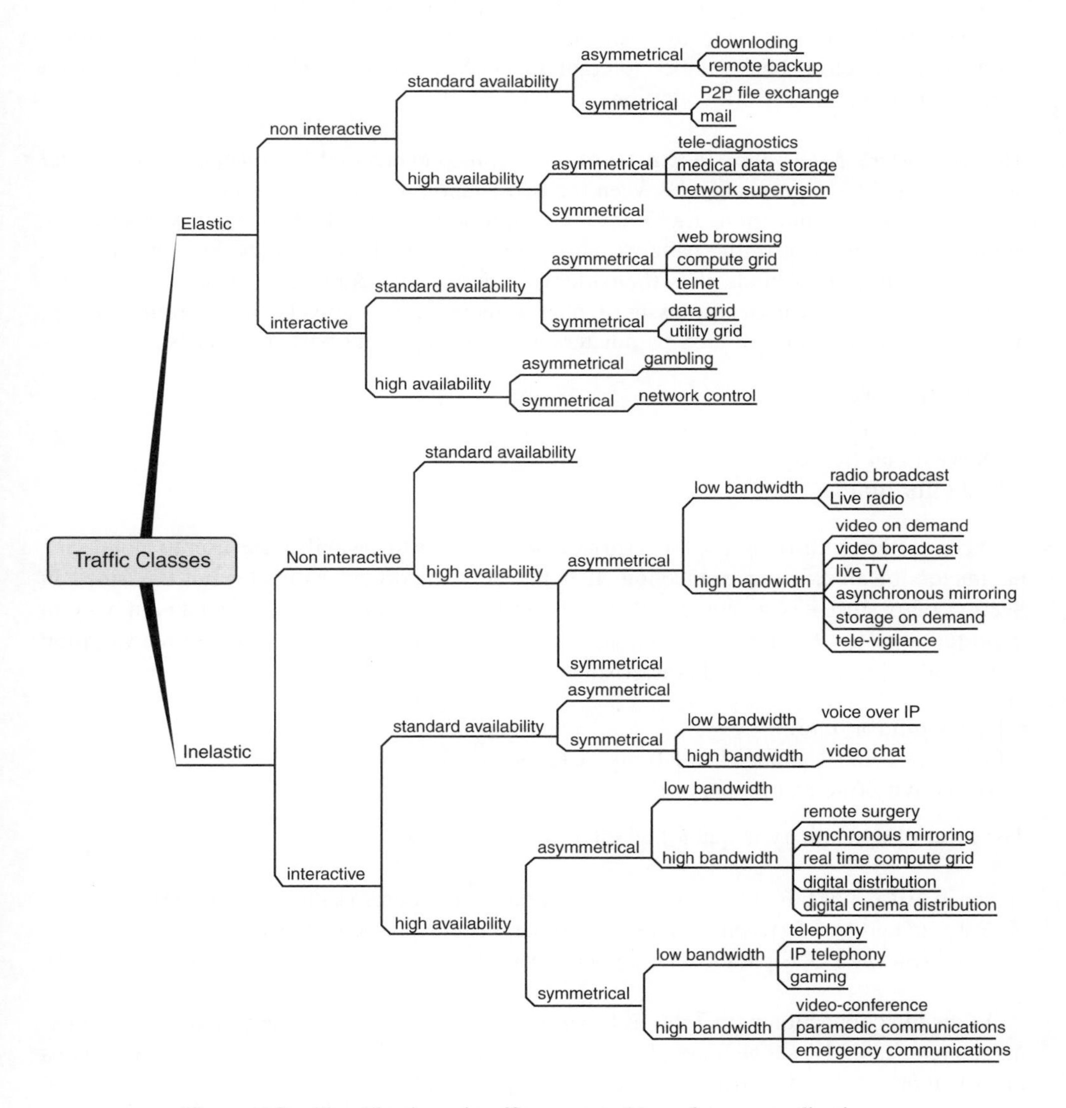

Figure 1.2 Classification of traffic generated by reference applications

Table 1.3 Qualitative classification of traffic types

	Elastic	Inelastic
Interactive	Transactional	Real time
Noninteractive	Best effort	Streaming

Elasticity refers to the level up to which the original traffic shape can be modified; the two main categories are as follows:

- **Inelastic traffic** (or stream traffic) is generated by applications whose temporal integrity overwhelms data integrity because they try to emulate virtual presence.
- **Elastic traffic** is generated by applications where data integrity overwhelms temporal integrity, therefore being rather tolerant to delays and being able to adapt their data generation rate to network conditions.

The term *interactivity* refers to a mode of operation characterized by constant feedback and an interrelated traffic exchange between the two endpoints of the connection.

To map users' applications traffic into the appropriate network services, it is essential to define a few classes of traffic patterns that share the main characteristics. For this purpose, Table 1.3 defines four kinds of traffic patterns in terms of QoS requirements.

Another important task is to assign QoS parameters quantitatively to the traffic classes. Table 1.4 sets the values of QoS parameters used to define four basic classes as:

- real-time traffic
- streaming traffic
- transactional traffic
- best-effort traffic.

In connection with Table 1.4, the term *dynamicity* refers to the ability of a user to modify the parameters of an existing connection. It is the only important parameter not described in Section 1.2.6, since it is not addressed directly by the classic QoS definition, but it is anyway an important quantity for application classification. The dynamicity refers to the time variation of the following connection characteristics:

- bandwidth (bit-rate);
- QoS parameters (latency, availability, data integrity);
- connectivity (the end-points of the connection).

The level of dynamicity is quantified on a three-state base:

- "none" (it is not possible to modify any parameters of an existing connection);
- "bit-rate and QoS" (when only these two parameters can be altered);
- "full" (bit-rate, QoS parameters, and connectivity modifications are allowed).

As seen in connection with Table 1.4, four traffic categories are defined based only on QoS parameters. Table 1.5 shows examples of applications belonging to each one of the four classes identified above, having different bandwidth requirements.

Table 1.4 Quantitative classification of QoS for traffic classes

	Blocking probability (%)	Network availability (%)	Set-up time (s)	Max. latency	Mean latency (ms)	Max. latency variation	Dynamicity	Packet loss rate
Real time	<0.1	>99.995	<1	<50 ms	*		Bit-rate, QoS and connectivity	<5 E-5
Streaming	<0.1	>99.99	<1	<1 s	*		None	<1 E-3
Transactional	<1	>99.9	<3	<1 s	<200		Bit-rate and QoS	<1 E-2
Best effort	*	*	*	*	*		Not applicable	*

Table 1.5 Traffic characterization based on bandwidth (BW) and QoS parameters and map of users' applications

QoS	Low BW (tens of kbit/s)	Medium BW (<2 Mbit/s)	High BW (>2 Mbit/s)
Real time	Legacy and IP telephony	Gaming	Video conference, grid computing
Streaming	UMTS	Remote backup, network supervision	TV and video broadcast, VoD[a]
Transactional	E-commerce	Telnet	SAN[b]
Best effort	E-mail, domotic, VoIP	p2p file exchange, data acquisition	p2p file exchange, data acquisition

[a]Video on demand.
[b]Storage area network.

Table 1.5 is useful to map most common users' applications into the four traffic classes (real-time, streaming, transactional, best-effort), taking also the bandwidth use into account.

Similar to the classification of user applications, network services are classified into five categories in association with the network services reported in Section 1.2.4. Thus, the network service map looks as follows:

- **L1 VPN**, which provides a physical-layer service between customer sites belonging to the same closed user group. These VPN connections can be based on physical ports, optical wavelengths, or TDM timeslots.
- **L2 VPN**, which provides a service between customer terminals belonging to the VPN at the data link layer. Data packet forwarding is based on the information contained in the packets' data link layer headers (e.g., frame relay data link circuit identifier, ATM virtual circuit identifier/virtual path identifier, or Ethernet MAC addresses).
- **L3 VPN**, which provides a network layer service between customer devices belonging to the same VPN. Packets are forwarded based on the information contained in the layer 3 headers (e.g., IPv4 or IPv6 destination address).
- **Public IP**, which is considered as the paradigm of best-effort network services. Namely, it is a generalized L3 VPN without restrictions to the user group, but with a consequently poor QoS.
- **Business IP**, which is included as a higher priority class that, for instance, can efficiently handle latency[2] in time-sensitive applications.

On top of this classification, further "orthogonal" categorizations are often introduced. VPN services are further subdivided into:

- **permanent VPNs**, to be provided on a permanent basis by the network service provider;
- **on-demand VPNs**, which could be controlled dynamically by the client user/network.

[2] See latency and other QoS defining parameters later in this section.

Table 1.6 Mapping network services groups to some applications (BW: bandwidth)

			real time	low BW	mid BW	high BW	streaming	low BW	mid BW	high BW	transactional	low BW	mid BW	high BW	best effort	low BW	mid BW	high BW
Public IP			■	░			■	≡	░		■	≡			■	≡	≡	
Business IP			■	░			■				■	≡			■		≡	≡
VPN - L3	permanent		■	░	░	░	■	░	░	░	■				■			
	on demand		■				■	≡	≡	≡	■				■			
VPN - L2	permanent	high availability	■	≡	≡	≡	■				■	░	≡	≡	■			
	permanent	low availability	■				■				■	░	≡	≡	■	≡	≡	░
	on demand	high availability	■	≡	≡	≡	■	≡	≡	≡	■	░	≡	≡	■			
	on demand	low availability	■				■				■				■	≡	≡	░
VPN - L1	permanent	high availability	■	░	≡	≡	■				■				■			
	permanent	low availability	■	░	≡	≡	■	░	≡	≡	■	░	░	░	■			
	on demand	high availability	■		≡	≡	■				■				■			
	on demand	low availability	■				■				■				■	░	░	░

L1 and L2 VPN services are also classified into high- and low-availability services. Table 1.6 provides a mapping between "user applications" and "network services": in this context, a stippled box means that that particular application may run over on this network service, but not very efficiently. The most efficient support to that application is designated with horizontal rows, whereas a white box should be interpreted as no support at all from this service to that application.

1.3.3 Network Architectural Requirements

This section gives an overview of the architectural requirements for transport networks supporting the services described above.

1.3.3.1 Network Functional Requirements

From an architectural point of view, data services have been traditionally transported over a wide set of protocols and technologies. For example, IP services are transported over the core network usually relying on SDH, ATM, or Ethernet transmission networks. A widespread alternative used in current convergent transport networks is to go towards a meshed network of IP/MPLS routers, interconnected through direct fiber or lambda connections and without any multilayer interaction.

This "IP-based for everything" approach was proved to be valid for the last decade, but with current traffic trends it would lead to scalability problems. Currently, backbone nodes need switching capacities of several terabits per second, and this need is predicted to double every 2 years. Routers are also very expensive and they are not optimized for high-bandwidth traffic transportation, while transport technologies such as SONET/SDH are not efficient enough for packet transport, due to a very coarse and not flexible bandwidth granularity.

On the other hand, a number of emerging services (e.g., new multimedia applications served over the Internet; i.e., real-time high-bandwidth video services) are imposing new requirements on the current "IP-based for everything" architecture in terms of bandwidth and QoS (end-to-end delay and availability). Moreover, mobility of users and devices and new traffic profiles (due to, for example, flash crowds and streaming services) require a network with an unprecedented dynamicity that is able to support unpredictable traffic patterns.

1.3.3.2 Network Scalability

The term *scalability* is a feature of a network architecture designating the ability to accommodate higher traffic load without requiring large-scale redesign and/or major deployment of resources. A typical (negative) example manifesting lack of scalability is an SDH ring where additional resources and manual configurations are mandatory in order to increase the capacity between two nodes. Thus, future transport networks should be scalable in order to support existing or yet-unknown clients and traffic volumes.

The lack of scalability is demonstrated in two distinctive ways. First, by means of an excessive deployment of network resources to accommodate higher traffic volumes. This inefficiency is leading to higher CAPEX and OPEX that are mainly attributed to the enduring very high cost of switching. Solving this issue requires the deployment of technologies able to transport traffic with a lower cost per bit. Second, it is associated with architectural and/or control plane scalability restrictions due to the excessive number of network elements to control (e.g., the number of paths in the network). To address this issue requires the adoption of layered architectures and aggregation hierarchies.

1.3.3.3 Network Reconfiguration Ability

Network reconfiguration ability refers to the ability of the network to change the status of some or all of the established connections, to modify the parameters of these connections (e.g., modify the amount of allocated bandwidth) or to modify the way the services are provided (for instance, changing the routing of a given connection to allow more efficient grooming on a different route or improve spare capacity sharing).

The interest in having a reconfigurable network comes from the fact that traffic profiles change very frequently, may be fostered by symmetrical traffic patterns, unexpected traffic growth, possible mobile data/multimedia services, varied geographic connectivity (e.g., home, work), and emerging services, such as user-generated content. All these facts make it reasonable to think in the future about a highly varying traffic profile in a network, thus meaning that reconfigurability would be a highly advantageous characteristic in data architectures.

1.3.3.4 Cost Effectiveness

Taking into account the fierce competition and the pressure upon network operators in the telecommunications market, as well as the descending cost per bit charged to the final user, the only solution for service providers to keep competitive is to reduce traffic transport costs. Therefore, cost effectiveness is the obvious requirement for any new technology. Basic approaches to achieve this cost reduction are to build networks upon cheap scale-economy technologies, adapted to the applications' bursty data traffic and specifically designed to keep functional complexity to a minimum. To facilitate this cost per bit reduction even in presence of unpredictable traffic growth, modular solutions are of paramount importance.

1.3.3.5 Standardized Solutions

Standardization of solutions is a key point, because it assures interoperability of equipment from different manufacturers and, as a consequence, it allows a multi-vendor environment.

This leads to the achievement of economies of scale that lower costs, since a higher number of suppliers use the same technology. Besides, standardization allows network operators to deploy networks with components from different suppliers, therefore avoiding dependence on a single manufacturer, both from a technological and an economical point of view.

1.3.3.6 Quality of Service Differentiation

As specified in Sections 1.2.6 and 1.3.2, a differentiating feature between the various applications consists in their dissimilar transport network requirements (e.g., minimum/maximum bandwidth, availability, security, delay, jitter, loss, error rate, priority, and buffering). For this reason, networks have to support QoS differentiation because their main goal is to assure a proper multi-service delivery to different applications. The intention of QoS specifications is to utilize network mechanisms for classifying and managing network traffic or bandwidth reservation, in order to deliver predictable service levels such that service requirements can be fulfilled.

1.3.3.7 Resilience Mechanisms

As reported in Section 1.2.5, an important aspect that characterizes services offered by telecommunication networks is service availability. Resilience mechanisms must be present in order to react to network failures, providing backup solutions to restore the connections affected by the failure. Typical resilience mechanisms provide full protection against all single failures; they distinguish in terms of how fast restoration is provided and on the amount of backup capacity required for protection, to fully support this single-failure event. Resilience schemes can also be characterized depending on their ability to provide various level of protection (e.g., full protection against single failures, best effort protection, no-protection, and preemption in the case of failure) and on their capability to provide very high availability services (e.g., full protection against multiple failures). For transport network clients, the important aspect is the resulting service availability, measured in terms of average service availability over a given period of time (e.g., 1 year) and of maximum service interruption time.

1.3.3.8 Operation and Maintenance

A fundamental requirement is to keep a proper control over the networking infrastructure: easy monitoring, alarm management, and configuration tools are required. The current trend for

OPEX reduction and maintenance simplification leads towards automated distributed control maintenance and operations.

Transport technologies or carrier-grade switching and transmission solutions differ from other technologies in the OAM features: it is important not only in administrating and managing the network, but also to provide services and to deal with its customers. Efficient operation tools and mechanisms must also be implemented within the transport networks.

Finally, it is important to consider the interoperability between different network layers that requires mutual layer independence; for this reason, the transport technology needs to be self-sufficient to provide its own OAM, independently of its client and server layers.

1.3.3.9 Traffic Multicast Support

A multicast transfer pattern allows transmission of data to multiple recipients in the network at the same time over one transmission stream to the switches.

A network with multicast capability must guarantee the communication between a single sender and multiple receivers on a network by delivering a single stream of information to multiple recipients, duplicating data only when the multiple path follows different routes. The network (*not* the customer devices) has to be able to duplicate data flows. There are only two degrees for the ability to support multicast transfer: able or unable (multicast is an on/off property).

Multicast distribution is considered a useful tool for transport technologies when dealing with IPTV and similar applications. However, it is pointed out that layer 2 multicasting is not the only solution to distribute IPTV.

1.3.3.10 Multiplicity of Client Signals

Previous sections highlighted that metro-core networks are supporting traffic from many different applications, such as business data, Web browsing, peer to peer, e-Business, storage networking, utility computing, and new applications such as video streaming, video conference, VoIP, and tele-medicine applications. The prevalence of multimedia services and the expansion of triple-play has an important role in traffic load and distribution in metro and core networks. A strong increase of broadband access penetration, based on a combination of different fixed and mobile access technologies, is expected for the next years, favoring the increase of terminal nomadism, which might introduce a more variable and unpredictable traffic, especially in the metro area. On the other side, corporate VPN services ranging from MPLS-based VPNs [10] to legacy services cope with the business telecom market.

From a technological standpoint, most services are migrating to packet-based Ethernet framing. This trend makes it mandatory for Core/Metro networks to support Ethernet client services. Nevertheless, many legacy networks are still based on other standards, such as SDH and ATM, and they still need to support these kinds of technology.

A transport infrastructure that can carry traffic generated by both mobile and fixed access is an important challenge for future transport networks.

Fixed and mobile applications present similar QoS requirements, and can be classified according to the four classes previously defined in Section 1.2.4. (i.e., best-effort, streaming, real-time, and transactional). However, current bandwidth requirements are lower for mobile applications than for fixed applications due to limitations in wireless access bandwidth and terminal screen size and resolution.

1.3.3.11 Transport Network Service Models and Client Interactions

Telecom networks have been upgraded with different network layer technologies, each providing its own set of service functionality based on its own switching paradigm and framing architecture. The GMPLS (Generalized Multi-Protocol Label Switching) protocol architecture paves the way for a convergence between transport and client networks reducing, thus, the overall control and management complexity. GMPLS can be configured to handle networks with dissimilar switching paradigms (on data plane) and different network management platforms (on control and management plane). This is made feasible by means of LSPs (Label Switched Paths) that are established between two end points. i.e. under the GMPLS protocol architecture the resources of the optical transport network are reserved based on the connectivity requests from a client packet-switched network.

The Overlay Model

The overlay model refers to a business model in which carriers or optical backbone (bandwidth) providers lease their network facilities to Internet service providers (ISPs). This model is based on a client–server relationship with well-defined network interfaces (or automatic switched optical network (ASON) reference points) between the transport network involved and client networks. The overlay model mandates a complete separation of the data client network control (that could be IP/MPLS based) and the transport network control plane (e.g., wavelength-switched optical networks/GMPLS). A controlled amount of signaling and restricted amount of routing messages may be exchanged; as a consequence, the overlay model is a very opaque paradigm. The IP/MPLS routing and signaling controllers are independent of the routing and signaling controllers within the transport domain, enabling the different networks to operate independently. The independent control planes interact through a user-to-network interface (UNI), defining a client–server relationship between the IP/MPLS data network and the wavelength-switched optical network (WSON)/GMPLS transport network.

Overlay network service models support different business and administrative classes (as developed in Section 1.5.3.) and preserve confidentiality between network operators. The connection services are requested from client networks to the transport network across distinct UNIs. When a connection is established in the transport network for a given client network, this connection can be used as a nested LSP or a stitched LSP to support the requirements of the client network.

The service interface in the overlay network model can be configured according to the level of trust of the two interacting structures. The interface can be based on a mediation entity such as an operation service support (OSS) or it can use the northbound interface of the network management system. Further, the interface between client network (higher layer network) and transport network (lower layer network) can operate a GMPLS signaling protocol, such reservation protocol with TE (RSVP-TE).

Peer Model

Compared with the overlay model, the peer model is built on a unified service representation, not restricting any control information exchanged between the transport network and the clients. This model is relevant and represents an optimal solution when a transport network operator is both an optical bandwidth provider and an ISP. In this case, the operator can optimally align the virtual topologies of its transport network with the network services required by its data network. The IP/MPLS control plane acts as a peer of the GMPLS transport network control plane, implying that a dual instance of the control plane is running over the data network (say, an IP/MPLS network) and optical network (say, a WSON/GMPLS network). The peer model entails the tightest coupling between IP/MPLS and WSON/GMPLS components. The different nodes are distinguished by their switching capabilities; for example, packet for IP routers interconnected to photonic cross-connects (PXCs).

Integrated Model

Compared with the peer model, the integrated model does not require different service interfaces between the different networks. The integrated model proposes the full convergence of data network control plane and transport network control plane. All nodes are label-switched routers (LSRs) all supporting several switching capabilities; say, wavelength, SDH and Ethernet. Each LSR is also able to handle several orders of the same switching capability as it happens; for example, with SDH. An LSR embeds one GMPLS control plane instance and is able to control simultaneously different switching capability interfaces. Only this model can handle a complete and global optimization of network resource usages through transport and client networks.

Augmented Model

The augmented model considers that the network separation offered by the overlay model provides a necessary division between the administrative domains of different network service providers, but also considers that a certain level of routing information should be exchanged between the transport network and the client networks. In a competitive environment, a complete peer network service model is not suitable because of the full exchange of topology information and network resource status between client and server optical networks, imposing on a network operator to control the resources of the client data networks and triggering the scalability issues of the management functions.

The augmented model provides an excellent support for the delivery of advanced connectivity services as might be offered from a multilayer network (MLN)/multiregion network (MRN). The capability, such as wavelength service on demand, integrated TE or optical VPN services, can require controlled sharing of routing information between client networks and the optical transport network.

User-to-network Interface

The UNI is a logical network interface (i.e., reference point) recommended in the "Requirements for Automatic Switched Transport Network" specification ITU-T G.807/Y.1302. The UNI defines the set of signaling messages that can be exchanged between a client node and a server node; for instance, an IP router and an SDH optical cross-connect (OXC) respectively. The server node provides a connection service to the client node; for example, the IP router can request TDM LSPs from its packet over SONET (PoS) interfaces. The UNI supports the exchange of authentication, authorization, and connection admission control messages, and provides the address space set of the reachable nodes to the client network. Different versions of the implementation agreement for a UNI have been produced by the Optical Internetworking Forum (OIF) since October 2001 as OIF UNI 1.0. The different OIF implementation agreement versions are supporting the overlay service model as well as the augmented service model. The signaling messages exchanged between the client node and the server node are focused on the LSP connection request, activation, deactivation, and tear down. The IETF specifies a GMPLS UNI that is also applicable for a peer model. Fully compliant with RSVP-TE, GMPLS UNI allows the end-to-end LSP handling from the ingress customer edge equipment to the egress customer edge equipment and at each intermediate LSR involved in the signaling sessions.

Network-to-network Interface

The network-to-network interface (NNI) is a logical network interface (i.e., reference point) recommended in the "Requirements for Automatic Switched Transport Network" specification ITU-T G.807/Y.1302. The NNI defines the set of both signaling messages and routing messages that can be exchanged between two network server nodes; for example, SONET OXC and SDH OXC. There are two types of NNI, one for intranetwork domains and one for internetwork domains: an external NNI (E-NNI) and an internal NNI (I-NNI) respectively.

- The E-NNI assumes an untrusted relationship between the two network domains. The routing information exchanged between the two nodes located at the edge of the transport

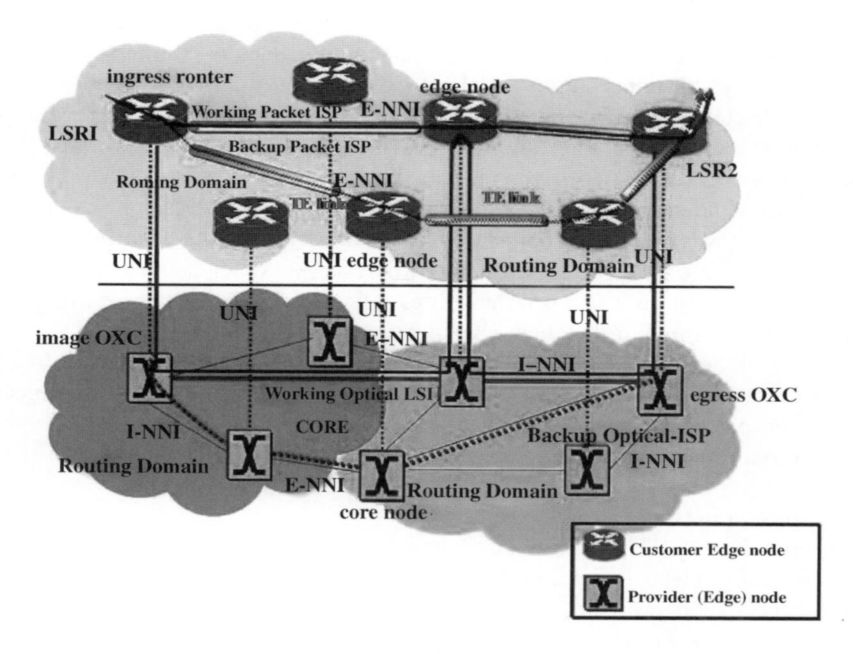

Figure 1.3 Customer nodes to public network link through server nodes by the UNI as defined in ITU architecture

network specified within the E-NNI is restricted. The control messages exchanged include reachable network addresses that are usually translated, authentication and connection admission control messages, and a restricted set of connection requests of signaling messages.

- The I-NNI assumes a trusted relationship between two network domains. The control information specified within the I-NNI is not restricted. The routing control messages exchanged include topology, TE link state, and address discovery. The signaling messages can allow controlling of the resources end to end between several network elements and for each LSP and its protection path.

1.3.4 Data Plane, Control Plane, and Management Plane Requirements

1.3.4.1 Data Plane Requirements

The challenges of the physical layer part of the data plane are covered in Chapters 3–6. In this section, two conceptual challenges of the data plane are addressed, namely the quest for transparency and the search for novel transport formats.

Transparency

During the last 20 years the cornerstone of transport network evolution has been the notion of "transparency." Today, there are two distinctive understandings of the term "transparency": bit-level transparency and optical transparency.

In the original idea, the two meanings were synonymous and they were based on the following simple concept. The advent of the erbium-doped fiber amplifier facilitated the proliferation of WDM (these topics are discussed in Chapters 3–6), which increased the product "bandwidth times length" to about two to three orders of magnitude, an event that eventually led to a significant curb of the transmission cost. As a result of this evolution, the cost of switching started dominating (and still does) the cost of a transport network. At the same time, there was a clear disparity between the data throughput that the fiber-optic systems could transmit and the amount of data that synchronous systems (SDH/SONET) could process, a phenomenon that was termed an "optoelectronic bottleneck." For these reasons, every effort was made to minimize electronic switching wherever possible, making use of optical-bypassing concepts for the transit traffic, namely avoiding transport schemes requiring frequent aggregation and grooming of the client signals through electronic switches.

The widespread deployment of applications based on the IP and the emergence of a "zoo" of other protocols (like Ethernet, ESCON, Fiber Channel, etc.) gave a renewed impetus to incentivize transparency in transport networks.

In more recent times, transparency went colored of different shades.

Bit-level Transparency

The transport network (OTN in particular) should convey client signals with no processing of the information content. This will minimize the aggregation/grooming used throughout the network, whilst it will provide a client/provider agnostic transportation (transparent bit-mapping into the transport frame). Here, transparency mainly indicates *service transparency*; that is, the minimization of bit-by-bit processing regardless of the technological platform that is used. This definition is shifting the interest from *technologies* to *functions*; hence, both all-optical and optoelectronic subsystems are of equal interest in building service-transparent networks.

Optical Transparency

Nevertheless, in conjunction with bit-level transparency, the initial notion of a transparent network is still of interest, where the optical/electrical/optical (O/E/O) conversions are minimized so the signal stays in the optical domain. The benefits from the reduction in the number O/E/O conversions include:

- reduction of a major cost mass by minimizing the number of transponders and large (and expensive) switching machinery;
- improved network reliability with a few in number of electronic systems;
- significant reduction in power consumption (from switching fabrics to cooling requirements).

In an optically transparent network, the routing of the signal is based on the wavelength and/or on the physical port of the signal. Framing takes place at the ingress to the optically transparent domain and it adds overhead information that makes it possible to detect errors – possibly occurring during transmission – at the egress node. Each standardized format has a specific frame, and several different frames are possible at the ingress to an optically transparent domain; for example, Ethernet frames, synchronous transport module frames, and G.709 [11] frames.

Ethernet as an Alternative Transport Platform

Ethernet is a frame-based technology that was defined in 1970s. It was originally designed for computer communications and for broadcasting, and since then it has been widely adopted. This was made possible thanks to two main competitive advantages. First, a successfully implemented switching capability besides the original broadcasting LAN technology. Second, because all generations of Ethernet share the same frame formats, making it feasible to support interfaces from 10 Mbit/s over copper to 100 Gbit/s over fiber (the latter still under standardization), thus ensuring seamless upgradeability.

Nowadays, Ethernet represents the most successful and widely installed LAN technology and is progressively becoming the preferred switching technology in metropolitan area networks (MANs). In the latter scenario, it is used as a pure layer 2 transport mechanism, for offering VPN services, or as broadband technology for delivering new services to residential and business users. Today, Ethernet traffic is rapidly growing and apparently it has surpassed SDH traffic. As mentioned in the previous section, bit transparent mapping is essential for a cost-effective transportation of data and it is a feature provided by Ethernet thanks to its framing format. As Ethernet is becoming the dominant technology for service provider networks and as 40/100 GbE interfaces will be standardized in the years to come, it is essential to keep Ethernet transport attractive and simple. Currently, under the Ethernet umbrella, three network layers are considered:

- **Network layer**, based on Metro Ethernet Forum (MEF) documents. The network services include E-Line, defining point-to-point connections, and E-LAN, defining multipoint-to-multipoint connections and rooted multipoint connections.

- **Layer 2**, which is also called the MAC layer. This provides network architectures, frame format, addressing mechanisms, and link security (based on IEEE 802.1 and IEEE 802.3).
- **Physical layer**, which includes the transmission medium (e.g., coaxial cable, optical fiber), the modulation format, and network basic topologies (based on IEEE 802.3).

The numerous networks where Ethernet is installed confirm that it is a valuable frame-based technology, capable of assuring an inexpensive physical stratum, providing high bit-rates, and allowing network architectures to offer emerging network services for distributing both point-to-point and p2mp variable bit-rate traffic efficiently.

There are several initiatives at standardization bodies that aim at a revision of Ethernet to make it valuable and deployable in transport networks. The work carried out at IEEE, IETF, and ITU-T is improving Ethernet with faster and more efficient resilience mechanisms and valuable OAM tools for fault localization and measurement of quality parameters to verify customers' service level agreements (SLAs).

Given this success in the access area and MAN and the simplicity and transparency it offers, Ethernet is stepping forward to the core network segment under the definition of carrier Ethernet which is investigated by the MEF as "a ubiquitous, standardized, carrier-class Service and Network." Carrier Ethernet improves standard Ethernet technology facing scalability issues (assuring a granular bandwidth increment from 1 Mbit/s to 10 Gbit/s). It also assures hard QoS mechanisms (allowing the transport on the same lambda of different traffic categories) and reliability (the network is able to detect and recover from failures with minimum impact on users). Carrier Ethernet aims to achieve the same level of quality, robustness, and OAM functions typical of circuit technologies (think of SDH or OTN) while retaining the Ethernet advantage in offering a cost-effective statistical aggregation.

1.3.4.2 Control Plane Requirements

The control plane is studied in detail in Chapter . Here, some important issues are highlighted.

Provisioning of End-to-end Connections over the Entire Network

The main function of the control plane is to set up, tear down, and maintain an end-to-end connection, on a hop-by-hop basis, between any two end-points. The applications supported from the transport network have specific QoS requirements (Section 1.2.6), which the control plane must uphold.

Unified Control Plane

In the quest to upgrade or build new integrated network infrastructures, a paradigm shift has been witnessed in network design principles. The focus has shifted from a layered-network model involving the management of network elements individually at each layer, to one of an integrated infrastructure able to provide a seamless management of packets, circuits, and light paths. The reasons for this industry trend towards a unified set of mechanisms (the unified control plane), enabling service providers to manage separate network elements in a uniform way, can be traced to the historical evolution of transport and packet networks. The IP became the uncontested platform for supporting all types of application and the associated, IP-based, GMPLS provides a single, unified control plane for multiple switching layers [12].

Horizontal Integration (Unified Inter-domain Control)
This issue refers to the way of multidomain interconnection at control plane level. Horizontal integration refers to the situation where, in the data plane, there is at least one common switching facility between the domains, whilst the control plane topology extends over several domains. For instance, the control plane interconnection between lambda-switching-capable areas defines a horizontal integration.

Control Plane and Management Plane Robustness
In the emerging optical network architectures, the interplay between the control plane and management plane is essential to ensure fast network reconfiguration, while maintaining the existing features of SDH/SONET like robustness against failures, which is essential for the preservation of traffic continuity.

Network Autodiscovery and Control Plane Resilience
Automated network discovery refers to the ability of the network to discover autonomously the entrance of new equipment or any changes to the status of existing equipment. This task is assigned to the control plane. Additional functions of the control plane are the automated assessment of link and network load and path computation process needed to substantially reduce the service provision time and the changes invoked in the network infrastructure to support these services. Moreover, the automation is essential to reallocate resources: as customers cancel, disconnect, or change orders, the network resources can be readily made available to other customers.

The term control plane resilience refers to the ability of the control plane to discover the existing cross-connect topology and port mapping after recovering from a failure of itself. For example, when only control plane failures occur within one network element, the optical cross-connects will still be in place, carrying data traffic. After recovery of the control plane, the network element should automatically assess the data plane (i.e., optical cross-connects), and reconfigure its control plane so that it can synchronize with other control plane entities.

Appropriate Network Visibility among Different Administrative Domains Belonging to Different Operators
Administrative domains may have multiple points of interconnections. All relevant interface functions, such as routing, information exchanges about reachable nodes, and interconnection topology discovery, must be recognized at the interfaces between those domains. According to ASON policy, the control plane should provide the reference points to establish appropriate visibility among different administrative domains.

Fast Provisioning
As part of the reliable optical network design, fast provisioning of optical network connections contributes to efficient service delivery and OPEX reduction, and helps reaching new customers with broadband services.

Automatic Provisioning
To achieve greater efficiencies, optical service providers must streamline their operations by reducing the number of people required to deliver these services, and reducing the time required to activate and to troubleshoot network problems. To accomplish these objectives,

providers are focusing on automated provisioning through a distributed control plane, which is designed to enable multi-vendor and multilayer provisioning in an automated way. Therefore, requests for services in the data network that may require connectivity or reconfiguration at the optical layer can happen in a more automated fashion. In addition, instead of provisioning on a site-by-site basis, the control plane creates a homogeneous network where provisioning is performed network-wide.

Towards Bandwidth On-demand Services

Providers can also set up services where the network dynamically and automatically increases/decreases bandwidth as traffic volumes/patterns change. If the demand for bandwidth increases unexpectedly, then additional bandwidth can be dynamically provisioned for that connection. This includes overflow bandwidth or bandwidth over the stated contract amount. Triggering parameters for the change may be utilization thresholds, time of day, day of month, per-application volumes, and so on.

Bandwidth on demand (BoD) provides connectivity between two access points in a non-preplanned, fast, and automatic way using signaling. This also means dynamic reconfiguring of the data-carrying capacity within the network; restoration is also considered here to be a bandwidth on-demand service.

Flexibility: Reconfigurable Transport/Optical Layer

A network operator may have many reasons for wanting to reconfigure the network, primarily motivated by who is paying for what. Flexibility of the transport layers means a fair allocation of bandwidth between competing routes dealing with bursts of activity over many timescales. Reconfigurability increases network flexibility and responsiveness to dynamic traffic demands/changes.

1.3.4.3 Interoperability and Interworking Requirements

Multidomain Interoperability

In many of today's complex networks, it is impossible to engineer end-to-end efficiencies in a multidomain environment, provision services quickly, or provide services based on real-time traffic patterns without the ability to manage the interactions between the IP-layer functionality of packet networks and that of the optical layer. According to proponents of ASON/GMPLS, an optical control plane is the most advanced and far-reaching means to control these interactions.

Another important issue is that of translating resilience classes from one domain to another. The ASON reference points UNI and I-NNI/E-NNI are abstracted functional interfaces that can resolve that topic by partitioning the transport network into sub-networks and defining accurately the exchanges of control information between these partitions. As recommended in Ref. [13], the UNI is positioned at the edge of the transport network as a signaling interface used by the customer edge nodes to request end-to-end connection services between client networks, with the explicit level of availability. Routing and signaling messages exchanged at the I-NNI concern only the establishment of connections within a network domain or across the subnetwork. The E-NNI is placed between network domains or sub-networks to carry the control message exchanges between these regions of different administration.

Multi-vendor Interoperability

The multi-vendor interoperability of metro and core solutions maximizes carrier performance and ensures the interoperability of legacy with emerging network architectures. One of the most important objectives of the development of a standardized ASON/GMPLS control plane is to contribute to interoperability, which validates the speed and ease of provisioning enabled by ASON/GMPLS in a live, multi-vendor network.

Seamless Boundary in between Networks

Given the vast amount of legacy SONET/SDH equipment, there is a clear need for an efficient interworking between traditional circuit-oriented networks and IP networks based on the packet-switching paradigm. For example, efficient control plane interworking between IP/MPLS and SONET/SDH GMPLS layers is indispensable and requires the specification of their coordination.

1.3.4.4 Management Plane Requirements

Easy-to-use Network

Emerging standards and technologies for optical networks allow for a significantly simplified architecture, easy and quick provision of services, more effective management, better interoperability and integration, and overall lower cost. In addition, it will be possible to provision services on these future networks such that global applications will be much more location independent.

Transparent for Applications: Hide Network Technology to Users

There are multiple separate service, technology, and technical considerations for networks depending on location, at the metro edge, metro core, aggregation points, long haul, and ultra-long haul. Next-generation optical networking has the potential to reduce significantly or eliminate all of these barriers, especially with regard to application and end users.

To some degree, one of the key goals in this development is to create network services with a high degree of transparency; that is, allow network technical elements to become "invisible" while providing precise levels of required resources to applications and services. To allow an optimal use of the optical network infrastructure interconnecting different types of application, network service management functions are required to establish automatically connection services with adequate amount of allocated network resources. The network service management layer can rely on the routing and signaling control functions.

Monitoring of End-to-end Quality of Service and Quality of Resilience

The requirement of integrated monitoring of the (optical) performance of connections, QoS, and fault management speeds up system installation and wavelength turn-up and simplifies ongoing maintenance. Furthermore, the management plane should be able to monitor end-to-end quality of resilience. That means the end-to-end type of transport plane resilience parameters (such as recovery time, unavailability, etc.) should be monitored and adhered according to the SLAs).

Connectivity and Network Performance Supervision

As networks run faster and become more complex, infrastructure, links, and devices must operate to precise levels in a tighter performance. As a result, a huge number of network problems stem from simple wiring and connection issues. Connectivity and performance supervision is at the heart of an efficient network management.

Network Monitoring

A monitoring system is dedicated to the supervision of the physical and optical layers of a network. Optical-layer monitoring should provide valuable, accurate information about the deterioration or drift with slow and small signal variations, helping to detect problems before they may become so serious to affect the QoS. It helps maintain the system from a lower layer's perspective.

Policy-based Management (Network and Local-basis)

Today's optical network architectures lack the proper control mechanisms that would interact with the management layer to provide fast reconfiguration. The problem of accurate intra-domain provisioning in an automated manner allows satisfying the contracts with customers while optimizing the use of network resources. It is required that a policy-based management system dynamically guides the behavior of such an automated provisioning through the control plane in order to be able to meet high-level business objectives. Therefore, the emerging policy-based management paradigm is the adequate means to achieve this requirement.

End-to-end Traffic Management (Connection Admission Control, Bandwidth Management, Policing)

Traffic management features are designed to minimize congestion while maximizing the efficiency of traffic. Applications have precise service requirements on throughput, maximum delay, variance of delays, loss probability and so on. The network has to guarantee the required QoS. For instance, the primary function of the connection admission control is to accept a new connection request only if its stated QoS can be maintained without influencing the QoS of already-accepted connections. Traffic management features are key elements in efficient networking.

Multi-vendor Interoperability

In the near future, network element management interfaces and OSS interfaces will be pre-integrated by control plane vendors. Indeed, independent control planes increase the performance of network elements and OSS, and reduce carriers' reliance on any single network element or OSS application. This eliminates the task of integrating new network elements into a mass of OSS applications.

Connection services (respectively, connectivity services) are described from the network infrastructure operator (respectively, the service customer) point of view, which is complementary for the connections implemented through the control functions at customer edge (CE) nodes. Provider VPN services offer secure and dedicated data communications over telecom networks, through the use of standard tunneling, encryption, and authentication functions. To reconfigure automatically the provisioning of VPNs, automated OSS functions are required to enhance existing network infrastructures for supporting networked applications sharing the optical infrastructures.

Network service functions can automatically trigger addition, deletion, move, and/or change of access among user sites. The description of each connection service includes the UNI corresponding to the reference point between the provider edge (PE) node and CE node. At a given UNI, more than one connection can be provisioned from the network management systems or automatically signaled from the control plane functions according to multiplexing capabilities. GMPLS controllers enable signaling of the connection establishment on demand, by communicating connectivity service end-points to the PE node. This operation can be assigned to an embedded controller to exchange the protocol messages in the form of RSVP-TE messages.

Support Fixed–Mobile Convergence

Fixed-mobile convergence means alliance of wired and wireless services and it is referring to single solutions for session control, security, QoS, charging and service provisioning for both fixed and mobile users. Fixed-mobile convergence is clearly on the roadmap of operators that want to create additional revenue streams from new value-added services.

1.4 New Transport Networks Architectures

Today's telecommunication networks have evolved substantially since the days of plain-telephony services. Nowadays, a wide variety of technologies are deployed, withstanding a substantial number of failures, supporting a broad range of applications based on diversified edge-user devices; they span an enormous gamut of bit-rates, and they are scaling to a large number of nodes.

In parallel, new services and networking modes (e.g., peer-to-peer) are emerging and proliferating very rapidly, modifying the temporal and spatial traffic profile in rather unpredictable ways. As has been discussed in the previous sections, it is widely recognized that the existing mind-set for the transport network architecture largely fails to accommodate the new requirements. However, the bottleneck is not only on the technology front. Architectural evolution presupposes a consensus between the many providers which, quite often, is hard to reach. This situation is exacerbating interoperability issues that, potentially, negate any competitive advantage stemming from architectural innovation. Market protectionism could stall technological advances.

Nevertheless, a major rethinking on network architectures is mandatory in the quest for a cost-effective, secure, and reliable telecommunications network. The research today is pivoted around notions on how network dynamicity can be significantly enhanced, how the cost of ownership can be reduced, and how the industrial cost of network services can be decreased. The scope of this section is to present plausible scenarios for the evolution of the core and the metropolitan transport networks, taking into account the data plane as well as the control/management planes. It is organized so as to provide snapshots of the current situation in both segments and for three discrete time plans:

- short term (2010)
- medium term (2012)
- long term (2020).

Figure 1.4 depicts the existing network architecture for metro/regional and core/backbone segments, which will be the starting point in the network evolution scenario. Today, the functionality requirements are dissimilar in the two network segments, leading to the adoption of different solutions, as was shown in Figure 1.3:

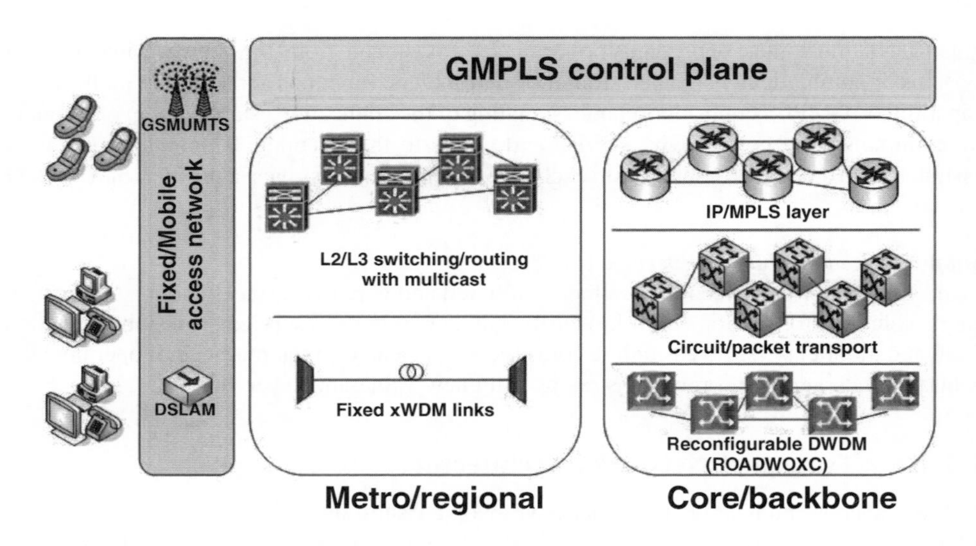

Figure 1.4 Existing metro and core network architecture

- In an MAN, the client traffic is transported from a "zoo" of protocols (IP, Ethernet, ATM, SDH/SONET, ESCON, Fiber channel, etc., to mention the most important instances only) whilst it is characterized from a low level of aggregation and grooming; this a problem exacerbated by the coexistence of unicast (video on demand, high-speed Internet and voice services) and multicast traffic (i.e., mainly IPTV). This environment postulates a highly dynamic networking, making packet-oriented solutions a necessity.
- In the core network, on the other hand, the efficient aggregation and grooming out of the MAN indicates a smoothed-out, slowly varying traffic profile, so that a circuit-switched solution is a good candidate for a cheaper switching per bit. These developments, for the core, are beefed up by the past and current developments of dense WDM (DWDM) and OTN technologies, to enhance the "bit-rate times distance" product significantly, by two to three orders of magnitude, compared to what was feasible in late 1980s, by shifting to transmission the balance for a lower cost per bit transportation.

Regarding the scenarios presented in the rest of this section, it is pointed out that their common denominator is progress in the following enablers:

- **Packet technologies** (in particular IP/MPLS and Ethernet) for a more efficient use of bandwidth due to the subsequent statistical multiplexing gains; that is, advanced aggregation and grooming.
- **Control plane** (currently dominated by ASON and GMPLS, which are further discussed in Chapter) to decrease the cost of provisioning dramatically and the possibility to have on-the-fly resilience mechanisms.
- **Optical transparency**, which, as explained in Section 1.3.4.1, aims at minimizing the level of bit-by-bit processing, simplifying client signal encapsulation, leading to transparent

bit-mapping in the transport frame and providing optical bypassing for the transit traffic. These are key functions for a reduction in CAPEX and OPEX.

As it emerges from simple inspection of the existing network architecture paradigm, efficiency and robustness today are achieved from the interplay between two, rather mutually exclusive, technologies: packets (mainly IP/MPLS and Ethernet) and circuits (SDH/SONET, OTN and WDM) do coexist in transport networks with a low level of interoperability and significant functionality duplication. Apparently, for an overall optimization, it is fundamental to increase the synergy between the layers and reduce the unnecessary functionality duplication. Thus, the emerging technologies (PTTs, see Section 1.2.9), which are in the standardization process, aim at combining the best features from both circuit and packet worlds. Therefore, features like OAM and control plane and resilience mechanisms are inherited from the circuit transport network, while frame format, statistical aggregation, and QoS support are similar to the corresponding features of packet technologies. Within the standardization bodies, two main technologies are currently under discussion: PBB-TE (IEEE802.1Qay [14] based on Ethernet) and MPLS-TP (developed in the ITU-T and IETF, starting from MPLS).

1.4.1 Metropolitan Area Network

The introduction of triple-play applications (voice, video, and high-speed Internet) has a strong impact on Metropolitan Area Network (MAN) traffic. The advances include improvements in residential access networks (whose traffic aggregates upwards to the MAN), multimedia distribution (which is using MANs in an intensive way) from points of presence point of presence (PoPs) to the home, and finally VPNs that are used for business and residential applications. The necessity to provide multicast services (i.e., to carry IPTV) and to add/release users to multicast groups very quickly is a strong driver towards packet solutions (IP, Ethernet, ATM, ESCON, Fiber channel, etc.).

In the MAN segment, the main advantages of circuits (low cost per switched bit per second, strong OAM, and efficient resilience mechanisms) are not essential: the bandwidth at stake is not really huge and the distances of the cables interconnecting the nodes are not very long, so that the probability of failure due to fiber cut is not that high to mandate a circuit-switched level of resilience.

However, using IP over Ethernet or pure Ethernet over WDM systems (architectural examples are available in Ref. [15]) presents some problems in terms of resilience, bandwidth guarantee, and OAM, because packet technologies currently do not have efficient mechanisms to cope with such functions. For these reasons, technologies with the ambition to couple circuit-like OAM and resilience mechanisms with packet-like flexibility, dynamicity, and granularity might represent the right candidates for next-generation networks.

Both the emerging PTTs (PBB-TE and MPLS-TP) currently have a lack of multicast traffic that, at the date of writing (July 2008), is still under study.

Figure 1.5 shows the possible evolution of the architecture for networks in the MAN or regional segment. The following sections describe in depth the concepts illustrated in this picture. At the moment, the most plausible scenario is a migrations towards technologies that, from one side, assure packet granularity and, from the other side, have "circuit-like" OAM

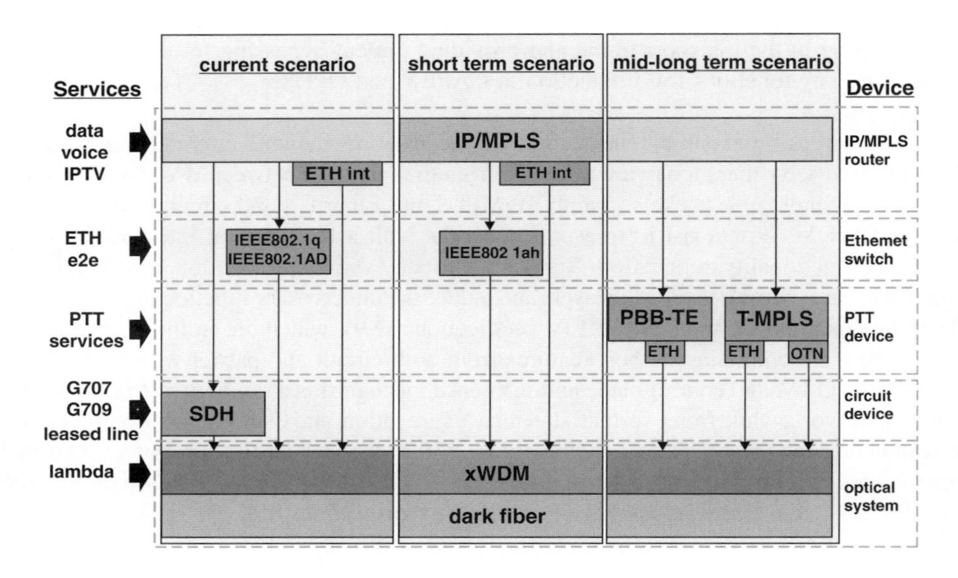

Figure 1.5 Evolution scenario for metropolitan/regional network architecture

and resilience mechanisms (capable of switching in times shorter than 50 ms after failure, a performance requested by the majority of applications).

1.4.1.1 Short Term

In the short term, there will probably be a progressive migration of fixed and mobile services to IP. This migration will speed up the increasing interest towards the Ethernet technology [16]. So, the roll out of native Ethernet platforms in the metro space is likely to start in the next time frame. Metro network solutions in the short term are expected to be mainly based on the Ethernet technology in star or ring topologies.

Nevertheless, in this phase, both packet (IP and Ethernet) and circuit (for the most part SDH) will coexist. In most cases, different kinds of traffic (with different quality requirements) will be carried on the appropriate platform (e.g., voice or valuable traffic) on SDH and the remainder on packet platforms.

No unified control plane is available for the transport layers. Further, the control plane is restricted to a single network domain and in most cases to a single layer within the network.

1.4.1.2 Medium Term

In the metro network, Ethernet will probably be the dominant technology in the medium-term scenario. The utilization of Ethernet is mainly driven by Ethernet conformal clients; however, non-Ethernet payloads, such as TDM, ATM, and IP/MPLS, will still exist for a long time and have to be adapted into an Ethernet MAC frame.

Therefore, any incoming non-Ethernet payload behaves as an Ethernet payload from a network perspective; the reverse operation is performed at the outgoing interface of the egress network node.

Since Ethernet networks are growing in dimension, moving from simple LAN or basic switched networks (e.g., in a campus behavior) towards a situation where an entire metropolitan (or regional) area is interconnected by an Ethernet platform, hundreds of thousands (or even millions) of MAC addresses would have to be learned by the switches belonging to the metro networks. To prevent this severe scalability problem, IEEE 802.1ah (PBB or MACinMAC) might be adopted. This evolution of the classical Ethernet allows layering of the Ethernet network into customer and provider domains with complete isolation among their MAC addresses.

Leaving the SDH technology, the main problems that still remain are related to the lack of efficient resilience mechanisms, present in SDH, but not mature with IP or Ethernet. In fact, traditional Ethernet (802.1q, 802.1ad, and 802.1ah) bases resilience on the "spanning tree" mechanism and its evolutions (for instance, VLAN spanning tree), which are inefficient for carrying traffic that has strong requirements in terms of unavailability.

The same argument might be argued if resilience is demanded at the IP level. In this case, routing protocols (e.g., open short path first, intermediate system to intermediate system) after a failure rearrange routing tables on surviving resources; also, this process assures stability after some seconds, a time that is often too long for voice or some video applications.

Innovative solutions to this problem might be represented by resilient packet ring (RPR) or optical circuit switching (OCS) rings. OCS rings are specially adapted to metro–core scenarios, as well as to metro access characterized by high-capacity flows between nodes (e.g., business applications and video distribution services), while RPR and dual bus optical ring network solutions fit better in scenarios with higher granularity and lower capacity requirements per access node.

At control plane level, the most important aspects that are expected for the medium-term scenario are the implementation of interfaces to make possible the exchange of information (routing and signaling) between control planes even between different domains and finally the vertical integration of the control planes of layer 1 and layer 2 technologies.

1.4.1.3 Long Term

The metro segment is composed of metro PoPs (GMPLS-capable LSR), some of which link the metropolitan network to the IP/optics core backbone (core PoP).

In this phase, the solutions based on Ethernet technology (that is, on 802.1ah (MACinMAC) and the IP/MPLS routing) will probably be replaced by innovative PTTs.

These technologies (MPLS-TP and PBB-TE) are connection-oriented transport technologies based on packet frames, enabling carrier-class OAM and fast protection. IP/MPLS should remain at the edge of the network (e.g., in the access), while the metro-core will be dominated by packet transport.

The reasons for a migration towards packet transport are a very efficient use of the bandwidth (due to the packet behavior of the connection) joint to OAM and resilience mechanisms comparable in efficiency to what is standardized in circuit-based networks.

In addition, PTTs keep the door open to the introduction of a fully integrated ASON/GMPLS network solution, which seems to be one of the most interesting approaches to meet network emerging requirements, not only overcoming the four fundamental network problems (bandwidth, latency, packet loss, and jitter) for providing real-time multimedia applications over networks, but also enabling flexible and fast provisioning of connections, automatic discovery, multilayer TE, and multilayer resilience, all based on an overall view of the network status.

1.4.2 Core Network

Consistent with the metro/regional description, Figure 1.6 depicts a possible migration trend for the architecture of the backbone network.

The current network architecture, depicted in the left side of the figure, is influenced by the long-distance traffic relationships that are currently covered by two networks: an IP/MPLS network (based on routers) and a transmission network based on optical digital cross-connects (in SDH technology).

In a first phase, the evolution is represented by the migration from legacy SDH to an OTN (based on ITU-T G.709 and its evolution); in parallel, Ethernet interfaces as routers' ports will substitute PoS ports. The next phase will be characterized by the adoption of PTTs for providing connectivity and substituting pass-through routers.

The following sections describe more deeply the concepts summarized in the figure.

1.4.2.1 Short Term

In the backbone segment, for an incumbent operator, the dominance of SDH carried on DWDM systems will be confirmed in the near future.

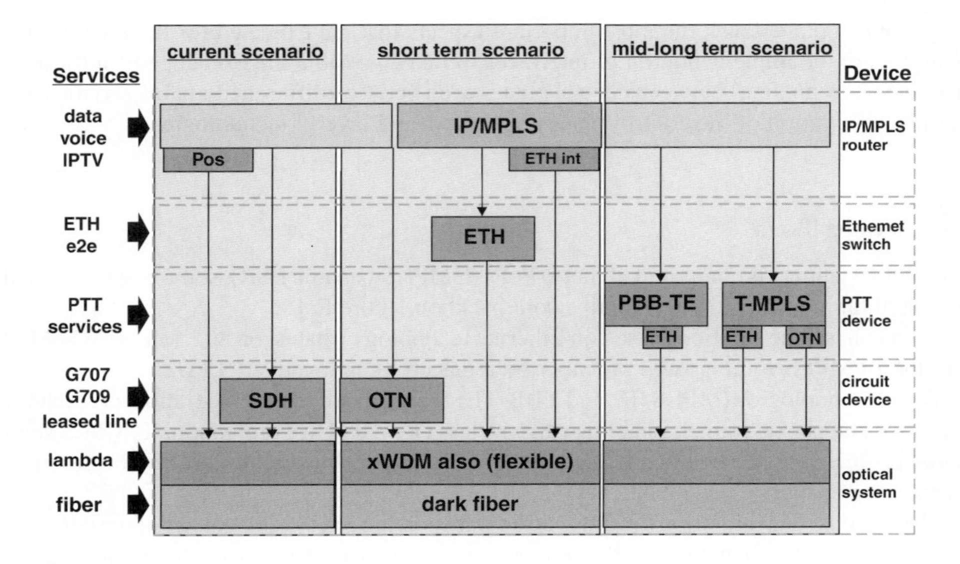

Figure 1.6 Evolution scenario for core/backbone network architecture

No unified control plane is available for the transport layers, yet. Furthermore, the control plane is restricted to a single network domain and in most cases to a single layer within the network.

Single-layer TE and resilience mechanisms will be still in use for quite a while. The full standardization of the ASON/GMPLS control plane is not yet complete. However, some vendors already provide optical switches equipped with a standard or proprietary implementation of the GMPLS control plane and make feasible control-plane-driven networking using the overlay network model. This suggests that the automatic switched transport network architecture using GMPLS protocols is being implemented and deployed together with different releases of the UNI and NNI.

1.4.2.2 Medium Term

In the core network, standard SDH (and to some extent OTH) is being introduced. The support for Ethernet-based services is increased. The use of the native Ethernet physical layer as layer 1 transport in the core will most likely depend on the availability of OAM functionality.

With the introduction of an intelligent SDH and OTH network layer, service providers can achieve significant cost savings in their backbones. IP over static L1 networks (e.g., IP over peer-to-peer links) should cope with a high amount of transit traffic in the core routers. As traffic increases, there comes a point where the savings in IP layer expenses realized by end-to-end grooming – where the bypass traffic is sent on the L1 layer without going back to the IP layer – compensate the extra expenses of introducing the intelligent layer 1 (SDH/OTH) switches needed.

The vertically integrated control plane refers to the underlying concepts that are called MLN/MRN at the IETF [17] and next-generation networks (NGNs) at the ITU-T. On the other hand, horizontal integration refers to the ability of control planes to provide support for service creation across multiple network domains.

The control plane will be aware of the physical layer constraints, which are important to consider, for instance, during routing in transparent/hybrid networks. Indeed, topology and resource information at wavelength level, as well as simplified signal quality/degradation information on links/wavelengths and nodes, is needed to allow the routing and wavelength assignment algorithm to place feasible paths efficiently into the network. In opaque networks, routing is based only on the overall path length constraint.

TE over domain borders between two or more domains will be a crucial topic.

1.4.2.3 Long Term

PTTs will also probably dominate the long-term scenario for the backbone segment, even if at later times than their adoption in the metro/regional network segment.

A probable architecture will consider some edge routers aggregating traffic and a network consisting of packet transport switches that will connect the edge routers.

The task of this packet transport network is to connect routers with dynamic connectivity (thanks to a control plane, probably of GMPLS type) and to assure multilayer resilience.

As shown in Figure 1.6, only some relationships might be confined at packet transport level, not the whole traffic.

For CAPEX reasons, the deployment of packet transport devices is more similar to that of an L2 switch than to L3 routers; as a consequence, the cost of switching (normalized per bits per second) is expected to be much lower than the current cost of switching in IP routers. For this reason, large bandwidth relationships (say, larger than 2 Gbit/s) should be carried more conveniently in connection-oriented mode. In a first phase, these circuit networks should be represented by the G.709 technology that, as said before, will probably dominate the medium-term scenario. Successively, the architecture of the backbone network will see the coexistence of G.709 and packet transport networks.

In general, integrated equipment can be assumed for the core network in the long-term scenario; this means that, within the core network, each equipment (LSR) would integrate multiple-type switching capabilities such as packet-switching capability (PSC) and TDM (utilizing SDH/SONET or OTH fabrics); or, in an even more evolutionary scenario, a solution where PSC and lambda switching capability (LSC) or where LSC and fiber switching capability coexist may be available.

The introduction of an integrated ASON/GMPLS network control plane solution might represent one of the most interesting approaches to meet network emerging requirements, both to overcome the four fundamental network problems (bandwidth, latency, packet loss, and jitter), to provide real-time multimedia applications over networks, and to enable flexible and fast provisioning of connections, automatic discovery, multilayer TE, and multilayer resilience, all based on an overall view of the network status.

As mentioned before, the control plane model considered for the long-term scenario is a fully integrated (horizontal and vertical) GMPLS paradigm, allowing a peer-to-peer interconnection mode between network operators, as well as network domains. Specifically, full integration means that one control plane instance performs the control for all the switching capabilities present in the network.

1.4.3 Metro and Core Network (Ultra-long-term Scenario)

Optical burst and/or packet switching might represent important technologies to face the flexibility demand of bandwidth in future networks. It is still unclear when some implementations of these technologies will be available, even if they could be developed (both at standard level and commercially) before 2012 to 2015, which seems very unlikely.

However, a dramatic increase of the traffic amount and the necessity of end-to-end QoS, in particular for packet-based network services, may open the door to PTTs as a new layer 1/layer 2 network solution that can overcome existing shortcomings.

The further evolution of PTTs may be represented by innovative solutions based on burst/packet switching, which would offer the following functionalities:

- Burst/packet switching will have the required dynamicity and flexibility already in layer 2, since an appropriate size of bursts/packets eliminates the grooming gap by offering a fine granularity, with less processing effort compared with short IP or Ethernet packets/frames.
- Reliability, monitoring, and QoS functionalities will be provided at layer 2, offering a solid carrier-class network service supporting higher layer network services at low cost.
- Hybrid circuit/burst/packet switching capabilities will be fully integrated into the GMPLS control plane philosophy (full vertical integration).

Specifically for core networks, layer 2 network technologies could consist of a hybrid circuit/burst/packet solution based on large containers carrying TDM, packet transport, and data traffic.

In metro networks – currently being dominated by Ethernet transport – the future architecture may be represented by the adoption of carrier-grade Ethernet protocols endowed with extensions on control, monitoring, and QoS. This implementation should also fit into the vertical integration strategy based on a GMPLS control plane and the horizontal integration with domain interworking providing end-to-end QoS.

From the control plane point of view in the ultra-long-term scenario, the ASON architecture based on GMPLS protocols is the most promising solution to integrate the TDM-based optical layer transport technologies (i.e., G.709) and the dominating packet-based data traffic using IP and IP over Ethernet protocols.

1.5 Transport Networks Economics

There is no unique infrastructure to support the required network services for the expected traffic; furthermore, not all plausible migration scenarios are cost effective for any of the network operators or within different market regulations. To analyze these differences, network operators use models that help them to evaluate how much a given network service implementation is going to cost, both in terms of CAPEX (the initial infrastructures roll out) and management and operation of the service (OPEX).

1.5.1 Capital Expenditure Models

CAPEX creates future benefits. CAPEX is incurred when a company spends money either to buy fixed assets or to add to the value of an existing fixed asset, with a useful life that extends beyond the taxable period (usually one financial year). In the case of telecommunications, operators buy network equipment to transmit, switch, and control/manage their infrastructures; this is part of CAPEX, but it also includes some more items:

- rights of way and civil works needed to set the equipment and deploy the lines;
- software systems (or licenses);
- buildings and furniture to house personnel and equipment;
- financial costs, including amortization and interest over loans (used to buy any of the former items).

A reasonable comparison among different solutions need not take all those items into account. However, two limiting approaches should be considered in evaluating investment and amortization:

- a "green field" situation, where a network operator starts to build their new network (or a part of it, as may happen for the access segment);
- upgrading the deployed network with new equipment or to add new functionality to the existing infrastructure.

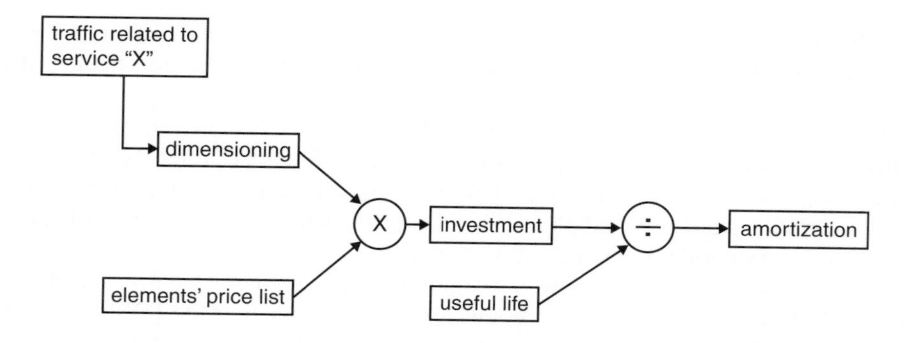

Figure 1.7 Process to evaluate the investment and amortization of a telecommunication network

As for the equipment deployment, the starting point naturally consists of dimensioning the requirements as a result of traffic estimation (Figure 1.7).

Amortization is the process of allocating one lump sum (CAPEX) to different time periods. Amortization can be calculated by different methods, but its concept describes the total expenses due to an asset over the time of its economic usefulness. As a coarse rule, the following list can be used to estimate the useful lifetime:

- network infrastructures (excavation, buildings, ...) 30 years
- optical fibers and copper wires 10 years
- network equipment 5 years
- software 3 years.

Different systems and technologies give rise to different components and, thus, different CAPEX analyses.

Figure 1.8 represents a general block model for most switching systems, including interface components, software, power supply, and common hardware elements. The cost model obviously arises from adding all individual prices for each component.

Sometimes, network operators prefer to establish a certain redundancy for some (common hardware) components so as to ensure service reliability.

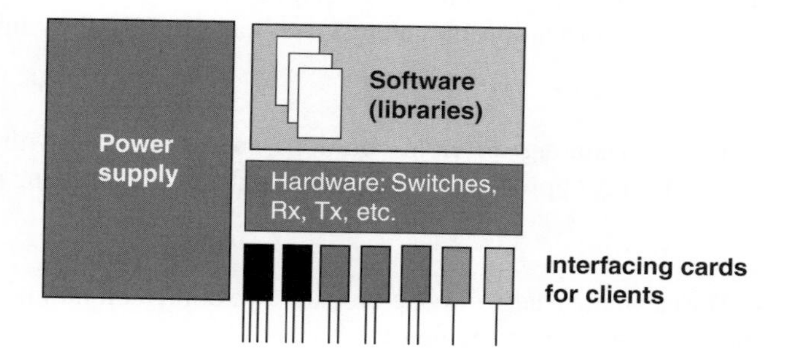

Figure 1.8 Simplified block model for a switching system

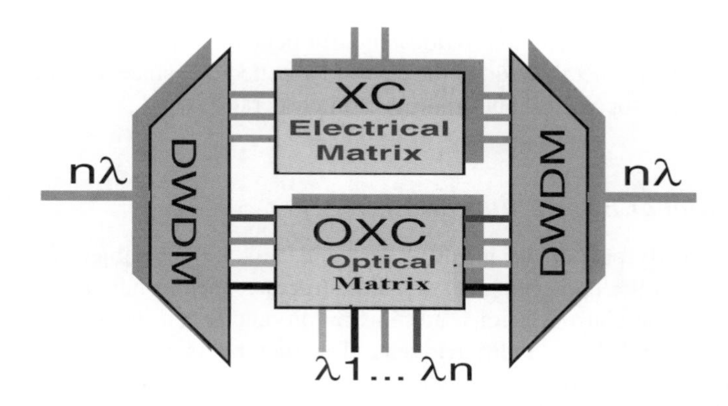

Figure 1.9 Hybrid SDH/lambda switching node

On the other hand, manufacturers also may introduce hybrid systems taking into account that common elements may be used by different technological solutions to accomplish a given function (switching packets, for instance).

Figure 1.9 shows the case for a transparent and opaque hybrid solution for a node where some lambdas are switched at the optical level, requiring no electrical regeneration (most lambdas are bypassed and transponders are only used to insert/extract lambda on/off the transmission line), whereas opaque switching (for low-granularity lines and to deal with the ingress/egress lambda of the transparent module) require electro-optical conversion. In fact, modular equipment is offered in a "pay as you grow" model to keep upgrading systems according to traffic growth for different (and interoperable) capacities like opaque/transparent hybrid nodes or L2/L3 multilevel solutions.

Aside from the amortization time schedule, network operators plan network infrastructures several years in advance. That is the reason why the introduction of new technologies must consider not only current prices, but also somewhat different ones for the future, taking into account certain conditions.

- Vendors offer discounts to network operators; discounts are very frequent because this is a way for vendors position their products as the *de facto* standard by massive installations.
- Equipment prices get lower after standardization agreements.
- Learning curves finally represent a significant price reduction as technologies mature. The empirical rule that states "as the total production of a given equipment doubles, the unit cost decreases in a constant percentage" can be combined with an initial estimate of market penetration to allow for predicting such a price reduction due to the maturity process.

These kinds of techno-economical prediction are usually carried out in combination with more general strategic considerations: whether network operators expect to expand their business outside present geographical limits or not, whether they will be able to reuse equipment for other purposes or places (from core networks to metropolitan ones, for example), or simply if it is possible to buy equipment from other companies, and so on.

On the other hand, cost models are always performed with a sensitivity analysis that highlights which elements are the most important to define a trend in the price evolution of a system and, as a consequence, to provide a tool for benchmarking it. This task must be done in combination with a general vision of the network architecture, since it is not straightforward to compare different topologies and multilevel interoperation.

1.5.2 Operational Expenditure Models

OPEX is not directly part of the infrastructure and, thus, is not subject to depreciation; it represents the cost of keeping the network infrastructure operational and includes costs for technical and commercial operations, administration, and so on. Personnel wages form an important part of the OPEX, in addition to rent infrastructure, its maintenance, interconnection (with other network operators' facilities) costs, power consumption, and so on.

Only considering specific OPEX derived from network services provision, the following list can be used as a guide to analyze its components:

- Costs to maintain the network in a failure-free situation. This includes current exploitation expenditures, like paying rents for infrastructures, power for cooling and systems operations, and so on.
- Operational costs to keep track of alarms and to prevent failures. This involves the main control activities to ensure QoS, namely surveying systems and their performance versus traffic behavior.
- Costs derived from failures, including not only their repair, but also economic penalties (in case an SLA states them for service failures).
- Costs for authentication, authorization, and accounting (AAA) and general management of the network.
- Planning, optimization, and continuous network upgrading, including software updating and QoS improvement.
- Commercial activities to enhance network usage, including new service offers.

Several approaches to analyzing OPEX can be used. To compare technologies and network architectures or different services implementations, a differential approach may be sufficient instead of considering all OPEX parts. However, if a business case-study requires knowledge of all costs and revenues, then a total OPEX calculation must be performed.

On the other hand, OPEX calculation is different for green-field and migration situations; for instance, bulk migration of customers or removal of old equipment will not be taken into account for a green-field scenario. Furthermore, bottom-up or top-down approaches can be used to calculate OPEX: the top-down method fits well to get a rough estimation of costs, as a starting point for a finer analysis of relative costs; the bottom-up approach is based on a detailed knowledge of operational processes.

Various approaches can be combined when dealing with OPEX calculations. In addition, OPEX and CAPEX may be balanced for accounting exploitation costs (e.g., buying or renting infrastructures) and some OPEX concepts can also be included in different items: salaries, for instance, can be considered as an independent subject or inside maintenance, commercial activities, and so on. A deep evaluation of OPEX is really important; in fact, it is possible that some technologies may offer high performances and perhaps at relative low CAPEX, but their

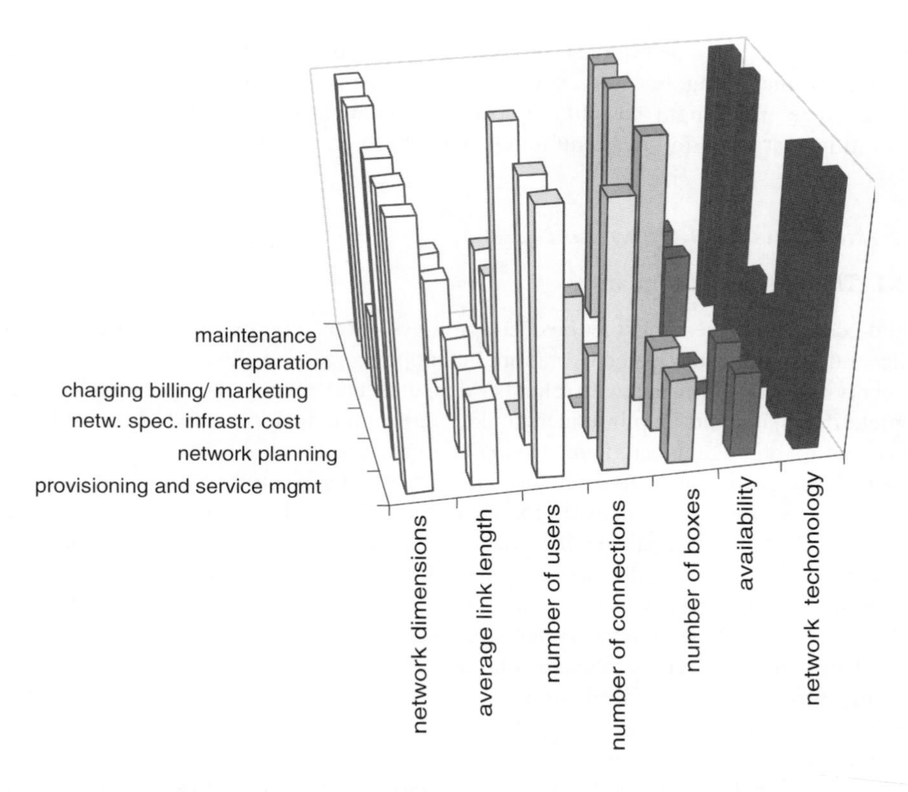

Figure 1.10 Network characteristics as cost drivers for OPEX calculations

complexity, software updating, power consumption, or personnel specialization may render them unaffordable.

Figure 1.10 shows, in arbitrary units, the dependence between OPEX components and network characteristics, considered as cost drivers so as to appreciate the impact of network technology on operational expenditure. It is clearly observed that network technologies determine the cost for maintenance, reparation, provisioning, service management, and network planning. They have less impact on charging/billing and commercial activities.

A more detailed analysis of Figure 1.10 gives the following information:

- The number of network elements (network components – e.g., routers, OXCs, PXCs) has an important impact on the cost for maintenance/reparation.
- Network technology determines not only network performance, but also some specific cost of infrastructure (more or less floor space and need of energy).
- Network dimension is important for all considered OPEX subparts, except AAA (run in centralized scheme) and marketing.
- The number of connections strongly influences the cost for provisioning and service operation and management (each connection needs to be set up), but it is less important for network planning.

- The number of users determines the cost for provisioning, network planning, charging/ billing, and marketing, but has a small impact on the cost of maintenance and reparation.
- The average link length has little impact on maintenance, but may be significant for reparation cost when failures require a technician to go on site.

1.5.3 New Business Opportunities

1.5.3.1 The Business Template

Techno-economic drivers must let business progress, since network services are no longer of public strategic interest, covered by national monopolies. Such a situation is not new, but it is still evolving in accordance with clients' demands and technical improvements. Just to complete the vision and help in understanding telecommunication network evolution, some ideas about market agents and their driving actions are presented here.

The technological evolution of network infrastructure leads it to becoming multifunctional. Hence, the old scheme of one network for one purpose and kind of client, in a vertical structure (see Figure 1.11), must be changed into a matrix scheme of services based on a unique transport infrastructure: all tasks related to network service provisioning are no longer repeated for every telecommunication business; for instance, AAA is common for IPTV, teleconferencing, or POTS, as well as QoS assurance systems, network configuration, or customer commercial issues. This scheme of cross-management can also lead to telecommunication companies exploding into several specialized companies that cover a part of the business (for more than one single network operator, perhaps). In addition, the market agents involved in the telecommunication business, from content providers to end customers, play their role freely in a cross-relational model (see Figure 1.12), where commercial interests should prevail.

So, companies have to redesign their business model along the guidelines discussed so far: a method of doing business by which telecom companies generate revenues, in order to set strategies and assess business opportunities, create or get profit of synergies and so align business operations not only for financial benefits, but also to build a strong position in the

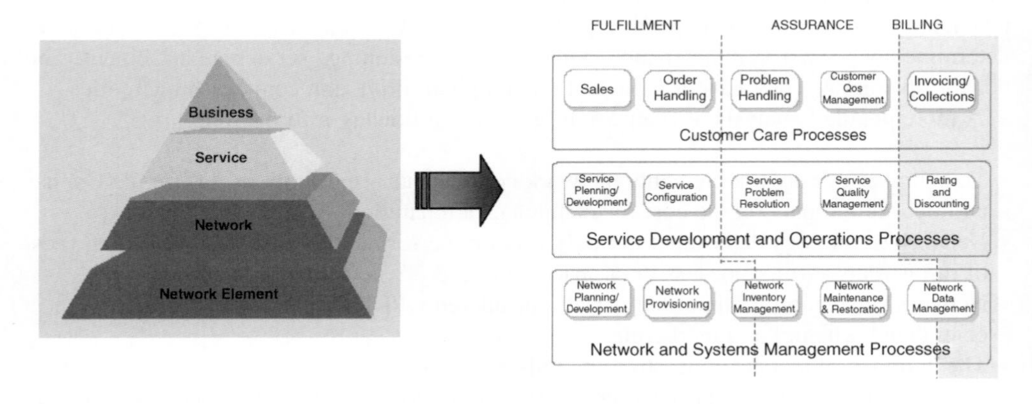

Figure 1.11 Network operator business model scheme adapted to NGN concept, from a pyramid structure to a matrix one

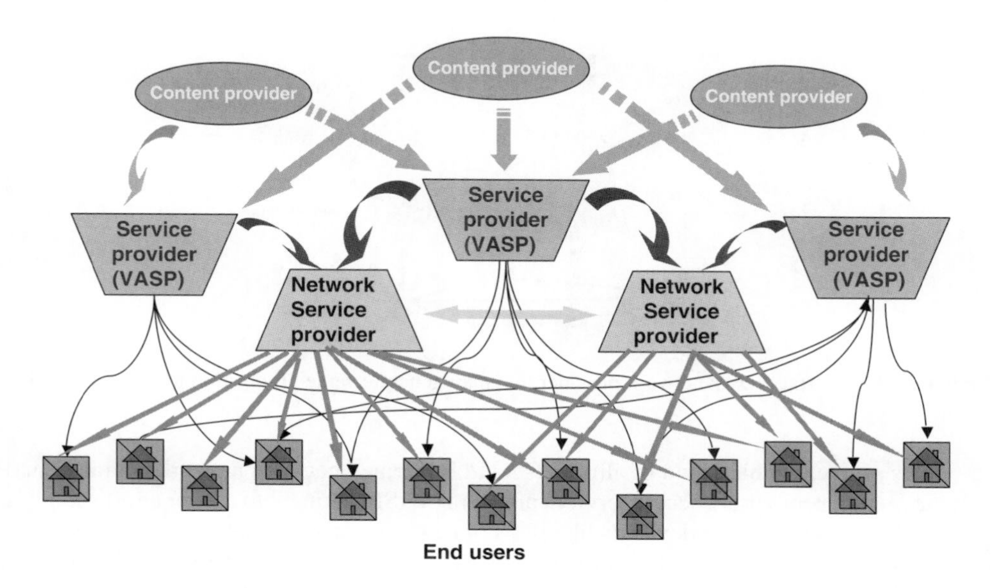

Figure 1.12 Telecommunication market agents and their cross-relationship just governed by commercial interests

market. In fact, network technology and market environment evolution affect all components of a business model, namely market segment, value proposition, value chain, revenue generation, and competitive environment. For these reasons, new business opportunities arise and the business model template has to be updated.

For the market segment, defined as the target customers group, it is clear that incumbent network operators may find virtual network operators as new clients and so the market liberalization generates new market segmentation. Each segment has different needs and expectations (e.g., business customers, banking) and companies create services for a specific types of customer;[3] in addition to customer types, geographical issues must also be considered for market segmentation.

The segmentation of the market, in turn, can be based on various factors, depending on the analysis to be performed; for example, different parts of a network can be shared by different market segments; every market segment will have its own behavior, reflected in its demand model.[4] Conversely, the way of using applications/services depends on the type of market segment to which they are focused (the same applications have different characteristics in residential and business areas, for example). Therefore, a proper market segmentation analysis should not only aim to map traffic demands onto transport services, but also tackle the heart of the network operator's business model: the question of how a network operator is designing

[3] Some broadband services target roughly the same segments, such as big enterprises (for instance, VPN services, virtual network operators, or regional licensed operators) and Internet service providers. But residential, business and public administration market segments are normally offered network services that specifically support applications for storage, grid computing, multimedia content distribution, and so on.

[4] This aspect includes the applications and the way they are used by those customers: frequency, time-of-the-day distribution, holding time, and so on.

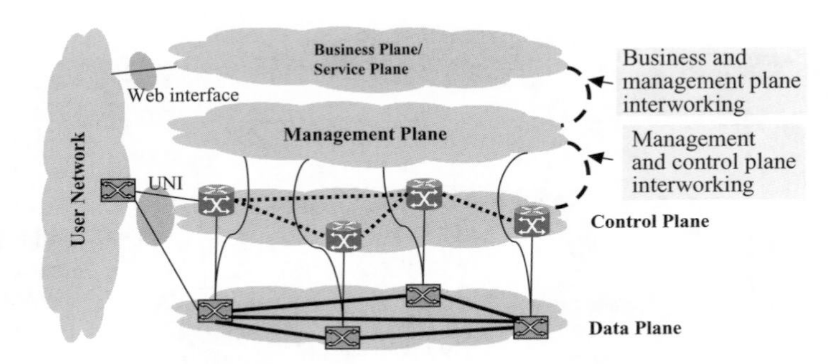

Figure 1.13 Interfaces user-network provider that may generate value proposition to different market segments

its access segment, which kind of alliance should be formed between network operators and content providers or virtual access services platform (VASP) taking into account traffic models derived from new customer demands like IPTV or new customer approaches to telecommunication issues like Web2.0. Finally, as far as network operators contract SLAs with their clients (availability and QoS), their profit depends on sharing their network infrastructure optimally.

The importance of properly segmenting the target market is critical in the development of a value proposition. The value proposition is identified by the combination of the customer needs, the services required to fulfill their needs, and the virtual value of the product.[5] NGNs allow network operators to design new products/services that can have value to some customers, like those derived from VPNs and temporary bandwidth leasing (BoD). In addition, customers are going to be able to access network services using new service interfaces and procedures. These interfaces can be provided on different planes (or any combination of them):

- **Service plane.** VASP may offer their products with (or without) a middleware to their customers using this interface to access network resources, either via previous reservation or in a dial-in mode (with a certain service availability).
- **Management plane.** This is the old way. Carriers establish SLAs and reserve network resources either for dedicated connections or by statistical knowledge of traffic. Internally, management/control plane interworking functions get the transport system (data plane) and set the required connections.
- **Control plane.** Network operators let some users directly enter into their network control system through the UNI. In this way (as established in an SLA), those customers can set up, modify, and tear down their own VPN once the required AAA process has been performed (by the management plane).

Thus, network operators' migration activities should take into account that deployment of new services and functionalities may attract new users and increase the operator position in the market. Such an approach is worth it even if the net cash balance remains unaffected due to extra costs for implementing the new services.

[5] The real value of the product is formed once customers select the service among available alternatives, according to its functionality and price.

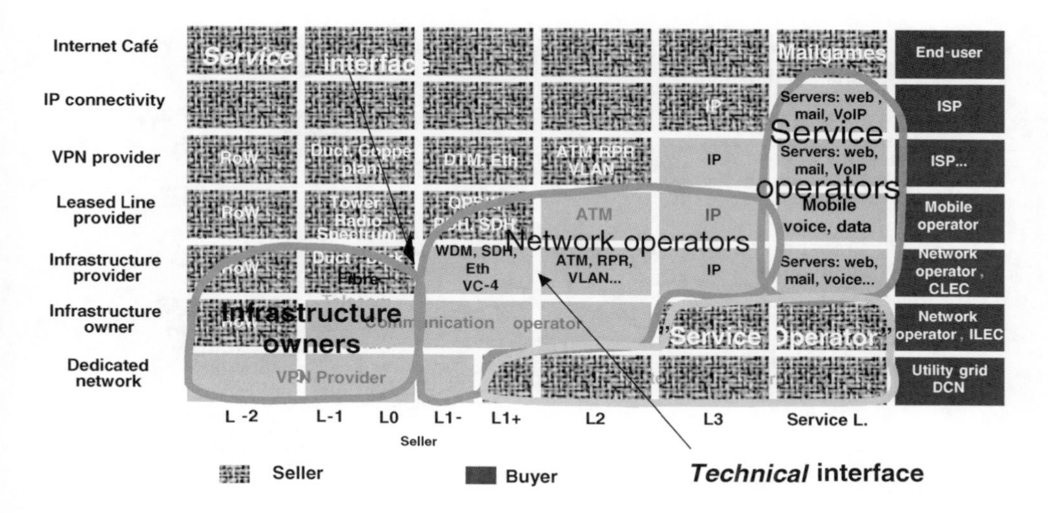

Figure 1.14 Market agents' roles mapped to the value chain by network layers to show service interfaces

On the other hand, virtual network operators (VNOs) may find easier ways of creating specialized services (TV distribution, for instance) without any handicap derived from the necessity of maintaining old services, QoS systems, or complex AAA mechanisms: their value proposition is clearly identified by a market segment. This is just the opposite case of incumbent telecommunication companies, composed of a number of departments that offer differently valued proposition services; then, synergies, market position, and customer fidelity are their main assets.

The creation of new services also modifies the value chain that represents a virtual description of the environment in which the market agents offer their services. If market desegregation implies unbundling the value chain to exploit a set of cross-relationships (Figure 1.14); also, tradeoffs and vertical alliances can be formed. These alliances are interesting whenever they produce added value to end customers because of technical improvement or service enhancement (Figure 1.14).

In fact, each box of the value chain represents an activity (Figure 1.15), and groups of activities are generally covered by independent market agents.[6] This value chain scheme illustrates the connections between the telecom network layers and the corresponding telecom services; layer handovers can be commercial transactions between different organizations or enterprises, or they can be internal (technical) interfaces without any real commercial transaction. For example, there may be fiber availability issues (with underlying ducts and rights of way); the access to this dark fiber is then a service (something to sell), and the seller is known as a "dark fiber provider." The organization that is purchasing this service – the buyer- has also to place the necessary equipment infrastructure (WDM, SDH) in order to become a Network Operator.[7] Other buyer–seller market profiles can be identified through the value chain in similar ways.

[6] Here is a key to the identification of some roles: customer, packager, connectivity provider, access network provider, loop provider, network service provider, application service provider, and content provider.

[7] The light boxes denote the seller and the dark boxes the buyer of the service at that interface.

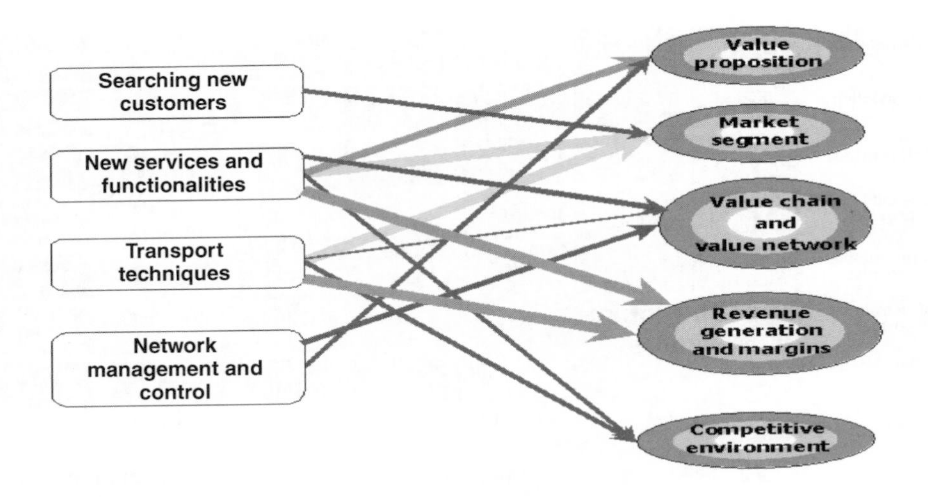

Figure 1.15 Impact of migration factors to business model components

In addition, the initial roles of the market agents can evolve according to commercial interests; for example, equipment vendors can develop network management activities, since their knowledge of the systems (hardware and software) allow them easily to propose value-added functions; in the same way, big media companies (content providers) may become service providers or, conversely, VASP may get exclusive rights on certain contents. Network operators can split old functionalities and even outsource some of them in order to concentrate their operations in either connectivity services ("bit transmission") or services of higher quality closer to the end customer (with higher margins); on the other hand, the FMC can have different consequences in the role that network operators would like to assume in the future.

The role of network operators within the value chain scheme given in Figure 1.15 is going to comprise other activities, within the value chain, apart from selling pure network services; for instance:

- Purchasing multimedia contents and distributing them throughout a nationwide platform for other (VASP) agents to deal with them.
- Leasing or buying underlying network infrastructure (right of way, network equipment, fiber) from utility companies or other vendors and renting part of it to other network operators as their network migration is carried on.
- Renting or selling buildings and complementary equipment (for cooling, for example) when they are no longer needed by new switching and transmission systems.
- Service provisioning to other VNOs.

The revenue amount is the main assessment to determine the network operator's ability to translate the value of the product into the money received. A typical revenue model is based on monthly subscription fees and costs of transactions, commissions, and services used by customers. A supplementary income can be achieved by selling or leasing a product/service to other, nonresidential companies. In each situation a network operator should consider an appropriate pricing model in order to maximize the revenue.

Moreover, the network migration strategy will be planned by a network operator, as long as this is possible, choosing the smallest cost of investment that may lead to greater profitability. Providing new services, after an initial investment, is a risk for several years to create new revenue-generation streams. Thus, increasing these margins is always an objective for any market agent before extending its role in the value chain or capturing new market. From the network operator point of view, it is always crucial to monitor the margins, since these are constantly change, and to plan network migration and to update the business plan, accordingly. This is not a straightforward operation; some puzzling considerations are described below:

- New services typically mean more profits from higher income; however, the higher the number of services, the more complex the management of them, aside from the effort required to search new customers. Then, the higher the diversification of services for value proposition and market segment capture, the bigger the problem with accounting and operating.
- Higher network quality and closer matching between customers' demands and network capabilities mean potentially more subscribed services and higher income. More subscribers, attracted by emerging, pioneer services lead to increasing income, provided that transport, control, and management do not overload network capabilities.
- In general, however, a higher number of customers means a faster return of investment, thus allowing reductions in the prices for services; and customizing existing services allows proposing more services and making them used more frequently, thus generating higher incomes.
- In addition, higher accessibility to the network services (through geographical expansion, higher availability derived from FMC or via interconnection facilities) and higher penetration due to combined service proposals increase the number of new subscribers and help to keep their fidelity.
- On the other hand, technical modifications affecting the traffic model will have consequences on the revenue streams. For example, the volume of metropolitan network traffic will be affected by the growth of social networks; network operators and VASP should then study carefully how IP networks are going to work in coordination with L1 and L2 transport layers and also analyze the importance of developing multicast-capable transmission systems, as well as the placement of service nodes in order to avoid bottlenecks. However, sometimes it can be proven that an extra initial investment (CAPEX) will be compensated by a consequent OPEX reduction: introduction of GMPLS for broadband standard networks will surely compensate the extra provision of bandwidth for some services unless network operators are obliged to share their transport capacity or extend it for the rest of their networks by the regulatory authority.
- Incumbent network operators must finally face the challenge of designing their network migration, not only to reach more revenue generation and gain market position, but also to keep backward compatibility. This decreases the risk of the investment and allows one to perform the infrastructure modernization in reasonable time schedules with the interesting possibility of reusing equipment.

In general, a higher value of the network and a higher position on the market allows financial advantages and new revenue streams for the telecommunication core business, like leasing offices and selling auxiliary equipment, since the eventually selected winning technology has lower requirements in terms of cooling, footprint and smaller real estate for the housing of the equipment.

Figure 1.15 summarizes the possible impact of the already analysed components on network migration factors and to NO's business model. Perhaps all factors involved in a migration strategy will have some influence on the company competitiveness, and so a migration strategy is always designed having as a goal to increase revenues and/or to gain a better position in the competitive environment. In the latter case, a network operator can deploy new services and functionalities to develop niche markets, thus attracting new customers, and achieve diversification of services so as to extend their market segment, attracting customers that prefer dealing with a unique network operator and get advantages of synergies for both network services and AAA tasks. The impact of regulatory framework on the migration strategies has, amongst other things, the following consequences from a NO point of view:

- Changes in the regulatory framework taken from the governments may either accelerate or discourage new entrant NOs, FMC or inter company (vertical) alliances.
- Vertical or horizontal alliances, even if they do not end up in one company's takeover, modify the competition environment (for suppliers and clients, respectively). On the other hand, the advancement in standardization, in contrast to exclusive vertical (in the value chain) dependencies, also modifies the competitive environment of network operators and other telecommunication market agents.

Network operator clients are also taking advantage of new network service implementations for end-to-end broadband communication, to be based on the NGN concept; thus, the competitive environment is becoming harder, as network operators' customers can trade off different network operator service proposals and use UNI to different transport infrastructures, so as to build up their own customized VPN (see Figure 1.16) or lease part of their bandwidth capacity, thus acting as VNOs.

1.5.3.2 Business Opportunities

VPN services, regardless to the association with FMC issues, seams to be the most important driver in telecommunication business plans in the short term. Currently, VPNs are mainly supporting legacy data services in circuit switched (CS) networks. However, the advances in

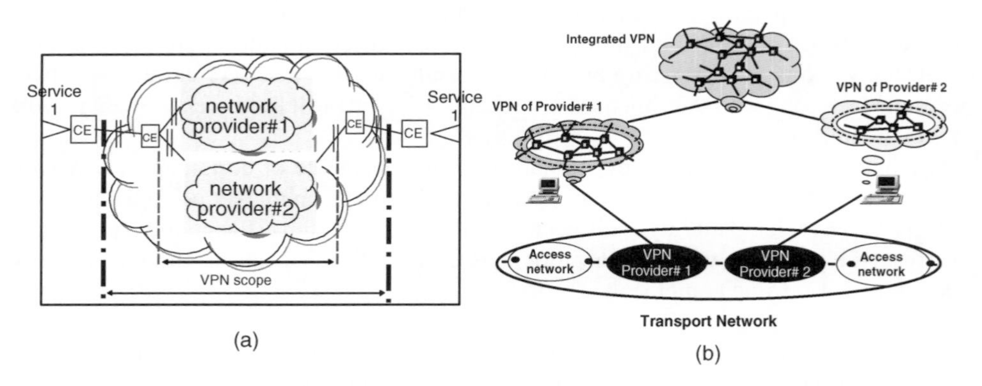

Figure 1.16 (a) Tradeoff network services from different network operators. (b) VPN concatenation to get extended services under a unified management. (CE: customer edge.)

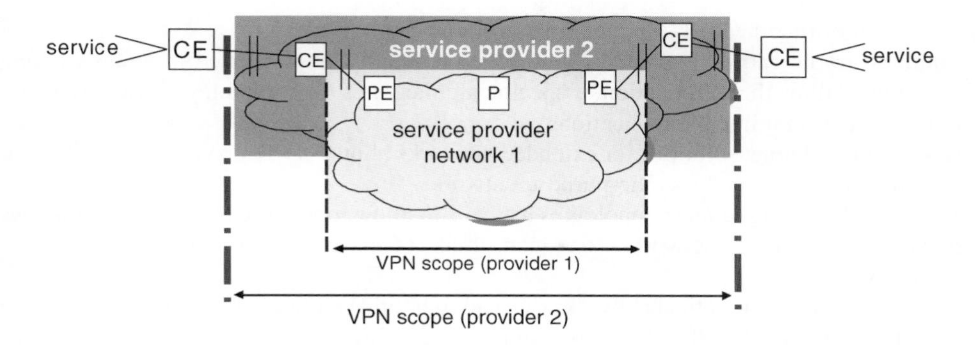

Figure 1.17 VPN cascade either to get dynamic network resources allocation or to be permanently used by VNOs (CE: customer edge; PE: provider edge; P: provider)

network technologies for VPNs are making it possible for NOs to operate a packet switched (PS) transport infrastructure like a CS one. But there is more than simulating CS networks over PS networks with the economical interest derived from statistical multiplexing and resource sharing. The currently perceived image of a "dummy" IP network will be reconsidered with the advances made possible in VPN technology: a NO may take advantage of the VPN technology to provision restricted connectivity using diffserv[8] -instead of over-provisioning network capacity- offering, thus, differentiated SLAs. Hence, ISPs and other VASPs could implement their preferred mechanisms to ensure application quality (for information integrity and the speed at which it reaches their customers).

A VPN may simply be used to implement VNO network resources (Figure 1.17). This fact, and the possibility of part-time network resource renting as BoD for any network operator client, is producing a radical change in the telecommunication business.

Moreover, incumbent network operators can evidently profit from operating their transport resources by means of a new developed TE approach. This policy can be based on the VPN technology, together with a network operator's ability to manage their capabilities and Ethernet network resources by reusing network elements (for the "virtuality" of the VPN paradigm) and make more scalable the management of the whole infrastructure by (virtually) dividing it and operating it as usual with the old pyramid scheme (Figure 1.17).

The expectations opened by new technologies and the market environment go beyond VPN-related topics. Improvements in network exploitation and new network service availability also give network operators new opportunities derived from:

- Cheaper and faster network upgrade (according to traffic demand) allowed by NGN. Furthermore, an NGN control plane makes it possible to adapt network resources quickly to client demands (a few minutes instead of days to reconfigure transport resources).

[8] For packet-switched networks, diffserv concepts of QoS can easily be understood as a mechanism of setting transmission preferences (at edge or core nodes) for determined sorts of packets (over a simple best-effort approach): those of a specific kind of application (real-time class, for instance) or those of a privileged customer, VPN, and so on. The concept is not new, and examples of its implementation may be found in standards of ATM services or RPR architecture. For IP networks, this non-neutral network operator performance, beyond a plain flux control, is a hot issue actively opposed by ISPs that promote ruling a net neutrality.

- Reduction of time needed to provide connections and new services.
- Multicast traffic carried over p2mp connection solutions, implemented in new switching equipment, allow the introduction of specific architectures for triple-play services (namely IPTV or other multimedia applications).
- Developing Ethernet solutions for extended networks allows carriers to reduce CAPEX for wide area networks and specific-purpose networks.
- Migration from ring to mesh topologies happens to allow more efficient resilient strategies for packet-switching networks: one can think of restoration versus protection, for example.
- Complete transparent optical networks (instead of hybrid solutions) for OSI level 1 are also expected to reduce CAPEX by eliminating undue O/E/O conversions. The possibility of directly managing wavelengths also has some OPEX advantages (wavelength-switched-oriented networks).
- Parallel development of the service layer, over the control plane, is often addressed to Ethernet, new applications like storage area networks, video on demand, grid computing, and so on.

Aside from these improvements in network exploitation and in the new services arising from NGN architecture and deployment of new technologies, some collateral new business must be mentioned too, since their economical impact is not negligible.

VoIP and migration to the IPv6 protocol are making possible a complete quadruple-play offer, as well as easier ways of managing networks to provide any kind of network service, including those for real-time applications, by means of the same packet-switched-oriented equipment.

As new equipment is smaller and need less cooling and DC generators, a considerable amount of real estate is made redundant in central offices, thus allowing NOs to get extra revenues from their selling or renting. This new real estate business opportunity should be perceived as in an integral part of the current trend in extending the role of NO's towards a value added service provider which will eventually turn them to their current antithetical pole i.e. the VASP could become a (V)NO offering specialized network services. Finally, the aforementioned SLA diversification supporting dynamic network service opportunities, as well as a complex trade off between the factors affecting it, may spawn the appearance of a new market agent; that of the extended service broker. This is an agent serving as the "middle-man" between suppliers and customers, operating not only in a direct NO-client scheme but also under a dynamic and multi-step (VPN cascade, for instance) network service leasing pattern.

Acronyms

AAA authorization, authentication, and accounting
ASON automatic switched optical network
ATM asynchronous transfer mode
BER bit error rate
BoD bandwidth on demand
CAPEX capital expenditure
E-NNI external NNI

FMC	fixed–mobile convergence
GMPLS	generalized multi-protocol label switching
IETF	Internet Engineering Task Force
I-NNI	internal NNI
IPTV	Internet protocol television
ITU	International Telecommunications Union
LSC	lambda switching capability
LSR	label-switched router
MAN	metropolitan area network
MEF	Metro Ethernet Forum
NGN	next-generation networks
NNI	network-to-network interface (see ASON)
OCS	optical circuit switching
O/E/O	optical/electrical/optical
OIF	Optical Internetworking Forum
OPEX	Operational Expenditures
OSI	open systems interconnection
OSS	operation service support
OTH	optical transport hierarchy
OTN	optical transport network
OXC	optical cross-connect
p2mp	point-to-multipoint
PBB	provider backbone bridge
PBB-TE	provider backbone bridge traffic engineering
PLR	packet loss rate
PoP	point-of-presence
PoS	packet over SONET
POTS	plain old telephone service
PSC	packet-switching capability
PXC	Photonic Cross-connect
QoS	quality of service
RPR	resilient packet rings (refer to IEEE 802.17)
RSPV-TE	reservation protocol with TE
SDH	synchronous digital hierarchy (refer to the ITU-T framework)
SLA	service level agreement
SONET	synchronous optical network (refer to the ANSI framework)
TDM	time-division multiplexing (see SONET and SDH)
UMTS	Universal Mobile Telecommunication System
UNI	user-to-network interface (see ASON)
VASP	virtual access services platform
VNO	virtual network operator
VoIP	voice over IP
VPN	virtual private network
WDM	wavelength-division multiplexing
WSON	wavelength-switched optical network

References

[1] Spargins, J.D., Hammond, J., and Pawlikowski, K. (1991) *Telecommunications: Protocols and Design*, Addison-Wesley.
[2] Dorf, R.C. (ed.) (1997) *Electrical Engineering Handbook*, 2nd edn, CRC Press.
[3] MEF 10.1 Technical specifications (November 2006) Ethernet services attributes Phase 2.
[4] IETF RFC 2702 (09/1999) Requirements for traffic engineering over MPLS.
[5] Davie, B. and Rekhter, Y. (2000) *MPLS Technology and Applications*, Morgan Kaufmann Phublishers.
[6] ITU-T Recommendation Y.1311 (03/2002) Network-based VPNs – generic architecture and service requirements.
[7] IETF RFC4026 (03/2005) Provider provisioned virtual private network (VPN) terminology.
[8] ITU-T Recommendation Y.1312 (09/2003) Layer 1 virtual private network generic requirements and architecture elements.
[9] ITU-T Recommendation Y.1313 (07/2004), Layer 1 virtual private network service and network architectures.
[10] Tomsu, P. and Wieser, G. (2002) *MPLS-based VPNs*, Prentice Hall.
[11] ITU-T Recommendation G.709 (03/2003) Interfaces for optical transport network (OTN).
[12] Comer, D.E. (2003) *Internetworking with TCP/IP Principles, Protocols and Architectures*, Prentice Hall.
[13] ITU-T Recommendations G.8080/Y.1304 (November 2001) Architecture for the automatically switched optical network (ASON).
[14] IEEE 802.1Qay (2007) Standard provider backbone bridge traffic engineering.
[15] Serrat, J. and Galis, A. (2003) *IP Over WDM Networks*, Artech house.
[16] Kadambi, J. Crayford, I., and Kalkunte, M. (2003) *Gigabit Ethernet Migrating to High Bandwidth LANs*, Prentice Hall.
[17] IETF RCF 5212 (07/2008) Requirements for GMPLS-Based Multi-Region and Multi-Layer Networks (MRN/MLN).

2

The Advances in Control and Management for Transport Networks

Dominique Verchere and Bela Berde

2.1 Drivers Towards More Uniform Management and Control Networks

We observe the convergence at different levels on the network given the application and service convergence as presented in Chapter 1 (and might be highlighted/summarized here); the convergence and the related Internet protocol "(IP)-orientation" in network services seems inevitable.

What is the network convergence? The converged IP-based service platform, despite the wide variety of network technologies, requires network control and management functions that tend to be uniform at the different switching layers in order to reduce operational expenditure (OPEX). The transition from "a network per a service" to "network integration of multi-services and multilayer with support for end-to-end quality of service (QoS)" concept enables one to get higher network resource utilization coupled with higher resiliency.

This section further elaborates the new role of the network control functions and network management functions relying on this integration at different transport layers by considering the network layer (i.e., IP layer or layer 3) with the convergence to IP-based services using the essentials of traffic engineering (TE) and QoS support, and how these requirements can be illustrated in IP or IP/multi-protocol label switching (MPLS) networks. Then we consider the data layer (i.e., layer 2) as essentially based on Ethernet with the transport MPLS (T-MPLS)[1]

[1] At the time this text was produced, key agreements have been defined between ITU-T SG15 and IETF leadership concerning T-MPLS/MPLS-TP evolution in December 2008. The agreements can be summarized in three statements: (i) there is no agreement, or proposal to cancel or deprecate the T-MPLS recommendations from ITU-T currently in force; (ii) ITU-T will not undertake any further work to progress T-MPLS; and (iii) it is possible that the ITU-T will have a future role in MPLS-TP standardization.

Core and Metro Networks Edited by Alexandros Stavdas

extensions, resilient packet ring empowered with TE and operation, administration, and maintenance functions. Finally, we consider the physical layer (i.e., layer 1) as essentially based on optical transmission networks with synchronous optical network (SONET)/synchronous digital hierarchy (SDH) and its next-generation SONET/SDH enhanced with data protocols such as generic framing procedures, virtual concatenation and link capacity adjustment schemes, G.709, and optical transmission technologies [33].

In circuit-switched layer transport networks, a layer (L1) path is constructed with physical links and ports, one or many optical wavelengths, or time-division multiplexing (TDM) timeslots, and an L1 path is established and dedicated to a single connection between the access points (APs) of the transport network for the duration of the applications using the connections.

In packet-switched networks, packets are forwarded hop by hop at each network element involved in the connection (e.g., IP routers or Ethernet switches) based on information in the packet header. An IP-switching-based network provides connectivity services while making efficient use of network resources by sharing them with many connections. It is based on the assumption that not all the provisioned connections need to use the resource all of the time. Packet-switched networks (PSNs) can be connectionless or connection oriented. PSNs can be based on the IP, such as IP networks (connectionless) or IP/MPLS networks (connection oriented), or can be based on the Ethernet protocol, such as Ethernet network (connectionless) or Ethernet/MPLS transport profile (MPLS-TP) networks (connection oriented). IP-based networks are said to be layer 3 and Ethernet-based networks are said to be layer 2 in reference to the Open Systems Interconnection (OSI) basic reference model [31].

In connectionless packet-switched networks, once the data packet is sent on the interface, the connection is available until further information is either sent or received at this same interface.

In connection-oriented PSNs, connections are established and maintained until connectivity is no longer required, regardless of whether data packet has been transmitted or not.

Network services are classified according to three assigned groups taking into account their switching capabilities, features, relations, and differences. The three groups identified are virtual private networks (VPNs) on transport layer 3, VPNs on transport layer 2, and VPNs on transport layer 1. Additional network services subclassification can be defined for the layer 3, such as public IP and business IP, extending the network service classification to five groups. The characteristics of each network service are usually described along with their performance parameters.

The different layers' VPNs are explained from the perspective of connectivity, control, scalability, and flexibility. Mechanisms enabling the network "confidentiality" are described, such as MPLS and tunneling in PSNs, and wavelength services enabled by L1-VPN services from optical networks. How network services should match the connectivity requirements of distributed applications is not developed in this chapter, but the connectivity requirements of on-demand storage, (grid) cloud computing, or multimedia-based distributed applications are developed in [32].

Three classes of network services are identified based on the related switching and transmission layers in which they are provided to the network customers edges (Table 2.1):

1. Layer 1 network services provide "physical layer" services between the network client and 5the provider network (server). The customer edge (CE) equipment belongs to the same L1-VPN as the other equipment of the provider network (provider edge (PE) nodes,

Table 2.1 Network service modes versus switching layers

Network service mode	Switch capability layer		
	Layer 1	Layer 2	Layer 3
Connectionless packet-switched	*Optical packet switching*	Ethernet	IP
Connection-oriented packet-switched		Ethernet/MPLS-TP, ATM, frame relay	IP/MPLS
Circuit oriented	OTUx, (OTN) STM-*n* (SDH), STS-*n* (SONET)		

provider (P) nodes). The connections can be established based on TDM timeslots (SONET/SDH), optical wavelengths, optical wavebands, or physical ports, such as Ethernet ports or fiber ports.
2. Layer 2 network services provide "data link layer" connection services to the CE equipment involved in the L2-VPN. At the interface, the forwarding of user data frames is based on the control information carried in the data link layer headers, such as media access control (MAC) addresses (for Ethernet), virtual circuit identifier/virtual path identifier (for asynchronous transfer mode (ATM)) or data link connection identifier (frame relay). The customer layer 2 data frames are associated with the right L2-VPN by each PE node.
3. Layer 3 network services provide "network layer" services between the CE equipment involved in the L3-VPN. At each user-to-network interface (UNI), the forwarding of data packets is based on the IP address information embedded in the layer 3 header; for example, IPv4 or IPv6 destination address.

The network services have been generically labeled VPN at L1, L2, and L3. For each of these three types of network service, typical performances are defined and usually referenced by network operators to allocate the connectivity service requests.

The layer 3 network service (such as IP) is, as already mentioned, divided into public and business IPs, where public IP is a "best-effort" service and business IP is a higher priority class of services that, for example, can handle latency-sensitive applications. "Business IP" is also presumed to guarantee higher bandwidth from CE to CE.

The VPN services on all layers, L1, L2, and L3, are divided into either a permanently configured network service (typically provisioned by a network management system (NMS)) or an on-demand service (typically triggered and signaled from a network controller) (Table 2.2). The permanent service is totally managed by the network operators, but the on-demand connectivity service can be triggered dynamically by the CE node through a suitable UNI.

L1 and L2 VPN services are further divided according to their availability; this can be high or low availability. The high-availability network services are normally configured with defined protection/restoration schemes [2], which offer an alternative way for carrying the traffic impacted by network failures. Different types of distributed application are identified and reported in [32], and for each case the application performance requirements are checked against the network services.

Table 2.2 Mapping applications into network services (light: application will run on this network service; dark: more efficient implementation, white: no support for the application)

		Storage	- Back-Up/ Restore	- Storage on Demand	- Asynchronous Mirroring	- Synchronous Mirroring	**Gridcomputing**	- Compute Grid	- Data Grid	- UtilityGrid	**Multimedia**	- Video on Demand	- VideoBroadcast(IP-TV)	- Video Download	- VideoChat	- Narrowband Voice,data(VoIP,...)	- Gambling	- Gaming	- Digitaldistributiondigitalcinema	- Videoconference	Tele-medicine/diagnostic
Public IP																					
Business IP																					
VPN - L3	*permanent*																				
	on-demand																				
VPN - L2	*permanent, Hi avail*																				
	permanent, Low avail																				
	on-demand, Hi avail																				
	on-demand, Low avail																				
VPN - L1	*permanent, Hi avail*																				
	permanent, Low avail																				
	on-demand, Hi avail																				
	on-demand, Low avail																				

It should be noted that the same transport layers could be used to provide multiple VPN services. For example, a transport layer based on SDH can be used to provide layer 1 (e.g., TDM), layer 2 (e.g., Ethernet), and layer 3 (e.g., IP) VPN services. For this reason, it is important to distinguish VPN clients and VPN transport networks (server part). Both VPN transport and VPN client each have their own set of connectivity inputs and outputs known as APs.

When the VPN client switching capability layer and VPN transport switching capability layer are different, the VPN client layer information must be adapted for transmission across the transport VPN. Examples of adaptation functions include multiplexing, coding, rate changing, and aligning (which may also include some form of fragmentation and sequencing if the transport layer traffic unit is smaller than the service layer traffic unit). Even if VPN transport layer trail/connectionless trail termination functions are VPN client layer independent, adaptation functions must exist for each transport–client pair defined [41]: adaptation functions are required between APs in the VPN transport layer and control plane planes/ forwarding planes at the VPN client layer.

2.2 Control Plane as Main Enabler to Autonomic Network Integration

Telecommunications equipment designers face a huge task in making the optical networking dynamically configurable according to the IP traffic demands of customers. Provisioning connections from NMSs or signaling connections from the network control planes based on real-time traffic patterns require the ability to manage the interactions between the IP-layer functionality of packet networks and of the lower transport layers (L2 or L1 networks).

End-to-end control issues in transport networks with multiple technology layers and multiple administrative or organizational domains with multi-vendor equipment are becoming common. This is the role of the control plane to bring an answer in the form of tools, algorithms, automated processes, and common management interfaces to these issues, with the automated connection provisioning capability.

2.2.1 Generalized Multi-Protocol Label Switching

Derived from MPLS and driven by the Internet Engineering Task Force (IETF) Common Control and Measurement Plane (CCAMP) working group, generalized MPLS (GMPLS) has drawn a lot of attention on enabling different network layers to interact automatically. GMPLS provides uniform control capabilities from the optical transport network (OTN) up to the IP client layer to permit open and scalable expansion for next-generation transport networks.

This section will examine the environment for GMPLS control architecture and describe the following topics:

- Primary drivers that compelled the developers of GMPLS to come up with a standard architecture.
- MPLS/GMPLS evolution.
- GMPLS architecture.
- Fundamental technology components – protocols (and others, such as path computation element (PCE)).
- GMPLS functionality – resource discovery, routing, signaling, and link management.
- Goals of GMPLS – operation automation, flexibility, scalability, restoration, path selection optimization, and TE capabilities.
- Primary current and future benefits and applications of GMPLS.
- What are competing or complementary standards to GMPLS and how do they compare?
- How do GMPLS-related equipment features and functions map to service provider requirements?
- Specific economic benefits of GMPLS with reference to capital expenditure (CAPEX) and OPEX.

2.2.1.1 Primary Drivers to GMPLS

In multilayer transport networks, where IP, ATM, Ethernet, SONET/SDH, and fiber/port switching devices must constantly cooperate, GMPLS was specified with the objectives to automate the connection service provisioning, and, especially, with reliability and TE capabilities. This allows the combining of different network layers for service delivery: IP and Ethernet (also with GMPLS), and so on, down to optical switching.

Service providers need to reduce the cycle times for service provisioning. Adopting a new architecture including GMPLS leads to a reduced length of provisioning times, which stems primarily from the requirements to manually set up interconnectivity between network partitions, such as SONET/SDH rings, and networks. For instance, the time period to provision in a traditional optical network has been estimated at an average of 15–20 days from the service order through to final test, where the actual provisioning of the physical network takes a major part. With the automated service subscription, one of the primary goals of GMPLS is to automate provisioning with cutting cycle times down to a few seconds [32].

Deploying and building out regional or national fiber infrastructure, or simply deploying expensive switching equipment, require operational cost reduction. Manually provisioned billed services with the incident service support do not fit this figure of expenditure. The reliability and protection guarantees for, especially, voice, TV, and video services over a multilayer network, not only SONET/SDH, pose the demand on operational machinery running at a reduced cost.

2.2.1.2 From MPLS Evolution to GMPLS Consolidation

MPLS, drafted in 1999 by the IETF for the convergence of IP, ATM, and frame relay layer technologies, provides higher speed data forwarding with the network internal options of TE and support for QoS. The predictability and reliability was then brought to IP networks with MPLS, with the efficiency of automation for connection establishment.

The routers, at the ingress of a network, inject a fixed-format label positioned at the layer 2 protocol header of a data packet as it enters the network. At every label-switched router (LSR), the pair of values built from the incoming label number and the incoming interface (i.e., port) determines the route. The complete path is actually predetermined at the beginning of the route at the ingress LSR. The data packet flows are carried over the transport network and signaled as what is called the label-switched path (LSP). Given that the LSRs perform a much less time-consuming label examination, and not a packet header-maximal matching forwarding, the LSPs enable significant speed improvements with reference to a traditional IP forwarding network.

Moreover, extending MPLS to other network technologies first requires separating the label control information from the packet header. Next, MPLS uses in-band control, meaning that the control information is sent with the actual data traffic flows. Also, the IP addressing schemes need to be adapted to other technologies. Another issue comes from the MPLS label formed as a number of up to 32 bits.

All these questions were addressed by the IETF in order to accommodate MPLS into the switching capabilities of other network layer, and that below the IP layer; that is, L2 and L1 networks. The result is the standard GMPLS architecture that allows generalizing the label switching to non-packet-switching technologies, such as time division (e.g., SONET/SDH, PDH, G.709), wavelength (lambdas), and spatial switching (e.g., incoming port or fiber to outgoing port or fiber).

2.2.1.3 Architecture to GMPLS

In the IETF standard [42], GMPLS addresses multiple switching layers and actually covers five groups of switching types. The switching type of a network element defines the data frames that the element can receive, switch, and control; it corresponds to the switching capability layer to which the network element can demultiplex the data signal from an input interface, to switch it and send it to the output interface. The switching types are ordered among switching capabilities as follows:

1. Packet-switch-capable (PSC) interfaces: interfaces that recognize packet boundaries and can forward data based on the content of the packet header. These devices, such as IP or ATM routers, can also receive routing and signaling messages on in-band channels.
2. Layer-2-switch-capable (L2SC) interfaces: interfaces that recognize frame/cell boundaries and can forward data based on the contents of the frame/cell. Examples include ATM, frame relay, Ethernet, and its evolution towards MPLS-TP.
3. TDM-capable interfaces: interfaces that forward data based on the data's time slot in a repeating synchronous cycle. Examples are SDH/SONET cross-connects and add–drop multiplexers.
4. Lambda-switch-capable (LSC) interfaces: interfaces that forward data based on the wavelengths on which data is received. Examples include photonic cross-connects (PXC) or optical cross-connects (OXC). These devices can operate either at the level of an individual wavelength or a group of wavelengths, called waveband-switching equipment.
5. Fiber-switch-capable (FSC) interfaces: interfaces that forward data based on the position of the physical interfaces. Examples are PXC or OXC equipment.

The importance of the hierarchy comes from the multi-switching capability that can occur on the same interface, or between different interfaces. As described in [42], a circuit can be established only between, or through, interfaces of the same type. Depending on the particular technology being used for each interface, different circuit names can be configured; for example, SDH circuit or optical trail. In the context of GMPLS, all these circuits are signaled with a common control entity named the LSP.

In a GMPLS-controlled OTN, the labels are physical resources and are used to signal the route of the data stream it carries. Each label is specific between connected nodes, and the values of the labels require to be controlled with the same meaning. In [44–46], labels are given a special encoding so that the referenced resource (SDH, SONET, G.709 or wavelength) can be deduced automatically without the need to be configured and negotiated through the link management protocol engine [47].

The importance of GMPLS, and especially in multilayer networks, comprising multiple but collaborative network technologies, comes from its TE capabilities. TE is defined as that aspect of network engineering dealing with the issue of data and information modeling, performance evaluation, and optimization of operational networks. TE encompasses the application of technology and scientific principles to the measurement, characterization, modeling, and control of Internet traffic [68]. An important objective of TE is to facilitate reliable network operations. These operations can be facilitated by providing mechanisms that enhance network integrity and by embracing policies emphasizing network survivability. This results in a minimization of the vulnerability of the network to service outages arising from errors, faults, and operational failures.

Using the word "multilayer" means that two (or more) switching-capability layers collaborate in the network. In particular, with GMPLS-based control functions, the switching-capability layers may be IP, Ethernet, SDH, and optical transport.

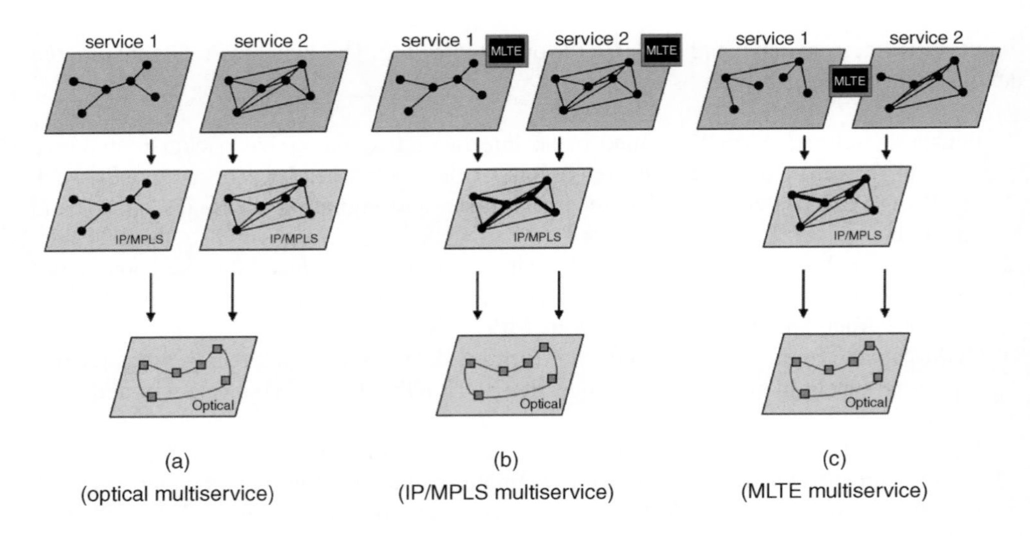

Figure 2.1 MLTE and service allocation modes

As an illustration for TE-like optimization, consider multiple IP/MPLS services provided to clients through an IP/MPLS over an optical network. The illustration presents three ways for an IP/MPLS services request to be allocated over optical network resources in Figure 2.1. It is considered that the optical connections (e.g., TDM LSP or higher switching capability type) are used as TE links and constitute a virtual network topology (VNT) for an IP/MPLS network. The hierarchy of the switching-capability layers (defined in Table 2.1) is used to define that the IP/MPLS layer is a client of the optical network layer.

The first allocation mode (case (a) in Figure 2.1) consists of designating separate network control instances for each network service request. This means that each IP/MPLS-based service request is built separately from a data flow and is routed and allocated through one IP/MPLS control instance. IP/MPLS service requests are sent independently to the optical network layer. The acceptance of one IP/MPLS service request by the optical connections may influence acceptance for the subsequent service requests from IP/MPLS network client.

The second allocation mode (case (b) in Figure 2.1) consists of combining IP/MPLS service requests into a common network control instance. This causes the service requests to be allocated at the IP/MPLS layer first; for example, in IP/MPLS routers, data packet flows associated with different IP/MPLS service requests will be processed in the same logical queues. Note that in this case the optical network server is still not optimally used because some optical connections can remain idle at a time when their capacity can be used in speeding the services of the other IP/MPLS connection requests. By handling the service requests onto a single IP/MPLS control plane, it is possible to recover some wasted capacity (e.g., due to nonoptimal light-path filling) and balance the service requests between the optical layer resource usage. To overcome this inefficiency, the TE at the IP/MPLS layer and at the optical layer are merged to produce a single multilayer TE (MLTE) control instance.

The third allocation mode (case (c) in Figure 2.1) uses MLTE joint optimization. A single MLTE control instance can optimize both IP/MPLS service requests, meaning that, for example, traffic flow measurements from several service requests are collected, combined, and optimized into one IP/MPLS connection. The measurement from this optimal combination IP/MPLS service is used in routing the service over the optical network. This is different from the previous allocation mode, where measurements (and MLTE actions) are performed separately. Since the traffic flows of several IP/MPLS services must be carried on a single IP/MPLS layer, a joint optimization will improve resources assignment, leading to the removal of idle connection servers in the optical layer. Contention may still exist when the joint IP/MPLS service requests are superior to the global optical network capacity.

2.2.1.4 Fundamental Technology Components

The GMPLS control plane is made of several building blocks. These building blocks are based on discovery, routing, and signaling protocols that have been extended and/or modified to support the TE capabilities of MPLS and then GMPLS. Each network element with a GMPLS controller in general needs to integrate a specialized link protocol for identifying and referring each data channel explicitly that the adjacent nodes can reference accurately. This complexity is due to the separation between the data transport network plane and the control network plane; consequently, there is no direct mapping between the data channels and the control channels. In order to support the network element discovering the data link capabilities and their identifiers, a control protocol for link management, called the link management protocol (LMP) [47] is defined.

However, fundamental technology building blocks to GMPLS include not only protocols (i.e., routing, signaling, and LMP), but also new elements developed to include brand new concepts. It is important, indeed, to describe, at least, two new notions for the GMPLS framework: the TE link and forwarding adjacency (FA) LSP concepts. A TE link is a representation of a physical link. The link state advertisements (LSA) is used by the routing protocol and stored into the link state database, also called the TE database (TEDB), to advertise certain resources, and their properties, between two GMPLS nodes.

This logical representation of physical links, called TE Links, is used by the GMPLS control plane in routing for computing and selecting the resources and in signaling parts for establishing LSPs. The TE extensions used in GMPLS corresponds, therefore, to the TE (link) information of TE links. GMPLS primarily defines, indeed, additional TE extensions for TDM, LSC, and FSC TE, with a very few technology-specific elements.

The FA-LSP corresponds to an LSP advertised in the routing domain and stored in the TE link state database as a point-to-point TE link; that is, to an FA. That advertised TE link no longer needs to be between two direct neighbors, as the routing protocol adapts the TE link information to that indirect neighborhood. Importantly, when path computation is performed, not only just conventional links, but also FA-LSPs are used.

From the point of view of protocol engines, extensions to traditional routing protocols and algorithms are needed to encode uniformly and carry TE link information. Therefore, for GMPLS, extended routing protocols were developed; that is, intermediate system to intermediate system (IS-IS)-TE and open short path first (OSPF)-TE. In addition, the signaling must be capable of encoding the required circuit (LSP) parameters into an explicit routing object (ERO). For this reason, GMPLS extends the two signaling protocols defined for MPLS-TE signaling; that is, resource reservation protocol (RSVP)-TE [50] and constraint-based routing–label distribution protocol (CR-LDP) [49] described in the signaling functional description [48]. GMPLS further extends certain base functions of OSPF-TE and IS-IS-TE and, in some cases, adds functionality for non-packet-switching-capable networks.

2.2.1.5 GMPLS Functionality

The GMPLS framework separates the network control plane containing the signaling and routing protocol. The fundamental functional building blocks of GMPLS can be structured as:

- resource discovery
- topology/state information dissemination
- path selection
- routing
- signaling
- link management
- restoration and protection
- management information base (MIB) modules
- other functionality.

GMPLS was designed for integrating MLTE and multilayer recovery mechanisms in the network. In addition, we are specifically interested in the promising alternatives of:

- Scalability, with reference to the number of protocol messaging. Practical limitations on information processing may exist in equipment.
- Stability, or more properly protocol convergence times. For MLTE, especially in upper layers that see logical topology updates, control plane traffic may be bursty, and large networks may see large convergence times.

Given the collaborative aspects in MLTE when running GMPLS protocols, the processing of the vast amount of information, which may potentially be emitted by network devices running GMPLS, the following sections briefly present base functions with reference to the fundamental building blocks.

Resource Discovery

The use of technologies like dense wavelength-division multiplexing (DWDM) may imply a very large number of parallel links between two directly adjacent nodes (hundreds of wavelengths, or even thousands of wavelengths if multiple fibers are used). Such a large number of links was not originally considered for an IP or MPLS control plane, although it could be done. Moreover, the traditional IP routing model assumes the establishment of a routing adjacency over each link connecting two adjacent nodes. Having such a large number of adjacencies does not scale well. Each node needs to maintain each of its adjacencies one by one, and the link state routing information must be flooded throughout the network.

To solve these issues, the concept of link bundling was introduced. Moreover, the manual configuration and control of these links, even if they are unnumbered, becomes impractical. The LMP was specified to solve the link management issues.

Topology/State Information Dissemination

The goal of information dissemination is to provide information on the TE link and its attribute information in order to allow LSRs to select an optimized path based on TE criteria. A primary goal of the GMPLS routing controller in handling this information is to deal with the problems of scalability that are essential to multilayer networks. The impact of control plane limitations is dependent on the "meshedness" of the topology formed in the control plane. This means that, for sparser topologies, greater (relative) changes occur in the GMPLS logical topology when traffic patterns shift.

Since the number of physical ports on an OXC may be large, GMPLS devises the concept of a bundled link, used to minimize, via aggregation, the amount of information that is propagated over the network. Bundle links are parallel links, equivalent for routing purposes, which share common attributes used for path selection.

When distributing topology information to other sub-networks, bundle links can be used and only aggregate information is provided. There is a trade-off between the granularity of the information disseminated and the scalability; balancing this trade-off is not trivial.

Link state information has both static and dynamic components. Examples of static information include neighbor connectivity and logical link attributes, while dynamic ones include available bandwidth or fragmentation data. Similar to the scalability issues regarding topology information, the amount of link state information to be conveyed has to be controlled. The premise is to distribute only what is absolutely necessary. Although all link state data must be set up in the initial database, only changing information needs to flood the network. There are rules used to set controls to determine when to send information: static information changes, exceeding of a threshold, or periodic refreshment of topology information are good examples.

Path Computation and Selection

Path selection means a constrained path computation process. Upon the LSP request, the LSR checks out all the request admission rules related to this request. These rules, plus some user-specified rules carried by the LSP request, are parsed into constraints that instruct the LSR only to retrieve the related resources from the TEDB and create a copy of the retrieved resource information in the memory. Constraint shortest path first (CSPF) computation is carried out on this reduced TE information to select a routing path.

Routing

MPLS developed the application of constraint-based routing, which added extensions to existing routing protocols such as OSPF and IS-IS. These extensions were designed to enable nodes to exchange control information about topology, resource availability, and policy constraints [50]. RSVP-TE and CR-LDP establish the label forwarding state used to compute the path. In optical networks, the dynamics of routing are made significantly more complex by the potentially huge number of numbered and unnumbered links (e.g., the impossibility of assigning IP addresses to individual fibers, lambdas, and TDM channels) and the difficulty in identifying physical port connectivity information.

One of the important extensions to MPLS to address routing issues in optical networks is the concept of the LSP hierarchy. LSPs of the same type (i.e., FSC, L2SC, LSC, TDM or PSC) that enter a network on the same node and leave a network on the same node can be bundled together; that is, aggregated and tunneled within a single LSP. This handles the fact that optical networks have discrete bandwidths that are largely wasted if they are transmitting a lower bandwidth stream from, for example, an IP network. Rather than use an entire 2.488 Gbit/s optical link for a 100 Mbit/s LSP data stream, the lower bandwidth stream can be tunneled through the optical LSP leaving the rest of the bandwidth for other data flows. There is a hierarchy that dictates the order in which these LSPs are aggregated or nested. The hierarchy orders the FSC interfaces at the higher order, followed by LSC interfaces, followed by TDM interfaces, with PSC interfaces at the bottom. The other important benefit of this hierachy [34] is the aggregation of information that would otherwise need to be disseminated in its original detail over the network.

Signaling

A generalized label request in GMPLS is signaled with RSVP-TE and it includes three kinds of information: (i) the LSP encoding type, which represents the nature of the LSP and it indicates the way the data are framed in the LSP (values represent packet, Ethernet, SDH or SONET, lambda, fiber or digital wrapper); (ii) the switching type used by the LSP that is being requested on a link, and this value is normally the same across all the links of an LSP; the basic GMPLS switching types are PSC, L2SC, TDM switch capable, LSC, or FSC; (iii) the generalized payload identifier (G-PID) is generally based on standard Ethertypes for Ethernet LSPs or other standard for non-packet payloads such as SONET/SDH, G.709, lambda encodings. The establishment of the LSP itself (i.e., the reservations of physical resources (interfaces) between adjacent nodes) is done through signaling messages [29].

Adjacent network elements have four basic interactions that involve signaling messages: create, delete, modify, and query messaging. Create means to create a hop that is within an end-to-end path. The create request requires information about the hop list, since the upstream node also suggests a label to the next node. Suggesting the label in advance means that the upstream node can begin to configure its hardware in advance, thus saving time and reducing latency. However, the suggested label can be rejected by the next node, in which case the upstream node will have to reconfigure itself. Since lower order LSPs can tunnel through higher granularity LSPs [34], the create request needs to include the ID of the higher order LSP in its hop list. Delete transactions deallocate a link, meaning it must be torn down. The modify request changes the parameters of the link and the query request asks for the status of a link within an LSP.

Link Management

To enable communication between nodes for routing, signaling, and link management, control channels must be established between a node pair. In the context of GMPLS, indeed, a pair of nodes (e.g., photonic cross-connects) may be connected by tens of fibers, and each fiber may be used to transmit hundreds of wavelengths in the case that DWDM is used. Multiple fibers and/or multiple wavelengths may also be combined into one or more bundled links for routing purposes.

Link management is a collection of procedures between adjacent nodes that provide local services such as:

- control channel management, which sets up the out-of-band control channel between the nodes;
- link verification, which certifies that there is connectivity between links;
- link property correlation, confirming that the mappings of interface IDs and aggregate links are consistent;

- fault management to localize which data link has failed;
- authentication, which confirms the identity of the neighboring node.

The LMP has been defined to fulfill these operations. The LMP has been initiated in the context of GMPLS, but it is a generic toolbox that can also be used in other contexts. Control channel management and link property correlation procedures are mandatory per LMP. Link connectivity verification and fault management procedures are optional.

The control channels that are used to exchange the GMPLS control information exist independently of the managed links.

Restoration and Protection

GMPLS enables a physically separate control channel from the data bearer channel for the signaling to be done out of band. It is especially relevant for protection and restoration, as one among the most important GMPLS extensions to MPLS. Through signaling to establish back-up paths, GMPLS offers the option of both protection against failed links and protection against a failed path. When a route is being computed, the ingress LSR also determines a back-up path through the network (also termed a secondary path [7,8]). When protecting on an end-to-end basis, the concept of shared risk groups (SRGs) can be used to guarantee that the secondary path does not share any physical links in common with the original one. The terminology used in GMPLS for the types of end-to-end protection are [9]:

- **1 + 1 protection.** simultaneous data transmission over two physically disjoint paths and a selector is used at the receiving LSR to choose the best signal.
- ***M*: *N* protection.** *M* pre-allocated secondary paths are shared between *N* primary paths. Data is not replicated on to the back-up path but is assigned to that path only in the case of failure.
- **1: *N* protection.** one pre-allocated secondary path is shared among *N* primary paths.
- **1: 1 protection.** one dedicated secondary path is pre-allocated for one primary path.

GMPLS allows per-segment-based path protection mechanisms [16].

Restoration is a reactive process, involving troubleshooting and diagnostics, dynamic allocation of resources, and route recalculation to determine the cause of a failure and route around it. The stages of restoration include failure detection, failure notification, traffic restoration, and post restoration. Failure detection can differ depending on the interface. In SONET/SDH, the change in performance of a physical signal may trigger an alarm indication signal. When the failure notification phase is complete at the node, the controller then triggers a restoration scheme.

GMPLS provides for line, segment, and path restoration. Paths can be dynamically restored; that is, they can be rerouted using an alternate intermediate route, if a path fails. The route can be between (i) the two node adjacent to the failure in the link restoration case, (ii) two intermediate nodes in the segment restoration case or (iii) the two edge nodes (ingress node, egress node) in the path restoration case.

GMPLS provides some unique options for restoration that allow flexibility. Mesh restoration, for example, means that instead of dedicating spare capacity for a back-up connection for each path, the originating node re-establishes the connection using available capacity after the failure event. Optimization algorithms help to restore more efficiently; for example, alternate routes can be precomputed by the originating node and cached in case of need. Restored paths may reuse original path nodes or new ones, but using existing ones is more efficient since the new nodes could be used as elements in other LSPs. In the section on GMPLS applications and benefits, the use of mesh protection in providing tiered services will be discussed. Although SONET/SDH voice channels require restoration in 50 ms or less due to the quality requirements of voice, some compelling arguments can be made for cost savings using mesh restoration techniques to use bandwidth more efficiently for data services.

MIB Modules

The introduction of GMPLS shifts some provisioning roles from network operators to the network service providers, since GMPLS controllers are managed by the NMS. The service provider should utilize an NMS and standard management protocols, such as the simple network management protocol (SNMP) [69–71] – with the relevant MIB modules – as standard interfaces to configure, monitor, and provision LSRs. The service provider may also wish to use the command line interface (CLI) provided by vendors with their devices.

SNMP MIB modules require additional flexibility, due to the versatility of the GMPLS control plane technology, to manage the entire control plane. Based on the Internet-standard management framework and existing MPLS-TE MIBs [72], various MIBs were developed for representing the GMPLS management information [73], and the work is still ongoing. As an example, the following is a summary of MIB objects for setting up and configuring GMPLS traffic-engineered tunnels:

- tunnel table (gmplsTunnelTable) for providing GMPLS-specific tunnel configuration parameters;
- tunnel hop, actual tunnel hop, and computed tunnel hop tables (gmplsTunnelHopTable, gmplsTunnelARHopTable, and gmplsTunnelCHopTable) for providing additional configuration of strict and loose source routed tunnel hops;
- performance and error reporting tables (gmplsTunnelReversePerfTable and gmplsTunnelErrorTable).

Other Functionality

An “LMP adjacency” is formed between two nodes that support the same LMP capabilities. Multiple control channels may be active simultaneously for each adjacency. A control channel can be either explicitly configured or automatically selected; however, LMP currently assumes that control channels are explicitly configured while the configuration of the control channel capabilities can be dynamically negotiated.

For the purposes of LMP, the exact implementation of the control channel is left unspecified. The control channel(s) between two adjacent nodes is no longer required to use the same physical medium as the data-bearing links between those nodes. For example, a control channel could use a separate wavelength or fiber, an Ethernet link, or an IP tunnel through a separate management network.

2.2.1.6 Advantages of GMPLS

Scalability

Optional configuration can be used to increase the scalability of GMPLS for large transport networks, and especially in the addressing and the routing. The concepts of unnumbered links and link bundling were introduced with the extensions to signaling (RSVP-TE and CR-LDP) through an ERO and record routing object (RRO) and routing (OSPF-TE and IS-IS-TE) protocols through the TE-link attributes, for combining the intelligence of IP with the scalability and capacity of optical transport layers. These two mechanisms can also be combined. In addition, the hierarchical LSP concept and the LMP both contribute to the improved scalability and, therefore, to the performance of the control plane, which should not depend on the scale of the network and should be constant regardless of network size.

Flexibility

GMPLS brings the mechanisms and functions to the control plane in the network in a way that it should have flexibility and provide full control and configurability in the task of optimizing the use of network resources by provisioning LSPs. In TE, GMPLS technical components provide and maintain flexibility so that the network-level coordination of the resource is optimal.

Operation Automation

From the operational cost savings due to the automation capability of GMPLS, the control plane relieves operators from unnecessary, painful, and time-consuming manual operations of network service provisioning. Reduction in the provisioning cycle of LSPs due to automation helps service providers – even running proprietary element and NMSs – to provision services at a reduced cost.

Adopting GMPLS means the automation and TE capabilities spanning multiple service provider domains. It means also that equipments from different vendors can be integrated in the same network.

Path Computation, Selection, and Optimization

This complexity of path computation usually exceeds the computational capabilities of ordinary LSR. A specialized network element has been proposed by the IETF to overcome this problem.

The PCE serves as a computing entity, specialized for constraint-based path computation. The network entities (routers, switches, NMS, or other service element) requesting the PCE service is called a path computation client (PCC) [38]. The protocol requirements between both entities are proposed in [37]. The PCC request includes source and destination node and additional path constraints. The PCE responds with a NO-PATH object if no path is found or it includes a list of strict or loose hops if a path has been found.

In this study we compared the newly proposed five different PCE approaches of the newly available IETF draft by Oki *et al.* [36]:

- single multilayer PCE
- PCE/VNT manager cooperation
- multiple PCEs with inter-PCE communication based on the PCE protocol
- multiple PCEs without inter-PCE communication
- multiple multilayer PCE.

Our comparisons are quantitatively in respect to the path setup delay in an SDH over wavelength-division multiplexing (WDM) network scenario. We assumed a reconfigurable wavelength-switched optical network as well as a reconfigurable TDM-switched network. Additionally, the single multilayer PCE approach is also experimentally evaluated in the MLTE demonstration scenario described in Section 4.4.

We found that the amount of multilayer paths, which legitimate the argument of PCEs, is very low in the considered SDH/WDM core network scenario. Thus, in our scenario, PCEs are legitimated more by the complexity of constraint-based path computation requests and by the reduced computation time than by extensive ML path computation.

Because of the small frequency of multilayer paths, the minimum and mean path setup times do not show much difference in our scenario. The expected path setup delay is in the order of tens of milliseconds. The maximum path setup delay (representing multilayer paths) triples the path setup delay in certain scenarios.

We found that, among all PCE deployment scenarios, one single multilayer PCE performs best. In all cases, path setup delays are far less than a second even in the case of multilayer paths. The small communication overhead and the reduced number of PCEs needed back up this decision.

Multilayer/Multiregion TE Support

The so-called multilayer/multiregion network (MRN)concept is closely linked to GMPLS [51]. A region is a control-plane-related notion and is based on interface switching capability [34]. A GMPLS switching type (PSC, L2SC TDM, LSC, FSC) describes the ability of a node to forward data of a particular data plane technology and uniquely identifies a network region. On the other hand, a network layer is a layer in the data plane; typically, based on client–server relationship, the usual layers are PSC over TDM, PSC over LSC, and PSC over PSC; that is, interface switching capabilities.

A network comprised of multiple switching types controlled by a single GMPLS control plane instance is called an MRN. The notion of LSP region is defined in [34]. That is, layers of the same region share the same switching technology and, therefore, need the same set of technology-specific signaling objects. In general, we use the term layer if the mechanism of GMPLS discussed applies equally to layers and regions (e.g., VNT, virtual TE-link), and we specifically use the term region if the mechanism applies only for supporting an MRN.

MLTE in GMPLS networks increases network resource efficiency, because all the network resources are taken into account at the same time. However, in GMPLS MRN environments, TE becomes more complex, compared with that in single-region network environments. A set of lower layer FA-LSPs provides a VNT to the upper layer. By reconfiguring the VNT (FA-LSP setup/release) according to traffic demands between source and destination node pairs of a layer, the network performance factors (such as maximum link utilization and residual capacity of the network) can be optimized.

Expectation of Service Provider

With control plane resilience, the network element can discover the existing cross-connects after recovering from the control plane protocol failure. For example, when control plane failure only occurs within one network element, the LSR, such as an OXC, will still be in place carrying data traffic. After recovery of the control plane, the network element should automatically assess the data plane (i.e., the OXCs here) and reconfigure its control plane so that it can synchronize with other control plane entities.

Flexibility of the transport layers means a fair allocation of bandwidth between competing routes dealing with bursts of activity over many timescales. Reconfigurability increases network flexibility and responsiveness to dynamic traffic demands/changes.

The service provider can also set up the service where the network dynamically and automatically increases/decreases bandwidth as traffic volumes/patterns change. If the demand for bandwidth increases unexpectedly, then additional bandwidth can be dynamically provisioned for that connection. This includes overflow bandwidth or bandwidth over the stated contract amount. The triggering parameters may be utilization thresholds, time of day, day of month, per-application volumes, and so on.

Bandwidth-on-demand (BoD) provides connectivity between two APs in a non-preplanned, fast, and automatic way using signaling of GMPLS. This also means dynamic reconfiguring of the data-carrying capacity within the network, routing, and signaling, and that restoration is also considered here to be a BoD service.

Economic Models and Benefits of GMPLS

To achieve ever-greater efficiencies, service providers must streamline their operations by reducing the number of people required to deliver services, and reducing the time required to activate and to troubleshoot network problems. To accomplish these objectives, they are focusing on automated provisioning through a distributed GMPLS control plane, which is designed to enable multi-vendor and multilayer provisioning in an automated way. Therefore, requests for services in the data network that may require connectivity or reconfigurations can happen in a more automated fashion. In addition, instead of provisioning on a site-by-site basis, the control plane creates a homogeneous network where provisioning is performed network-wide.

2.2.2 Evolution in Integrated Architectures

2.2.2.1 The Path Computation Engine

The interior gateway protocols (IGP) in IP networks relies on fully distributed routing functions. Each network element that is part of the routing domain has its own view of the network stored inside the IGP routing table (link-state database). For scalability, performance, and security reasons, link-state routing protocols (e.g., IS-IS-TE and OSPF-TE) are used in today's carrier networks. Constraint-based path computation, typically using CSPF [27], is a fundamental building block for TE systems in MPLS and GMPLS networks. Path computation in large, multidomain, multiregion, or multilayer networks is highly complex, and may require special path computational components and efficient cooperation between the different network domains.

A PCE is an entity that is capable of computing paths in a network graph, applying computational constraints [38]. A PCE supports a network with a distributed control plane as well as network elements without, or with rudimentary, control plane functions. Network nodes can query the PCE for calculating a usable path they want to set up via signaling. The PCE entity can be seen as an application on a network node or component, on a separate out-of-network server. PCEs applied to GMPLS networks are able to compute paths by interfacing with one TEDB fed by a routing controller engine, and considering the bandwidth and other constraints applicable to the TE-LSP service request enhances this definition for GMPLS networks: "In GMPLS networks, PCE provides GMPLS LSP routes and optimal virtual network topology reconfiguration control, and assesses whether a new higher switching capability LSP should be established when a new LSP is triggered. PCE also handles inter-working between GMPLS and IP/MPLS networks, both of which will coexist at some point during the migration process." The IETF defines a region or an LSP region which refers to a switching technology domain; interface switching capabilities construct LSP regions. The deployment of a dedicated PCE can be motivated under several circumstances:

1. **CPU-intensive path computation** – for example, considering overall link utilization; computation of a P2MP tree; multicriteria path computation, such as delay and link utilization; integrated multilayer path calculation tasks.
2. **Partial topology knowledge** – the node responsible for path computation has limited visibility of the network topology towards the destination. This limitation may, for example, occur when an ingress router attempts to establish an LSP to a destination that lies in a separate domain, since TE information is not exchanged across the domain boundaries.
3. **Absence of the TEDB** – the IGPs running within parts of the network are not able to build a full TEDB; for example, some routers in the network do not support TE extensions of the routing protocol.
4. **A node is located outside the routing domain** – an LSR might not be part of the routing domain for administrative reasons.
5. **A network element lacks control plane or routing capability** – it is common in legacy transport networks that network elements do not have controller. For migration purposes, the path computation can be performed by the PCE on behalf of the network element. This scenario is important for interworking between GMPLS-capable and non-GMPLS-capable networks.
6. **Backup path computation for bandwidth protection** – a PCE can be used to compute backup paths in the context of fast reroute protection of TE-LSPs.

The main driver for a PCE when it was born at the IETF was to overcome particular problems in the path computation in the inter-area environment; that is, inter-area/autonomous system (AS) optimal path computation with nodes having partial visibility of other domains only and computation of inter-area/AS diverse paths with nodes having partial visibility of other domains only. However, a PCE is seen as suitable for performing complex path computation in single-domain scenarios; for example, for MLTE concepts or in migration scenarios with non-GMPLS-capable nodes.

2.2.2.2 The Role of the Management Plane: Provisioned Connection

NMS functions manage a set of network equipment that is inventoried in the resource inventory database. This resource inventory database can be compounded with tunable and reconfigurable optical add–drop multiplexers (R-OADMs), transport service switches, IP/MPLS routers, or other switching network elements. These managed network elements embed controller functions that are classified in different agents, such as GMPLS agents, RSVP-TE subagents, OSPF-TE subagents and LMP subagents. These agents report the information in a repository attached to the NMS that gathers and manages the information about the network devices. The subagents report the network element management information to the NMS periodically or in a burst mode when there are changes on the network infrastructure due to upgrades with new equipment or maintenance periods.

Network management functions are composed of a set of activities, tools, and applications to enable the operation, the administration, the maintenance, and the provisioning of networked systems providing connectivity services to the application users. Administration includes activities such as designing the network, tracking the usages, and planning the infrastructure upgrades. Maintenance includes diagnosis and troubleshooting functions. According to automatic switched transport network (ASTN) recommendations [40], connection provisioning concerns the settings of the proper configuration of the network elements so that the connection is established by configuring every network element along the path with the required information to establish an end-to-end connection. Connection provisioning is provided either by means of the NMS or by manual intervention. When an NMS is used, an access to the resource inventory database of the network is usually required first, to establish the most suitable route, and then to send commands to the network elements that support the connection. When a connection is provisioned by the NMS, it is usually referred to as a permanent connection. The ITU-T introduced the fault, configuration, accounting, performance, and security (FCAPS) framework. This framework for network management is part of a bigger model that is the Telecommunications Management Network model of the ITU-T Recommendation series M.3000.

The management plane and control plane functions offer complementary functions, which can be used to construct an optimal provisioning control approach that is efficient, fast, and cost effective.

The functional components of the NMS are:

- network resource management inventory
- network resource provisioning – that is, network element configuration
- network resource performance management
- support for resource trouble management
- manage network service inventory
- network service configuration and activation.

By opposition to permanent connection, signaled connections are established on demand by communicating end points within the control plane using a dynamic protocol message exchange in the form of signaling messages (e.g., RSVP-TE protocol and messages). These messages flow across either the internal network-to-network interface (I-NNI) within the control plane instance or UNI/external network-to-network interface (E-NNI) between two distinct control plane instances. GMPLS-based connection establishment is referred to as a switched connection and it requires network naming and addressing schemes and control protocol functions.

2.2.2.3 Hybrid (Control-Management) Solutions

Architectures in production networks in the short term would essentially combine certain control plane and management plane components into a hybrid scheme. This type of connection establishment exists whereby the transport network provides a permanent connection at the edge to the client network. And it utilizes a signaled connection within the network to provide end-to-end connections between the permanent connections at the network edges; that is, CE node and PE node. Within the transport network, the optical connections are established via control signaling and routing protocols. Permanent provisioning, therefore, is only required on the edge connections and, consequently, there is usually no UNI at the edge nodes. This type of network connection is known as a soft permanent connection (SPC).

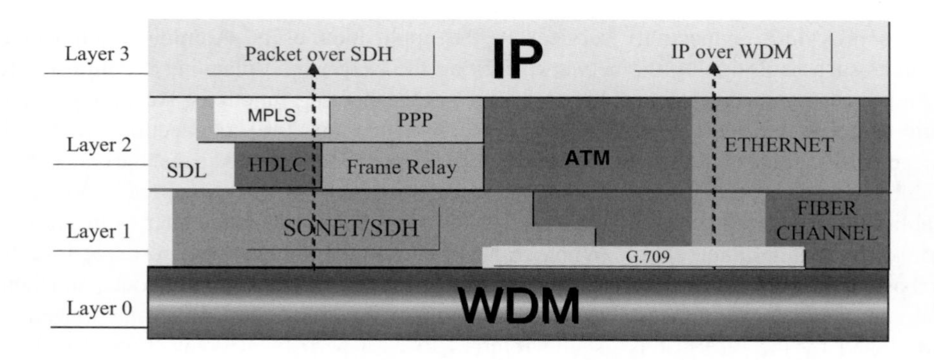

Figure 2.2 Two examples of multilayer network architecture models

From the perspective of the end points, an SPC appears no different than a management-controlled connection; that is, a permanent connection.

2.3 Multilayer Interactions and Network Models

This section develops on the different options of network control plane configurations over a multilayer transport network from the interactions between several control networks to the integration of one uniformed control network instance.

2.3.1 Introduction

Operator networks have been upgraded with different network technologies, each providing its own set of functionalities, defining its own switching capability and framing. Enabled with the GMPLS protocol architecture, the functional interactions between the different network layers can become more uniform and allow a reduction in the complexity of the global control and management of the network. With GMPLS, the data network (typically L3 or L3 networks) and transport network (typically OTNs) layer convergence can be addressed by providing the end-to-end LSPs (i.e., GMPLS-controlled connections) integrating the requirements expressed by the users of the connections.

The network service layer introduced in Section 2.1 can be referenced with the OSI communications model as illustrated in Figure 2.2. Each network service layer has the property that it only uses the connectivity service offered by the server layer (i.e., the layer below) and only exports functionality to the layer above. The different network service layers and their interaction models express how the transport network layer server should interact with the client network layer to establish connections (permanent connections, soft-permanent connections, or signaled connections [40]) in support of the user-network services.

Two classic ways of transporting IP traffic by optical networks are packet over SONET/SDH (PoS) and IP over WDM. Within a GMPLS-controlled network, each network layer[2] $i-1$ can carry another network layer i as it corresponds to the concept of nesting LSPs; that is, LSPs

[2] A network layer, also referred as a "switching layer," is defined as a set of data links with interfaces that have the same switching and data encoding types, and switching bandwidth granularity. Examples of network layers are SDH VC-4, SONET STS-3, G.709 OTU-1, Ethernet, IP, ATM VP, and so on.

originated by other LSRs at layer i into that LSP at layer $i-1$ using the label stack construct, defined by the LSP hierarchy [34]. For example for $i=2$, L2-LSPs (Ethernet) can be carried by TDM LSPs or G.709 LSPs.

Nesting can occur between different network layers within the same TE domain, implying interfaces switching capability information to be controlled in a hierarchical manner as shown in Figure 2.2. With respect to the switching capability hierarchy, layer i is a lower order switching capability (e.g., Ethernet) and layer $i-1$ is the higher order switching capability (e.g., SONET/SDH).

Nesting of LSPs can also occur at the same network layer. For example at the layer 1, a lower order SONET/SDH LSP (e.g., VT2/VC-12) can be nested in a higher order SONET/SDH LSP (e.g., STS-3c/VC-4). In the SONET/SDH multiplexing hierarchy, several levels of signal (LSP) nestings are defined.

2.3.1.1 Overlay Model

The overlay model refers to telecom carriers or optical backbone (bandwidth) providers who lease their network infrastructure facilities to Internet service providers (ISPs). This model is based on a well-defined client–server relationship with controlled network interfaces (or reference points) between the provider networks and customer networks involved. The overlay model mandates a complete separation of the client network control (e.g., based on MPLS architecture [27]) and the transport network control plane (e.g., based on GMPLS [42]). Only a controlled amount of signaling messages may be exchanged. As a consequence, the overlay model is very opaque. The client network routing and signaling controllers are independent of the routing and signaling controllers within the transport network domain. The two independent control planes interact through a UNI [39], defining a client–server relationship between the customer network and the transport network.

2.3.1.2 Peer Model

Compared with the overlay model, the peer model does not restrict any control routing information exchanged between the network switching layers. This model is relevant and can be optimal when a carrier network is both a network infrastructure provider (NIP) and an ISP. In this case, the provider networks can align the topological design of their transport network with the service operations of their data network, but they might be in conflict with some national or international policies.[3] The client network control plane acts as a peer of the GMPLS transport network control plane, implying that a dual instance of the control plane is running over the data network (e.g., IP/MPLS) and optical network (e.g., SDH/GMPLS), as illustrated in Figure 2.3. The peer model entails the tightest coupling between the customer networks and the transport network. The different nodes (CE, PE or P) can be distinguished by their switching capabilities; for example, PSC for IP routers interconnected to GMPLS PXCs.

[3] The European Commission, in its Green Paper, has regulated that a formal split be made within telecoms in network operating departments and service provider departments with a formal supplier relationship that is equal to an external vendor/buyer relationship. This relationship must be nondiscriminatory. Similarly, requirements of the FCC Ruling on Interconnection in the USA are encouraging companies to formally separate their network provider and service provider business.

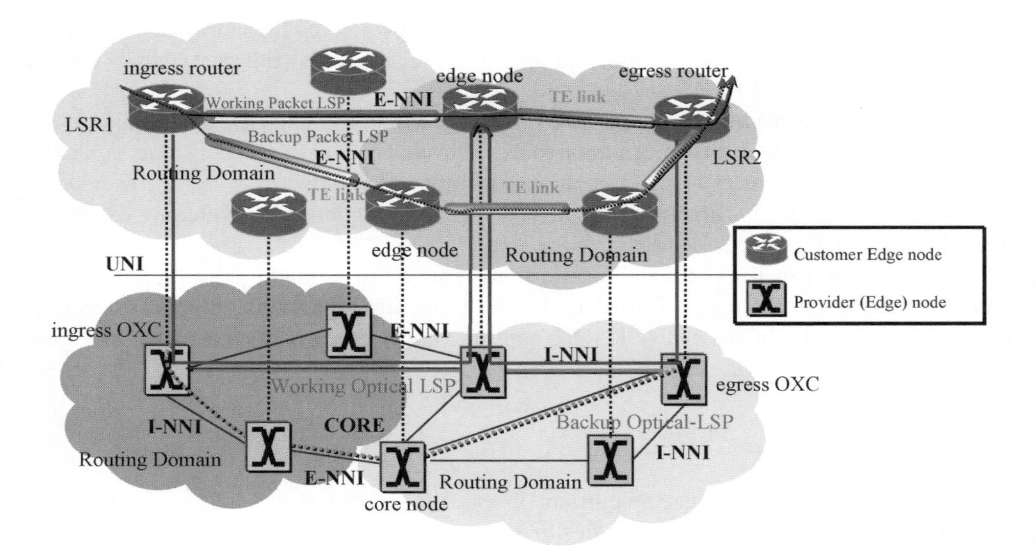

Figure 2.3 Network models and reference points [39]

2.3.1.3 Integrated Model

Compared with the peer model, the integrated model does not require different control plane interfaces between the network layers and different TE domains. The integrated model proposes one single instance of a control network for the customer networks and the provider network. All the nodes are LSRs and they are not classified in different network domains due to the administration they belong to or their interface switching capabilities. Each LSR is an integrated platform able to handle several orders of switching capabilities: for example, IP, Ethernet, TDM, and wavelength. An LSR embeds one GMPLS control plane instance and is able to control different switching-capability interfaces simultaneously. On the one hand, only this model can handle a global optimization of the network resource usages through the network; for example, packet, layer 2 (Ethernet, ATM, etc.), TDM (SONET, SDH, etc.), lambda (G.709), and fiber switching capabilities; on the other hand, this model has to face the scalability challenges to integrate the control of TE-links belonging to different switching capabilities and to control their states in a very reactive manner.

2.3.1.4 User-to-Network Interface

The UNI is a logical network interface (i.e., reference point) introduced in the requirements for the ASTN specification [40] and recommended in [39]. The UNI defines the set of signaling messages that can be exchanged between a node controller of the client network and a server node of the transport network. The server node provides a connection service to the client node; for example, the IP router can signal TDM LSPs on its PoS interfaces. The UNI supports the exchange of authentication and connection admission control messages and provides to CE nodes the address space set of the reachable nodes. The first implementation agreement for a UNI was produced by the Optical Internetworking Forum (OIF) in October 2001 [57]. This OIF implementation agreement is for an overlay model. The signaling messages exchanged between each client node and server node are restricted to LSP connection request/tear down only. The IETF specifies a GMPLS UNI that is applicable for a peer model [50]. Fully compliant with RSVP-TE, the GMPLS UNI allows the end-to-end LSP handling along LSR signaling paths. Some recent contributions at OIF integrate the alignments of RSVP-TE within OIF UNI 2.0 [58].

2.3.1.5 Network-to-Network Interface

The network-to-network interface (NNI) is a logical network interface (i.e., reference point) recommended in the "Requirements for Automatic Switched Transport Network" specification ITU-T G.807/Y.1302. The NNI defines the set of both signaling messages and routing messages that can be exchanged between two server nodes; for example, between two GMPLS-controlled PXCs. There are two types of NNI: one for two different TE domains and one for intra-domain TE: i E-NNI and ii I-NNI respectively.

- An E-NNI assumes an untrusted relationship between the two network domains. The information exchanged between the two nodes located at the edge of the transport network specified within the E-NNI is restricted. The control messages exchanged include the reachability network addresses that are usually summarized, authentication and connection admission control messages, and a set restricted to connection requests of signaling messages. Some contributions have been initiated at the OIF concerning the signaling message exchanges [59], but the project plan for producing an OIF E-NNI Routing 2.0 implementation agreement [60], which started in November 2007, is without significant progress despite some solid contributions from NOBEL2.
- An I-NNI assumes a trusted relationship between two network domains and is usually implemented in the same TE domain or administrative domain. The control information specified within the I-NNI is not restricted. The routing control messages exchanged include connection service topology, LSA (from IGP), and node discovery. The signaling messages can allow controlling of the resources end to end for the two networks of each LSP and its protection.

Network integration usually removes the restrictions imposed by administrative network domains. It covers interworking capabilities and optimized control functions for multiple switching layers running GMPLS with unified signaling and routing approaches for connection (i.e., LSP) provisioning and recovery. Integrated network architectures can include (i) network equipment hosting multiple switching capabilities that are controlled by a single instance of the GMPLS control plane and (ii) seamless collaboration between network domains (e.g., routing areas, autonomous systems [28]) of the network.

Hence, the multilayer interaction network models have been categorized into a horizontal integration and vertical integration framework (Figure 2.4):

- **Vertical integration** Collaborative switching layers are controlled by a single control plane instance: the UNI reference point is coalescing with the I-NNI control interface.

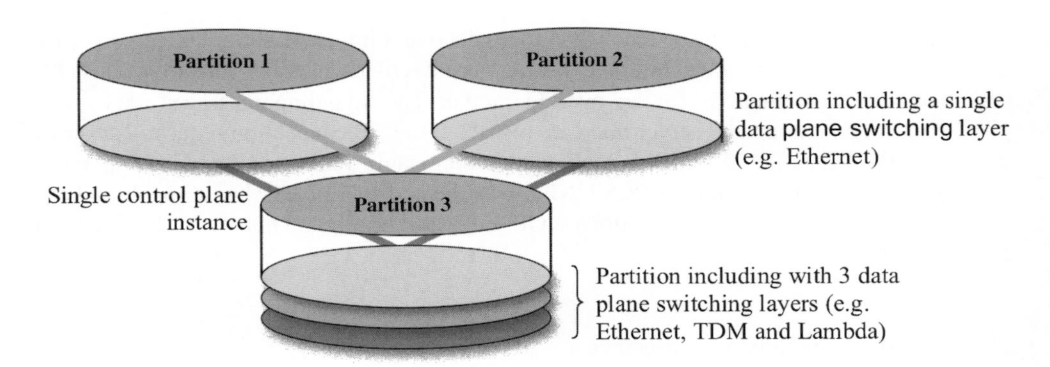

Figure 2.4 Horizontal and vertical integration of network domains with reference to the GMPLS control plane instances (not shown)

- **Horizontal integration** Each administrative entity constituting a network domain is controlled by one single control plane instance usually inferring one common (data plane) switching capability and the control plane topology extends over several routing areas or autonomous systems. This horizontal integration includes control functions such as: (i) signaling and routing in multi-domain transparent networks, (ii) interface adaptation between MPLS and GMPLS control domains, and (iii) proposed multi-domain signaling protocol extensions for the end-to-end service.

2.3.2 Vertical Integration and Models

The GMPLS control network is more complex when more than one network layer is involved. In order to enable two network layers to be integrated, interface adaptation capabilities are required.

2.3.2.1 Towards the IP/Ethernet Convergence

The Ethernet MAC is one of the main (if not the only) layer 2 technology that is going to remain really attractive in the long run. Together with pure layer 3 IPv4/IPv6 packet forwarding, these are the two fundamental data plane building blocks of any future access, metro, and backbone network. Starting from local area network (LAN) environments, Ethernet is under deployment in metropolitan area networks (MANs), and its extensions to metro–core and core network environments are foreseen.

2.3.2.2 Integrating Several Transport Layers

The role, limitations, and strengths of synchronous networking (SONET/SDH) and its association with WDM. How statistical multiplexing and framing can be further enhanced during the migration process to "packetized" and "flatter" multilayer networks with optimized costs of transport equipment.

2.3.3 *Horizontal Integration and Models*

When a single instance of a network controller covers several routing areas or autonomous systems, each having its own TE domain, the model is said to be horizontally integrated. Usually, horizontal integration is deployed within a single network layer spanning several administrative network domains [53], whereas vertically integrated models involve multiple network layers.

2.3.3.1 Multi-Domain Interoperability

In many of today's complex networks, it is impossible to engineer end-to-end efficiencies in a multi-domain environment, provision services quickly, or provide services based on real-time traffic patterns without the ability to manage the interactions between the IP-layer functionality of PSNs and that of the optical transmission layer. According to proponents of automatically switched optical network (ASON)/GMPLS, an optical control plane is the most advanced and far-reaching means of controlling these interactions.

The multi-domain issue and evolution in interconnecting network domains.

Currently, GMPLS is evolving to cover the end-to-end service establishment in multi-domain configuration.

A control plane consists inherently of different control interfaces, one of which is concerned with the reference point called the UNI. From a historical point of view, each transport technology has defined UNIs and associated control mechanisms to be able to automatically establish connections. NOBEL WP4 has examined the currently available standardized UNI definitions, and points out the differences between them as illustrated in Figure 2.5.

The UNIs most applicable for ASON- and GMPLS-controlled networks are being actively worked on by the IETF, OIF, and ITU-T. They show a significant protocol reuse for routing and

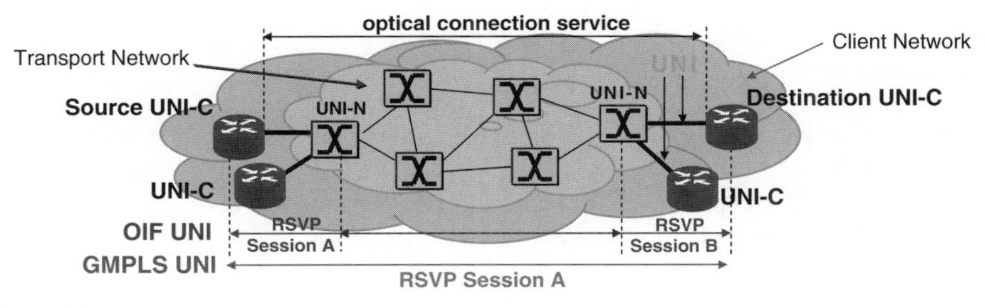

OIF UNI (Overlay Network)

Path message triggered w/o EXPLICIT_ROUTE:

- UNI Session A Tunnel add.: source UNI-N node ID
- If RSVP-TE: Session Tunnel add.: dest. UNI-N node ID
- UNI Session B Tunnel address: dest. UNI-C node ID

Source OXC UNI-N ERO/RRO Processing

- ERO: all OXC hops up-to destination UNI-N (strict)
- Source UNI-N computes the path to reach dest. UNI-N and creates an ERO in the outgoing Path messages

GMPLS UNI (Augmented Network)

Path message triggered with EXPLICIT_ ROUTE:

- Session Tunnel address: destination UNI-C node ID

Source LSR UNI-C ERO/RRO Processing:

- ERO: ingress UNI-N (strict) and dest. UNI-C (loose)
- Source UNI-N computes path to reach dest. UNI-C, updates the ERO and includes it in the outgoing Path messages

Figure 2.5 Network models compared in NOBEL: OIF UNI versus GMPLS UNI

signaling in the different standards bodies. Each standards body has initially used identical fundamental protocols, which have been further developed and modified to fulfill the standards body's requirements, including the European carrier network requirements (i.e., telecom network operators). Thus, the initially fundamental protocols have all evolved in slightly different directions, as at the time there were no significant established liaisons requesting cooperation between them.

The differences were driven by user requirements and are significant enough to lead to incompatibility between the UNIs recommended by different standards bodies. Although the IETF and ITU-T/OIF are both based on RSVP-TE, the requirements are different.

This section reports on the analysis and comparison of the currently available UNI definitions of the IETF, OIF, and ITU-T at the time of the NOBEL project analysis. Furthermore, this section provides a fully GMPLS-compliant UNI definition, and the corresponding technical description, for the support of end-to-end signaling for fast connection provisioning with different options for connection recovery. This definition, called GMPLS UNI, is one of the contributions of NOBEL to achieve convergence towards future high-performance networks providing high-impact, end-to-end services for optical networks (TDM, wavelength or port switching).

A UNI is a reference point over which predefined (i.e. standardized) messages are exchanged with the objective of requesting services from a network. An UNI reference point is logically associated with a physical interface on the interdomain link between the user network and the provider network. The messages exchanged between the user and the provider require a physical bearer. The physical network infrastructure designed for that purpose – that is, to relay control and network management information – is called a data communication network [39].

The following information and procedures are typically managed over UNI:

- endpoint name/address registration – directory service [57]
- authentication, authorization, and connection admission control (CAC)[4]
- connection service messages
- connection request/release + confirmation
- resource discovery
- grade of service (GoS) parameters selection (which are typically captured in service level agreements (SLA)s)
- handling of error and exception conditions during the recovery process.
- non-service affecting attribute modification
- query of attribute values for a given circuit.

There is typically no routing protocol information; that is, link state flooding [28] exchanged over the UNI. This is fundamentally different from other interfaces, such as I-NNI and E-NNI. However, under certain policies, the CE node can request and maintain optical connection services by signaling explicitly an explicit route with, for example, an explicit route object [29,48,59].

[4] A user's connection request towards a carrier network can be specified with a specific time of the day and/or day of the week. The -N controller agent verifies whether connection requests are received during the contracted hours. All information required to make the policy decision (time-of-request receipt) is contained in the UNI signaling message.

The concept of "user" in this context is considered to be the connection requestor, that is, the entity that requests to utilize from the network particular network infrastructural resources, with the typical aim to exchange information with one or more remote entities. The UNI can be single-ended end-users or corporate LAN interconnections, metro network connections, and so on. From the description of "user," a multitude of scenarios can be envisioned under which the UNI needs to operate. This obviously puts significant pressure on the UNI, as it needs to be able to operate under a broad spectrum of different applications areas, requiring the UNI to be able to support all the different types of connection and physical infrastructures. It equally shows the importance of the UNI functionality evolving towards the user requesting automatically the connectivity services offered by the dynamic optical network infrastructure enabled by GMPLS control functions.

There is mainly one view on the recommendations of the UNI that views the UNI as a client–server relation The following nonexclusive list presents some points:

- there is a client–server relation between the connectivity service user and the NIP;
- routing information is not exchanged between the parties involved and it is based on an overlay network interaction model;
- there is no trust relation between the two parties involved, but there is a (commercial) agreement;
- there is a business relationship between an end-user and a provider;
- this business relationship is based on an SLA (i.e., a commercial contract);
- this business relationship is typically transport technology dependent.

In this document we focus solely on the control and management functions and specific extensions required for the UNI client–server relationship.

2.3.3.2 The User-to-Network Interface

The UNI is an asymmetrical interface and is divided into two parts: UNI-C on the client side and UNI-N on the network side, as shown in the Figure 2.5.

The different functional requirements on the UNI client and network sides necessitate the two distinct parts constituting the UNI,; that is, the client-specific part and network-specific part. On UNI-N, an additional routing interface for connection control has, for example, to be provided [39]; that is, the UNI-N has a routing controller that participates in the network-related routing. However, the UNI-N does not distribute that routing information over the UNI towards the UNI-C.

The UNI supports, as a minimum, the following information elements or functions to trigger SPCs or switched connections [39,40]:

- authentication and admission control
- endpoint name and addressing
- connection service messages.

The control plane functions of the UNI-C side are mainly call control and resource discovery. Only limited connection control and connection selection are necessary at UNI-C. The following functions reside at the UNI-N side [39]:

- call control
- call admission control
- connection control
- connection admission control
- resource discovery and connection selection.

The optical network layer provides transport services to interconnect clients such as IP routers, MPLS LSRs, carrier-grade Ethernet switches, SDH cross-connects, R-OADMs, and so on. In its initial form, the OTN uses SONET/SDH as the interface switching-apable interfaces, and the network is migrating to Ethernet framing clients in the future.

The OIF defines the *UNI as the ASON control interface between the transport network and client equipment*. Signaling over the UNI is used to invoke connectivity services that the transport network offers to clients.

The purpose of the OIF UNI is to define interoperable procedures for requesting, configuring, and signaling dynamic connectivity between network equipment clients (e.g., Ethernet switches or IP routers) connected to the transport network. The development of such procedures requires the definition of a logical interfaces between clients and the transport network, the connectivity services (specified as a call in Ref. [39]) offered by the transport network, the signaling protocols used to invoke the services, the mechanisms used to transport signaling messages, and the autodiscovery procedures that aid signaling.

We have the following definitions:

- **Client/user:** network equipment that is connected to the transport network for utilizing optical connection.
- **Transport network:** an abstract representation, which is defined by a set of APs (ingress/egress) and a set of network services.
- **Connection:** a circuit connecting an ingress transport network element (TNE) port and an egress TNE port across the transport network for transporting user signals. The connection may be unidirectional or bidirectional [61].
- **UNI:** The logical control interface between a client device and the transport network.
- **UNI-C:** The logical entity that terminates UNI signaling on the client network device side.
- **UNI-N:** The logical entity that terminates UNI signaling on the transport network side.

2.3.3.3 The OIF or Public UNI

The OIF UNI Connection Services

The primary service offered by the transport network over the UNI is the ability to create and delete connections on demand. A connection is a fixed-bandwidth circuit between ingress and egress APs (i.e., ports) in the transport network, with specified framing [61]. The connection can be either unidirectional or bidirectional. Under OIF UNI 1.0, this definition is restricted to being a TDM connection of payload bandwidth 50.112 Mbit/s (e.g., SONET STS-1 or SDH VC-3) and higher.

The properties of the connection are defined by the attributes specified during connection establishment. Four activities are supported across the UNI, as listed below and illustrated with RSVP-TE [29,48]:

- connection establishment (signaling) – Path/Resv exchange sequences;
- connection deletion (signaling) – PathTear/ResvTear exchange sequences;
- connection notification (signaling) – PathErr/ResvErr/Notify exchange sequences;
- status exchange (signaling) – discovery of connection status;
- autodiscovery (signaling) – discovery of connectivity between a client, the network, and the services available from the network.

Actual traffic (usage of the established connections) takes place in the data plane, not over the service control interface.

For each activity there is a client and a server role.

The OIF UNI Signaling Sequences

UNI signaling refers to the message exchange between a UNI-C and a UNI-N entity to invoke transport network services. Under UNI 1.0 signaling, the following actions may be invoked:

1. **Connection creation:** This action allows a connection with the specified attributes to be created between a pair of APs. Connection creation may be subject to network-defined policies (e.g., user group connectivity restrictions) and security procedures.
2. **Connection deletion:** This action allows an existing connection to be deleted.
3. **Connection status enquiry:** This action allows the status of certain parameters of the connection to be queried.

OIF UNI Supporting Procedures

UNI Neighbor Discovery (Optional)

The neighbor discovery procedure is fundamental for dynamically establishing the interface mapping between a client and a TNE. It aids in verifying local port connectivity between the TNE and the client devices. It also allows the UNI signaling control channel to be brought up and maintained.

Service Discovery (Optional)

Service discovery is the process by which a client device obtains information about the available connectivity from the transport network, and the transport network obtains information about the client UNI signaling (i.e., UNI-C) and port capabilities.

Signaling Control Channel Maintenance

UNI signaling requires a control channel between the client-side and the network-side signaling entities. Different control channel configurations are possible, as defined in the OIF UNI specification [57]. OIF UNI supports procedures for maintenance of the control channel under all these configurations.

There are two service invocation models, one called *direct* invocation and the other called *indirect* invocation. Under both models, the client-side and network-side UNI signaling agents are referred to as UNI-C and UNI-N respectively. In the direct invocation model, the UNI-C functionality is present in the client itself. In the indirect invocation model, an entity called the *proxy* UNI-C performs UNI functions on behalf of one or more clients. The clients are not required to be collocated with the proxy UNI-C.

A control channel is required between the UNI-C and the UNI-N to transport signaling messages. The OIF UNI specification supports an *in-fiber* signaling transport configuration, where the signaling messages are carried over a communication channel embedded in the data-carrying optical link between the client and the TNE. This type of signaling applies only to the direct service invocation. An *out-of-fiber* signaling transport configuration is also supported, where the signaling messages are carried over a dedicated communication link between the UNI-C and the UNI-N, separate from the data-bearing optical links. This type of signaling applies to the direct service invocation model as well as the indirect service invocation model.

Discovery Functions (Optional)

The neighbor discovery procedures allow TNEs and directly attached client devices to determine the identities of each other and the identities of remote ports to which their local ports are connected. The IP control channel (IPCC) maintenance procedures allow TNEs and clients to continuously monitor and maintain the list of available IPCCs.

The protocol mechanisms are based on the LMP.

Service discovery in OIF UNI is optional, and it can be based on OIF-specific LMP extensions [57]. Service discovery is the procedure by which a UNI-C indicates the client device capabilities it represents to the network, and obtains information concerning transport network services from the UNI-N; that is, the signaling protocols used and UNI versions supported, client port-level service attributes, transparency service support, and network routing diversity support.

OIF UNI Extensions [58]

The primary service offered by the transport network over the UNI is the ability to trigger the creation, the deletion, and the modification of optical connections on demand. In the context of the NOBEL project, a connection is a fixed-bandwidth circuit between ingress and egress APs (i.e., ports) in the transport network, with specified framing. The connection can be either unidirectional or bidirectional. Under UNI 2.0, this connection can be a SONET service of bandwidth VT1.5 and higher, or an SDH service of bandwidth VC-11 and higher, an Ethernet service, or a G.709 service. The properties of the connection are defined by the attributes specified during connection establishment.

The following features are added in OIF UNI 2.0 [58]:

- separation of call and connection controllers as recommended in [39]
- dual homing for diverse network infrastructure provider routing
- nondisruptive connection modification through rerouting traffic-engineered tunnel LSPs; that is, implementing make-before break [29]
- $1:N$ signaled protection ($N \geq 1$) through segment LSP protection [8] or end-to-end LSP protection [7]
- Sub-TDM signal rate connections (SONET/STS-1, SDH/VC-12, SDH/VC-3, etc.)
- transport of carrier-grade Ethernet services [22,26]
- transport of wavelength connection services as recommended with G.709 interfaces
- enhanced security.

UNI Abstract Messages

This section describes the different signaling abstract messages. They are termed "abstract" since the actual realization depends on the signaling protocol used. OIF UNI describes LDP and RSVP-TE signaling messages corresponding to the abstract messages that can be exchanged between a pair of CE nodes implementing the UNI-C function and the PE node implementing the UNI-N functions. Abstract messages comprise connection create request (Path), connection create response (Resv), connection create confirmation (ResvConf), downstream connection delete request (PathTear), upstream connection delete request (ResvTear), connection status enquiry, connection status response, and notification (Notify).

The attributes are classified into identification-related, signaling-related, routing-related, policy-related, and miscellaneous. The encoding of these attributes would depend on the signaling protocol used.

2.3.3.4 The IETF GMPLS UNI

The following section describes the UNI from the IETF point of view, termed a private UNI [2]. In Ref. [62] the IETF describes the signaling messages exchanges that can be configured between IP network clients of optical network servers. In the NOBEL project, the network model considered consists of IP routers[5] attached to a set of optical core networks and connected to their peers over dynamically signaled optical channels. In this environment, an optical sub-network may consist entirely of transparent PXCs or OXCs with optical–electrical–optical (OEO) conversions. The core network itself is composed with optical nodes incapable of switching at the granularity of individual IP packets.

With reference to [62], three logical control interfaces are differentiated by the type and the possible control of information exchanges: the client–optical internetwork interface (UNI), the internal node-to-node interface within an optical network domain (I-NNI), and the E-NNI between two network domains. The UNI typically represents a connection service boundary between the client packet LSRs and the OXC network [15]. The distinction between the I-NNI and the E-NNI is that the former is an interface within a given network under a single administration (e.g., one single carrier network company), while the latter indicates an interface at the administrative boundary between two carrier networks. The I-NNI is typically configured between two sets of network equipments within the same routing area or autonomous system. The I-NNI and E-NNI may thus differ in the policies that restrict routing information flow between nodes. Ideally, the E-NNI and I-NNI will both be standardized and vendor-independent interfaces. However, standardization efforts have so far concentrated on the E-NNI [59,60]. The degree to which the I-NNI will become the subject for standardization is yet to be defined within a roadmap.

The client and server parts of the UNI are essentially two different roles: the client role (UNI-C) requests a service connection from a server; the server role (UNI-N) can trigger the establishment of new optical connections to fulfill the QoS parameters of the connection request, and assures that all relevant admission control conditions are satisfied. The signaling messages across the UNI are dependent on the set of connection services defined across it and the manner in which the connection services may be accessed.

[5] The routers that have direct physical connectivity with the optical network are referred to as "edge routers" with respect to the optical network.

The service available at this interface can be restricted, depending on the public/private configuration of the UNI. The UNI can be categorized as public or private, depending upon context and service models. Routing information (i.e., topology and link state information) can be exchanged across a private UNI. On the other hand, such information is not exchanged across a public UNI, or such information may be exchanged with a very explicit routing engine configuration.

Connection Service Signaling Models

Two service-models are currently defined at the IETF, namely the unified service model (vertically) and the domain services model (horizontally). Under the unified model, the IP and optical networks are treated together as a single integrated network from a routing domain point of view. In principle, there is no distinction between the UNI, NNI, and any other control interfaces.

The optical domain services model does not deal with the type and nature of routing protocols within and across optical networks. An end-system (i.e., UNI-C and UNI-N) discovery procedure may be used over the UNI to verify local port connectivity between the optical and client devices, and allows each device to bootstrap the UNI control channel. This model supports the establishment of a wavelength connection between routers at the edge of the optical network. The resulting overlay model for IP over optical networks is discussed later. Under the domain service model, the permitted services through the UNI are as follows:

- lightpath creation
- lightpath deletion
- lightpath modification
- lightpath status enquiry
- service discovery, restricted between UNI-C and UNI-N.

Routing Approaches

Introduction

The following routing approaches are closely related to the definition of the UNI and the interconnection models considered (overlay, augmented, peer).

Under the peer model, the IP control plane acts as a peer of the OTN control plane (single instance).

Under the overlay model, the client layer routing, topology distribution, and signaling protocols are independent of the routing, topology distribution, and signaling protocols within the optical domain. The two distinct control planes interact through a user network interface defining a separated client–server relationship. As a consequence, this model is the most opaque, offers less flexibility, and requires specific rules for multilayer routing.

Finally, under the augmented model, there are separate routing instances in the IP and optical domains, but certain types of information from one routing instance can be passed through to the other routing instance.

Integrated Routing (GMPLS)

This routing approach supports the peer model with the control from a single administrative domain. Under the integrated routing, the IP and optical networks are assumed to run the same instance of an IGP routing protocol (e.g., OSPF-TE) with suitable TE extensions for the "optical networks" and for the "IP networks." These TE extensions must capture optical link parameters and any routing constraints that are specific to optical networks. The virtual topology and link state information stored in the TEDB and maintained by all nodes' (OXCs and routers) routing engines may be identical, but not necessarily. This approach permits a router to compute an end-to-end path to another router considering the link states of the optical network.

The selection of the resources in all layers can be optimized as a whole, in a coordinated manner (i.e.,, taking all layers into account). For example, the number of wavelength LSPs carrying packet LSPs can be minimized.

It can be routed wavelength LSPs that provide a virtual topology to the IP network client without reserving their bandwidth in the absence of traffic at the IP network client, since this bandwidth could be used for other traffic.

Domain-Specific Routing

The domain-specific routing approach supports the augmented interconnection model. Under this approach, the routing processes within the optical and IP domains are separated, with a standard border gateway routing protocol running between domains. IP inter-domain routing based on the border gateway protocol (BGP) is usually the reference model.

Overlay Routing

The overlay routing approach supports the overlay interconnection model. Under this approach, an overlay mechanism that allows edge routers to register and query for external addresses is implemented. This is conceptually similar to the address resolution mechanism used for IP over ATM. Under this approach, the optical network could implement a registry that allows edge routers to register IP addresses and VPN identifiers. An edge router may be allowed to query for external addresses belonging to the same set of VPNs that it belongs to. A successful query would return the address of the egress optical port through which the external destination can be reached.

IETF GMPLS UNI Functionality [50]

Routing information exchange may be enabled at the UNI level, according to the different routing approaches above. This would constitute a significant evolution: even if the routing instances are kept separate and independent, it would still be possible to dynamically exchange reachability and other types of routing information.[6]

Addressing

The IETF proposes two addressing schemes. The following policies are relevant:

- In an overlay or augmented model, an end client (edge node) is identified by either a single IP address representing its node-ID, or by one or more numbered TE links that connect the client and the core node.
- In the peer model, a common addressing scheme is used for the optical and client networks.

[6] Another more sophisticated step would be to introduce dynamic routing at the E-NNI level. This means that any neighboring networks (independent of internal switching capability) would be capable of exchanging routing information with peers across the E-NNI.

Table 2.3 IETF models and approaches

Signaling control	Interconnection model	Routing control	UNI functionality
Uniform end-to-end	Peer	Integrated (I-NNI)	Signaling + common link state database (TEDB)
Overlay (edge-to-edge)	Augmented	Separated (E-NNI)	Signaling + inter-layer routing
	Overlay	No routing	Signaling

Signaling through the UNI

The IETF proposes to use standard GMPLS signaling for the UNI, which can be configured for a client–server model in the case of a per domain signaling model and for end-to-end integrated signaling in the case of a unified service model. A comparison of the different UNI signaling features is shown in Tables 2.3 and 2.4.

Overlay Service Model

The signaling for UNI considers a client–server relationship between the client and the optical network. Usually the switching capability of the client network is lower than the transport network. The source/destination client addresses are routable, and the identifier of the session is edge-to-edge significant. In principle, this implies several signaling sessions used throughout the UNI, I-NNI, and E-NNI that are involved in the connection.

The starting point for the IETF overlay model (IETF GMPLS UNI) is the use of the GMPLS RSVP-TE protocol specified in Ref. [10]. Based on that protocol, the draft GMPLS specifies mechanisms for UNI signaling that are fully compliant with the signaling specified in Refs. [10,48]. There is a single end-to-end RSVP session for the user connection. The first and last hops constitute the UNI, and the RSVP session carries the LSP parameters end to end.

Furthermore, the extensions described in GMPLS address the OIF UNI shortcomings and provide capabilities that are required in support of multilayer recovery.

Unified Service Model

In this model, the IP and optical networks are treated together as a single integrated network from a control plane point of view. In principle, there is no distinction between the

Table 2.4 Comparison between the OIF UNI and the IETF UNI

UNI and Service Model	OIF UNI	IETF UNI (overlay)	IETF UNI (unified)
Signaling	Direct and indirect	Direct	Direct
Symmetry/scope	Asymmetrical/local	Asymmetrical/ edge-to-edge	Symmetrical/end-to-end
Routing protocol	None	None/optional	Link state preferred
Routing information	None	UNI-N may reveal reachability based on policy	Reachability (augmented) and TE attributes
Address space	Must be distinct	Can be common in part	Common
Discovery	Optional	Optional	Through routing
Security	No trust	Limited trust	High trust

UNI, NNI, and any other router-to-router interfaces. The IP client has the possibility to compute an end-to-end path to another client across the optical network. The signaling protocol will establish a lightpath across the optical domain. The IETF proposes to use the concept of "forwarding adjacency," which is described hereafter. Applied in GMPLS, it takes into account the characteristics of the lightpath, such as the discrete bandwidth, which can be reused. GMPLS-capable nodes may advertise a lightpath as a virtual link (TE link) into the link-state protocol responsible for routing this lightpath. Such a link is referred to as an FA, and the corresponding LSP is termed an FA-LSP. An FA-LSP is both created and used as a TE link by exactly the same instance of the GMPLS control plane. The usage of the FA provides a mechanism for improving lightpath reservation; that is, the optical interface reservation. When creating an FA-LSP, the bandwidth of the FA-LSP can be greater or equal to the bandwidth of the LSP that induced its creation. Thus, it is essential that other routers realize the availability of the free excess bandwidth.

2.3.3.5 The GMPLS UNI

The objective of this section is to explain the ASON/GMPLS concepts and the GMPLS UNI described in NOBEL, and how they enable interactions between the layers of a resilient multilayer network, for which multiple reference points of information exchange are defined:

- **UNI:** the reference point between administrative domain and user;
- **E-NNI:** the reference point between administrative domains, and between routing areas within an administrative domain;
- **I-NNI:** the reference point between controllers within areas.

This work was derived from the parallel (but complementary) ITU-T and IETF efforts to define UNI signaling interface standards. These standards and approaches are complementary in the following sense:

- ITU-T defines overall/high-level control plane architecture (Q12/SG15) and functional models (Q14/SG15);
- IETF defines protocol suite/extensions to meet these requirements (driven by CCAMP WG interacting with MPLS WG, OSPF WG, IS-IS WG);
- OIF assesses carrier requirements and multi-vendor interoperability issues and works in coordination with the ITU-T and IETF to resolve them.

The ITU-T ASON definition contributes G.8080/Y.1304 "ASON Architecture," G.807/Y.1301 "ASTN Requirements," G.7714/Y.1705 "Generalized Automatic Discovery Techniques," and G.7713/Y.1704 "Distributed Call & Connection Management." The IETF GMPLS definition contributes RFC3471 "GMPLS Signaling Functional Description," signaling protocol extensions to RSVP-TE, routing protocol extensions to OSPF-TE, and the LMP.

The GMPLS UNI is particularly suited for intra-carrier networks; that is,, the network element exchanging signaling messages belongs to the same or different routing areas or autonomous system but belongs to the same administrative entity (e.g., network operator). Opting straight for the GMPLS UNI provides enhanced coordination, enabling optimized usage of the overall network resources. The GMPLS UNI is beneficial, since the functionality offered by this interface is a superset of those of the OIF UNI. The GMPLS UNI can be configured to operate exactly like the OIF UNI; that is,, it can enforce a strict separation between the different signaling sessions of the overlay-based network (see Figure 2.5). Since it is upgradeable towards better control plane integration, it is also more future-safe. From these observations, the GMPLS UNI, which is backward compatible with IETF RFC 2205/2210/2961/3209/3477/3471 and compliant with RFC 3473, is definitively the interface to be considered. Additional work is dedicated to further extending this interface by allowing endpoint reachability information to be exchanged across it. As such, the GMPLS UNI is not the endpoint but an important step in the continued evolution for delivering intelligent and unified network solutions (see Table 2.4).

In order to quantify the added value of the GMPLS UNI in the context of IP/MPLS over OTNs in terms of resource requirements, notification time, and recovery time, simulation studies have been performed under NOBEL activities. Related to recovery, it is essential to have an interface that facilitates functional sequences and coordinations between layer control planes and minimizes the overall resource requirements and the delays inferred by the restoration of the traffic affected by link failure event cuts in the transport network.

Three advanced IP/MPLS over SDH/GMPLS recovery scenarios were designed and simulated:

- **scenario 1:** SDH sub-connection recoveries handled by ingress OXC;
- **scenario 2:** end-to-end SDH connection recoveries handled by ingress LSR;
- **scenario 3:** IP/MPLS connection recoveries handled by ingress LSR.

The aim is to compare and classify the performance (according to the parameters defined in the following section) of these scenarios on the proposed COST266 European reference network topology.

At the reference point between the IP/MPLS LSR interface and the SDH/GMPLS LSR (i.e., OXC) interface, a signaling interface is defined to establish the TDM LSPs. This signaling interface is referred to as a UNI between the IP/MPLS client domain and SDH/GMPLS transport network domain. Two versions of UNIs exists, and in different standardization bodies. For the three multilayer recovery scenarios developed, scenario 2 is an end-to-end SDH connection recovery handled by an ingress LSR requiring that the edge LSR can signal a connection within the transport network. Then a rich functional interface UNI is required, such as defined in Ref. [2].

Performance evaluations were derived from network simulations using a tool based on data flows generation. These control plane engine behaviors were simulated by considering the different stages of processing message durations inside the nodes (LSR and OXC) and the transmission delays along the route between the different nodes: edge nodes and core nodes.

The processing time and transmission duration are implemented for the signaling-like RSVP-TE messages and routing OSPF-TE-like messages.

By modeling the corresponding LSP provisioning, LSP recovery sequence diagrams, the signaling of end-to-end protection LSP is simulated. Some indications will be given for the end-to-end LSP reversion.

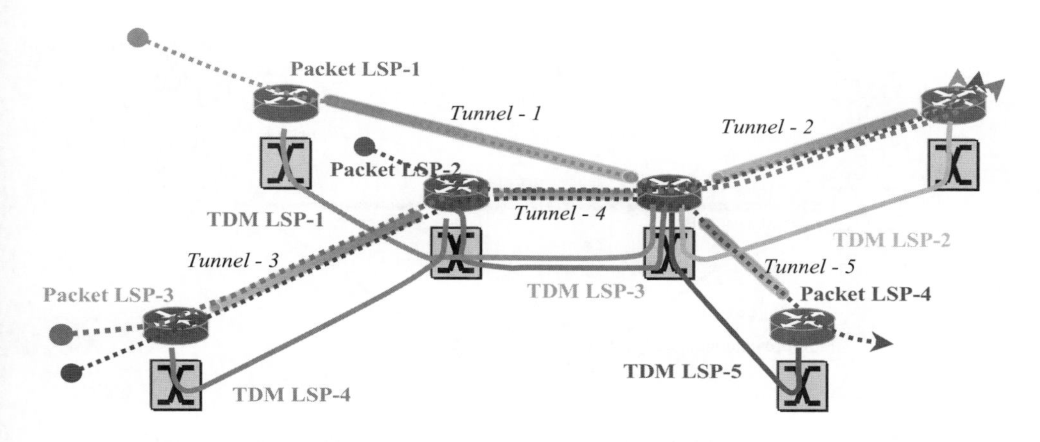

Figure 2.6 Logical topology (virtual topology) representation for the IP/MPLS network

The establishment of the protected TDM LSP – that is, the establishment of the working TDM LSP and its protecting TDM LSP – mandates that the set of ducts used by the working TDM LSP has no intersection with the set of links used by the protecting TDM LSP. Each set of links constituting a "shared risk link group" (SRLG) is unique and is attached to each physical link used by either a working LSP or a protecting LSP. For the given input traffic matrix, each working TDM LSP is established SRLG diverse from its protecting LSP. The SRLG information for each TDM LSP is handled in order to route the protecting LSP.

Just as the physical topology is the support for the routing and establishment of TDM LSPs, the virtual topology is the support for the routing and establishment of packet LSPs. The representation of the Virtual Topology (VT) is to consider it as being a set of TE tunnels connecting the edge nodes; that is, IP/MPLS routers (Figure 2.6). These tunnels should be viewed as bundled TDM LSPs and should have attributes enabling the routing of packet LSPs over them.

The routing of the packet LSPs over the virtual topology makes use of SRLG information that is attached to the TDM LSPs that carry the packet LSP. This routing information is a summary of the control information brought back from the OTN in order for the detours packet LSP to be physically (and not only virtually) diverse from their protected packet LSP.

Routing packet LSPs over the virtual topology takes into account the length of the path of the packet LSP. Several costs can be defined: the hop length at the IP/MPLS layer and the hop length of the tunnel itself, information that also is summarized back from the transport network when establishing TDM LSPs that compose the tunnel.

Routing packet LSPs over the virtual topology takes into account the cost induced by the need to establish a new TDM LSP along a tunnel before nesting the packet LSP in the tunnel. So information about the available bandwidth of the tunnel LSP is reported.

The above listed routing constraints enable the definition of some of the attributes of a tunnel. Other attributes may be necessary for other purposes.

The bundling policies of TDM LSPs combined with the routing policy in the transport network domain have an important impact on the ability to find physically disjoint routes for protected packet LSPs and their detours.

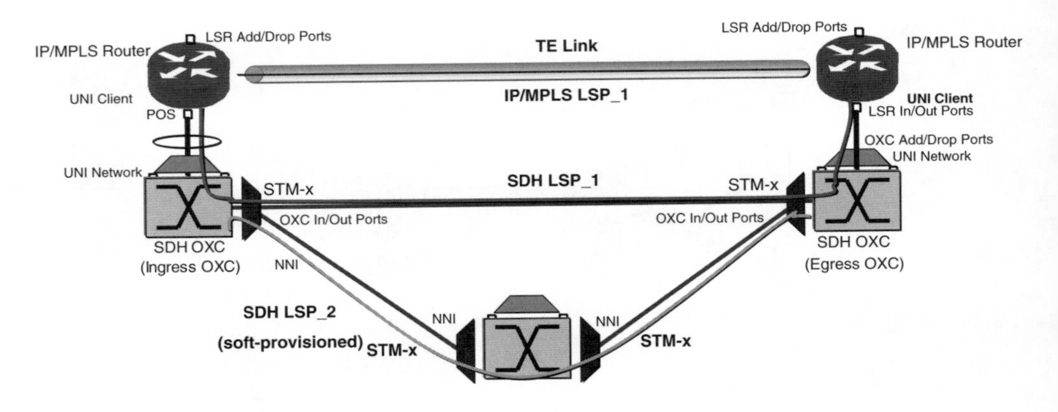

Figure 2.7 Scenario 1: SDH connection recoveries handled by ingress OXC

2.3.3.6 SONET/SDH Sub-Connection Recovery

A packet LSP is established between the ingress LSR and the egress LSR across the logical IP/MPLS topology. Each virtual link in the logical topology corresponds to a connection in the transport network that is recoverable using the shared meshed restoration. Figure 2.7 depicts the segment mesh restoration in the SDH transport layer; it is noted that, in principle, any available transport layer recovery mechanism can be used.

A packet LSP is transported by a protected TDM LSP previously triggered by the OXC adjacent to the ingress LSR. The establishment between the same pair of end-OXCs of a working TDM LSP and its protecting LSP that is link/node/SRLG is disjoint from the working LSP. However, in this case the protecting LSP is not fully instantiated; thus, it cannot carry any extra traffic (note that this does not mean that the corresponding resources cannot be used by other TDM LSPs). Therefore, this scenario protects against working TDM LSP failure(s), but it requires activation of the protecting LSP after the link failure occurrence. The TDM LSP provisioning and its protection configuration are fully handled by the ingress OXC.

In the case of a link failure, the recovery switching from the working TDM (LSP_1) to the protecting TDM LSP (LSP_2) occurs between the edge cross-connects (ingress and egress cross-connects). When an intermediate OXC detects a data plane failure, it sends a Notify message to the ingress OXC.

The ingress OXC detects the data plane failure and initiates a notification timer. The ingress LSR detects the data plane failure and initiates a notification timer and a hold-off timer.

When the ingress OXC receives the Notify message from the intermediate OXC or when its notification timer expires, the ingress OXC allocates the resources for the protecting TDM LSP in order to switch the packet traffic by triggering a new Path message along the protecting TDM LSP.

2.3.3.7 End-to-End SONET/SDH Connection Recovery

This second scenario (Figure 2.8) should be considered an extension of scenario 1, where the recovery switching point is moved at the ingress routers. By comparison with scenario 1, a

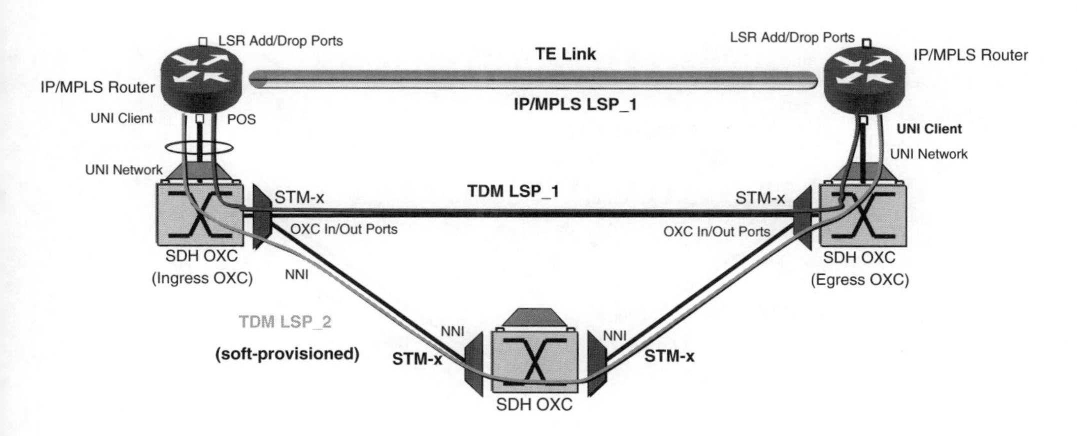

Figure 2.8 Scenario 2: SDH connection recoveries handled by ingress LSR

better sharing of resources can be configured because the TDM LSP protection can be done end-to-end. The protecting TDM LSP corresponding to the recovery SONET/SDH connection (TDM LSP_2) used to protect the working SONET/SDH connection (TDM LSP_1) can be shared with the local detour protecting another working packet LSP.

When an intermediate OXC detects a data plane failure on its interface and it sends a Notify message to the ingress LSR. The ingress LSR detects the data plane failure and initiates a notification timer and an MPLS hold-off timer to give precedence to GMPLS recovery. When the ingress LSR receives the Notify message from the intermediate OXC or when the notification timer expires, the ingress LSR allocates the data plane resources for the protecting LSP.

This scenario requires the UNI to enable the ingress LSR (client side) to establish a signaling session end to end; that is, the Path messages created by the ingress LSR indicate the session tunnel address as the egress LSR (UNI-C node ID) and carries an ERO. The explicit route information allows the ingress LSR to establish end-to-end link/node/SRLG disjoint working TDM LSPs and their protecting LSPs.

When a link failure occurs, the adjacent intermediate OXC detects a data plane failure on its interface and sends a Notify message directly to the ingress LSR (that is, the sender of the signaling session).

The ingress LSR detects the data plane failure and initiates a notification timer and a hold-off timer. When the ingress LSR receives the Notify message from the intermediate OXC or when the notification timer expires, the ingress LSR allocates the necessary bandwidth for recovering the working TDM LSP in order to switch the packet LSPs by triggering a new Path message along the protecting TDM LSP.

2.3.3.8 End-to-End IP/MPLS Connection Recovery

In this scenario (Figure 2.9), a packet LSP is protected end to end by means of an end-to-end recovery LSP at the MPLS layer. The SONET/SDH connections are unprotected. Recovery switching in this case occurs at the source and destination of the packet LSP; that is, responsibility for recovery is shifted completely to the packet domain. The only requirements

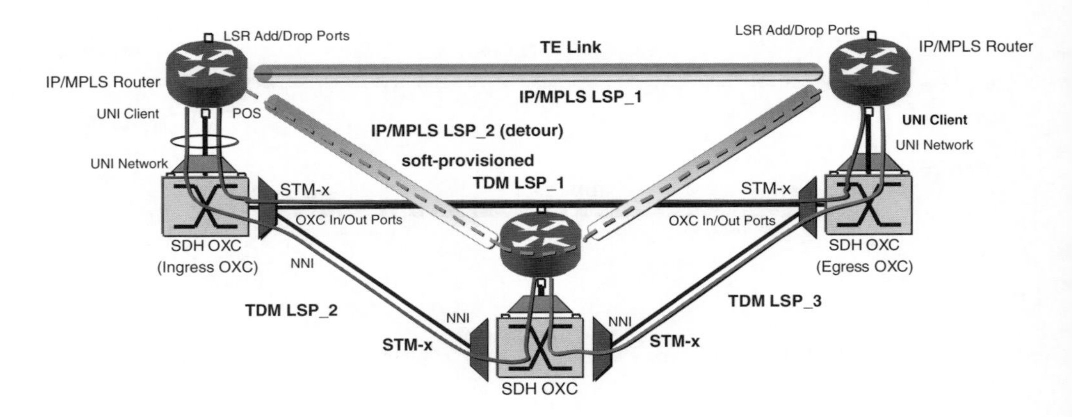

Figure 2.9 Scenario 3: packet connection recoveries handled by ingress LSR

for the SONET/SDH domain are the support of fast notification and soft provisioning of the recovery packet connections (i.e., detours).

When an intermediate OXC detects a data plane failure it sends a Notify message to the ingress LSR. The ingress LSR detects the data plane failure and initiates a notification timer and an MPLS hold-off timer.

When the ingress LSR receives the Notify message from the intermediate OXC, or if the connection is still in the down state when the notification timer expires or when the connection is still down when the MPLS hold-off timer expires, the ingress LSR activates its backup as MPLS fast-reroute recovery.

When the link between the ingress OXC and ingress LSR is protected by an automatic protection switching (APS) mechanism, the MPLS hold-off timer should be configured such that the APS is given sufficient time before the fast-reroute mechanism is activated in the case that this link fails.

Table 2.5 provides a comparison pointing out the strengths and weaknesses of each of these three recovery scenarios.

2.3.3.9 Performance Parameter Definition

Following the definition provided in Ref. [9], two parameters are measured. These parameters have been identified as outputs of network simulations in order to compare the different resiliency mechanisms:

- Network resource requirements in terms of number of synchronous transport module (STM)-n interfaces for the network topology at 2.5 Gbit/s ($n = 16$) and 10 Gbit/s ($n = 64$).
- Mean time to recovery (MTTR) restoration times.

As described in the previous section, the recovery functional complexity in the different scenarios is derived, as it will impact the implementation and the deployment costs of the solutions proposed in a real operator network environment.

Table 2.5 Strengths/weaknesses of each of the three recovery scenarios

	Scenario 1	Scenario 2	Scenario 3
	SONET/SDH Sub-connection recovery	End-to-end connection recovery	End-to-end packet LSP recovery
Responsible layer	Transport (SDH/GMPLS)	Transport/data	Data (IP/MPLS)
Recovery resource/router driven extra traffic or sharing of recovery resources	Pre-established (no router-driven extra traffic) or soft-provisioned (sharing of resources)	Pre-established (with extra traffic) or soft-provisioned (sharing of resources)	Pre-established (with extra traffic) or soft-provisioned (sharing of resources)
Protection of link between edge LSR	Dedicated mechanism	Inherent	Inherent
Protection of edge devices	No	No (edge OXCs in case of dual homing)	Yes (except for source/destination devices)
Complexity of implementation	Low/intermediate (if sharing of resources)	Intermediate	High
Resource efficiency	Low (if no cross-connect-driven extra traffic)/intermediate (if sharing of resources)	Intermediate/high (if good sharing of protection resource) s	High
Recovery granularity (assuming granularity of LSP bandwidth is lower than connections)	Coarse (per connection)	Coarse (per connection)/fine (per LSP)	Fine (per LSP)
Required UNI	(Public)/private	Private	Private
Link bundling required	No	Yes	Yes

2.3.3.10 Network Resource Requirements

The results obtained prove that the proposed multilayer recovery scenario 2, namely end-to-end TDM LSP recovery, is more efficient in terms of interface utilization.

It is important to underline that a phased introduction of the different multilayer recovery scenarios is possible; that is, a gradual introduction of the capabilities required for the different scenarios. This limits the initial effort required in terms of development (vendor side) and deployment (operator side). In addition, some of the features required in support of the enhanced multilayer recovery scenarios have a value on their own. This is especially true for the private UNI.

Figure 2.10 depicts the number of interfaces (both LSR and cross-connect interfaces) for STM-16 (left) and STM-64 (right) interfaces. The difference in number of interfaces between scenario 1 and scenario 2 is mainly due to the interconnection between the OXC and its attached LSR.

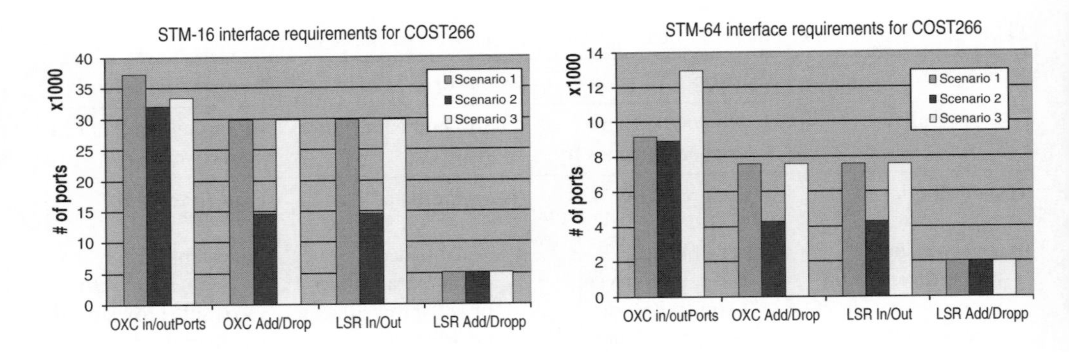

Figure 2.10 The number of interfaces for STM-16 (left) and STM-64 (right) interfaces

In scenario 1, the LSR-OXC interface required is redundant with either permanent $(1 + 1)$ or non-permanent $(1 : 1)$ bridging. By contrast, in scenario 2, the LSR-OXC interface is inherently protected by the $1 : N$ SONET/SDH shared protection mechanism applied, reducing the number of interfaces. In scenario 3, the working packet LSP and its detour are routed over link-diverse SONET/SDH connections requiring to be fully provisioned.

2.3.3.11 Mean Time to Recovery Performance

The recovery time analytical model is presented in Ref. [13] and used for scenarios 1 and 3. It was adapted for scenario 2 using the end-to-end signaling for recovery. The results reported compare their recovery time delays. For each time requiring an exchange of control messages (e.g., Path/Resv messages for the resource request/allocation or Notify message for the failure notifications) a reception/emission time is included for each hop. The recovery time t_r is composed of four terms:

1. the link failure alarm detection time t_1
2. the notification time t_2
3. the activation of the protecting LSP along its route t_3 and the cross-connection time of the egress node t_4.

The model is based on the assumption that the Notify messages of the TDM LSPs affected by the link failure are forwarded sequentially at each node interface along the route of the failed TDM LSPs.

Figure 2.11 shows for each scenario the recovery time delay using two possibilities for the failure notification based on RSVP-TE: either PathErr or Notify messages. For each of the three scenarios, the restoration ratio (i.e., the percentage of LSPs that were affected by the failure and that have been recovered) is plotted as a function of time. Let us denote the recovery time, defined as the amount of time that elapses between the moment the failure is detected $(T = 0)$ and the time at which the affected traffic is completely restored, by $T^{(i)}$ and the recovery time of the first LSP that is recovered first by $T_0^{(i)}$ (where i identifies the recovery scenario applied).

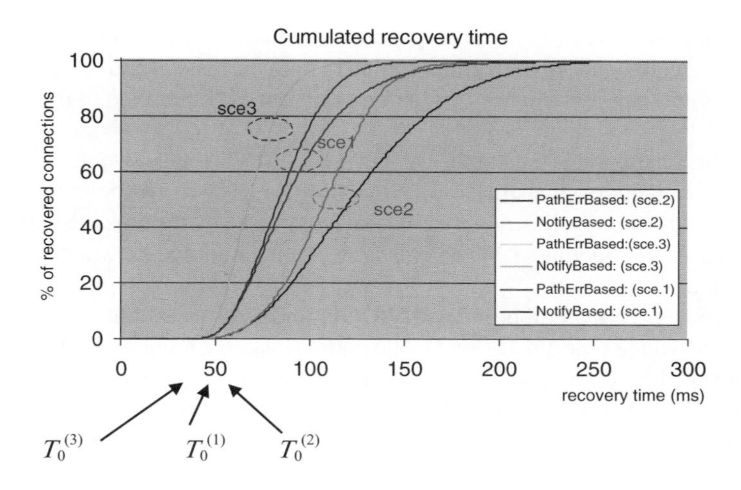

Figure 2.11 Recovery time comparison for IP/MPLS over SDH/GMPLS

The total recovery time is the recovery time of the LSP that is recovered last (more accurately, the total recovery time is the amount of time that elapses between the moment at which the failure is detected and the time at which the last recovery action is performed). From Figure 2.11, it is clear that the recovery speed (which is inversely proportional to the total recovery time) in the case of the SONET/SDH connection-based recovery (scenarios 1 and 2) is lower than in the case of packet-LSP-based recovery (scenario 3), as can be seen both from the slope of the curves at the inflection point and the fact that $T_0^{(3)} \leq T_0^{(1)} \approx T_0^{(2)}$.

This is a consequence of the soft provisioning applied in scenarios 1 and 2; that is, in this case the resources of the protecting SONET/SDH connections need to be allocated before IP data traffic from the client network can be switched onto them. The slight difference in recovery speed between scenarios 1 and 2 (about 5 ms) is mainly due to additional hop in the path of the RSVP-TE Notify message. Note that if an IP/MPLS-based recovery mechanism were to be used as a fallback mechanism in case SONET/SDH based recovery is not successful, in this example, a hold-off time of about 130 ms or more should be applied, before which the IP/MPLS layer should not initiate any recovery attempt.

A high degree of the client network connectivity enables the transit traffic to be bypassed by the OXC nodes. In these scenarios, each LSR is used to aggregate the access traffic and to nest it into TDM LSPs. The aggregating function of the LSR and moving the transit traffic to OXC interfaces are both optimal in scenario 2. In the scenario 2 case, the protection resources have a better sharing ratio thanks to the end-to-end signaling capability provided by GMPLS UNI between an LSR and an OXC. In scenario 2, only the working TDM LSPs are cross-connected; the soft-provisioned protecting TDM LSPs are handled end to end by the ingress LSR and can be shared between several end-to-end diverse working TDM LSPs. Typically, packet LSPs between LSRs have a smaller bandwidth granularity than TDM LSPs. Each ingress LSR can optimize both the path followed by each TDM LSP and the nesting function of packet LSPs into the different TDM LSPs.

In scenario 3, increasing the number of TE links (logical link controlled by OSPF6TE) – that is, TDM FA LSP between the two edge LSRs – improves the routing diversity of packet LSPs and their detours. But this strategy does not allow an intelligent nesting of packet LSPs into TDM LSPs and clearly leads to overdimensioning of both the need of add–drop SDH interfaces in the transport network and the need for an input/output PoS interface in the IP/MPLS data network.

In the IP/MPLS network, if the number of TDM FA LSPs is reduced to the adjacent point of presence (PoP) neighbors, then the utilization ratio of each corresponding TDM LSP is increased. With this strategy, the transit traffic is not bypassed by the intermediate OXCs. If the average number of intermediate LSR hops is large, then the global network cost can become very significtant.

The COST266 network is composed with 28 PoPs and 41 physical links ($n = 28$ and $l = 41$), the average node degree is around 3 (2.93). The optimal average number of TE links per LSR is between 9 and 10. This corresponds to the optimal trade-off for the IP traffic for bypassing the LSR (through TDM LSPs) or using the available bandwidth (on TE links for which the unreserved bandwidth advertised by the opaque LSA is not equal to zero) at the interface of the LSR.

Owing to its opacity and its limited extensibility, the OIF UNI imposes coordination restrictions on the recovery mechanisms between the two client nodes and the server nodes that can be delivered. With this observation in mind, an enhanced UNI has been developed by the IETF: the GMPLS UNI. The GMPLS UNI advanced signaling interface provides the capabilities required in support of three IP/MPLS over SONET-SDH/GMPLS networks that require recovery scenarios designed to meet the increasing need for more efficient, functional, robust, and deterministic recovery. The simulation results prove the benefits of the proposed recovery solutions; that is, the GMPLS UNI enables:

1. Deterministic recovery through enhanced signaling coordination. The end-to-end signaling capability allows for exchanges of RSVP-TE notification messages that aggregate failure indications of multiple connections. This allows coordination of all or part of the recovery actions to the edge LSRs connected to the SONET/SDH/GMPLS transport network. The OIF UNI would only indicate such failure on a per connection basis based on PathErr message exchanges.
2. Failure notifications to reach the edge LSR. The coordination of the recovery actions is thus possible between network layers. This faster recovery behavior is illustrated in Figure 2.11, where scenario 3 shows that the recovery ratio reaches 100% when scenarios 1 and 2 are still around 50%; that is, the former is twice as fast.
3. Improved control plane scalability. With respect to scenario 1, the OIF UNI requires at least four RSVP-TE LSPs per end-to-end SONET/SDH protected connection (two at the source and destination UNI and two within the network, one for the working LSP and one for the protecting LSP). By contrast, the GMPLS UNI requires only two RSVP-TE LSPs (one for the end-to-end protected and one for the network protecting LSP). This leads to a difference of 50% in terms of control plane scalability.
4. Improved network resource usage efficiency. With respect to scenario 2, the OIF UNI shows a difference of 30% with respect to the number of LSR-OXC links compared with the numbers obtained for the GMPLS UNI. The reason is that the OIF UNI does not maintain the semantics of the client-initiated connection within the network. Therefore, when using the OIF UNI, it is not possible for the network to discriminate a soft-provisioned protecting from a working connection. On the contrary, the GMPLS UNI allows the performing of such

a distinction because protecting connections can be soft-provisioned within the network. The soft-provisioned resources for the protecting connections can be shared among multiple lower priority connections. In the case of failure of the working connection, the lower priority connection(s) are preempted to recover the higher priority working connections(s). In scenario 3, a difference of 10–15% in resource usage can be observed that results from the full provisioning of any connection compared with scenario 2. Moreover, since the resource selection algorithm uses a "first-fit strategy" by selecting paths that reduce the number of hops traversed by each connection, these results can be further improved by using an algorithm that optimizes the sharing ratio of the recovery resources.

Taking this one step further, the GMPLS RSVP-TE engine has been implemented to demonstrate the feasibility of these recovery mechanisms when using the GMPLS UNI in NOBEL WP4 Deliverable 30. These recovery mechanisms have to be based on the GMPLS UNI to benefit fully from their performance advantages that provide a high available network with an optimal network resource usage and the fastest and deterministic recovery sequences. This confirms the advantages of implementing the GMPLS UNI for deploying end-to-end recovery configurations.

Public UNI and Overlay Model

The strict control plane separation imposed by the public UNI is both its main strength in the cases of multi-domain networks and its main weakness in the case of a single network operator, depending on the operational requirements imposed. It limits the amount of control information that needs to be exchanged between the two control plane instances. The transport network assigned (TNA) addresses identify UNI connection end points; that is, the links connecting UNI-C and UNI-N. The TNA address is a public external address that is globally unique, and is assigned and exposed to the clients by the transport network. It forms a third address space that complements the client and transport network address spaces, while the internal transport network address space is completely hidden from clients. These TNA addresses are carried in UNI signaling messages, within the GENERALIZED_UNI object (see below) to identify uniquely the end points of a connection. OIF UNI TNA addresses are formed using:

- values assigned by the network that identifies (bundled or unbundled) TE links using IPv4/IPv6/NSAP
- corresponding "names" assigned to TE links (i.e., <Local Link_Id, Remote Link_Id>)
- local resource identification; that is, <TNA, label>, with label: = <Link_Id, Label> with Label as per [48].

At source (Node A, UNI-C), the signaling Path message includes:

• source IP address (header)	node A IP controller address
• destination IP address (header)	node B IP controller address
• tunnel endpoint address	node B IP controller address
• tunnel ID	16 bits (selected by the sender)
• extended tunnel ID	node A IP controller address
• tunnel sender address	node A IP controller address
• LSP ID	16 bits (selected by the sender)

- GENERALIZED_UNI object
  - source address (connection) endpoint A eddress (IPv4/v6/NSAP)
  - destination address (connection) endpoint D address (IPv4/v6/NSAP)[7]

where endpoint A address and endpoint D address are assigned by the transport network.

On the other side, at source (Node C, UNI-N) the Path message includes:

- source IP address (header) node C IP controller address
- destination IP address (header) node D IP controller address
- tunnel endpoint address node D IP controller address
- tunnel ID 16 bits (selected by the sender)
- extended tunnel ID node C IP controller address
- tunnel sender address node C IP controller address
- LSP ID 16 bits (selected by the sender)
- GENERALIZED_UNI object
  - source address (connection) endpoint A address (IPv4/v6/NSAP)
  - destination address (connection) endpoint D address (IPv4/v6/NSAP)

The GENERALIZED_UNI object contains the connection endpoint selection information. This enables a strict separation in addresses, both address types and address spaces, between the client and transport networks. Specifically, the RSVP-TE session endpoints (tunnel endpoints) are not the same as the connection endpoints. The Transport Network Assigned (TNA) address and GENERALIZED_UNI reflect that although the OIF UNI is based on GMPLS signaling, a number of OIF-specific (non-GMPLS-compliant) extensions were introduced to meet its goals. Furthermore, in order to guarantee strict independence between the client and the transport network, three separate signaling sessions must exist:

- one ingress session between the client edge node (node A) and the server edge core node (node B) (at left);
- at least one network session between the ingress edge core node (node B) and the egress edge core node (node C) (in the middle);
- one egress session between the server edge core node (node C) and client edge node (node D) (at right).

Having three separate signaling sessions introduces additional complexity due to the coordination they require, compared with implementations where the same functionality is obtained using one single signaling session (contiguous signaling session).

Therefore, the client–server signaling and overlaid routing of the OIF UNI leads to the results where two single-hop LSPs must be established between UNI-C and UNI-N peers.

For establishing an end-to-end connection in the transport network, the OIF UNI induces at least the establishment of three separate signaling sessions that are stitched together. Splitting into multiple RSVP sessions per connection increases the number of RSVP sessions to be controlled per connection and limits the scalability: $N-1$ (in the case of N nodes configured for the

[7] Network service access point.

connections) instead of one. Furthermore, it decreases the interoperability level and, therefore, one of the key distributed control plane objectives does not help in increasing flexibility or in providing more features. It also implies additional processing at edges; for example, the tunnel end-point address is different from the connection destination address, so the RSVP-TE needs to process an extra look-up and, therefore, substantially impacts performance. Since the client considers the server (i.e., the transport network) as a "black box," only very limited coordination is possible between the IP/MPLS and transport network control planes. In addition, the strict separation of the control planes necessitates per-layer TE, which can lead to inconsistent network states [54].

Over time, these OIF UNI limitations will become more apparent. Although the OIF is presently defining a new version of its UNI, due to backward compatibility reasons with the current version, some of these deficiencies will be removed. Moreover, since there are no significant deployments and the constraints imposed by the OIF public UNI may be relaxed when both IP/MPLS and transport networks belong to the same administrative domain, it is preferable to design a new interface that is fully GMPLS compliant and that is more flexible and extensible than the OIF UNI.

NOBEL GMPLS UNI and Overlay Networks

It is precisely with these observations in mind that a new user-to-network signaling interface is proposed: the NOBEL GMPLS UNI. Table 2.4 shows the positioning of this interface with respect to the control plane interconnection models described in the previous section. The addressing scheme can be configured according to GMPLS TE links and bundles [63] between UNI-C and UNI-N:

- bundled TE links are numbered or unnumbered;
- a bundled TE link represents a set of component links (numbered or unnumbered);
- component or bundled Link_id; that is, <Local Link_Id, Remote Link_Id> with Link_Id being either Interface_Id or IP address;
- local resource identification; that is, <Bundled Link_Id, Component Link_Id, Label> with Label as per Ref. [48].

One of the main differences from the OIF UNI is that GMPLS UNI offers direct addressing, even though internal data plane addresses may be used within the transport network.

The major differences between the public (OIF) UNI and the (NOBEL) GMPLS UNI signaling interface can be summarized as follows (also see Table 2.6):

1. The OIF UNI restricts endpoint identification to numbered (IPv4/IPv6) or unnumbered interfaces. This means that the UNI-C (node A) has only access to the end-points (such as node D) of the data network adjacent to the transport network. For example, an ingress router (node A) having an OIF UNI-C can only address connections towards its attached cross-connect (node B), or other routers defined in the TNA space provided by the ingress cross-connect in the transport network without network address translation [56], and potentially cannot establish it to the real end-point of the connection.
2. The NOBEL GMPLS UNI allows client-driven explicit routing (typically loose routing in this context) through the transport network. The OIF UNI does not support that.
3. The NOBEL GMPLS UNI maintains a single end-to-end signaling session, instead of three signaling sessions. This simplifies the RSVP-TE procedures inside the transport network, but with no signaling message exchange restriction at the client side.

Table 2.6 Comparison between the OIF UNI and the GMPLS UNI signaling interfaces

Function	OIF public UNI	NOBEL GMPLS UNI
Routing	No routing information (overlay model) is reported from the core node to the attached edge node.	No routing information (overlay model) Some reachability information (augmented model) Reachability, topology, and resources information, including FAs (peer model)
Signaling	RSVP signaling session is local across the UNI Client-driven explicit routing is not supported	RSVP signaling session is client-to-client Client-driven explicit routing (loose and strict) is supported
Addresses	UNI addresses (TNA) logically separate from client internal addresses and from network internal addresses UNI addresses can be IPv4, IPv6, or NSAP	A common address space covering network, UNI, .and clients The addresses can be IPv4, IPv6
Symmetry/scope	Asymmetrical/local	Asymmetrical/end-to-end
Security	No trust	Limited trust

The NOBEL GMPLS UNI signaling sessions are defined end to end, meaning that the connection is uniquely identified (i.e., the combination of the SESSION and SENDER_TEMPLATE/FILTER_SPEC object in GMPLS-RSVP-TE) includes only end-to-end significant information.

The situation is explained by the following. At source (node A) the Path message includes:

- source IP address (header) — node A IP controller address
- destination IP address (header) — node B IP controller address
- tunnel endpoint address — node D IP controller address
- tunnel ID — 16 bits (selected by the sender)
- extended tunnel ID — sender specific (0 by default)
- tunnel sender address — node A IP controller address
- LSP ID: — 16 bits (selected by the sender)

While at source (node C) the Path message contains:

- source IP address (header) — node C IP controller address
- destination IP address (header) — node D IP controller address
- tunnel endpoint address — node D IP controller address
- tunnel ID — 16 bits (selected by the sender)
- extended tunnel ID — sender specific (0 by default)
- tunnel sender address — node A IP controller address
- LSP ID — 16 bits (selected by the sender)

Addressing

A client edge node (node A) is identified by either a single IP address representing its node-ID (and referred to as the edge node IP address), or by one or more numbered TE links that connect the edge node to the core nodes. Core nodes are assumed to be ignorant of any other addresses associated with an edge node; that is, addresses which are not used in signaling connections through the core network.

An edge node (typically node A) only needs to know its own address (i.e., the edge-node IP address), a control plane address of the adjacent core node (referred to as the edge core-node IP address), and to know or be able to resolve the control plane address of any other edge node to which it wishes to connect (typically node D).

Explicit Routing
This greatly simplifies provisioning operations and allows for a straightforward reuse of widely available control plane source-code. To provision a SONET/SDH connection, the ingress edge node may rely on explicit routing by means of the Explicit Route object (ERO) that specifies the route as a sequence of abstract nodes.

If the initiator of the connection has knowledge of a path that is likely to meet the connection requirements in terms of, for example, bandwidth or efficient use of resources, or satisfies some policy criteria, then it can insert an ERO into the Path message. Thus, in contrast with the OIF UNI, the ingress edge node is allowed to include an ERO in Path messages. A core node receiving such a Path message may, depending on its local policy, either reject or accept and process. In the former case (or if the path is not feasible), it returns a PathErr message to the sender. In the latter case, it verifies the route against its topology database before forwarding the Path message across the network. To support explicit label control on the edge node interface, an ERO can contain only the ingress core node and the egress edge node. In this case, loose routing is expected between the ingress and egress core nodes, allowing the transport network to splice existing internal connections to deliver the requested end-to-end service.

Diversity
In the context of the GMPLS UNI, diversity with respect to nodes, links, and SRLGs can be achieved using the EXCLUDE_ROUTE object (XRO). Whereas RSVP-TE only allows for explicit inclusion of the abstract nodes listed in the ERO of the Path message, the XRO allows them to be explicitly excluded. Two cases are considered. In the first case, a list of abstract nodes or resources (referred to as the exclude route list) is specified that should not occur anywhere on the path. The exclude route list is carried in a new RSVP-TE object: the XRO.

In the second case, certain abstract nodes or resources should not occur on the path between a specific pair of abstract nodes present in the ERO. Such specific exclusions are referred to as an explicit exclusion route and are supported by introducing Explicit eXclude Route sub-object(s) in the ERO [52].

End-to-End Recovery
For connection recovery [7,8], the GMPLS UNI allows the usage of the PROTECTION object. This allows a client to specify explicitly the desired, for example, SONET/SDH connection end-to-end recovery level (thus starting at the ingress edge node and terminating at the egress edge node). The following protection types have been defined: dedicated protection, protection with extra traffic, (preplanned) rerouting without extra traffic, and dynamic rerouting.

Fast Notification
The Notify message provides a fast mechanism to notify nonadjacent nodes (that have been previously configured with a NOTIFY_REQUEST object) of LSP-related events such as LSP failures. Notify messages are only generated when explicitly requested by the node controller during LSP establishment. This is done by carrying the NOTIFY_REQUEST object in the

Path message, an object that includes the destination IP address of the intended receiver. The Notify message differs from the other error messages (i.e., PathErr and ResvErr) in that it can be "targeted" to a node other than the immediate upstream or downstream neighbor. In addition, Notify messages do not have to follow the same path as the Path/Resv messages that were used to establish the corresponding LSP, and are only forwarded and not processed at intermediate nodes, if an alternative path is available.

A NOTIFY_REQUEST object can be inserted into Path or Resv messages sent by the client LSR, indicating the IP address of a node that should be notified upon LSP failure. Notifications may be requested in the upstream and/or downstream directions by including a NOTIFY_REQUEST object in the Path and/or Resv message respectively. Any node receiving a message containing a NOTIFY_REQUEST object stores the notify node address in the corresponding RSVP state block: Path State Bock or Resv State Block [10,64]. If the node is a transit node, it also includes a NOTIFY_REQUEST object in the outgoing Path or Resv message. The outgoing Notify Node Address may be updated based on local policy. This means that a core node, upon reception of a NOTIFY_REQUEST object from the edge node, may modify the notify node address (by replacing it with its own address).

Comparative NOBEL UNI and OIF UNI

The major differences between the OIF (public) UNI and the NOBEL (GMPLS) UNI signaling interface can be summarized as in Table 2.6.

2.3.4 Conclusions on UNI Definitions from ITU-T, OIF, IETF, and OIF UNI: GMPLS UNI Interoperability Issues

Section 2.3 has depicted the UNI definitions suitable for ASON/GMPLS from the ITU-T recommendations, IETF protocol specifications, and OIF implementation agreements and performed a comparison on their functional scope and compatibility. These different standards bodies show a significant protocol reuse for signaling. Each standards body has started from the same RSVP-TE protocol specifications, which they have further developed and modified to fulfill their requirements, leading to the fact that the protocols have all evolved in slightly different directions, as at that time there existed no significant cooperation between the different organizations. Typically, the differences are not extensive, but significant enough to lead to incompatibility between the UNI signaling protocols mentioned. Although, for example, both IETF and ITU-T use RSVP-TE, the perspective is different.

The most important open issue with UNI definitions at the moment is the interoperability between OIF UNI and IETF UNI. The IETF UNI has evolved from protocols that are extensively deployed, and with good documented operational experiences. Three such protocols create the major basis for the IETF GMPLS UNI:

- RFC3473 "Generalized Multi-Protocol Label Switching (GMPLS) Signaling Resource ReserVation Protocol-Traffic Engineering (RSVP-TE) Extensions."
- RFC3209 "RSVP-TE: Extensions to RSVP for LSP Tunnels."
- RFC2205 "Resource ReSerVation Protocol (RSVP) – Version 1 Functional Specification."

Even though this Internet draft has not formally been published as an RFC yet, it has been approved by the Internet Engineering Steering Group to become a proposed standard, and has been implemented by several vendors and is deployed. This deployment has been with the understanding that the standards work is finished, as a major rework into something incompatible would cause major problems.

A key design decision for the GMPLS UNI was to run a signaling protocol over the UNI with an optimal set of messages exchanges between nodes (UNI-C and UNI-N) within the peer network, which means that there is no need to perform protocol mappings and address translation. It is also designed to be possible to run without having any routing protocol information exchanges over the UNI. However, there may be a routing protocol interaction between a core node and an edge node for the exchange of reachability information to other edge nodes. Depending on the interactions, it can be based on BGP extensions [65] or OSPF extensions [3].

As recommended, the ITU-T (ASON) UNI gives several options for the implementations of the signaling protocol engine. One of the options is the OIF UNI v1.0 Release 2. The OIF UNI is based on the RSVP-TE, but some new objects (e.g., GENERALIZED_UNI objects) have been added to the UNI signaling. These extensions were integrated to cover the following concerns:

- The ASON architecture makes no assumptions about the protocol within the peer network. This means that the node at the UNI-N side has to provide a mapping between the UNI signaling protocol and the intra-network signaling protocol.
- It is not possible to interconnect a node implementing an OIF UNI to a GMPLS core because of the protocol incompatibilities described above.
- It is not possible for a node implementing the UNI as defined by the ITU-T architecture to interoperate with a node implementing the UNI as defined by the IETF without translating a number of key signaling messages and objects.

The reason is that the added objects make an implementation following the IETF standard incompatible with an implementation according to the OIF specification. An adjacency may be established, but when the new objects are received from the OIF side, the IETF-compliant implementation has no way of dealing with these objects (in principle, there is no problem in the other direction, since only the OIF definition uses proprietary protocol extensions). The UNI signaling will be stalled. And even with OIF UNI implementations on each side of the UNI, end-to-end signaling adjacency is not possible because of breaking the signaling session between an OIF UNI-C node and OIF UNI-N node.

The IETF UNI does not currently include support for calls, and in fact a large part of the difference between the OIF UNI version 2.0 and the IETF UNI is the support for calls (CALL object). The IETF is working on this; however, the mechanism described in the draft GMPLS specification is not the same as in OIF UNI 2.0. The work to solve the incompatibility has started and is now one of the topics to be discussed in a series of meetings between the IETF and ITU-T.

2.4 Evolution of Connection Services and Special Cases of Optical Networks

2.4.1 Evolution in Network Services

Since emerging high-speed networks are expected to integrate a wide variety of applications with different traffic characteristics and service requirements, to help with QoS prioritization,

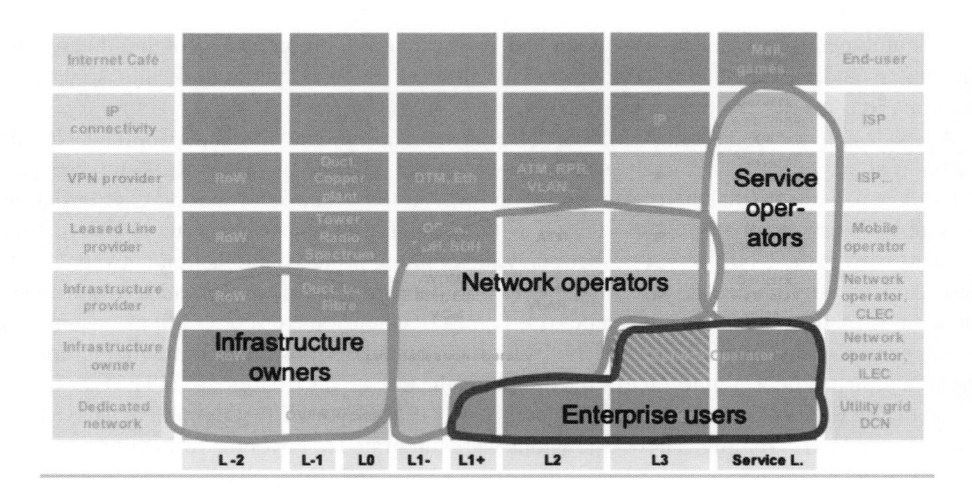

Figure 2.12 Various service categories and distinct groups of players

applications can be classified into four categories with respect to their requirements: delay sensitive and loss sensitive (e.g., interactive video, gaming), delay sensitive but tolerant of moderate losses (e.g., voice), sensitive to loss of data but tolerant to moderate delays (e.g., interactive data), relatively tolerant to both delay and some limited loss of information (e.g., file transfer).

Service structuring with reference to network technology is shown in Figure 2.12 and was elaborated in Figure 1.13.

2.4.2 Virtual Private Networks

VPNs are defined from a transport plane viewpoint. For the ITU-T, a network-based VPN is that part of a network which provides connectivity among a limited and specific subset of the total set of network resources operated by the network provider. A VPN has the appearance of a network that is dedicated specifically to the user's within the network infrastructure. This dedication is achieved through logical/virtualization rather than physical means, hence the use of the word virtual. Users within a VPN cannot communicate, via the VPN provider, with users not included in the specific VPN subset and vice-versa. As we will see, these transport plane characteristics of VPNs put very particular demands on the control and management planes, which become then essential components of any VPN deployment.

NOBEL deliverables provide the basic requirements and features for the realization of the VPN at different layers in a number of scenarios. VPNs at layer 3, layer 2, and layer 1 are considered, and a timeline for their deployment is drawn based on the challenges remaining to be solved and the identification of applications potentially benefiting from them. It is envisioned that VPNs at all three layers will get increasingly dynamic over time, and that control plane importance will significantly grow as a mechanism to support integration among layers. As [74] concludes, "large-scale service deployment, such as L3-L2-L1 VPNs in the NOBEL network, involves collections of control plane collaborations between networks where the required resources are widely distributed".

Among the functions that the control plane, in a VPN context, is expected to perform, the most representative are:

- setting up connectivity at the server layer
- advertising topology within the VPN
- discovery of VPN members
- admission control.

The management plane's main roles are, on the one hand, to configure and provision each VPN instance, and, on the other hand, to provide per VPN visibility on its status, connectivity, and topology, both towards the provider and towards the customer. As resources are dedicated in VPNs on a logical level, security is an important issue in VPNs.

Although building a common framework for control and management of VPNs is a key NOBEL objective, the specific aspects of the layer 1, layer 2, and layer 3 VPNs have been studied with an emphasis on L1-VPNs, as they lead to notably different requirements from the control and management perspectives. It is also worth differentiating between the three networking modes (connectionless-packet switching, connection-oriented packet switching and connection-oriented circuit switching) and their combinations in a client–server realization, which may result in up to nine types of VPN. Note that not all of them are advisable in practice.

Another classification of VPNs that greatly impacts its control and management is based on the transport plane resource allocation policy. Resources may be dedicated to one VPN instance or shared, which leads to the following sub-cases:

- **Dedicated resources VPN**
  - Physical-port-based VPN: the management plane assigns physical network points as the NNI side to one VPN, and the control plane enforces this assignment.
  - Logical-port-based VPN: a defined part of a physical port (e.g., a channel range) is exclusively assigned to a VPN instance.
  - Node-based VPN: a VPN customer has exclusive access to entire nodes; other VPNs are not allowed to traverse these nodes.
- **Shared resources VPN**
  - Connection-based VPN: the control (or management) plane dynamically assigns logical ports to a VPN instance whenever the VPN connectivity is rearranged (by a sequence of connection requests).
  - Specific shared resources VPN, or closed user group: all VPNs use the same network resources to establish connections; the focus here is reachability/connectivity among client subsets – that is, by means of defined VPNs, it can be controlled which clients communicate with others.

These cases may be combined into hybrid models, where some resources are exclusively assigned to one VPN while others are publicly available or shared by a limited number of VPNs.

Control plane resources may also be shared or allocated to individual VPN instances. Use is made of the terms dedicated control-plane private networks (DCPN) and shared control-plane private networks (SCPN) for referring to these cases.

- A DCPN is a VPN with dedicated control plane instances; that is, different control plane instances are created to support different VPNs.
- An SCPN is a VPN in which control plane instances are shared; that is, the same control plane instance may be used for the control of multiple VPNs.

The following subsections discuss in more detail the requirements and issues associated with VPN control and management.

2.4.2.1 Control Plane Requirements and Issues

The basic difference between a regular control plane and the control plane for VPNs is that the latter must support and multiple logical networks on top of the same network infrastructure. This typically means the control plane must keep a differentiated view on the topology, connectivity, and resource allocation of all VPNs instantiated in the network. If a DCPN model were adopted, it would require multiple control plane instances to run simultaneously and independently.

Routing of connections is also affected, since allocation policies restrict which resources may be traversed by connections belonging to a given VPN. The control plane must be aware of which resources have been assigned to which VPNs, and respect these assignments when connections are set up. For this to be enforced end to end, and particularly if routing decisions can be performed at intermediate nodes, a VPN identifier needs to be signaled with each connection request and known all along the route.

CE devices may also participate in routing decisions and exchange topology information with the provider control plane, both to learn the VPN topology within the provider network (if the VPN service supports it) and to make routing announcements that the provider control plane will propagate to other VPN members.

Another requirement is the ability to maintain per VPN address spaces and to map them to unique identifiers within the provider network.

2.4.2.2 Management Plane Requirements and Issues

For any type of VPN it is recommended to have a management platform where the VPN-related information can be collected and managed. The service and NMS would centralize information related to instances of a VPN and allow users to configure and provision each instance from a central location.

A service provider must be able to manage the capabilities and characteristics of their VPN services, and customers should have means to verify fulfillment of the VPN service they subscribed to. This involves engineering, deploying, and managing the switching, routing, and transmission resources supporting the service, from a network perspective (control plane and network element management), to manage the VPNs deployed over these resources (network management), and to manage the VPN service (service management).

Two main management functions are identified:

- **A customer service management function:** This function provides the means for a customer to query, configure, and receive customer-specific VPN service information. Customer-specific information includes data related to contact, billing, site, access network,

addressing, routing protocol parameters, status, service quality, faults, and so on. In addition, the SLA may give the customer the ability to directly perform configurations on the resources subscribed. In this case, the customer service management function must provide the respective functionality.
- **A provider network management function:** This function is responsible for planning, building, provisioning, and maintaining network resources in order to meet the VPN SLAs outlined in the SLA offered to the customer. This mainly consists of (i) setup and configuration of physical links, (ii) provisioning of logical VPN service configurations (including resource assignments), and (iii) life-cycle management of VPN service, including adding, modifying, and deleting VPN configurations.

It is worth noting that there have to be relationships between the customer service management function and the provider network management function as the provider network is managed to support/realize/provide the customer service.

2.4.3 Layer 1 VPN

Optical networks currently deployed are based SONET/SDH, controlled and managed typically by a proprietary element management system and NMS. This architecture precludes flexibility and network automated service provisioning. Associated with the importance of packet-based network services, and the acceleration of IT application services requiring higher bandwidth, this traditional type of network operation is at risk of not only increased operational cost, but also the loss of business opportunities for network operators.

New advances in technologies, such as R-OADMs, and network control functions have matured with the progress of NOBEL contributions. A standardized GMPLS-based protocol architecture enables network operators to automate the service provisioning operations according to the status of the transport resources with multi-vendor interoperability.

Network operators can now provide multiple network services, such as layer 3 VPN services, layer 2 VPN services (see section below), point-to-point and multipoint private line services. The service networks and the transport network infrastructures are often owned and operated by different administrative organizations. L1-VPNs are still maturing to support on-demand wavelength network services with optimal network infrastructure sharing [3,4].

As recommended in Ref. [5], L1-VPNs define node clients to access and share automatically the transport network infrastructures as if controlled by them. Consequently, L1-VPNs enable network service providers to enrich their service offerings, creating new revenue opportunities and making optical network infrastructures more profitable. This section develops on the network architecture evolution and describes the business models for L1-VPNs and network use cases. To address the objectives of L1-VPNs, the introduction of GMPLS-based control functions in terms of service publication, discovery, and signaling are presented along with the related standardization activities at the IETF.

2.4.3.1 Layer 1 Network Services and GMPLS Motivations

Network operators are evolving to offer different levels of network services from new optical technologies and architectures for the purpose of meeting the demanding distributed application requirements. On the network infrastructure side, an important requirement of next-generation networks is the capability to deploy advanced services while reducing their developments and integrations and their related operations. Optical networks, such as TDM or wavelength-switching networks, provided guaranteed bandwidth but only fixed point-to-point connectivity; and as network services were static, manual provisioning was sufficient. This mode of operations was directly impacted from the capabilities of equipment functionality, and also in service offerings, operations, and standards.

SONET/SDH service provisioning and equipment management capabilities ranged from use of manual provisioning (craft terminal interfaces) to support of semi-automatic functions via a carrier's proprietary management interface (e.g., command line interfaces) between the network elements and the network operations system. As the transport network expanded geographically and increased in capacity, network operations became more complex. The existing approaches of using proprietary management interfaces were simultaneously CAPEX and OPEX costly for network operators to offer new services. Standards bodies took on the task of specifying three network functional layers: network control functions, EMS, and NMSs. Bodies such as the ITU-T and IETF are still progressing on open management interfaces in order to enable network operators to be much less dependent on their existing operation systems, in which GMPLS is playing an important role as a distributed (IP-based) control plane for their transport equipment to support the main management capabilities, including automated inventory, autodiscovery, end-to-end dynamic wavelength setup, and flexible segment recovery schemes.

The IETF's CCAMP has focused on providing a comprehensive, base set of standards to support network equipment inter-working from different vendors. CCAMP is currently extending its work to support multiple network domain applications, including UNI, recovery schemes, inter-domain (e.g., ASON) routing and signaling, and L1-VPNs. As with any new technology, L1-VPN deployment in existing optical networks will be gradual, conditioned by multiple weighting factors and network service applications varying for each network service provider.

L1-VPN architectures define a network service interface with connection control from CE devices and management. In the following, the layer 1 services provided by L1-VPN elements to CEs are summarized with their related merits, and examples of use cases are presented.

Elements of L1-VPN Framework and Motivations

This section presents the different elements of L1-VPNs and the architectural models that can be used to offer point-to-point network services (Figure 2.13). Transport networks and client networks are built as follows:

(i) Different client sites and their networks are connected to a shared transport network; for example, on the one hand, a carrier's MPLS network supporting IP-VPN services and, on the other hand, their IP network carrying application data traffic are enabled over a single wavelength-switched optical network.

(ii) The transport network and each client network can be owned and operated by different business organizations that are separate companies (e.g., other carriers) or departments within the same company.

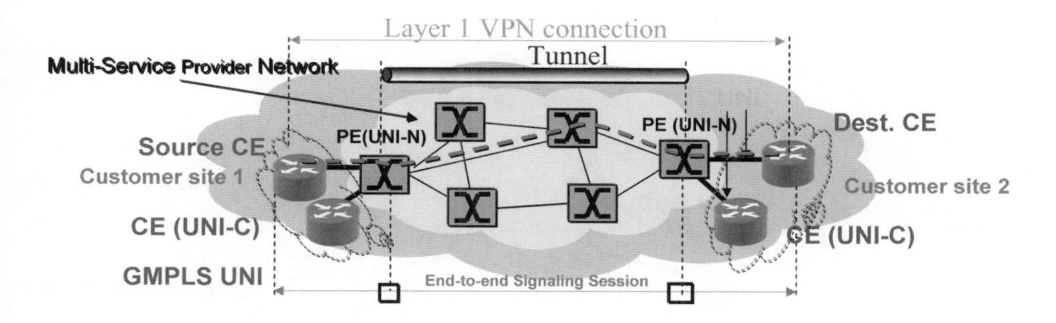

Figure 2.13 Layer 1 VPN service components and signaling scope

The first point is obvious in terms of network multiplexing efficiency. The shared transport network provides bandwidth and capacity to its customer networks, and each client networks use a portion of the transport network capacity and can increase their service requirements as needed.

The second point is a common approach for network operators for several reasons. First, network service deliveries and network operations are separated and adapted to different client networks. The network equipment capabilities, the control protocol configuration, including requirements such as recovery time, and service scalability are very different.

Moreover, network isolation is required for scalability and simplified operations from the transport network service provider (i.e., the network operators). For example, client[8] network#1 must not share any topology information or failure information with client network#2, and it also does not need to know full topology information or detailed failure information of the provider network. The configuration of the different network services from the NIP to its customers should be implemented to avoid interferences in terms of performance, security, and confidentiality. Company clients can operate their virtual networks without the need to deploy their own network infrastructure, which can reduce significantly globally the CAPEX of the network infrastructure. Further, it can create several opportunities for creating business models based on virtual network operators (VNO) and NIPs.

The solutions related to multiple network operator domains in the context of different organizations can be largely simplified by using GMPLS as a unified control plane extended to support multiple domains [12,53], but additional requirements, such as virtual network security, still need to be addressed, as pointed in Ref. [4].

In order to address this network architecture, L1-VPNs define a service interface, which enables dynamic layer 1 (e.g., SDH network) OTN connection provisioning across this service interface. A client network is a customer network of L1-VPN services, while a transport network becomes a provider network of L1-VPN services.

As illustrated in Figure 2.14, L1-VPNs enable one to separate the operations between the NIP (usually owned by large telecom network operators) and client networks. L1-VPNs provide a logically, secure separate network to each VNO customer. Customers can control connections and view related information within their VPN, but not other VPNs.

[8] In this model, it is assumed that each client network is served by the same provider network.

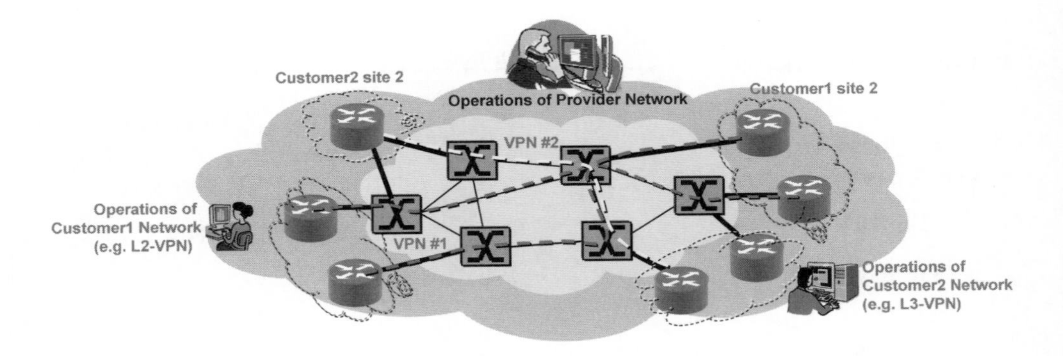

Figure 2.14 Infrastructure sharing concepts based on L1-VPN architectures

According to Ref. [1], the advantages of L1-VPNs for customer networks and provider networks are:

(i) Customer networks are virtually deployed over the transport network and are operated as if the transport network were dedicated to them. The company customer does not own the transport network infrastructure (CAPEX advantages). At the same time, the management of the transport network is under the responsibility of the provider network operator (OPEX advantages). Provider network operators can provision layer 1 connections on demand according to the traffic increase between any pair of customer sites.
(ii) The provider can deliver on-demand layer 1 network services, by reducing the complexity of the manual provisioning procedures to automatic connection control. They can support internal VNO clients or rent their resources more dynamically to create new revenue.

Traditional provisioning is based on application data traffic estimation for each pair of sites, from which new equipment is ordered and installed. Then, CE to PE links are provisioned, followed by connection provisioning between the PE nodes and then the end-to-end connectivity service activation. As client network traffic is increasing very rapidly and bandwidth needs can change sporadically, traffic load estimation is becoming more complex, and requiring spare equipment needs to be configured in advance. L1-VPNs enable a better allocation of connectivity requests over the transport network resources. A layer 1 connection is provisioned between any arbitrary pair of sites (i.e.,, mesh connectivity) by using spare equipment based on more up-to-date TE status. As CE-to-PE links continue to decrease in price (e.g., with the deployments of carrier-grade Ethernet), coupled with the optical transmission multiplexing capabilities (e.g., WDM or optical channelized links) continuing to decrease in price, L1-VPNs will provide an innovative service management option for customers and providers.

Network Use Cases

There are various network usage scenarios where L1-VPNs can be manually provisioned or automatically signaled [40]. Depending on customer connectivity requirements, GMPLS control within the customer network, and trust relationship between the (VNO) customer and the infrastructure provider, connection services may be accessed through different methods and different functionalities may be given to customers. Three major service models based on ASTN requirements, as shown in Figure 2.15, can be deployed.

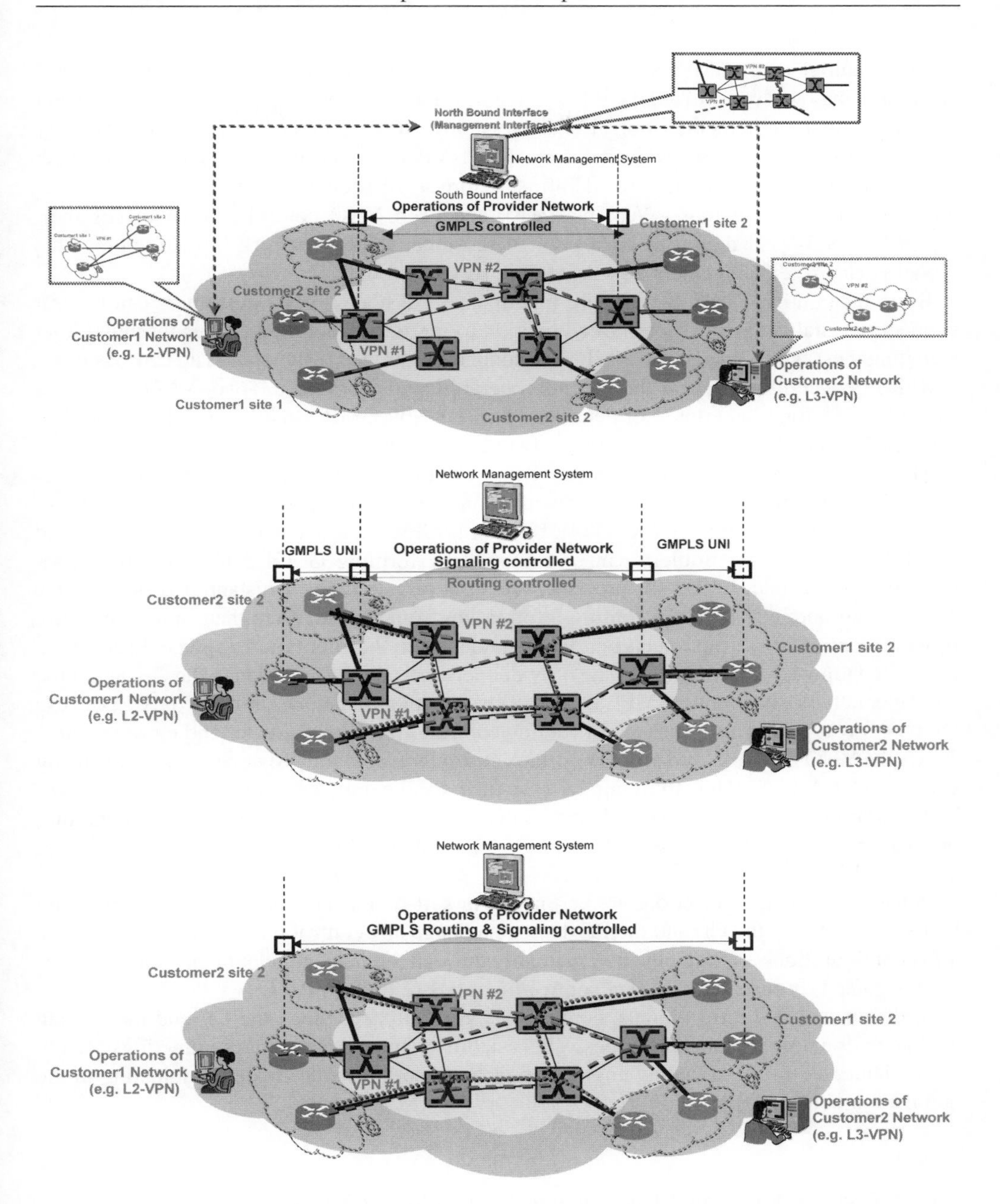

Figure 2.15 The different L1-VPN service models

As an initial step, GMPLS is not required to be implemented in customer networks; customers may wish to utilize L1-VPN services on-demand. For example, they may not want to change their implemented operational procedures to new practices. In such a case, by enhancing the NMS in the provider network, the L1-VPN service can be manually provisioned. VNO customer accesses to the network infrastructure are restricted by the provider via a service management interface (e.g., Web interface), and each VNO controls and manages their L2-VPN or L3-VPN deployed on a pool of resources of the network infrastructure. This model is called a management-based network service model.

By implementing GMPLS signaling functions at CE nodes, VNO customers can request connection establishment/deletion/modification to the infrastructure provider through direct signaling message exchanges. In GMPLS, a connection service is signaled as an LSP. By using GMPLS between the CE and the PE, interoperability is guaranteed. Also, VNO customers can subscribe for fast failure notifications of an LSP by GMPLS Notify mechanisms [15]. This model is called a signaling-based network service model.

The first and second L1-VPN service models are built on an overlay approach, where CEs set up routing adjacencies over provider network connections. This opaque approach can lead to a so-called *N*-square routing problem [54]. If not only CEs, but also customer sites are operated by GMPLS, then this problem can be removed by forming GMPLS routing adjacencies between the CE and the PE (or precisely L1-VPN private context instantiated on the PE). The CE can obtain remote site routing information from the PE. Note that, in this scenario, provider routing infrastructure is completely hidden from the client routing plane. In addition, abstracted topology information of the provider network may be provided to the CE, and if the customer network is running GMPLS, then all customer internal nodes, not just the CE, can use the L1-VPN topology information for routing control. This allows greater and more effective TE on the VNO networks and NIP network. A customer can seamlessly operate its VPN using end-to-end GMPLS. This third model is called a signaling and routing service model. Note, forming routing adjacencies between the CE and the PE precludes any confidentiality options; thus, use of this model will depend on the relationship between the VNO customer and the NIP.

In the first model, connections are permanent or soft-permanent because they are triggered by the NMS. Since the CE and the PE do not exchange any control plane protocol messages, VPN configurations are implemented manually through the management plane, not automatically signaled through the GMPLS control functions (e.g., RSVP-TE).

In the second and third models, connections are switched. Since the CE and the PE can exchange signaling protocol messages, VPN establishment and maintenance functions need to be implemented in the signaling engines of each network provider node participating in the connection configurations.

2.4.3.2 GMPLS Functional Components Applied for L1-VPNs

There are defined functions configuring L1-VPN services based on GMPLS protocols. In order to establish a new VPN connection, VPN reachability information has be propagated [3,65], followed by connection routing and signaling sequences to establish the connections. Reachability information propagation is accomplished by using an autodiscovery mechanism, such as BGP when multiple network domains are involved [65], or by using a flooding protocol such as OSPF for intra-domain reachability [24], while connection establishment is accomplished using GMPLS signaling, such as RSVP-TE [10].

One or more TE links connect a pair of CE and PE nodes where the UNI signaling sequences take place while the PE node has access to the provider network status through the TEDB. Each TE link connecting the CE to the PE is associated with a unique identifier within a given VPN (i.e., the customer equipment identifier (CE_id) of the network element port) and with a unique identifier of the provider network equipment (i.e., the provider equipment identifier (PE_id) of the network element port). CE_id and PE_id can be numbered or unnumbered [55]. The CE IP address must be VPN-wide unique and the PE IP address must be provider network-wide unique.

In addition to its PE_id, each TE link terminating on a PE also has an identifier that is unique within the VPN; that is, the VPN-PE_id. The PE IP address used for the VPN-PE_id need not be the same as the PE IP address used for the PE_id and need not be unique within the provider network.

At least one of the TE links between any CE–PE pair then includes at least one channel allowing IP connectivity between the CE and the PE. This channel, allowing for exchanging control information between the CE and the PE, is referred to as an IPCC. The CE's and the PE's IP addresses of this channel are required to be unique within the VPN they belong to, but are not required to be unique across multiple VPNs.

With the developments in customer-based VPNs, architectures such as MPLS-based layer 2 and layer 3 VPNs [21–23] enable one to reduce the operational complexity in managing and configuring these VPN services over the provider network infrastructure. An important component is called autodiscovery capabilities [3,24,65]. The main objective of a VPN autodiscovery mechanism is to allow VPN members to auto-learn from the provider network appropriate reachability information to be used for setting up intersite connectivity services. The VPN autodiscovery mechanism limits the configuration of addition/changes of a new site to only those devices that are attached to that site. The autodiscovery mechanism is responsible for flooding the new information (i.e., addition of a new site) to all the PE devices that need to know about this information. This reachability information flooding is accomplished with minimal manual interventions from the network provider operator.

Using an auto-discovery mechanism in the context of L1-VPNs had been presented [1] but the work has been stopped at the IETF. Then autodiscovery mechanisms were developed based on OSPF-TE in Ref. [24] for L1-VPN. This mechanism enables PE devices using OSPF to learn dynamically about the existence of each other, and the attributes of configured CE links and their associations with L1-VPNs. The VPN autodiscovery will then distribute this information to all relevant PEs in the network that have at least one site in common with that VPN. If this is the case, then each PE will learn the remote site information and then may, through local policies, decide to initiate connectivity services across the provider network between the customer sites.

In the context of L1-VPN connectivity that is initiated by the CE (i.e., case 3 with switched connections establishment), each CE can auto-learn the set of VPN members attached to the provider network. In Ref. [3], OSPF is extended to propose L1-VPN autodiscovery capabilities at the CE. The CE-based discovery takes advantage of the provider-based autodiscovery implemented within the transport network. Once a PE learns the set of remote CE_ids, it will pass that information to the attached CE. The CEs use this information to select which CEs they want to initiate connectivity to. Protocols that can convey to the CE the set of remote CE reachability information can also be BGP or LMP, as described previously in Ref. [20].

Connections between a pair of CEs are point-to-point and can be soft-permanent (i.e., established by NMS between local CE_id and remote CE_id and maintained by the signaling engines between local PE_id and remote PE_id), or switched (i.e., established between local CE_id and remote CE_id by the GMPLS-based routing and signaling engines). For customer-driven SPC, the provider network is responsible for the establishment and maintenance of the switched connection. An L1-VPN can comprise a set of SPCs and/or switched connections. In the following, switched connections are considered (i.e., case 3). With switched connection capabilities, VPNs can be configured dynamically. Once a local CE obtains the information about the remote CE_ids belonging to the same L1-VPN, the CE triggers the connection establishment using GMPLS RSVP-TE signaling, one (or more LSPs) to the desired CE_id. To support private address assignment, one of two approaches may be used (Figure 2.16):

1. **Contiguous:** Information carried in RSVP-TE messages identifying an LSP (i.e., SESSION and SENDER_TEMPLATE objects) is translated at the ingress and egress PEs.
2. **Nesting:** When a Path message arrives at the ingress PE, the ingress PE checks whether there is appropriate resources between the pair of PEs involved. If there is not, then the ingress PE initiates a PE-to-PE FA-LSP. This FA-LSP is advertised as a TE link by the PE nodes. Through this TE link, CE-to-CE lower order LSPs can be established.

LSP stitching and LSP nesting have similar signaling sequences. One LSP segment can be stitched to one and exactly one LSP segment such that exactly one end-to-end LSP is established; that is, the CE-to-PE LSP, the PE-to-PE LSP, and the PE-to-CE LSP correspond exactly to one LSP. This implies that no label exchange occurs between the head- and the tail-end of the LSP stitching segment compared with the mandatory label exchange over the FA that happens in the LSP nesting case.

In case of LSP stitching, instead of a hierarchical-LSP (H-LSP) [34], an LSP segment (S-LSP) is created between two nodes (CE or PE nodes). An S-LSP for the stitching approach is by analogy equivalent to an H-LSP for the nesting approach. An S-LSP created in one layer (e.g., at SDH VC-4 layer) provides a data link to other LSPs in the same switching layer. Similar to an H-LSP, an S-LSP can be advertised as a TE link by the corresponding IGP. If so advertised, the PCE nodes can use the corresponding TE link during their path computation. While there is an FA between end-points of an H-LSP (see Ref. [34] Section 5), there is no forwarding adjacency between end-points of an S-LSP.

Procedures for signaling a contiguous LSP are described as follows: the Path message originated by the CE node contains the CE_id of the ingress CE and the CE_id of the egress CE. The ingress CE_id value (or TE Router_Address of the ingress CE, for unnumbered CE_id [55]) is inserted into the SENDER_TEMPLATE object and the egress CE_id value (or TE Router_Address of the egress CE, for unnumbered CE_id) is inserted in the Tunnel Endpoint Address field of the SESSION object.

When the PE node attached to the ingress CE (originator of the Path message) receives the request, the PE processes a look-up on the table of the PE nodes to get the address PE_id of the egress PE node, and to find out the associated CE_id of the egress CE.

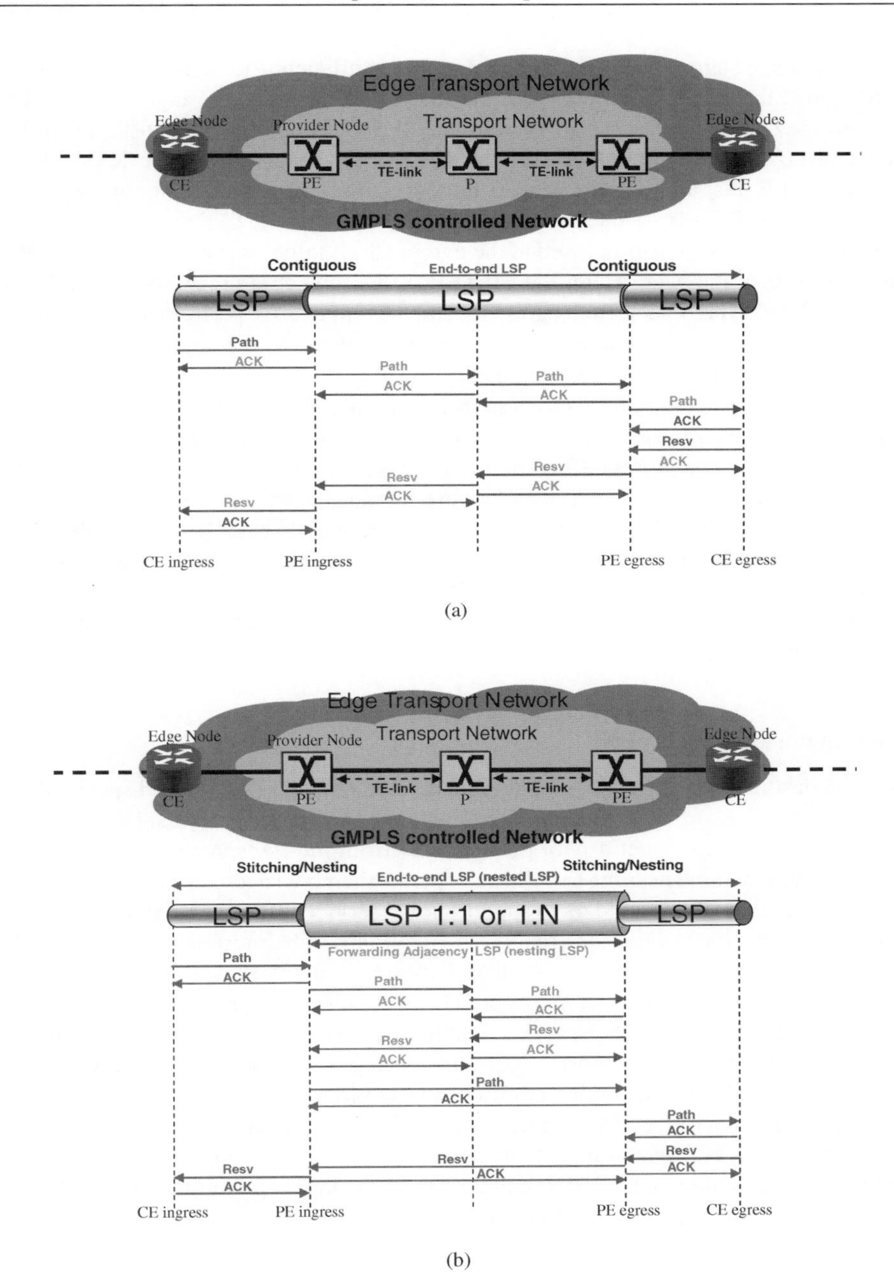

Figure 2.16 (a) Contiguous approach; (b) stitching/nesting approach

The PE_id (of the egress PE node) is necessary and sufficient for the ingress PE to signal a new LSP. Once the mapping is retrieved, the ingress PE replaces the ingress/egress CE_id values with the corresponding PE_id values. The SESSION and the SENDER_TEMPLATE objects (and possibly the sub-objects of the ERO object) included in the RSVP-TE Path message carry PE_ids, and not CE_ids. At the egress PE, the reverse mapping operation is performed; that is, PE_ids → CE_ids mapping.

When the Path message is processed at the egress CE, it initiates towards the egress PE the exchange of a Resv message. Here, the FILTER_SPEC object is processed similarly to the SENDER_TEMPLATE object. Both egress PE and then ingress PE perform the same mapping operation as with the corresponding Path message. Once the Resv message reaches the ingress CE, the signaled connection is established.

When the provider network needs to hide their internal topology, it may only accept a Path message that contains a loose ERO with a sequence of: ingress PE (strict mode), egress CE CE_id (loose mode). The ingress PE computes or gets a path to reach the egress PE, and then inserts this information as part of the ERO before forwarding the Path message downstream.

Furthermore, an ingress PE or an egress PE may remove/filter the RRO from the received message before forwarding it. Filtering an RRO consists of editing its content and includes only the sub-objects based on a local or global NIP policy.

By appropriately filtering RRO, the ingress/egress CE can be aware of the selected links and labels on the egress/ingress CE side respectively.

2.4.3.3 Conclusion on L1-VPN

This section describes the L1-VPN architecture defined at the ITU-T [5] and specified by the IETF [1]; this architecture is evolving to flexibly define shared optical NIPs. L1-VPNs enable client networks (VNOs) to get a pool of resources from the transport network as if it was dedicated to them, without needing to build and operate their own dedicated network. L1-VPNs are also evolving to enable providers to support novel service offerings to create new revenue opportunities. Here, the main issues related to L1-VPN will be presented and analyzed.

L1-VPNs are an enabler for a virtualized transport network serving multiple client network operators, either owned or operated by external operators. L1-VPNs enable dynamic, on-demand connection services, enabling customers to utilize a layer 1 network, and enabling providers to support VNOs, triggering new opportunities through these advanced services. With the implementation and deployment of the full set of L1-VPN functionalities, thanks to GMPLS control plane protocols, not only network management systems, but also operational support procedures will be less complex. With these new capabilities and advantages, L1-VPNs can be one of the key applications of GMPLS in the next-generation multiservice optical networks.

2.4.4 Layer 2 VPN

The L2-VPN architecture and its functional components are presented in this section. L2-VPN provides an end-to-end layer 2 connection (e.g., Ethernet) between two CEs (e.g., a company department) over a service provider's IP/MPLS or IP core networks. The Pseudo Wire Reference Model with Ethernet services will be introduced as an example of applications.

An L2-VPN provides typically an end-to-end Ethernet connection to a set of network users that are outside the transport networks. Each network user can be owned by an enterprise office, the department of a large organization, or be a residential network. In the context of NOBEL, the scope is mainly on "provider provisioned VPNs," referring to VPNs for which the network service provider participates in the management and the provisioning of the VPNs. The IETF lists the layer 2 connection requirements in Ref. [22], in which the network services considered are mainly Ethernet services. The attributes and the parameters of the network services can be aligned with the recommendations listed in Ref. [26] and are further described in the next section.

2.4.4.1 The MEF Ethernet Virtual Circuit

The MEF proposes three types of Ethernet services: a point-to-point Ethernet virtual circuit (EVC), a multipoint-to-multipoint EVC, and a rooted multipoint EVC [26]. The definition of an MEF network service is in terms of what is seen by each CE node. The description of the EVC services includes the UNI corresponding to the physical reference point between the node under the responsibility of the network service provider (which can be operated by a network infrastructure operator [67]) and the node under the responsibility of the user (typically the enterprise attached to the provider network [67]).

Each UNI is dedicated to a single company (that is, a subscriber charged for the L2 network services). The EVC specifies connectivity between UNIs and is an association of two or more UNIs. At a given UNI, more than one EVC can be supported according to the multiplexing capabilities. The connectivity multiplexing service can be configured with any combination of point-to-point EVCs and multipoint-to-multipoint EVCs.

As illustrated in Figure 2.17, CE nodes B, C, and D are attached to the service provider metro Ethernet network via one single 10 Gbit Ethernet interface (UNI). In this example, the UNI of

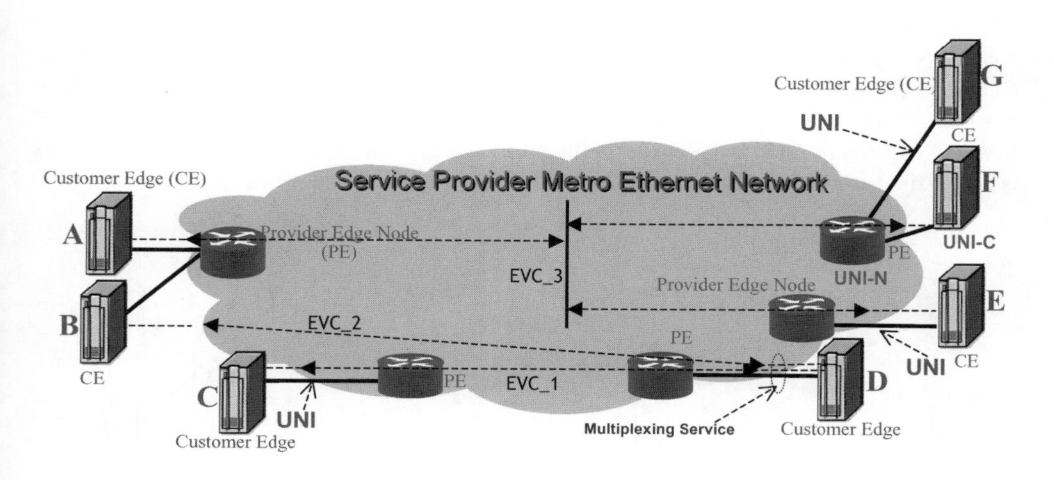

Figure 2.17 Metro Ethernet services: point-to-point and multipoint-to-multipoint

CE node D supports service multiplexing and it allows elimination of the need for two different physical interfaces. Since only one EVC is established at the UNIs A, B, C, E, and F, these UNIs need not support service multiplexing. These UNIs may or may not support virtual LAN (VLAN) tags, depending on what the service supports and the subscriber requires. For example, on EVC_1 from service multiplexed UNI_D that supports tagged services frames to UNI_B that does not support tagged service frames, the network would remove the CE-VLAN tags from Ethernet frames sent from UNI_D to UNI_B and add the CE-VLAN tags for frames sent from UNI_B to UNI_D.

Different options can be supported at one UNI (between CE node and PE node) to map the customer traffic flows to the EVC. Ethernet frames can be untagged or tagged with 802.1Q tags. It is worth noting that, between two UNIs, an EVC may support one UNI side with only untagged service frames while the other UNI side may support only tagged service frames. In this case the service provider network handles the integration or the removal of the VLAN tags.

In Ref. [66], two point-to-point EVCs and one multipoint-to-multipoint EVC are represented. A multipoint-to-multipoint EVC with two UNIs is different from one point-to-point EVC because one or more additional UNIs can be added.

Different types of Carrier Grade Ethernet service can be deployed. The network operator can decide to deploy network services according IETF L2-VPN specifications as described in the next section.

2.4.4.2 IETF L2-VPN

As specified by the IETF [22], a VPN is a service that provides secure and private data communications over a network infrastructure operated by a telecom operator, through the use of standard tunneling, encryption, and authentication techniques.

There are different types of L2-VPN service that a network operator can offer to its customers. These service types include point-to-point L2 connections named virtual private wire services (VPWSs) and multipoint-to-multipoint L2 connections named virtual private LAN services (VPLSs) over a PSN such as carrier-grade Ethernet networks.

A VPN can be contrasted with L1 leased lines that can only be used by one company (see Section 2.4.3.1.1). The main purpose of a VPN is to give the company the same capabilities as private leased lines using the shared transport network infrastructure.

A VPWS (Figure 2.18) is a point-to-point circuit (link) connecting two CE devices. The link is established as a logical through a PSN. The CE in the customer network is connected to a PE

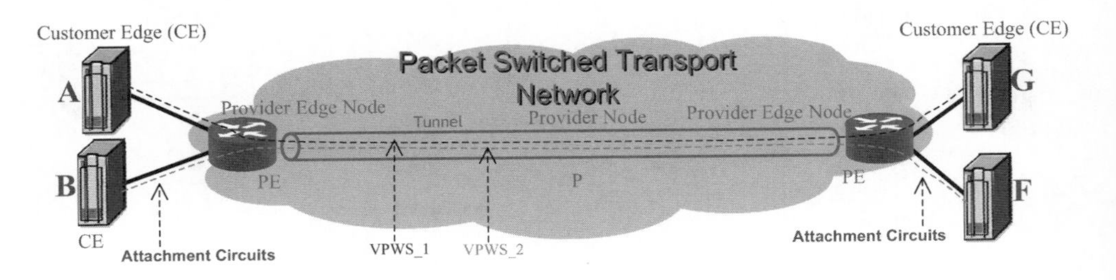

Figure 2.18 Entities of the VPWS reference model

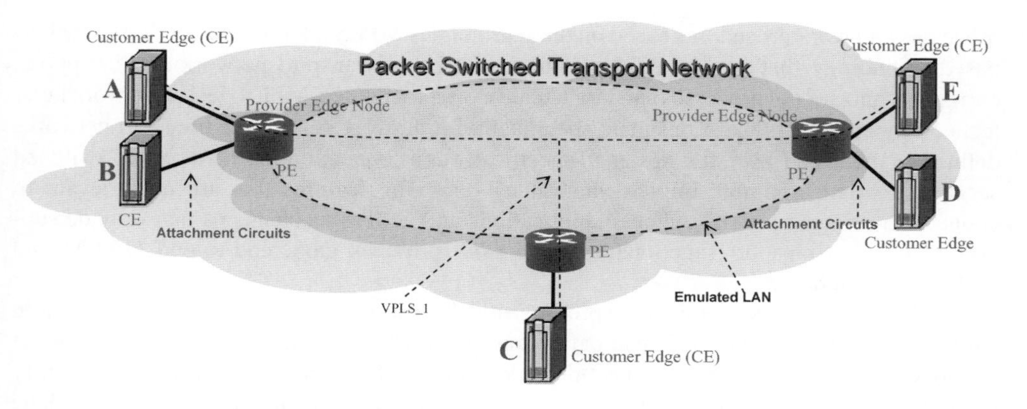

Figure 2.19 Entities of the VPLS reference model

in the provider network via an attachment circuit (AC), where the AC is either a physical or a logical circuit. The PEs in the core network are connected via a pseudo-wire (PW). The CE devices can be routers, switches, or end-user hosts. In some implementations, a set of VPWSs is used to create a multisite L2-VPN.

A VPLS (Figure 2.19) is a full-mesh connection of multiple sites using Ethernet access in a single bridged domain over a provider-managed PSN (e.g., MPLS-based PSN). All customer sites in a VPLS instance appear to be on the same LAN. An Ethernet interface at the customer simplifies the LAN/WAN boundary and allows for rapid and flexible service provisioning.

In a VPLS, the provider network emulates a learning bridge, and forwarding decisions are taken based on MAC addresses or MAC addresses and VLAN tags.

The IETF defines the PW as a logical circuit established over an L2 PSN. A PW is an emulated connection between two or more PEs. The establishment and maintenance of the state information of each PW is under the responsibility of the two PEs, which are its endpoints. In this specification, point-to-point PWs are always considered bidirectional and multipoint-to-point and point-to-multipoint PWs are always considered unidirectional. Each packet is carried from the ingress AC to egress AC through one PW. In the case of heterogeneous L2 transport networks, L2 inter-working functions should be applied, as further specified in Ref. [22].

2.4.4.3 ITU-T T-MPLS/IETF MPLS-TP

MPLS is a matured L2 switching technology based on labels and starts playing an important role in transport networks for connection establishments. However, not all of MPLS's capabilities as specified in Ref. [27] and the related documents [28,29] are needed and are consistent with transport network operations. There is, therefore, the need to adapt MPLS and to customize it to define an MPLS-TP in order to support the operations through combining the packet experience of MPLS with the operational experiences of TDM switching technologies. MPLS-TP uses the similar architectural principles of layered networking that are used in SONET/SDH and OTN. Service providers have already developed network management processes and work procedures based on these principles MPLS-TP provides a reliable packet-based technology that is also aligned with circuit-based transport networking, it supports current organizational processes and large-scale work procedures. MPLS-TP is a network layer technology at the data plane designed specifically for network services in transport networks defined by the ITU-T [30]. It is designed specifically as a connection-oriented packet-switched service. It offers a simple implementation by removing features that are not relevant to connection-oriented packet-switched applications and adding mechanisms that provide support of critical transport functionality. MPLS-TP provides QoS, end-to-end OA&M and protection switching.

MPLS-TP creates an application profile with the use of labels. The MPLS forwarding paradigm is based on labels. These data plane functions are complemented with transport network functionality, such as connection service and performance monitoring, survivability (including protection switching and restoration), and network management and control plane that can be based on ASON and GMPLS respectively.

An MPLS-TP layer network can operate independently of its clients and its associated control networks (i.e., management and signaling). This independence enables network operators the freedom necessary to design packet transport networks for their own use and to transport customer traffic (transparently). MPLS-TP trails can carry a variety of client traffic types, including IP/MPLS and Ethernet data traffic.

2.4.5 *Layer 3 VPN*

An IP layer network interconnects a set of network elements based on IP addresses as described in Figure 2.20 and it refers to the IP for the network connections between a set of sites. The different connections making use of a shared network infrastructure as described in the Figure 2.20 are termed IP-VPN. In this chapter, an IP network is also termed an L3-VPN.

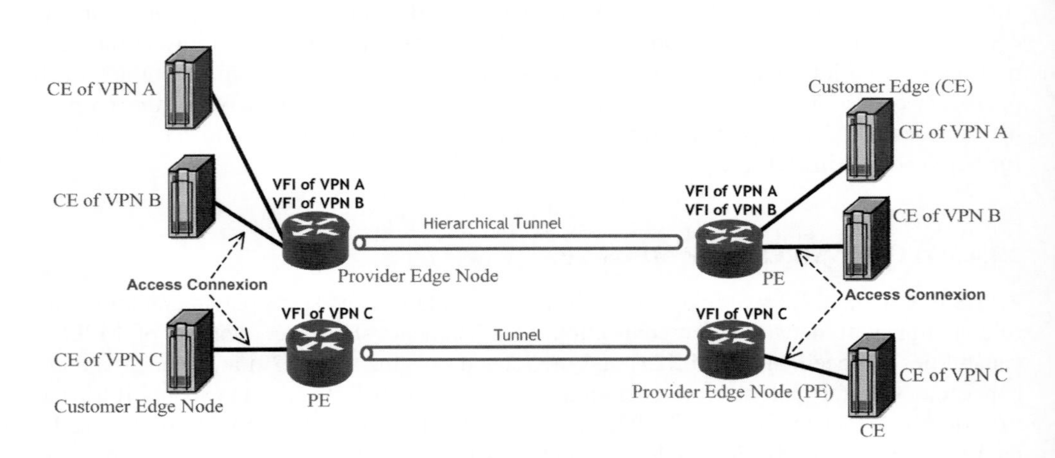

Figure 2.20 PE-based L3-VPNs

Each L3-VPN is composed of the following elements:

- CE equipment is part of the user networks located at the faces at a customer site. The CE has an access connection to a PE node. CE equipment can be a router or a switch that allows users at a "customer site" to communicate over the access network with other sites interconnected by the L3-VPN. In the case of a CE-based L3-VPN, also termed a provider-provisioned CE-based VPN, the network service provider manages (partially) the CE node. At the customer site, the network operator manages at least the access connection part of the PE node.
- PE node: each PE node faces the client transport network on one side and attaches via an access connection over one or more access networks (the networks of the customers) to one or more CE devices.[9] It participates in performing routing and forwarding functions.
- P device: a device within a transport network (referenced as provider network) that interconnects PE (or other P) devices but does not have any direct attachment to CE devices. The P node does not control any VPN states and is VPN unaware.

According to Ref. [21], the transport network is PSN through which the tunnels supporting the L3-VPN services are set up. The service provider network is a set of interconnected PE and P devices administered by a single service provider (which can be a network operator owning the infrastructure) in one or more autonomous systems (ASs).

Different deployment scenarios can be considered according to the network service provider point of view: (i) a single service provider and a single AS in which the L3-VPN sites are attached to one service provider within the scope of a single AS, (ii) a single service provider with multiple ASs in which the VPN sites are attached to one service provider within the scope consisting of multiple ASs, and (iii) cooperating service provider in which VPN sites are attached to the transport networks of different service providers that cooperate with each other to provide the L3-VPN.

The networks service providers have many requirements (in part expressed by their users) in common, including network service requirements for security, privacy, manageability, and interoperability. In the case of grid application service deliveries, the projections on the number of the customer VPNs over the next several years can be critical, as well as the rate of changes of the VPN services and the complexity of deploying these services over the transport networks.

There are two major classes of L3-VPN networks: (i) PE-based L3VPNs and (ii) CE-based L3-VPNs [23]. In a PE-based L3-VPN service, the PE controls a VPN forwarding instance (VFI) for each L3-VPN that it serves. A VFI is a logical entity that resides in a PE that includes the router information base and forwarding information base for a VPN. The VFI terminates tunnels for interconnection with other VFIs and also terminates access connections for accommodating CEs. The VFI contains information regarding how to forward data received over the CE–PE access connection to VFIs in other PEs supporting the same L3-VPN.

2.5 Conclusion

Through the NOBEL project, the different deliverables reported the major results on advanced TE and resilience strategies that can be possible with more uniform network management and

[9] The definitions of CE and PE do not necessarily match the physical deployment of network equipment on customer premises or a provider PoP.

control functions enabled with generalized multiprotocol label switching control applied for the different switching layers and corresponding services: L1-VPN, L2-VPN and L3-VPN as presented in this chapter.

Multiple case-studies on different aspects of multilayer, multiservice, and multi-domain resilience and TE mechanisms have been carried out from which these case-studies have been summarized to provide guidelines in metro and core transport networks.

The NOBEL transport network trends foresee an evolution towards a unique multiservice network able to support any kind of service offered by network infrastructures operated by uniform control and management functions. The main drivers for a uniform and integrated network control are the CAPEX and OPEX savings achieved by simplifying the implementation of dynamic connection provisioning according to the networked applications (Table 2.2).

In that respect, current optical network infrastructures, often based on hierarchical and uncoordinated layered architectures with static point-to-point links between routers, are presenting scalability problems, mainly due to the increase of long connections and their associated pass-through traffic. In order to solve this problem, operators have started deploying "more autonomous" transport networks based on TDM and OTN equipment with GMPLS-based routing and signaling controllers. Thus, with the appearance of IP traffic over GMPLS-controlled networks; the number of self-reconfiguration alternatives for resilience is increased. While resilience is exclusively managed at the IP layer in pure IP networks, both TE and resilience mechanisms can be carried out in different transport layers (L1/L2/L3) in IP over GMPLS. Furthermore, such dynamic resilience can be configured to be executed in a coordinated way with the availability requirements of the services.

Accordingly, some key challenges will arise in the short, medium and long term. Such challenges have been identified and technically evaluated in NOBEL2 deliverables of work-package 2 and some key contributions have been reported in this chapter.

References

[1] Takeda, T. *et al.* (April 2007) Framework for layer 1 virtual private networks, IETF RFC4847.

[2] Papadimitriou, D. and Dominique, D.(July 2005) GMPLS user-to-network interface in support of end-to-end rerouting. *IEEE Communications Magazine*, **43**, 35–43.

[3] Bryskin, I. *et al.*, L1-VPN working groups (December 2007) OSPF based layer 1 VPN auto-discovery, work in progress at IETF.

[4] Takeda, T. *et al.*, L1-VPN working groups (October 2007) Applicability statement for layer 1 virtual private networks (L1VPNs) basic mode, work in progress at IETF.

[5] ITU-T (July 2004) Layer 1 virtual private network service and network architectures, ITU-T Rec. Y.1313.

[6] A. Autenrieth, *et al.* (December 2006) WP2 final report on Traffic engineering and resilience strategies for NOBEL solutions (resilience part). NOBEL phase 1, Deliverable D27.

[7] Lang, J.P. *et al.* (May 2007) RSVP-TE extensions in support of end-to-end generalized multi-protocol label switching (GMPLS) recovery, IETF RFC 4872.

[8] Berger, L. *et al.* (May 2007) GMPLS based segment recovery, IETF. RFC 4873.

[9] Mannie, E. *et al.* (March 2006) Recovery (protection and restoration) terminology for generalized multi-protocol label switching (GMPLS), IETF RFC 4427.

[10] Berger, L. *et al.* (January 2003) Generalized multi-protocol label switching (GMPLS) signaling resource reservation protocol-traffic engineering (RSVP-TE) extensions, IETF RFC3473.

[11] Pan, P. *et al.* (May 2005) Fast reroute extensions to RSVP-TE for LSP tunnels, IETF RFC 4090.

[12] Farrel, A. *et al.* (September 2007) Inter domain multiprotocol label switching (MPLS) and generalized MPLS (GMPLS) traffic engineering – RSVP-TE extensions, work in progress at IETF, draft-ietf-ccamp-inter-domain-rsvp-te-07.txt.

[13] Tapolcai, J. *et al.* (June 2006) Joint quantification of resilience and quality of service. IEEE International Conference on Communications.

[14] Song, Q. *et al.* (2005) Performance evaluation of connection setup in GMPLS IP optical network. Optical Fiber Communication Conference, OFC/NFOEC.

[15] Verchere, D. *et al.* (October 2004) Multi-layer recovery enabled with end-to-end signalling. Optical Networks and Technologies, IFIP TC6 Proceedings, pp. 246–253.

[16] Tapolcai, J. *et al.* (October 2005) A novel shared segment protection method for guaranteed recovery time. International Conference on Broadband Networks BROADNETS, Boston, MA, USA, pp. 127–136.

[17] Verchère, D. and Papadimitriou, D.(October 2004) Prototyping the GMPLS UNI implementation for end-to-end LSP re-routing. MPLS 2004 International Conference.

[18] Späth, J. and Foisel, H.-M.(March 2006) MUPBED network architecture, interoperability topics and relevance for NREN/GEANT2. Workshop, Bonn, http://www.ist-mupbed.eu.

[19] Reference Networks, http://www.ibcn.intec.ugent.be/INTERNAL/NRS/index.html.

[20] Takeda, T. *et al.* (2005) Layer 1 virtual private networks: driving forces and realization by GMPLS. *IEEE Communications Magazine*, **43** (7), 60–67.

[21] Andersson, L. *et al.* (March 2005) Provider provisioned virtual private network (VPN) terminology, IETF RFC4026.

[22] Andersson, L. *et al.* (September 2006) Framework for layer 2 virtual private networks (L2VPNs), IETF RFC4664.

[23] Callon, R. *et al.* (July 2003) A framework for layer 3 provider-provisioned virtual private networks (PPVPNs), IETF RFC4110.

[24] Bryskin, I. *et al.* (November 2008) OSPFv3-based layer 1 VPN auto-discovery, work in progress at IETF, in draft-ietf-l1vpn-ospfv3-auto-discovery-00.txt.

[25] Bryant, S. *et al.* (March 2005) Pseudo wire emulation edge-to-edge (PWE3) architecture, IETF RFC3985.

[26] Metro Ethernet Forum (June 2004) Ethernet service definition – Phase 1, Technical Specification, MEF 6.

[27] Rosen, E. *et al.* (January 2001) Multi-protocol label switching architecture, IETF RFC3031.

[28] Katz, D. et al. (September 2003) *Traffic Engineering (TE) Extensions to OSPF Version 2*, IETF.

[29] Awduche, D. *et al.* (December 2001) RSVP-TE: extensions to RSVP for LSP tunnels, IETF RFC3630.

[30] ITU-T (undated) Operation and maintenance definition mechanism for T-MPLS networks, ITU-T Recommendations Y.1711.1.

[31] ITU-T (July 1994) Open systems interconnection (OSI): basic reference model, ITU-T Recommendations X.200.

[32] IST NOBEL (Next generation Optical network for Broadband European Leadership) (Sept. 2004) Preliminary definition of drivers and requirements for core and metro networks supporting end-to-end broadband services for all, Deliverable 6.

[33] ITU-T (March 2003) Interfaces for the optical transport network (OTN), ITU-T Recommendations .G.709/Y.1331.

[34] Kompella, K. *et al.* (October 2005) Label switched paths (LSP) hierarchy with generalized multi-protocol label switching (GMPLS) traffic engineering (TE), IETF RFC4206.

[35] Colle, D. *et al.* (September 2004) NOBEL dictionary: resilience related terminology NOBEL report.

[36] Oki, E. *et al.* (November 2009) Framework for PCE-based inter-layer MPLS and GMPLS traffic engineering. Works in progress at IETF PCE working group (draft-ietf-pce-inter-layer-frwk-10.txt).

[37] Atlas, A.K. *et al.* (September 2006) Path computation element (PCE) communication protocol generic requirements, IETF RFC4657.

[38] Farrel, A. *et al.* (August 2006) A path computation element (PCE)-based architecture, IETF RFC 4655.

[39] ITU-T (June 2006) Architecture for the automatically switched optical network (ASON), ITU-T Recommendation G.8080/Y.1304.

[40] ITU-T (July 2001) Requirements for automatic switched transport networks (ASTN), ITU-T Recommendation G.807/Y.1302.

[41] ITU-T (March 2000) Generic functional architecture of transport networks, ITU-T Recommendation G.805.

[42] Mannie, E. *et al.* (October 2004) Generalized multi-protocol label switching (GMPLS) architecture. IETF RFC3945.

[43] Kompella, K. *et al.* (October 2005) Routing extensions in support of generalized multi-protocol label switching (GMPLS), IETF RFC4202.

[44] Mannie, E. *et al.* (August 2006) Generalized multi-protocol label switching (GMPLS) extensions for synchronous optical network (SONET) and synchronous digital hierarchy (SDH) control, IETF RFC4606.

[45] Papadimitriou, D. *et al.* (January 2006) Generalized multi-protocol label switching (GMPLS) signaling extensions for G.709 optical transport networks control, IETF RFC4328.
[46] Otani, T. *et al.* (March 2009) Generalized labels for G.694 lambda-switching capable label switching routers, work in progress at IETF CCAMP working group.
[47] Lang, J. *et al.* (October 2005) Link management protocol (LMP), IETF RFC4204.
[48] Berger, L. *et al.* (January 2003) Generalized multi-protocol label switching (GMPLS) signaling functional description, IETF RFC3471.
[49] Ashwood-Smith, P. *et al.* (January 2003) Generalized multi-protocol label switching (GMPLS) signaling constraint-based routed label distribution protocol (CR-LDP) extensions, IETF RFC3472.
[50] Swallow, G. *et al.* (October 2005) Generalized multiprotocol label switching (GMPLS) user-network interface (UNI): resource reservation protocol-traffic engineering (RSVP-TE) support for the overlay model, IETF RFC4208.
[51] Shiomoto, K. *et al.* (July 2008) Requirements for GMPLS-based multi-region and multi-layer networks (MRN/MLN), IETF RFC5212.
[52] Lee, C.-Y. *et al.* (April 2007) Exclude routes – extension to resource reservation protocol-traffic engineering (RSVP-TE), IETF RFC4874.
[53] Farrel, A. *et al.* (November 2006) A framework for inter-domain multiprotocol label switching traffic engineering, IETF RFC4726.
[54] Colle, D. *et al.* (2002) Data-centric optical networks and their survivability. *IEEE Journal on Selected Areas in Communications*, **20** (1), 6–20.
[55] Kompella, K. *et al.* (January 2003) Signalling unnumbered links in RSVP-TE, IETF RFC3477.
[56] Srisuresh, P. *et al.* (January 2001) Traditional IP network address translator (traditional NAT), IETF RFC 3022.
[57] The Optical Internetworking Forum (OIF) (October 2001) User network interface (UNI) 1.0 signaling specification.
[58] The Optical Internetworking Forum (OIF) (February 2008) User network interface (UNI) 2.0 signaling specification: common part.
[59] The Optical Internetworking Forum (OIF) (February 2009) OIF E-NNI signaling specification (OIF E-NNI 2.0), working document (OIF2005.381.30).
[60] The Optical Internetworking Forum (OIF) (November 2007) OIF E-NNI routing 2.0 implementation agreement, draft document (OIF2007.379.01).
[61] ITU-T (March 2000) Generic functional architecture of transport networks, ITU-T Recommendation G.805.
[62] Rajagopalan, B. *et al.* (March 2004) IP over optical networks: a framework, IETF RFC 3717.
[63] Kompella, K. *et al.* (October 2005) Link Bundling in MPLS traffic engineering (TE), IETF RFC 4201.
[64] Braden, B. *et al.* (September 1997) Resource reservation protocol (RSVP) – Version 1 message processing rules, IETF RFC 2209.
[65] Ould-Brahim, H. *et al.* (June 2008) BGP-based auto-discovery for layer-1 VPNs, IETF RFC 5195.
[66] Metro Ethernet Forum (October 2006) Ethernet services attributes phase 2, Technical Specification MEF 10.1.
[67] CARRIOCAS, Distributed computing over ultra-high bit rates optical internet network, http://www.carriocas.org.
[68] Awduche, D. *et al.* (September 1999) Requirements for Traffic Engineering Over MPLS IETF RFC 2702.
[69] Case, Jeffrey *et al.* (December 2002) Introduction and Applicability Statements for Internet Standard Management Framework IETF RFC 3410.
[70] Harrington, David *et al.*, (December 2002) An Architecture for Describing Simple Network Management Protocol (SNMP) Management Frameworks IETF RFC 3411.
[71] Presuhn, Randy *et al.* (December 2002) Version 2 of the Protocol Operations for the Simple Network Management Protocol (SNMP) IETF RFC 3416.
[72] Srinivasan, Cheenu *et al.* (June 2004) Multiprotocol Label Switching (MPLS) Traffic Engineering (TE) Management Information Base (MIB) IETF RFC 3812.
[73] Nadeau, Thomas D. *et al.* (February 2007) Definitions of Textual Conventions for Generalized Multiprotocol Label Switching (GMPLS) Management IETF RFC 4801.
[74] Preliminary definition of network scenarios and solutions supporting broadband services for all IST Integrated Project NOBEL - Work Package 1 - Deliverable 11, Jan. 2005.

3

Elements from Telecommunications Engineering

Chris Matrakidis, John Mitchell and Benn Thomsen

3.1 Digital Optical Communication Systems

This chapter presents the basic mathematical constructs that are required to describe and evaluate signals in digital optical systems and networks. Optical communications systems, as illustrated in Figure 3.1, are almost exclusively used for the transmission of digital signals. However, in order to transmit the digital signal efficiently it must first be encoded onto the optical carrier in a process known as modulation. Current optical communication systems use a very simple form of modulation known as on–off keying, which is a form of digital amplitude modulation where a 'one' is represented as a bright pulse and a 'zero' is represented as no light. At low bit-rates (e.g. up to 2.5 Gbit/s), this can be easily performed by directly modulating the current of a semiconductor laser and thereby controlling the emitted optical power. However, for high-speed modulation (e.g. at 10 and 40 Gbit/s), external modulators are required. These modulators act as some kind of fast switch that is controlled by the electrical data and either blocks or transmits the incoming continuous-wave light, depending on the value of the electrical data to be transmitted. The encoded optical signal that is transmitted is now, of course, an analogue representation of the original digital signal. The purpose of an optical communication system is to transmit this analogue signal with as little distortion as possible and effectively reconstruct the original digital signal at the receiver.

3.1.1 Description of Signals in the Time and Frequency Domains

Here, we present a brief summary of the description of signals in the time and frequency domains, with a particular emphasis on those of interest to optical communications. The term signal is used in a telecommunications context to describe the way information is transmitted. In general, a signal is some function of at least one variable (usually time) that represents some

Core and Metro Networks Edited by Alexandros Stavdas

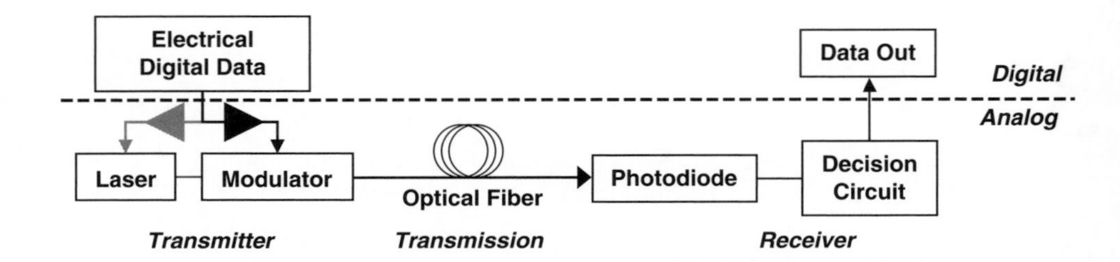

Figure 3.1 Optical digital communication system

measurable physical quantity that we are interested in. As an example, the speech signal is the variation of air pressure that results from the vocal chord movement; similarly, the electrical current coming out of a microphone as a result of these air pressure variations is another signal.

In communication systems there are several types of signal that are of interest. *Continuous time signals*, where the free variable (time) is a real value, and *discrete time signals*, when time takes values from a discrete set (e.g. integer values). Examples of continuous and discrete time signals are shown in Figure 3.2.

An important signal distinction is whether they are *periodic* or *aperiodic*. Periodic signals (Figure 3.3) are all signals $x(t)$ that have the property that there is a real number T for which $x(t) = x(t + T)$ for all values of t. T is called the period of the signal. The smallest value of T is called the fundamental period, symbolized by T_0. Aperiodic signals are those where no period exists.

An example of a periodic signal is $x(t) = \sin(at + b)$ with fundamental period $T_0 = 2\pi/a$.

Another common distinction is between *causal* and *noncausal* signals (Figure 3.4). A causal signal is one where its value is zero for all negative time, while a noncausal signal has nonzero values for both positive and negative time.

In communications systems, signals can either be *deterministic* or *random*. A deterministic signal is a signal in which there is no uncertainty about its value at any time and so can be modeled as a completely specified function of time. Conversely, the value of a random signal is uncertain before it actually occurs. Random signals in communication systems are represented using techniques from probability theory.

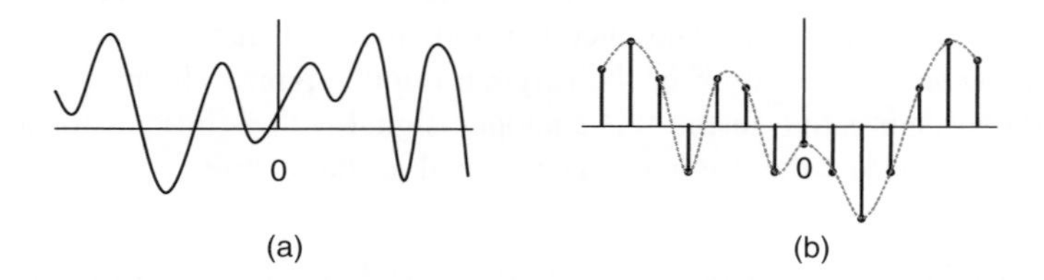

Figure 3.2 (a) Continuous-time and (b) discrete-time signals

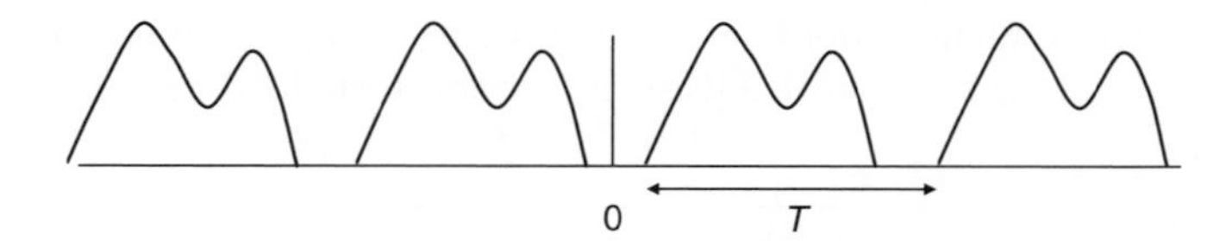

Figure 3.3 Example of a periodic signal

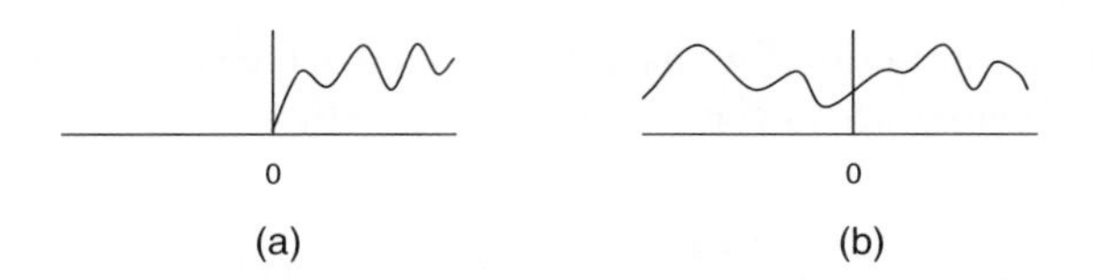

Figure 3.4 (a) Causal and (b) noncausal signals

An important property of all signals is that they can be decomposed into a sum of an odd and an even signal. An even signal is one having the property that $x(t) = x(-t)$ for all values of t, while an odd signal has $x(t) = -x(-t)$. The decomposition of an arbitrary signal $g(t)$ can be done in the following way:

$$g(t) = \frac{1}{2}(g(t) + g(-t)) + \frac{1}{2}(g(t) - g(-t))$$

with $\frac{1}{2}(g(t) + g(-t))$ being an even signal and $\frac{1}{2}(g(t) - g(-t))$ being an odd signal (Figure 3.5).

3.1.1.1 Frequency Domain Representation of Signals

In many situations, the representation of signals in the frequency domain is useful. This is the case when we want to study their bandwidth and their frequency spectrum. For this purpose the Fourier series is used to represent periodic signals and the Fourier transform is employed to convert between the time- and frequency-domain signal representations for both periodic and aperiodic signals.

A signal $f(t)$ that has a finite number of minima and maxima in an interval (t_0, t_1) and a finite number of discontinuities in the same interval can be represented as a Fourier series if it is absolutely integrable in this interval; that is, if $\int_{t_0}^{t_1} |f(t)|\, \mathrm{d}t < \infty$. The Fourier series is given by

$$f(t) = \sum_{k=-\infty}^{\infty} a_k \, \mathrm{e}^{\mathrm{j}k\omega_0 t} \tag{3.1}$$

with fundamental frequency $\omega_0 = 2\pi/(t_1 - t_0)$ and Fourier series coefficients

$$a_k = 1/(t_1 - t_0) \int_{t_0}^{t_1} f(t) e^{-\mathrm{j}k\omega_0 t} \mathrm{d}t$$

The Fourier coefficients are, in general, complex numbers.

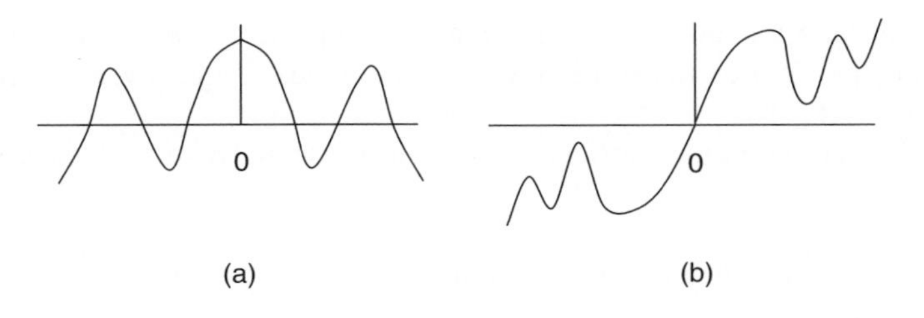

Figure 3.5 (a) Even and (b) odd signals

For periodic signals with period T we have $T = t_1 - t_0$ and then the fundamental frequency is $\omega_0 = 2\pi/T$ and the Fourier series coefficients are given by

$$a_k = 1/T \int_{-T/2}^{T/2} f(t)\mathrm{e}^{-\mathrm{j}k\omega_0 t}\,\mathrm{d}t$$

Some useful properties of Fourier series are the following:

1. For a real-valued signal $f(t)$ the Fourier coefficients have the property that $a_k = a^*_{-k}$
2. For an even real signal $f(t)$ the Fourier coefficients a_k are real numbers.
3. For an odd real signal $f(t)$ the Fourier coefficients a_k are imaginary numbers.
4. For a periodic signal $f(t)$, Parseval's theorem states that

$$\frac{1}{T}\int_{-T/2}^{T/2} |f(t)|^2\,\mathrm{d}t = \sum_{k=-\infty}^{\infty} |a_k|^2 \tag{3.2}$$

A signal $g(t)$ that has a finite number of minima and maxima in any finite real number interval and a finite number of discontinuities in any finite real number interval can be represented as a Fourier series if it is absolutely integrable for all real numbers; that is, if $\int_{-\infty}^{\infty} |g(t)|\,\mathrm{d}t < \infty$. The Fourier transform is given by

$$G(\omega) = \int_{-\infty}^{\infty} g(t)\mathrm{e}^{-\mathrm{j}\omega t}\,\mathrm{d}t \tag{3.3}$$

and the original signal can be obtained using its Fourier transform by

$$g(t) = \frac{1}{2\pi}\int_{-\infty}^{\infty} G(\omega)\mathrm{e}^{\mathrm{j}\omega t}\,\mathrm{d}\omega \tag{3.4}$$

Commonly, to indicate that $G(\omega)$ is the Fourier transform of $g(t)$ we use the notation

$$G(\omega) = F(g(t)) \tag{3.5}$$

and to indicate that $g(t)$ is the inverse Fourier transform of $G(\omega)$ we use

$$g(t) = F^{-1}(G(\omega)) \tag{3.6}$$

While the Fourier series gives a discrete frequency representation of a periodic (or a time-limited) signal, the Fourier transform gives a continuous frequency representation of an aperiodic signal.

Some useful properties of the Fourier transform are presented here, with the assumption that $G(\omega) = F(g(t))$:

1. If $g(t)$ is real for all t, then the Fourier transform has the property that $G(\omega) = G^*(-\omega)$.

2. For a $g(t)$ that is an even real function, the Fourier transform is an even real function and can be calculated by

$$G(\omega) = 2\int_{0}^{\infty} g(t)\cos(\omega t)\,\mathrm{d}t$$

3. For a $g(t)$ that is an odd real function, the Fourier transform is an odd imaginary function and can be calculated by

$$G(\omega) = -2\mathrm{j}\int_{0}^{\infty} g(t)\sin(\omega t)\,\mathrm{d}t$$

4. Parseval's theorem version for the Fourier transform of a function is

$$\int_{-\infty}^{\infty} |g(t)|^2\,\mathrm{d}t = \frac{1}{2\pi}\int_{-\infty}^{\infty} |G(\omega)|^2\,\mathrm{d}\omega$$

5. Linearity

$$F(a_1 g_1(t) + a_2 g_2(t)) = a_1 G_1(\omega) + a_2 G_2(\omega), \text{ with } G_n(\omega) = F(g_n(t))$$

6. Duality

$$g(\omega) = F(G(-t))$$

7. Time shift

$$\mathrm{e}^{-\mathrm{j}\omega t_0} G(\omega) = F(g(t - t_0))$$

8. Scaling

$$\frac{1}{|a|} G\left(\frac{\omega}{a}\right) = F(g(at))$$

9. Convolution

$$F(g_1(t) * g_2(t)) = G_1(\omega)G_2(\omega), \text{ with } G_n(\omega) = F(g_n(t))$$

3.1.1.2 Bandpass Signals

In optical communications the signals we are interested in are an example of bandpass signals. A bandpass signal is one whose frequency representation is nonzero in the area around some high frequency F_c which is known as the carrier frequency (Figure 3.6). This is obviously the case with an optical signal, where light is modulated and, therefore, the carrier frequency is of the order of hundreds of terahertz, while the modulated signal bandwidth is several orders of magnitude lower.

In order to simplify their analysis, we often use the baseband representation of the signal. To obtain this, we take the positive part of the frequency representation and multiply it by two. Then we shift it towards zero frequency by $-F_c$. The end result is a complex signal that we can use instead of the bandpass signal. An important thing to note is that the frequency F_c does not have to be the real center frequency of the bandpass signal, but can be any arbitrary frequency.

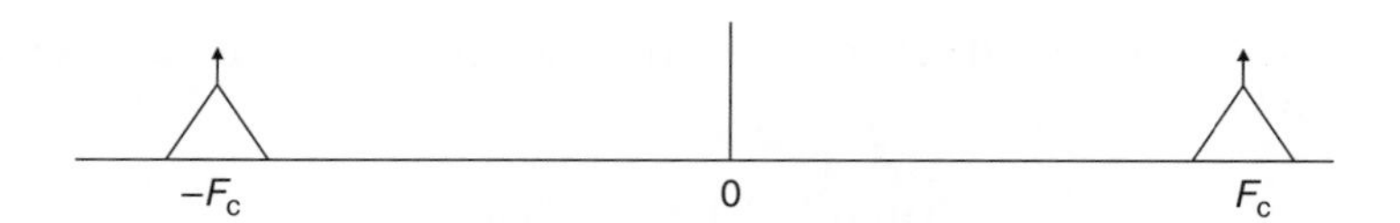

Figure 3.6 Frequency representation of a bandpass signal

Regarding the complex baseband signal representation, we can separate it into real and imaginary parts. The real part is called the in-phase component of the signal, while the imaginary part is called the quadrature component.

3.1.2 Digital Signal Formats

Most optical communication systems transmit digital signals rather than analogue signals, as digital signals are far more resilient to noise. In a digital system, which transmits ‘ones’ or ‘zeros’, the receiver simply has to decide whether the received signal is less than or greater than some threshold. The most commonly transmitted optical digital signal format is known as nonreturn to zero (NRZ), shown in Figure 3.7. In this format, a digital ‘one’ is represented by a pulse of light and a ‘zero’ by the absence of light. The pulses occupy the entire bit slot and, therefore, signal transitions only occur when the digital data transitions from $1 \rightarrow 0$ and vice versa.

The ideal NRZ data sequence is defined as a sum of rectangular functions $p(t)$:

$$\begin{aligned} s(t) &= \sum_n a_n A p(t - nT_b) \\ p(t) &= \begin{cases} 1 & 0 \leq t < T_b \\ 0 & \text{otherwise} \end{cases} \end{aligned} \tag{3.7}$$

where T_b is the bit period, A is the signal amplitude and a_n are the data bits, either 0 or 1. If the data sequence is infinitely long, then the spectrum of this signal is simply the Fourier transform of the rectangular function; that is

$$\begin{aligned} S(f) &= AT_b \frac{\sin(\pi T_b f)}{\pi T_b f} \\ &= AT_b \sin c(T_b f) \end{aligned} \tag{3.8}$$

We note that this ideal NRZ signal occupies an infinite spectral bandwidth, as shown in Figure 3.8, and thus is not ideal for transmission over a bandwidth-limited channel.

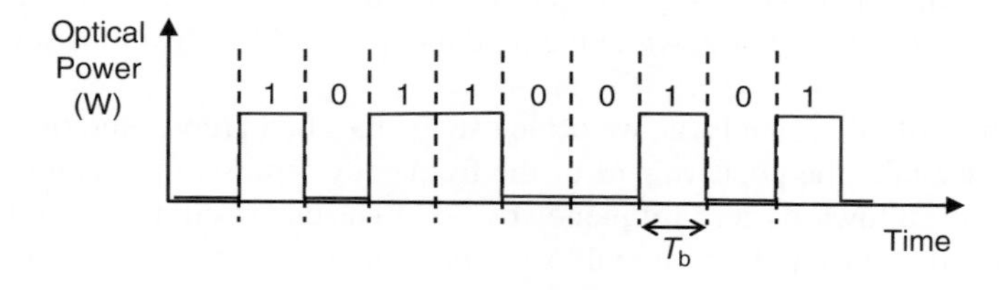

Figure 3.7 Ideal NRZ signal

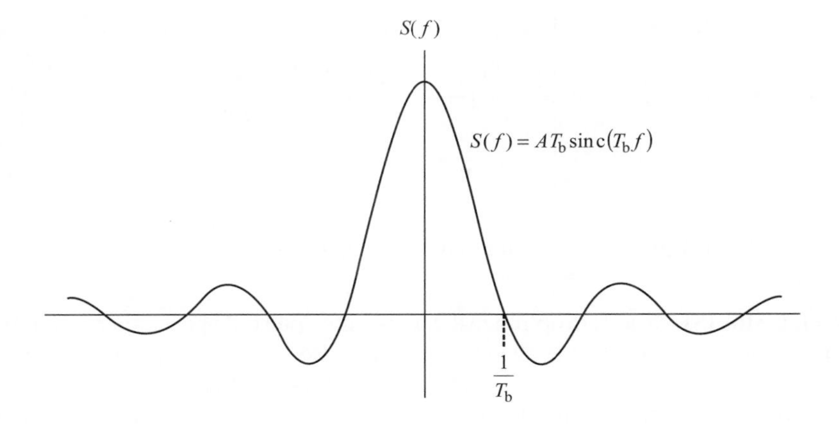

Figure 3.8 Spectrum of ideal NRZ signal

To optimize transmission over any channel we need to consider the impact of two things: the channel frequency response and the noise introduced by the channel. The passband of the channel frequency response must be sufficiently wide so as not to attenuate the frequency content of the transmitted data significantly. In optical fibers, this condition is easily met; however, the phase response of an optical fiber is not constant across this passband and, thus, they are dispersive. Dispersion of the transmitted signal results in spreading of the pulse energy across adjacent bit slots and leads to intersymbol interference (ISI) at the receiver. Both ISI and noise give rise to errors in the reconstructed data stream after the decision element at the receiver.

In order to minimize the impact of noise on the system it is necessary to maximize the signal-to-noise ratio (SNR) before the decision element. If we consider the communications system model shown in Figure 3.9, where additive white Gaussian noise (AWGN) $n(t)$ is added to the transmitted signal pulse $g(t)$, this implies that we need to design a receiver filter $h(t)$ that matches the transmitted data signal. It can be shown (see Ref. [1]) that the optimum filter, the so called *matched filter*, in the presence of AWGN is given by a delayed and time-reversed version of the transmitted signal pulse with an appropriate scaling factor k.

$$h(t) = kg(T-t) \tag{3.9}$$

Or equivalently in the frequency domain:

$$H(\omega) = kG^*(\omega)\mathrm{e}^{-\mathrm{j}\omega T} \tag{3.10}$$

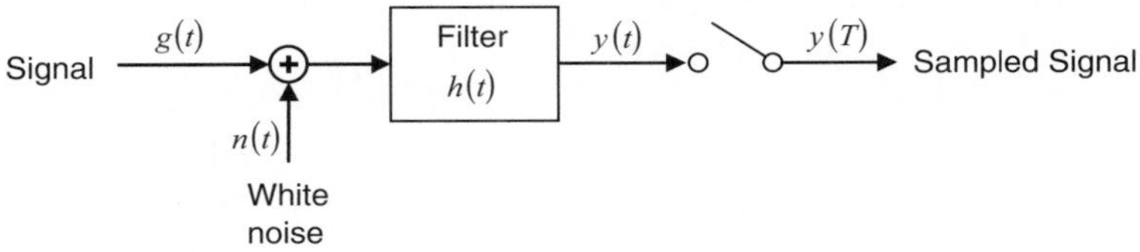

Figure 3.9 Communications channel noise model with matched filtering

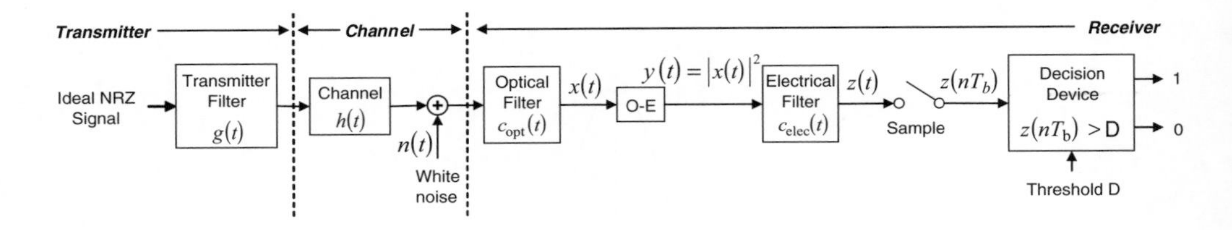

Figure 3.10 Transmit and receiver filtering to minimize ISI and noise

To minimize the ISI we need to optimize the transmitted pulse shape in order to minimize the effects of dispersion in the channel. That is to say, we need to band limit the signal. This is achieved by designing transmit and receive filters as shown in Figure 3.10.

In an optical fiber channel the dominant source of dispersion is first-order chromatic dispersion; thus, the fiber channel is modeled as

$$H(\omega) = \mathrm{e}^{-\mathrm{j}(\beta_2 L/2)\omega^2} \tag{3.11}$$

where β_2 is the group velocity dispersion parameter, which is related to the dispersion parameter by $D = -2\pi c\beta_2/\lambda^2$, which has units of ps/(nm km) and L the fiber length. In order to optimize the transmit $g(t)$ filter design we wish to maximize the *energy transfer ratio* ETR, which describes the energies within a single per bit slot of the received signal $s_\mathrm{o}(t)$ as a ratio of the input signal $s_\mathrm{i}(t)$ [2]:

$$\begin{aligned} \mathrm{ETR} &= \frac{\displaystyle\int_{-T_\mathrm{b}/2}^{T_\mathrm{b}/2} |s_\mathrm{o}(t)|^2\,\mathrm{d}t}{\displaystyle\int_{-T_\mathrm{b}/2}^{T_\mathrm{b}/2} |s_\mathrm{i}(t)|^2\,\mathrm{d}t} \\ s_\mathrm{o}(t) &= h(t)\otimes g(t)\otimes s_\mathrm{i}(t) \end{aligned} \tag{3.12}$$

Essentially, this means designing a filter that maintains as much of the original pulse energy within the bit slot as possible. This criterion will also maximize the SNR and, thus, meet the matched filtering requirement discussed previously. Gaudino and Viterbo [2] used this criterion to design the optimum transmitter filter for an NRZ transmission format. Here, the optimal transmitted pulse has an amplitude profile that varies from Gaussian-like when the channel dispersion is small to raised cosine-like as the channel dispersion increases. The optimal phase profile is parabolic, as would be expected from inspection of Equation 3.11. In practice, however, most optical NRZ transmitters do not control the phase and so opt for a *raised cosine* amplitude profile and a constant phase. This bandwidth limits the NRZ signal to the data rate and minimizes the ISI. The raised cosine NRZ signal has a temporal amplitude defined as

$$p(t) = \begin{cases} 0 & t \le -\dfrac{\alpha T_\mathrm{b}}{2},\ T_\mathrm{b} + \dfrac{\alpha T_\mathrm{b}}{2} < t \\ \dfrac{1}{2}\left[1 + \sin\left(\dfrac{\pi}{\alpha T_\mathrm{b}} t\right)\right] & -\dfrac{\alpha T_\mathrm{b}}{2} < t \le \dfrac{\alpha T_\mathrm{b}}{2} \\ 1 & \dfrac{\alpha T_\mathrm{b}}{2} < t \le T_\mathrm{b} - \dfrac{\alpha T_\mathrm{b}}{2} \\ \dfrac{1}{2}\left\{1 - \sin\left[\dfrac{\pi}{\alpha T_\mathrm{b}}(t - T_\mathrm{b})\right]\right\} & T_\mathrm{b} - \dfrac{\alpha T_\mathrm{b}}{2} < t \le T_\mathrm{b} + \dfrac{\alpha T_\mathrm{b}}{2} \end{cases} \tag{3.13}$$

where α is the rolloff factor, which varies between 0 and 1. When the rolloff factor approaches zero, Equation 3.13 tends to the ideal NRZ waveform of Equation 3.7. Typical optical systems use a rolloff factor of 0.8, such that the central 20% of the bit slot is represented by the flat portion of the raised cosine waveform. This representation, which is convenient for simulation purposes, also closely matches the temporal optical waveform that is generated by a typical optical transmitter, as shown in Figure 3.1, where the bandwidth of the signal is limited by the electrical bandwidth of the driver amplifier and the optical modulator. The frequency spectrum of the raised cosine waveform is given by the Fourier transform of Equation 3.13 as

$$P(f) = \mathrm{sin}\,c(T_\mathrm{b}f)\left[\frac{\cos(\pi\alpha T_\mathrm{b}f)}{1-4\alpha^2 T_\mathrm{b}^2 f^2}\right] \tag{3.14}$$

The bandpass optical filter $c_\mathrm{opt}(t)$ at the receiver is designed to match the bandwidth of the transmitted waveform in order to suppress amplified spontaneous emission (ASE) – ASE beat noise and minimize ISI. Typically, the optical filter is an arrayed waveguide grating (AWG) device which also acts as the channel demultiplexer in a wavelength-division multiplexed (WDM) system. These have either a Gaussian or flat top passband (often modeled as a super-Gaussian). At 10 Gbit/s these AWGs have a typical FWHM bandwidth of 0.5 nm (60 GHz at 1550 nm); however, we see from Equation 3.14 that better noise performance, with some degradation in ISI, can be achieved with an optical filter bandwidth that approaches $f_\mathrm{FWHM} = 2/T_\mathrm{b}$, which at 10 Gbit/s is 20 GHz (0.16 nm at 1550 nm). The filtered optical signal is then converted into the electrical domain using a square-law detector. This is then followed by a lowpass electrical filter to limit the noise bandwidth further. In practice, the receiver filter transfer function is characterized by a fourth-order Bessel–Thomson response according to [3]

$$\begin{aligned} C(p) &= \frac{1}{105}(105 + 105y + 45y^2 + 10y^3 + y^4) \\ p &= \mathrm{j}\frac{\omega}{\omega_\mathrm{r}} \\ y &= 2.1140p \end{aligned} \tag{3.15}$$

where f_o is the bit rate and the 3 dB reference frequency is $f_\mathrm{r} = 0.75f_\mathrm{o}$.

Return-to-zero (RZ) signal formats, where the signal level returns to zero within each bit slot, can be described in a similar fashion to NRZ signals. RZ signals are often modeled as Gaussian, Sech, or raised cosine functions. RZ signal formats are used to minimize ISI especially in differential phase encoded formats; however, the increased spectral bandwidth over NRZ that is required to support the shorter pulses means that RZ signals require broader receiver filters and are less tolerant to the channel dispersion.

3.2 Performance Estimation

The successful development of complex and large-scale optical networks relies heavily on the ability to model and simulate the end-to-end performance of the system with a high degree of accuracy. In this section we will look at a number of the techniques available to evaluate the performance of an optical link and will demonstrate the regime of validity of the most commonly used methods.

3.2.1 Introduction

The most common measure of system performance in digital communications at the physical layer is the bit error rate (BER), determined from the probability that a data bit is received in error. The receiver circuit integrates the energy over a bit period and the resulting signal is compared with a given level, termed the decision threshold, to determine whether the received bit is a '0' or '1' symbol. However, random noise and distortion of the signal may result in errors in the output if it causes it to cross the decision threshold.

It is typical to assume in an intensity-modulated direct-detection (IM-DD) system that the received signal current will be as follows:

$$I_{\mathrm{D}}(t) = \begin{cases} i_1 & \text{for the symbol 1} \\ 0 & \text{for the symbol 0} \end{cases} \tag{3.16}$$

Bit error rate evaluation is relatively simple in principle, if we know certain characteristics of the system. First, we define the probability of error:

$$P_{\mathrm{e}} = p_0 \Pr\{N > D \,|\, 0\} + p_1 \Pr\{N \leq D \,|\, 1\} \tag{3.17}$$

If we know the probability of each symbol being transmitted, p_0 and p_1, then all we need to find to calculate the probability of error is the likelihood of a '0' being detected when a '1' was transmitted ($\Pr\{N \leq D \,|\, 1\}$) and the likelihood of a '1' being detected when a '0' was transmitted ($\Pr\{N > D \,|\, 0\}$), where D is the decision threshold and N is the signal level. The reason for misdetection of symbols is noise and/or distortion on each symbol level that may cause the symbol energy to cross the decision threshold. This can be seen from Figure 3.11, which shows the digital signal at the input to the decision circuit and the corresponding probability densities. In most instances it is assumed that the probability of transmission for each symbol is 1/2. This is denoted in the equal area under each of the density functions. In Figure 3.11, the density functions are represented as two equal Gaussian (normal) distributions; however, in a real system, their exact shapes will be dependent on the noise statistics present on the 1s and 0s.

To calculate the probability of error for the noise distribution, the evaluation of infinite integrals is almost always required. If we define the probability density function (PDF) of the noise on the 1s and the 0s to be $f_1(x)$ and $f_0(x)$ respectively, then the error probability when both symbols are equiprobable is given by

$$P_{\mathrm{e}} = \frac{1}{2}\int_{D}^{\infty} f_0(x)\,\mathrm{d}x + \frac{1}{2}\int_{-\infty}^{D} f_1(x)\,\mathrm{d}x \tag{3.18}$$

The result is the BER, although often a more useful metric which can be derived from it is the power penalty. This is a measurement of the additional power required to maintain a given BER or SNR, when a certain noise contribution is considered. To calculate the exact nature of the noise distribution at the optical receiver can be extremely difficult; and for certain noise types, even if the distributions are known, direct error probability calculation is not always forthcoming. Therefore, a number of techniques have been devised that consider the use of elementary statistical measures, such as the mean and the variance, to facilitate the calculation of approximate performance values. This section details a number of the evaluation methods that are applicable to these calculations.

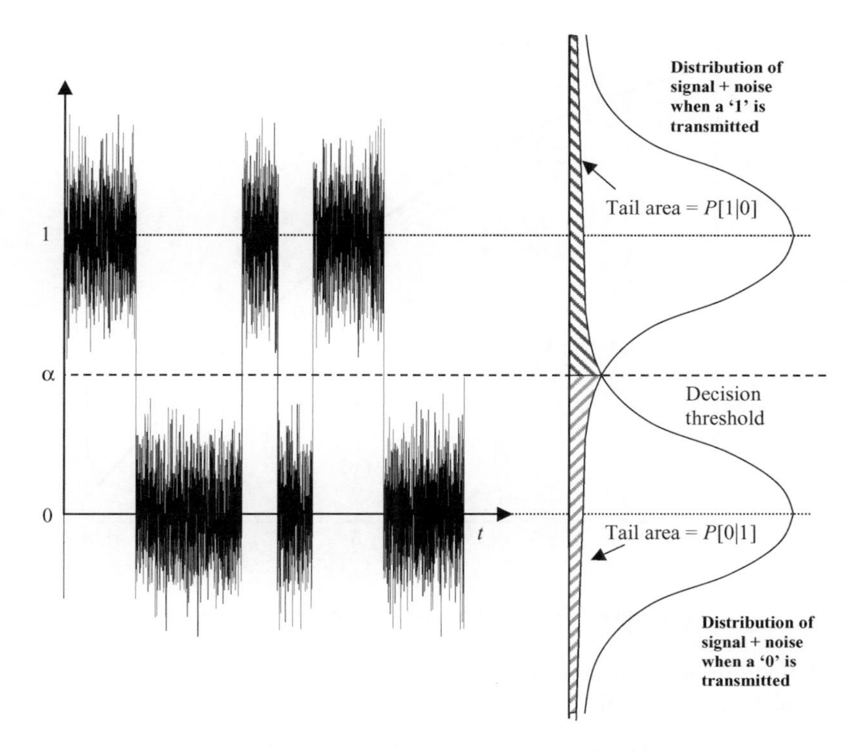

Figure 3.11 Noise distribution on signal bits

3.2.1.1 The Gaussian Approximation

Although the basic statistical nature of optical detection is a combination of noise processes, a Poisson process for shot noise and a Gaussian process for receiver thermal noise, it has been common practice for a Gaussian probability density to be used to approximate the error probability. A Gaussian approximation (GA) involves treating the total impairment as if it were characterized by a Gaussian PDF which is defined as

$$P(k|N) = \frac{1}{\sigma\sqrt{2\pi}} \mathrm{e}^{-[(k-N)^2]/2\sigma^2} \qquad P_{\mathrm{g}}(x) = \frac{1}{\sqrt{2\pi}} \mathrm{e}^{-x^2/2} \tag{3.19}$$

Therefore, to calculate the error probability, as shown in Section 3.2.1, it is required that we measure the area of the PDF that crosses the decision threshold z:

$$Q'(z) = \frac{1}{\sqrt{2\pi}} \int_{z}^{\infty} \mathrm{e}^{-x^2/2}\, \mathrm{d}x \tag{3.20}$$

Notice that $Q'(z)$ is the probability of exceeding value z and is, therefore, the complementary cumulative density function of the Gaussian distribution. This integral cannot be evaluated in

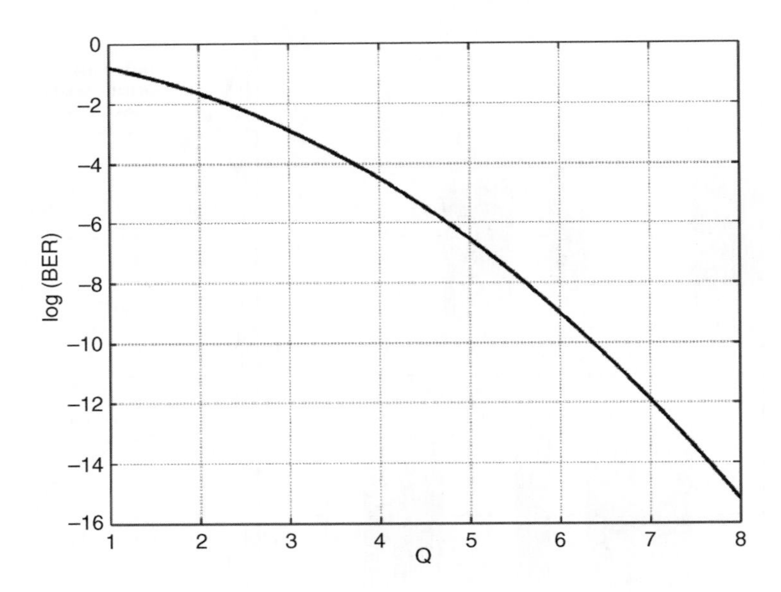

Figure 3.12 Relationship between Q and BER

closed analytic form, so it is usually approximated using the first term of its Taylor expansion, which is termed the 'Q' function:

$$Q'(z) \approx Q(z) = \frac{1}{z\sqrt{2\pi}} e^{-z^2/2} \quad \text{for } z > 3 \tag{3.21}$$

The relationship between Q and BER can be seen in Figure 3.12.

This function is related to the mathematical function erfc (complementary error function) as follows:

$$\mathrm{erfc}(z) = \frac{2}{\sqrt{\pi}} \int_z^{\infty} e^{-x^2}\, dx \approx \frac{e^{-z^2}}{z\sqrt{\pi}} \tag{3.22}$$

$$Q(z) = \frac{1}{2}\mathrm{erfc}\left(\frac{z}{\sqrt{2}}\right) \tag{3.23}$$

This approximation is frequently adopted in view of its simplicity; it makes use of only the minimum of statistical information about the noise and interference, the mean and variance, to obtain a standard approximation of the tail integral.

In its simplest form it can be given as

$$\mathrm{BER} \approx \frac{1}{2}\left[Q\left(\frac{\langle P_1\rangle - D}{\sigma_1}\right) + Q\left(\frac{D - \langle P_0\rangle}{\sigma_0}\right)\right] \tag{3.24}$$

where D is the decision level and σ_1 and σ_0 are the noise variances on the '1' and '0' symbols respectively. The noise variances will typically be made up from a number of contributions in optically amplified systems. Here, the noise variance becomes the square root of the sum of the squares of all the contributing terms:

$$\sigma = \sqrt{\sum_i \sigma_i^2} \tag{3.25}$$

We will now consider some of the terms that may be included in this calculation.

3.2.1.2 Calculating the Noise Variance

Thermal noise occurs due to the random motion of electrons that is always present at a finite temperature. It has been shown that the Gaussian distribution provides a good approximation to the true statistics:

$$\sigma_{\text{Thermal}}^2 = \frac{4K_{\text{B}}T}{R}B_{\text{e}} \tag{3.26}$$

where B_{e} is the electrical bandwidth of the receiver. Shot noise (or quantum noise) represents the Poisson statistics that characterize the arrival of photons at the photodiode. However, the use of the Poisson distribution is not straightforward when combined with the Gaussian model of thermal noise. Fortunately, in most practical cases where the mean photocarrier rate is high, the Gaussian distribution will provide a reasonable approximation. For a *pin* photodiode this will result in a variance that is given by

$$\sigma_{\text{shot}}^2 = 2e\hat{I}B_{\text{e}} \tag{3.27}$$

If an avalanche photodiode is used, where M is the average avalanche gain and $F(M)$ is a noise figure due to the random fluctuations of the avalanche process, then the variance of the shot noise current will be

$$\sigma_{\text{shot}}^2 = 2eIB_eM^2F(M) \tag{3.28}$$

The use of optical amplifiers also introduces a number of extra noise terms. These are due to resultant terms produced when the ASE from the amplifier falls on a square law reception device. This will result in terms due to the beating of the noise with the signal and with itself. These are termed signal–spontaneous beat noise, due to ASE noise beating with the signal:

$$\sigma_{\text{sig-spon}}^2 = 4\Re^2 GP_{\text{s}}S_{\text{ASE}}B_{\text{e}} \tag{3.29}$$

and spontaneous–spontaneous beat noise will results in a mean-square noise current of:

$$\langle\sigma_{\text{sp-sp}}^2\rangle = 4\Re^2 S_{\text{ASE}}^2 B_{\text{o}}B_{\text{e}} \tag{3.30}$$

where S_{ASE} is the noise spectral density of the ASE noise and B_{o} is the bandwidth of the optical filter. This additional optical noise power will also add to the shot noise, giving a total shot noise of

$$\sigma_{\text{shot}}^2 = \sigma_{\text{shot-signal}}^2 + \sigma_{\text{shot-ASE}}^2 = 2e(\hat{I}_{\text{signal}}B_{\text{e}} + \hat{I}_{\text{ASE}}B_{\text{e}}) \tag{3.31}$$

where $\hat{I}_{\text{ASE}} = \Re S_{\text{ASE}}B_{\text{o}}$ and

R = resistance (often 50 Ω)
T = temperature

K_B = Boltzmann's constant (1.38×10^{-23} J/K)
B_e = electrical bandwidth
G = amplifier gain
f_c = optical frequency
e = charge on an electron (1.6×10^{-19} C)
$\hat{I}$ = average photocurrent generated ($=\Re P_{rx}/2$)
$\Re$ = responsivity
P_{rx} = peak optical power at the receiver
h = Planck's constant (6.63×10^{-34} J/Hz)
P = power entering the amplifier
γ = population inversion factor
N_{sp} = amplifier spontaneous emission factor
S_{ASE} = power spectral density of the ASE noise.

Power Penalty Calculations

When making a design of a system, a calculation of the BER yields some useful information on the performance of the system. However, in this form it gives no information as to possible improvements in system performance. We may think of each noise or distortion contribution as a power penalty, in that it will require an increase in the signal power required to maintain the BER at the desired level. We can define the power penalty PP quite simply as the ratio of the optical power required for a given BER with the impairment included P'_{rx} over the optical power required without the impartment P_{rx}:

$$\mathrm{PP} = -10\log\left(\frac{P'_{rx}}{P_{rx}}\right) \tag{3.32}$$

This can also be related to other measures that are included in the design of optical links; for example, it is often suggested that an acceptable amount of pulse spreading due to dispersion is 49% of the total bit period. This figure comes from the calculation that a 49% spread results in a 2 dB power penalty due to dispersion [4] (however, this justification is often omitted).

To demonstrate the calculation of the power penalty we consider the fact that in all practical transmitters the power level of the zero symbol will not be represented by a complete absence of light. This may either be deliberately imposed to reduce transient effects in directly modulated lasers or due to component tolerance in external modulators. In this instance the ratio of the power of the 1s to the power in the 0s is defined as r, termed the extinction ratio and given by

$$r = \frac{P_0}{P_1} \tag{3.33}$$

The increase in the 0 level leads to a penalty at the receiver as the difference between the 1 and 0 level is reduced. The penalty can be defined as

$$\mathrm{PP}_{\mathrm{extinction}} = -10\log\left(\frac{r-1}{r+1}\right) \tag{3.34}$$

Many other examples of the calculation of the power penalty in various circumstances and with various assumptions can be found in the literature.

3.2.1.3 Calculating the Decision Threshold

The decision threshold is (nominally) set such that $P[1 \mid 0] = P[0 \mid 1]$. If noise is equal on 1s and 0s, then this will be in the center of the eye. This is appropriate when the dominant noise contribution is from signal-independent noise. For signal-dependent noise, with perfect extinction of the '0' symbols the normalized decision threshold corresponds to equal 'margins' for 0s and 1s of F and $(1-F)$ relative to the respective root-mean-square noise values σ_0 and σ_1, such that

$$\frac{F}{\sigma_0} = \frac{1-F}{\sigma_1} \Rightarrow F = \frac{\sigma_0}{\sigma_0 + \sigma_1} \tag{3.35}$$

Signal-dependent noise requires a definition of Q that accommodates the optimum threshold <0.5. For a peak signal level of $2s$ the threshold is optimally placed at $\sigma_0/(\sigma_0 + \sigma_1)$:

$$Q = \frac{2s}{\sigma_0 + \sigma_1}; \quad \mathrm{BER_{opt}} = \mathrm{erfc}(Q) \tag{3.36}$$

Voltages at the decision instants are Gaussian random variables with means V_0 and V_1 and variances σ_0^2 and σ_1^2.

3.2.2 *Modeling*

Here, the reader will be presented with an insight into techniques used to construct a modeling environment for optical signals. The concept of Monte Carlo simulation will be introduced, as well as more advance techniques for the evaluation of noise within an optical system based on the consideration of PDFs and other higher order statistics.

3.2.2.1 Monte Carlo Simulation

Optical communication systems are most often modeled directly in the time domain. In this way the optical signal is represented as a discrete time-sampled signal and all operations such as optical amplification, fiber propagation, filtering, and so on operate directly on the sampled optical signal. To estimate the system performance it is necessary to simulate not only the deterministic effects, but also the effects of noise on the system. A stochastic 'Monte Carlo' simulation provides the most straightforward way of assessing the effects of noise in complex optical communication systems. A Monte Carlo model is used to determine the likely outcome of a deterministic system by iteratively evaluating the deterministic model with random inputs. In the case of optical systems, the noises that arise from the laser transmitter, optical amplification, and the electrical receiver are modeled as stochastic processes and are added to the signal where it occurs in the system. This means that the component models are then applied simultaneously both to the signal and noise fields. Thus, it is easy to model the impact of filtering and nonlinear signal–noise interactions.

A basic Monte Carlo model of an optical communications system is shown in Figure 3.13. A pseudorandom bit sequence is transmitted through the system so that pattern-dependent effects are simulated. The problem with Monte Carlo simulations is that it is often necessary to simulate a large number of bits in order to build up sufficient statistical confidence in the results. A typical simulation of a 10 Gbit/s system might simulate 2^N bits (where N varies from 7 to 10)

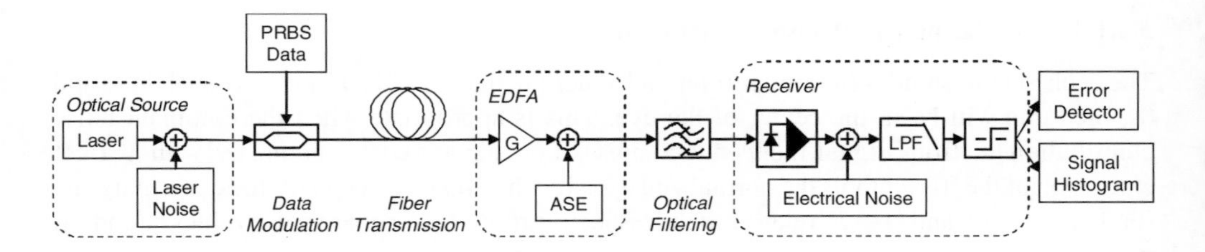

Figure 3.13 Monte Carlo optical system model

with 16 samples per bit, corresponding to a simulation bandwidth of 160 GHz. The number of samples per bit is chosen to ensure that the simulated bandwidth is sufficient to represent the signal waveform accurately and larger than the bandwidth of the optical and electrical filters present in the system.

Most optical communication systems use transmitters based on semiconductor laser sources, where the laser noise is dominated by spontaneous emission. Spontaneous emission results in intensity and phase fluctuations in the emitted optical field that vary over a time scale as short as 100 ps. For directly modulated laser transmitters it is often necessary to simulate the output field by directly solving the laser rate equations [5]. In this case the noise is modeled by adding Langevin forces to the rate equations, where the forces are modeled as Gaussian random processes [6].

In an optically amplified system the dominant noise source is ASE from the erbium-doped fiber amplifier (EDFA), and often this is the only source of noise that is modeled. In most simulations of a communications system the EDFA is modeled using a black-box model that is characterized by its gain$G$, saturated output power, and noise figure F_n [7]. The noise is added at each amplifier and is modeled as an additive white Gaussian process whose power spectral density is given by

$$S_{\mathrm{ASE}}(\nu) = \frac{1}{2} G F_{\mathrm{n}} h \nu \tag{3.37}$$

When adding the noise to the optical field it is important to ensure that the noise power is spread equally over both quadratures and polarization states.

At the receiver the impact shot noise and thermal noise from the photodiode and electrical amplifiers is modeled by a white Gaussian process with zero mean and whose variance is given in Section 3.2.1.2.

The system performance is generally determined from the signal distributions at the output of the simulator, as it is not usually possible to simulate sufficient numbers of bits to allow for direct error counting. The simplest measure of system performance is obtained assuming Gaussian statistics using the Q factor defined in Equation 3.56; however, it is also possible to fit more realistic distributions to the simulated results, such as those discussed in the following sections.

3.2.2.2 Techniques Using Higher Order Statistics

It is possible to formulate a power penalty by adjusting P_s to equate the quality factors (argument of $Q[\bullet]$) of a signal with and without noise terms other than the receiver thermal noise [8]. However, in doing this, two important implicit assumptions are involved. First, that the noise on signal 0s may be neglected. Second, that equating the quality factors of corrupted and uncorrupted signals, which simply scales the $Q[\bullet]$ function without changing its form, is a reasonable approximation. This will depend on the exact nature of the distribution being considered. Both of these may be broadly acceptable for the simplistic case involving approximately Gaussian noise terms; however, for analysis of more complex systems with, for example, a strictly non-Gaussian composite noise plus interference distribution, their validity cannot be taken for granted.

From central limit theorem arguments one would expect that for a large number of independent terms this will provide a good approximation; and, as we shall see, this is ultimately the case. However, if the constituent PDFs are markedly different from Gaussian, then the convergence under the central limit theorem can be very slow indeed [9].

One of the most studied examples in optical communications arises due to optical crosstalk on same-wavelength signals in wavelength-routed WDM networks. In this example the PDF of a single incoherent interfering term corresponds to an arc-sinusoidal distribution [10–12]. This can be shown to be strictly bounded; therefore, the GA will be a poor descriptor for this PDF. Even with several such interferers, convergence will be slow, especially in the tails of the distribution, which are of critical importance for BER considerations [9]. Owing to the simple formulation of the GA, it is understandable that it is a convenient tool for a first approximate assessment and can be applied more generally to instances when the central limit theorem can be relied upon. However, inaccurate results may be found in many instances or erroneous trends may be predicted [13]. Accordingly, it should be applied with care, recognizing the wisdom and implicit caution in the observation 'The central limit theorem is more commonly applied by incantation than by proof!' [14].

3.2.2.3 Calculation of Probability Density Functions

If we require more accurate evaluation of BERs than those demonstrated in Section 3.2.1.1 then we need to consider the exact statistics of the signal and noise components, the stochastic element being the sum of the random variables involved. The summation of independent random variables corresponds to the convolution of the PDFs of the random variables, which for many of the distributions encountered is a nontrivial operation. A number of papers have demonstrated results formed from the numerical convolution of the PDFs [15,16], but this has proved to be a computationally time-consuming task.

3.2.2.4 Techniques using the Moment-Generating Function

Another solution is to consider the moments of the PDF, which can be utilized in a number of different ways. The simplest form is, in reality, an extension of the GA, as the mean and variance are indeed the first and second moments of the PDF. The literature has demonstrated a number of series expansions, including Gram–Charlier series [17] and Hermitian polynomials [18], which, although more detailed, still represent a truncated set of moments. It is possible to use the entire set of moments by taking advantage of the properties of the moment-generating function (MGF). The MGF's most useful property is that the summation of independent random variables, although it is equivalent to the convolution of PDFs, is equivalent to the multiplication of MGFs [9], which is a much simpler mathematical operation. Evaluation methods making use of the MGF, for example [19,20], can overcome many of the shortcomings of approximate methods based on limited statistics, albeit sometimes at the expense of increased complexity.

We start by introducing the definition of the MGF of random variable X as

$$M_{X(s)} = E[\mathrm{e}^{sX}] \tag{3.38}$$

where $E[\bullet]$ denotes the statistical expectation. Care must be taken not to confuse the moment-generation function with the characteristic function, which is similar in definition [21] but complex:

$$\Phi(\theta) = E[\mathrm{e}^{\mathrm{j}\theta X}] \tag{3.39}$$

As an example of the application of this technique we formulate the MGF for first a number of intra-channel (same wavelength) crosstalk terms, often called interferometric noise in the literature. This type of noise is created by the beating of a desired optical signal with an interfering optical term that is close enough in wavelength for the beat term to fall within the passband of the receiver, which introduces a non-Gaussian interference component (arc-sinusoidal in distribution) in addition to the additive Gaussian noise.

The PDF of this contributing non-Gaussian noise term X, which is known to be arc-sinusoidal [14,15], for a single interfering term is given by

$$X : f_X(t) = \frac{1}{\pi\alpha\sqrt{1-\left(\frac{x}{\alpha}\right)^2}} \tag{3.40}$$

where α is the power of the interfering term, given by $\alpha = 2P_\mathrm{s}\sqrt{\varepsilon}$, with $\varepsilon = P_\mathrm{s}/P_\mathrm{i}$ and where P_s and P_i are the signal and interfering power respectively. Using the formulation of Equation 3.41, the MGF of the random variable X may be found as

$$M_X(s) = E[\mathrm{e}^{sX}] = I_0(s2P_\mathrm{s}\sqrt{\varepsilon}) \tag{3.41}$$

where $I_0(\bullet)$ is the modified Bessel function of the first kind, zero order.

If we consider a network where N interfering terms, each of power P_k, occur, this corresponds to a summation of N random variables, giving rise to convolution of their PDFs and, hence, multiplication of MGFs. Therefore, the total MGF for N interferers, representing the maximum interfering signal, is given by

$$M_{X,N}(s) = \prod_{k=1}^{N} I_0(s2P_\mathrm{s}\sqrt{\varepsilon_k}) \tag{3.42}$$

However, as the beat noise is signal dependent, symbol conditioning is required. To simplify matters we consider the case of all optical sources exhibiting perfect extinction. This results in beating only occurring when both the signal and the interfering term are transporting a data '1'. When either or both are data '0', no beating term will appear. To include this in the derivation of the MGF we introduce the statistical likelihood of each data combination. Therefore, considering a priori symbol probabilities, $P_0 = P_1 = 1/2$, the total MGFs for signal '1' and signal '0' respectively are defined as

$$M_{X,N,1}(s) = \prod_{k=1}^{N} \frac{1}{2}[1 + I_0(2P_\mathrm{s}\sqrt{\varepsilon_k s})] \tag{3.43}$$

$$M_{X,N,0}(s) = 1 \tag{3.44}$$

The expansion in Equation 3.43 is introduced because, for each interfering symbol, half of the time the interfering symbol will be zero and no beating will occur (MGF = 1) and is valid providing all interferers are of equal power.

The *Chernoff bound* is a well known and commonly used bounding mechanism which, if applied appropriately, can often produce acceptable results. However, we can improve on its performance by appreciating the particular scenario often encountered in optical communications of a noise term formed of an additive zero-mean Gaussian random variable and another, non-Gaussian, random variable term. The relationship between the Chernoff bound and its modified version can be expressed as

$$\mathrm{MCB}(s) = \frac{\mathrm{CB(s)}}{\sqrt{2\pi s\sigma}} \tag{3.45}$$

The modified Chernoff bound (MCB) was originally proposed by Prabhu [22] and extends the Chernoff bound to exploit directly the presence of an additive Gaussian noise component. It was refined for optical communications [23] and a simplified derivation has recently been produced [14]. Under the MCB, the average probability of error, for probability of transmission $p_0 = p_1 = 1/2$, is then

$$P_\mathrm{e} \leq \frac{\mathrm{e}^{s^2\sigma^2/2}}{2s\sigma\sqrt{2\pi}}\left[\mathrm{e}^{-sD}M_{X|0}(s) + \mathrm{e}^{sD}M_{X|1}(s)\right]; \quad s > 0 \tag{3.46}$$

where $M_{X|0}$ and $M_{X|1}$ are the total MGF for the '0' and '1' symbols respectively. This is the modified Chernoff (upper) bound (MCB), developed for optical communications by direct manipulation of conditional error probability expansion [23]. Equation 3.46 may be minimized to give the tightest form of the bound by selecting s appropriately, either as a common parameter or separately as s_1 and s_0 for data 1s and 0s respectively.

To consider the *saddlepoint approximation* (SPA) we return to the original problem of finding the probability of error by considering the probability that a random variable described by a PDF $f(x)$ exceeds a decision threshold denoted by D:

$$P_\mathrm{e} = \int_{-\infty}^{D} f(x)\,\mathrm{d}x \tag{3.47}$$

As shown previously, a convenient description of this probability function that describes the sum of noise terms is the MGF $M_x(s)$. We can, therefore, replace the PDF $f(x)$ in Equation 3.47 with the inverse Laplace transform of the MGF. The derivation of this can be found in [24] and results in

$$P_\mathrm{e} = \frac{1}{2\pi}\int_{-\infty}^{\infty} \mathrm{e}^{\psi(c=\mathrm{j}x)}\,\mathrm{d}x \tag{3.48}$$

where $\psi(s) = \ln(M_x(s)) - \ln(-s) - s\alpha$ and $c < 0$ is a real parameter. In order to obtain the SPA, a Taylor series expansion of the exponent of Equation 3.48 is performed with c chosen so that $\phi'(c_0) = 0$. Typically, only terms up to the second derivative of the series are used, which gives the SPA as defined in [24]:

$$P_{\mathrm{e}} \approx \frac{1}{2}\left[\frac{\mathrm{e}^{\psi_0(s)}}{\sqrt{2\pi\psi_0''(s)}} + \frac{\mathrm{e}^{\psi_1(s)}}{\sqrt{2\pi\psi_1'(s)}}\right] \tag{3.49}$$

where

$$\begin{aligned} \psi_{0(s_0)} &= \ln[M_{\mathrm{G}|0} M_{X|0}] - s_0 D - \ln|s_0| \\ \psi_{1(s_1)} &= \ln[M_{\mathrm{G}|1} M_{X|1}] - s_1 D - \ln|s_1| \end{aligned} \tag{3.50}$$

where D is the normalized decision threshold and s_0 and s_1 are determined by the positive and negative roots of $\psi_i'(s) = 0$ ($i = 0$, 1 respectively). M_{G} is the MGF of the Gaussian noise component as given in Equation 3.45. The error probability is determined by finding the saddlepoint of the contour integral of the complex MGF. As described previously, the terms given in Equation 3.49 are second-order terms from the respective Taylor series expansion; greater accuracy can be obtained if higher order terms are included.

3.2.3 *Comparison of Techniques*

In this section, the regimes of validity of a number of the evaluation techniques discussed will be considered. In particular, we show the areas in which the GA can be considered to be valid and where more advanced evaluation techniques (such as those discussed in Section 3.2.3) should be used. This is presented in terms of the dominant noise terms within the network to enable the reader to choose accurately the correct evaluation method for the link under consideration.

The three most widely used techniques in this area are the GA, the Chernoff bound and the SPA. Figure 3.14 compares the results produced by the MCB with those obtained using a very simple GA, the SPA, and the conventional Chernoff bound and demonstrates the inadequacy of the GA and of the conventional Chernoff bound [14].

We will further look at two particular BER assessment techniques making use of the MGF: the SPA [24] and the modified Chernoff bound [23]. The close relationship between these has been shown previously [25]. We have described these two MGF-based evaluation methods which offer increased rigor compared with GA-based methods, which are applied here to interferometric noise. To illustrate this, let us consider just the simple case of a single interfering term with symbol conditioning and perfect extinction. For interferometric noise as described in [10] and considering only a single interfering term, we demonstrate the full formulation for both techniques:

$$P_{\mathrm{e}} \leq \mathrm{MCB} = \frac{\exp[(s\sigma)^2/2]}{2s\sigma\sqrt{2\pi}}\left\{\exp(-sD) + \left[\frac{1 + I_0(2s\sqrt{\varepsilon}P_{\mathrm{s}})}{2}\right]\exp[-s(V - D)]\right\}; \quad s > 0 \tag{3.51}$$

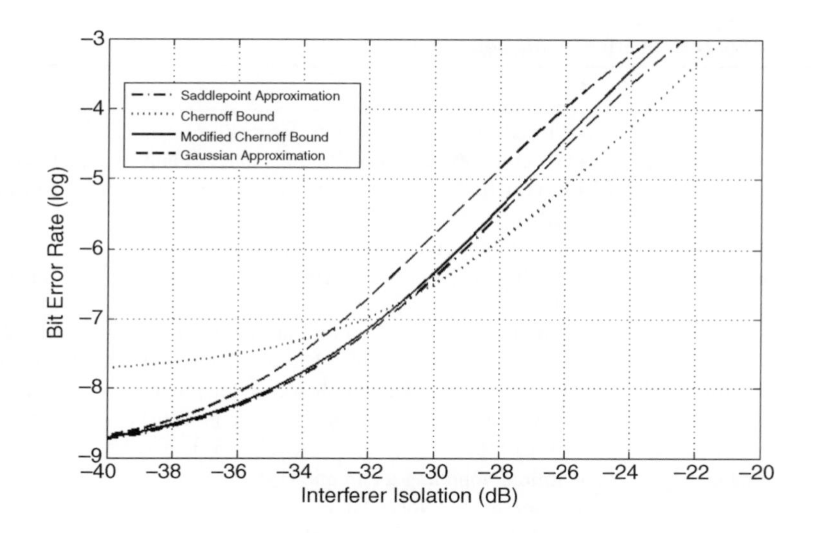

Figure 3.14 Comparison of the GA, SPA, Chernoff bound, and MCB

$$P_e \approx \mathrm{SPA} = \frac{\exp\left[\frac{(s\sigma)^2}{2} - sD\right] \Big/ |s|}{2\sqrt{2\pi\left(\sigma^2 + \frac{1}{s^2}\right)}} + \frac{\exp\left[\frac{(s\sigma)^2}{2} I_0(2s\sqrt{\varepsilon P_s}) - sD\right] / |s|}{2\sqrt{2\pi\left(4\varepsilon P_s^2\left\{\frac{I_0(2s\sqrt{\varepsilon P_s}) + I_2(2s\sqrt{\varepsilon P_s})}{2[1+I_0(2s\sqrt{\varepsilon P_s})]} - \frac{[I_1(2s\sqrt{\varepsilon P_s})]^2}{[1+I_0(2s\sqrt{\varepsilon P_s})]^2}\right\}\right) + \sigma^2 + \frac{1}{s^2}}}; \quad s > 0 \tag{3.52}$$

It is clear to see that, compared with the SPA, the initial formulation of the MCB is easier and more compact despite the relatively simple form of the MGFs being evaluated. The optimizing process (selection of 'best' s value) for both is similar, although the complexity of the SPA will complicate this. Since the formulation of the SPA involves the second differential of the equation in Equation 3.50, it is clear that once the full expressions for the MGFs are included the result is an expansive, complicated final formulation.

It is not surprisingly to see that the additional analytic complexity of the SPA leads to a similar increase in the computation requirement. Figure 3.15 compares the run time (measured in number of floating-point operations) required to minimize and evaluate both these expressions. We see that for a low number of independent terms where a rigorous approach is considered necessary the MCB demonstrates significantly better computational efficiency. However, although computational efficiency is important, numerical accuracy must also be maintained. Figure 3.16 shows the results of the MCB and the SPA evaluations performed for the simple case demonstrated above, as well as extended scenarios with more interferers. This demonstrate that the tolerable crosstalk value calculated using each method differs by <0.16 dB, showing efficiency is maintained. A comparison of the main attributes of the most common techniques is given in Table 3.1

Table 3.1 Summary of interferometric noise evaluation methods [26]

Technique	Comments
The GA	• Closed-form power penalty estimation formula. • Accurate for noise distributions that are approximately Gaussian. • Central limit theorem suggests accuracy increases with the number of terms. • Typically a good first approximation [27,28]
Inner eye closure	• Pessimistic penalty estimate (upper bound) from the inner-eye closure. • Simple but quickly becomes inaccurate if the range of the interfering variable is much greater than the standard deviation [10,29].
The Gauss quadrature rule	• Uses an appropriate number of moments of the noise PDF. • It was observed that it provided a similar level of accuracy to the SPA [32].
The Chernoff bound	• Simple bounding technique [14] • Based on the Noise MGF. • Although very easy to apply it is not always a very tight bound.
The MCB	• Uses the presence of a Gaussian component in the noise to improve the Chernoff bound. • Provides very accurate results under most conditions [22,23].
The SPA	• Finds the saddlepoint value of the integrand of the MGF to approximate the error probability. • Typically accurate but can be mathematically expansive for complex noise distributions [24].
Gram–Charlier series approximation and other series expansions	• Essentially a correction of the GA [17]. • It expresses an arbitrary PDF as an infinite series whose leading term is a Gaussian distribution. • All the series coefficients can be evaluated from the moments of the random variable. • Provides good results even for a small number of interferers [33].
Hermetain polynomials	• Series expansions based upon Hermetian polynomials • Closed-form BER formula, although truncation may limit accuracy [18].
Simulation	• Detailed simulation of a system modeled. • Accurate characterization of all network components required. • Care must be taken in the BER estimation part of the simulated model so that it does not implicitly estimate the BER by assuming that the noise is Gaussian. A quasi-analytical or semi-analytical method must be used where the noiseless signal prior to the receiver decision variable is simulated. At the decision threshold the Gaussian receiver noise is then added to the sampled signal to compute the BER, this being the analytical part [34].
Adaptive importance sampling	• Weights the noise contribution so that Monte Carlo simulation techniques can be used with low error rates [35].
Numerical convolution	• Numerical convolution of the interferometric noise PDFs. • Provides accurate results, but is hard to apply while maintaining numerical accuracy (due to the need to integrate infinite tails) especially as the number of interferers increases [30,31].

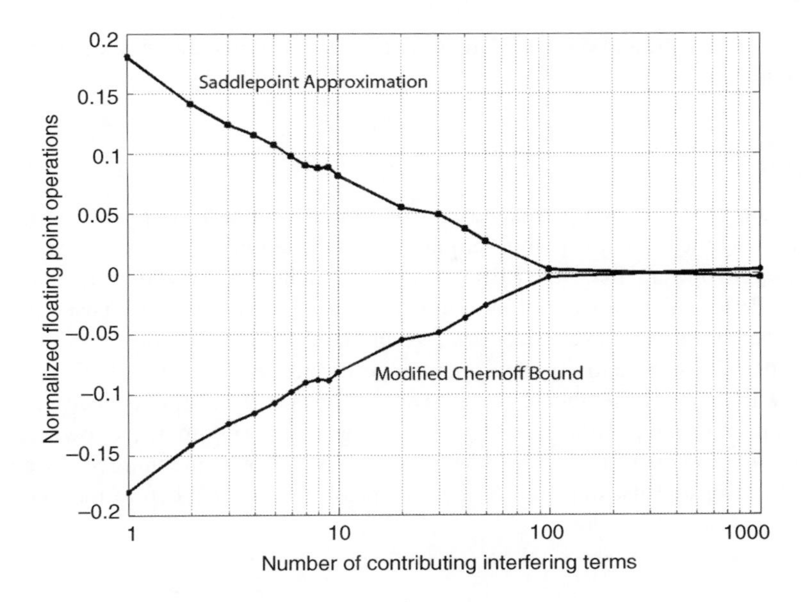

Figure 3.15 Comparison of floating-point operations required for MCB and SPA [14]

3.2.4 Standard Experimental Measurement Procedures

To complete the coverage of performance evaluation, this final section relates the theory presented for the analytical evaluation of an optical system relates to the practical measurement of system performance. The standard laboratory techniques for the evaluation of signal quality,

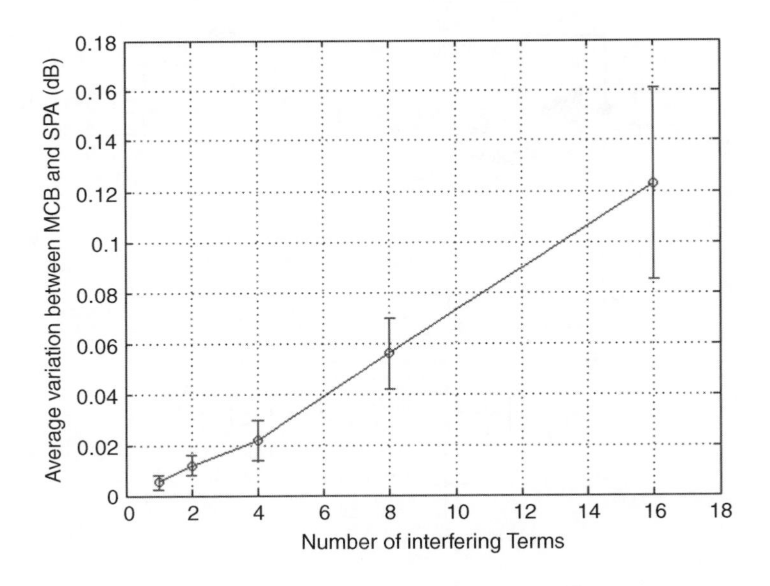

Figure 3.16 Comparison of the performance of the MCB and SPA [14]

namely BER, Q-factor, and optical SNR (OSNR), are presented and related to their mathematical equivalents.

3.2.4.1 BER

The BER is the most commonly used measure of optical communication system performance. When measuring the BER we are most commonly interested in how a system parameter, such as noise or signal power, affects the BER. In this section, the standard experimental techniques that use the BER as the measure of system performance are described. These are known as receiver power sensitivity, swept decision threshold, and OSNR penalty.

The receiver power sensitivity measures the performance in terms of BER of the optical receiver as a function of the optical power. The receiver can be a simple electrical-to-optical converter consisting simply of a photodiode, transimpedance amplifier, and lowpass electrical filter followed by a decision circuit, as shown in Figure 3.17a, or, as in Figure 3.17b, an optically amplified receiver that uses an EDFA and bandpass optical filter before the photodiode to

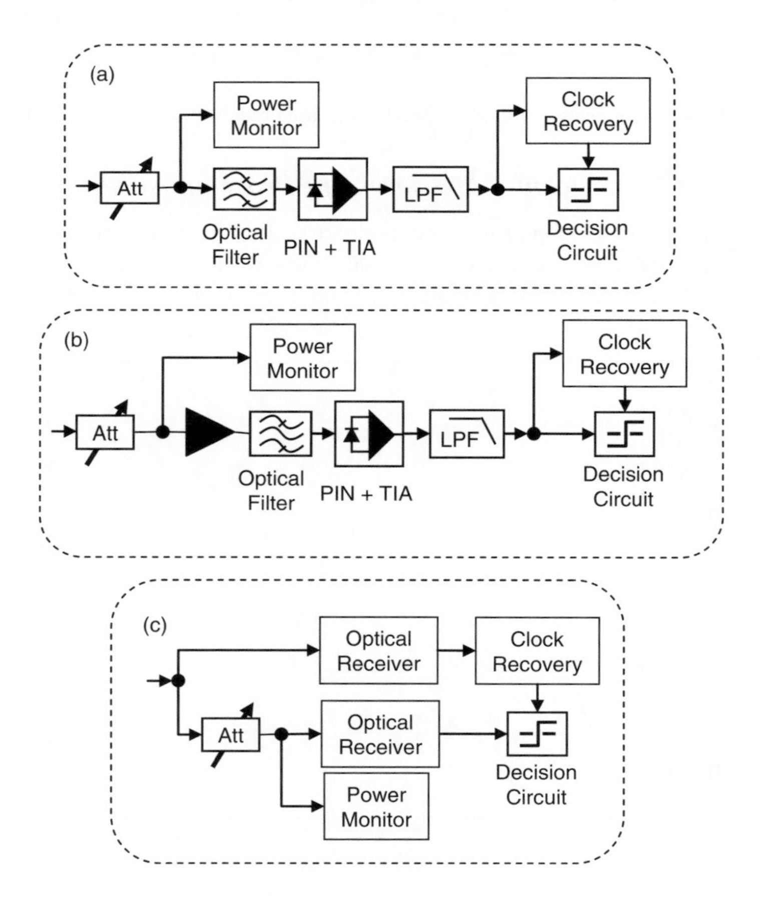

Figure 3.17 (a) Simple optical receiver; (b) optically amplified receiver; (c) decoupled clock recovery

increase the sensitivity. In both cases the receiver is characterized by attenuating the optical power before the receiver and recording the BER. The BER is measured by transmitting a known pseudo-random binary sequence and comparing this with the received signal after the decision circuit, in which case the BER is simply the ratio of the numbers of errors counted in the received sequence per unit time. In these measurements, the decision threshold is optimized for each received optical power level. In addition to optimizing the decision threshold, it is also necessary to determine the optimum sampling time. This process is carried out by the clock recovery unit shown in Figure 3.17. When characterizing the receiver, the performance of the clock recovery is often decoupled from the measurement, as shown in Figure 3.17c, by tapping off part of the optical signal before the attenuator. However, to obtain a true measure of the receiver performance, the clock recovery should not be separated from the receiver.

Figure 3.18 shows the result of a typical receiver sensitivity measurement using a commercial pin photodiode and transimpedance amplifer with and without an optical preamplifier. It is standard practice on the vertical scale to plot the $-\log(-\log(\mathrm{BER}))$ and on the horizontal scale the optical power in dBm in order to obtain a linear relationship between the BER and the optical power. This is a direct consequence of the assumption in Section 3.2.1.1 that the noise statistics are Gaussian. The Q in decibels calculated using the equation from the measured BER is often used rather than plotting the BER and is shown on the right-hand axis in Figure 3.18.

The receiver sensitivity is defined as the required optical power to achieve a specific BER. The BER chosen depends on the performance requirements, and a $\mathrm{BER} = 1 \times 10^{-9}$ (corresponding to a $Q = 15.5$ dB) is most commonly used; however, $\mathrm{BER} = 1 \times 10^{-12}$ is also used for higher specification systems. Thus, the receiver shown here has a receiver sensitivity of

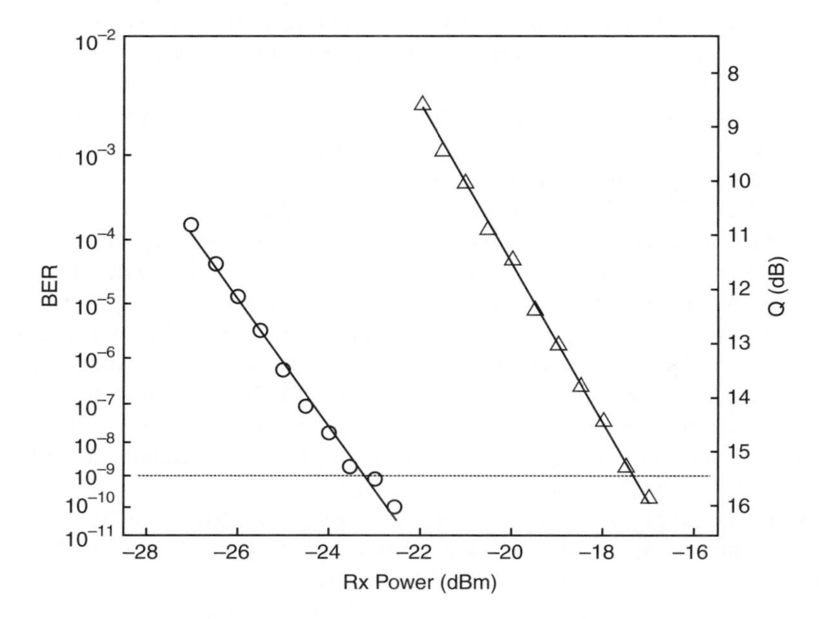

Figure 3.18 Receiver sensitivity as a function of the received optical power of commercially available receiver with and without an optical preamplifier

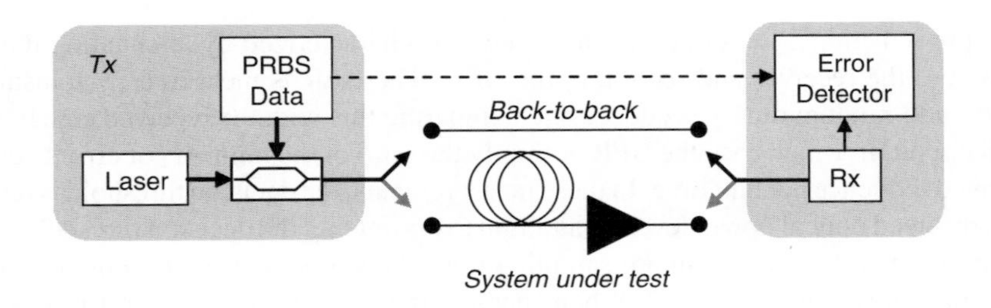

Figure 3.19 System performance characterization based on receiver sensitivity penalty

-17.5 dBm and -23 dBm at BER $= 1 \times 10^{-9}$ without and with optical preamplification. In addition, the slope of the BER curve depends on the dominant source of the noise in the receiver. Without optical preamplification the dominant noise source is the thermal noise arising from the photodiode and electrical amplifiers; however, when optical preamplification is used, then the dominant noise source is the ASE–signal beat noise.

Optical systems are often characterized by their impact on the receiver sensitivity, in what is know as the receiver sensitivity penalty. A typical measurement scenario is illustrated in Figure 3.19. The transmitter and receiver are first connected in what is known as a back-to-back configuration to measure the baseline receiver sensitivity. The system under test, in this case an amplified transmission span, is then placed between the transmitter and the receiver and the receiver sensitivity is measured again. The power penalty is then simply defined as the difference between receiver sensitivities at a specified BER.

In many optical systems it is not practical to measure the errors directly when the BER is less than 1×10^{-12}. Even at a bit rate of 10 Gbit/s it would take 1 min 40 s to record a single error when the BER $= 1 \times 10^{-12}$; thus, recording sufficient errors to produce a reliable measurement takes too long. In order to circumvent this, Bergano *et al.* developed a measurement technique, known as a swept decision threshold measurement, which allows the BER to be extrapolated from a measurement of the distributions of the 0s and the 1s [36]. In this technique, the decision threshold is swept across the eye from the signal rail that represents 0s to the signal rail that represents the 1s and the BER is recorded as a function of the decision voltage. Figure 3.20 shows the measured Q in linear units as a function of the threshold voltage. The Q in linear units is used as this allows for a linear extrapolation to the assumed Gaussian signal distributions. In order to ensure that the linear assumption holds, it is necessary only to fit to the curves where BER $< 1 \times 10^{-3}$ ($Q > 3$). The system BER or Q in this case is then obtained by calculating the crossing point of the linear fits, as shown in Figure 3.20, to obtain the optimum threshold and Q.

Recently, with the advent of forward error correction (FEC), optical system designers have become interested in system performance at much higher BER, such as 1×10^{-3}. Enhanced FEC provides sufficient coding gain to increase the BER of uncorrected data with a BER of 1×10^{-3} to 1×10^{-15} (ITU standard G.975.1) [37]).

To measure system performance in this region, optical noise loading before the receiver is used to enable direct counting of errors. This method of determining system performance was first proposed by Mazurczyk *et al.* [38]. In this technique, the system performance is measured indirectly by increasing the BER to measurable levels. The BER is degraded by noise loading using a broadband optical noise source before the receiver to increase the BER in a controlled way. Usually, ASE from an optical amplifier is used as the optical noise source. For a single

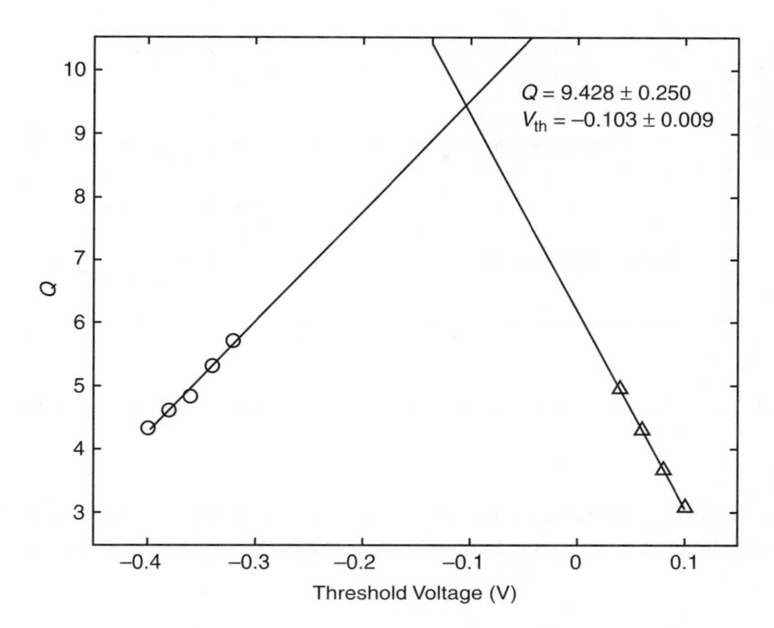

Figure 3.20 Swept threshold measurement. The circles show the measured Q as the threshold approaches the mean of the marks whilst the triangles are obtained as the threshold approaches the mean of the spaces

measurement point of system performance with respect to the independent variable, such as system dispersion, the OSNR directly before the receiver is decreased until the required BER reference point (1×10^{-3}) is reached and then this value of required OSNR is recorded. In a similar fashion to the power penalty, the OSNR penalty of a system can be defined.

This system performance metric is extensively used to characterize receivers that employ digital signal processing, as the more conventional Gaussian extrapolation technique [36] is no longer valid after nonlinear digital signal processing.

3.2.4.2 Eye Diagram

In digital communications, an eye diagram is one of the most useful and easy to use signal performance indicators. An eye diagram is constructed by triggering an oscilloscope using the symbol timing clock. This creates an overlay of the digital signal that can be used to indicate degradation in both vertical and horizontal directions.

The NRZ eye diagram show in Figure 3.21a is obtained by digitally sampling the electrical signal after optical-to-electrical conversion using a digital sampling oscilloscope that is triggered by a clock signal that is synchronized to the data. The measured points are overlaid on a single or few bit periods to build up a time-averaged image of the received signal over many bit periods. Thus, the eye diagram is a three-dimensional representation of the signal where the distribution over many bits in both amplitude and time is displayed. Most measurements on the eye diagram involve looking at the histogram of a slice through the

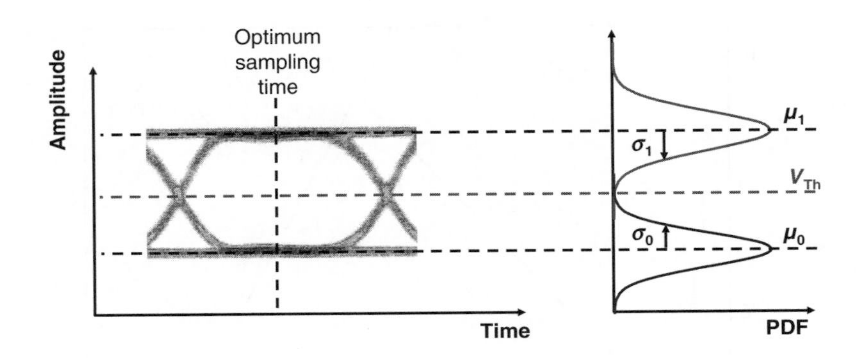

Figure 3.21 (a) NRZ eye diagram. (b) Histogram of vertical slice at the optimum sampling point

eye, such as the histogram shown for a vertical slice at the optimum sampling point show in Figure 3.21b. The parameters that can be obtained from the eye diagram shown in Figure 3.21 are defined below.

a. **Optimum sampling time.** The time at which the eye has the largest opening.
b. **1 level/0 level.** The means μ_1 and μ_0 of the logic 1 and 0 levels obtained from the histogram for a vertical slice of the eye diagram at the optimum sampling point.
c. **Amplitude distortion.** The standard deviations σ_1 and σ_0 of the logic 1 and 0 levels obtained from the histogram for a vertical slice of the eye diagram at the optimum sampling point.
d. **Rise time/fall time.** Defined as the time taken for the signal transition to go from 10 to 90% of its final amplitude.
e. **Timing jitter.** The standard deviation σ_T at the eye-crossing obtained from the histogram for a horizontal slice of the eye diagram at the amplitude where eye transitions cross.
f. **Eye width or timing margin.**

$$(t_2 - 3\sigma_T) - (t_1 + 3\sigma_T) \tag{3.53}$$

g. **Eye opening.** Also known as eye height or amplitude margin. The vertical histogram used to calculate the mean and standard deviations is usually averaged over a timing window that is 20% of the bit slot to allow for timing jitter.

$$\mathrm{EO} = (\mu_1 - 3\sigma_1) - (\mu_0 + 3\sigma_0) \tag{3.54}$$

h. **Mask margin.** Defined as a square mask that represents a two-dimensional margin, where there is both a valid timing margin and amplitude margin.
i. **Eye closure/opening.**

$$\mathrm{ECP} = \frac{\mathrm{EO}}{2\bar{P}} \tag{3.55}$$

where $\bar{P}$ is the average signal power [40].

j. **Q factor.** Calculated from the means and variances of the histogram for a vertical slice of the eye diagram at the optimum sampling point:

$$Q = \frac{\mu_1 - \mu_0}{\sigma_1 + \sigma_0} \tag{3.56}$$

k. **Extinction ratio.** Assuming a DC coupled receiver is used, then the extinction ratio is defined as

$$\mathrm{ER} = \frac{\mu_1}{\mu_0} \tag{3.57}$$

It should be realized that whilst the eye diagram provides an intuitive and straightforward way to measure a number of signal parameters, the measurement accuracy is limited by the bandwidth and storage depth of the oscilloscope. This is particularly so with Q measurements obtained in this way. Estimating signal quality using the Q-factor should also be taken with caution, since any strong signal distortion (due to dispersion, for example) may lead to erroneous interpretation. For instance, heavy presence of ISI causes pessimistic Q factor estimation [39]. Assuming the main source of noise is the signal–spontaneous beat noise and a large duty cycle intensity-modulated format is used, the Q factor is related to the OSNR by the following expression [40]:

$$\mathrm{OSNR} = \frac{Q^2 B_e}{B_o} \frac{1+r}{(1-\sqrt{r})^2} \tag{3.58}$$

where B_o is the optical bandwidth, B_e the electrical bandwidth, and r is the extinction ratio.

The eye diagram also provides a qualitative way to assess the impact of transmission impairments on the optical signal.

Figure 3.22 shows the ISI on the eye diagram that occurs as the dispersion is increased.

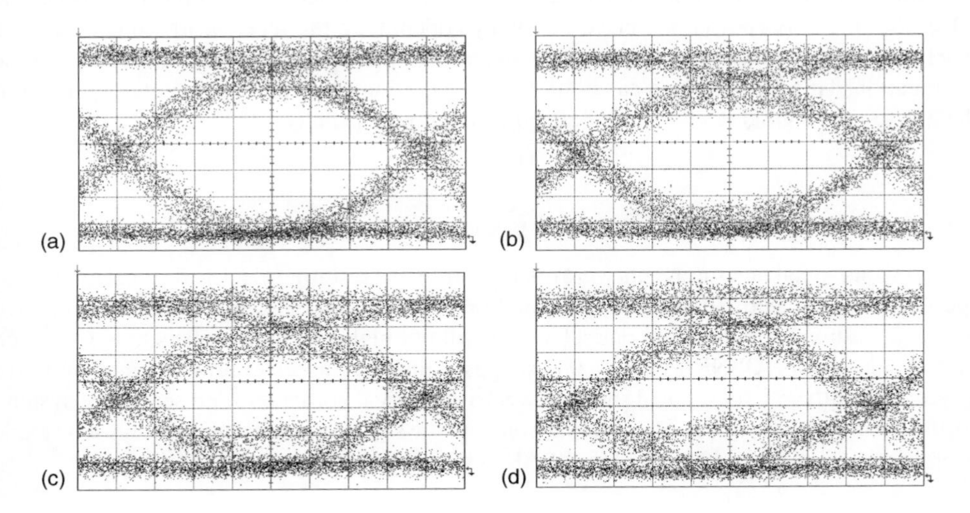

Figure 3.22 Impact of dispersion on the eye diagram of a 10 Gbit/s NRZ signal obtained using a receiver with a 7.5 GHz bandwidth after transmission through fiber with a dispersion of (a) 200 ps/nm (b) 400 ps/nm (c) 600 ps/nm, and (d) 800 ps/nm

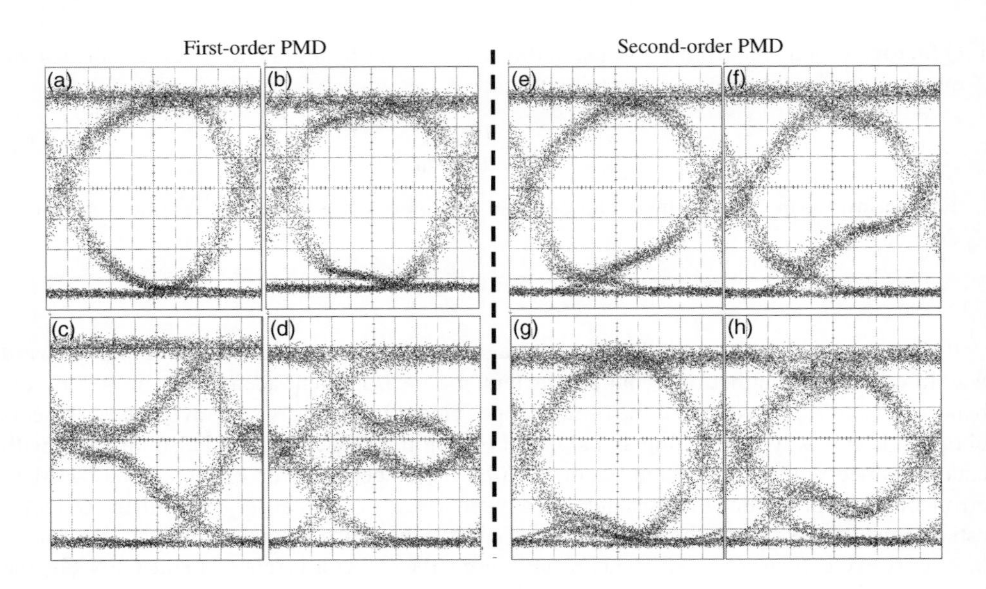

Figure 3.23 Effects of first-order PMD on the eye diagram: (a) $\Delta\tau = 50\,\text{ps}$; (b) $\Delta\tau = 60\,\text{ps}$; (c) $\Delta\tau = 80\,\text{ps}$; (d) $\Delta\tau = 100\,\text{ps}$. Effects of second-order PMD on the eye diagram: (e) $\Delta\tau = 62\,\text{ps}$, (f) $\Delta\tau = 70\,\text{ps}$, (g) $\Delta\tau = 85\,\text{ps}$, (h) $\Delta\tau = 90\,\text{ps}$

The effects of both first- and second-order polarization mode dispersion (PMD) are also observed on the eye diagrams shown in Figure 3.23.

A simple yet effective comparative technique for performance evaluation is to consider the effect that the noise power will have on the eye opening at the receiver. If we consider the maximum values of the noise, then we can form a qualitative measure of the worst-case degradation. This then allows a calculation of the additional power level required to 'reopen' the eye to the original level (this is defined as the power penalty).

3.2.4.3 OSNR

The previous signal measurements were obtained from the resulting electrical signal after high-speed optical-to-electrical conversion. However, it can be advantageous to be able to derive quality measurements from the optical signal without high-speed photodetection. One such measurement is the OSNR. As the name suggests, this measurement provides a quantitative assessment of the relative optical signal power to the optical noise power present in the system. As the noise will be broadband in comparison with the signal, this measurement must be made over a specific noise bandwidth. The OSNR is defined as

$$\text{OSNR(dB)} = 10\log\left(\frac{P_{\text{signal}}}{P_{\text{noise}}}\right) \tag{3.59}$$

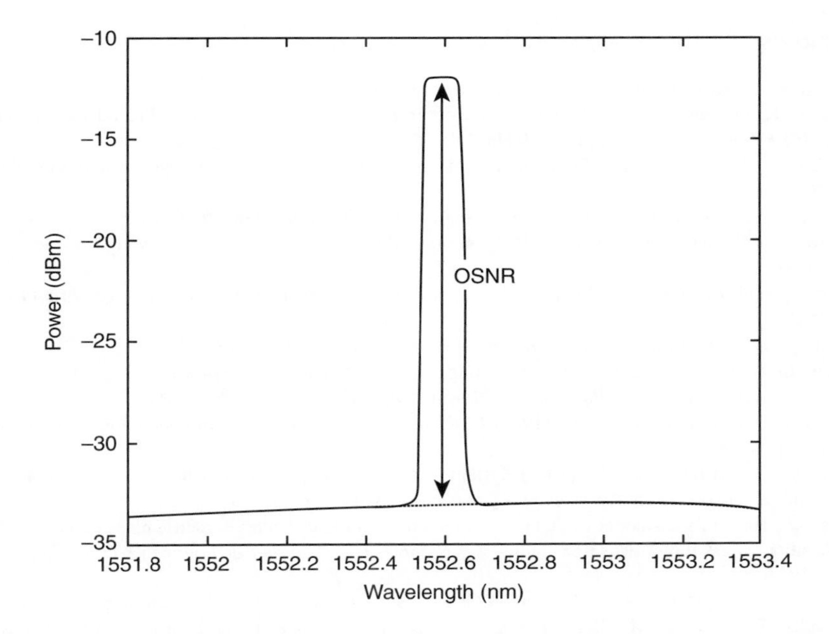

Figure 3.24 Optical spectrum showing OSNR measurement taken with a resolution of 0.1 nm. Here, the measured OSNR is 20 dB

The appropriate noise bandwidth depends on the data rate. For 10 Gbit/s NRZ systems this is usually 0.1 nm or 12.5 GHz, which is just slightly larger than the optical signal bandwidth. If the OSNR is required for a different bandwidth, then it is straightforward to convert using the following formula under the assumption that the optical noise is white:

$$\mathrm{OSNR(dB)_N} = \mathrm{OSNR(dB)_{REF}} + 10\log\left(\frac{B_{\mathrm{REF}}}{B_{\mathrm{N}}}\right) \tag{3.60}$$

The OSNR is most commonly obtained from the optical spectrum measured using an optical spectrum analyzer (OSA), as shown in Figure 3.24. In order to measure the entire signal power and the noise power within a 0.1 nm bandwidth, the resolution of the OSA is set to 0.1 nm. The signal power is then determined from the power within the resolution bandwidth, whilst the noise power within the signal bandwidth is interpolated from the noise power immediately outside the signal bandwidth, as shown in Figure 3.24. The interpolation is most commonly based on a quadratic fit, as shown by the dashed line in Figure 3.24.

This measurement technique becomes more challenging in WDM systems, especially when the channel spacing becomes small, as it is often no longer possible to identify the noise floor between channels. Other techniques, such as polarization nulling [41], which relies on the fact that the signal is polarized whilst the noise is unpolarized to distinguish between the signal and noise, can be used to overcome this.

References

[1] Haykin, S. (1994) *Communcations Systems*, 3rd edn, Wiley.

[2] Gaudino, R. and Viterbo, E. (2003) Pulse shape optimization in dispersion-limited direct detection optical fiber links. *IEEE Communications Letters*, **7**(11), 552–554.

[3] ITU-T (May 2006) Optical interfaces for equipments and systems relating to the synchronous digital hierarchy. ITU-T Recommendation G.957.

[4] Bellcore. (1995) SONET transport systems: common generic criteria. GR-253-CORE.

[5] Agrawal, G.P. and Dutta, N.K. (1993) *Semiconductor Lasers*, 2nd edn, Van Nostrand Reinhold, New York.

[6] Henry, C.H. (1982) Theory of the linewidth of semiconductor lasers. *IEEE Journal of Quantum Electronics*, **18**, 259.

[7] Agrawal, G.P. (2002) *Fiber-Optic Communication Systems*, 3rd edn, Wiley.

[8] Takahashi, H., Oda, K., and Toba, H. (1996) Impact of crosstalk in an arrayed waveguide multiplexer on $n \times n$ optical interconnection. *IEEE Journal of Lightwave Technology*, **14**(6), 1097–1105.

[9] O'Reilly, J.J. and Cattermole, K.W. (1984) *Problems of Randomness in Communication Engineering*, vol. **2**, Pentech, London.

[10] O'Reilly, J.J. and Appleton, C.J. (1995) System performance implications of homodyne beat noise effects in optical fibre networks. *IEE Proceedings–Optoelectronics*, **142**(3), 143–148.

[11] Legg, P.J., Tur, M., and Andonovic, I. (1996) Solution paths to limit interferometric noise induced performance degradation in ask/direct detection lightwave networks. *IEEE Journal of Lightwave Technology*, **LT-14**(9), 1943–1953.

[12] Moura, L., Karafolas, N., Lane, P.M. *et al.* (1996) Statistical modelling of the interferometric crosstalk in optical networks: The Race II MUNDI network. *Proceedings of European Conference on Networks & Optical Communications\NOC'96, Technology, Infrastructure, WDM Networks*, volume **3**, pp. 72–79.

[13] Mitchell, J.E., Lane, P.M., and O'Reilly, J.J. (1998) Statistical characterisation of interferometric beat noise in optical networks. *Proceedings of Conference on Optical Fiber Communication (OFC 98), San Jose, February*, Paper WD3.

[14] O'Reilly, J.J. and Mitchell, J.E. (2005) Simplified derivation of the modified Chernoff upper bound. *IEE Proceeding –Communications*, **152**(6), 850–854.

[15] Cornwell, W.D. and Andonovic, I. (1996) Interferometric noise for a single interferer: comparison between theory and experiment. *IEE Electronic Letters*, **32**(16), 1501–1502.

[16] Eskildsen, L. and Hansen, P.B. (1997) Interferometric noise in lightwave systems with optical preamplifiers. *IEEE Photonics Technology Letters*, **9**(11), 1538–1540.

[17] Ho, K.-P. (1999) Analysis of homodyne crosstalk in optical networks using Gram–Charlier series. *IEEE Journal of Lightwave Technology*, **17**(2), 149–154.

[18] Ho, K.P. (1998) Analysis of co-channel crosstalk interference in optical networks. *IEE Electronic Letters*, **34**(4), 383–385.

[19] Monroy, I.T. and Einarsson, G. (1997) Bit error evaluation of optically preamplified direct detection receivers with Fabry–Perot optical filters. *IEEE Journal of Lightwave Technology*, **15**(8), 1546–1553.

[20] Moura, L., Darby, M., Lane, P.M., and O'Reilly, J.J. (1997) Impact of interferometric noise on the remote delivery of optically generated millimeter-wave signals. *IEEE Transactions on Microwave Theory and Techniques*, **45**(8), 1398–1402.

[21] Whittle, P. (1970) *Probability*, Penguin Books Ltd, Harmondsworth, Middlesex, England.

[22] Prabhu, V.K. (1982) Modified Chernoff bounds for PAM systems with noise and interference. *IEEE Transactions on Information Theory*, **IT-28**(1), 95–100.

[23] O'Reilly, J.J. and Da Rocha, J.R.F. (1987) Improved error probability evaluation methods for direct detection optical communication systems. *IEEE Transactions on Information Theory*, **IT-33**(6), 839–848.

[24] Helstrom, C.W. (1978) Approximate evaluation of detection probabilities in radar and optical communications. *IEEE Transactions on Aerospace and Electronic Systems*, **AES-14**(4), 630–640.

[25] Schumacher, K. and O'Reilly, J.J. (1990) Relationship between the saddlepoint approximation and the modified Chernoff bound. *IEEE Transactions on Communications*, **38**(3), 270–272.

[26] Attard, J.C., Mitchell, J.E., and Rasmussen, C.J. (2005) Performance analysis of interferometric noise due to unequally power interferers in passive optical networks. *IEEE/OSA Journal of Lightwave Technology*, **24**(4), 1692–1703.

[27] Tur, M. and Goldstein, E.L. (1989) Dependence of error rate on signal-to-noise ratio in fiber optic communication systems with phase inducted intensity noise. *Journal of Lightwave Technology*, **7**(12), 2055–2058.

[28] Gimlett, J.L. and Cheung, N.K. (1989) Effects of phase to intensity noise conversion by multiple reflections on gigabit per second DFB laser transmission systems. *Journal of Lightwave Technology*, **7**(6), 888–895.

[29] Rosher, P.A. and Hunwicks, A.R. (1990) The analysis of crosstalk in multichannel wavelength division multiplexed optical transmission systems and its impact on multiplexer design. *IEEE Journal Selected Areas in Communications*, **8**(6), 1108–1114.

[30] Silva, C.F.C., Passy, R., Von der Weid, J.P. *et al.* (1999) Experimental and Theoretical investigations of interferometric noise power penalties in digital optical systems. *MTT-S IMOC'99 Proceedings*, pp. 562–564.

[31] Eskilden, L. and Hansen, P.B. (1997) Interferometric noise in lightwave systems with optical preamplifiers. *IEEE Photonics Technology Letters*, **9**(11), 1538–1540.

[32] Cancela, L.G.C. and Pires, J.J.O.(April 2001) Rigourous evaluation of crosstalk requirements for large optical space switches based on directional couplers, Conftele 2001, Figueira da Foz, Portugal.

[33] Boston, R.C. (1971) Evaluation of Gram–Charlier coefficients. *IEE Electron Letters*, **7**, 492.

[34] Jeruchim, M.C., Balaban, P., and Shanmugan, K.S. (December 1992) *Simulation of Communications Systems (Application to Communication Theory)*, Plenum Pub Corp., ASIN 0306439891.

[35] Remondo, D., Srinivasan, R., Nicola, V.F. *et al.* (2000) Adaptive importance sampling for performance evaluation and parameter optimization of communications systems. *IEEE Transactions on Communications*, **48**, 557–565.

[36] Bergano, N.S., Kerfoot, F.W., and Davidsion, C.R. (1993) Margin measurements in optical amplifier system. *IEEE Photonics Technology Letters*, **5**(3), 304–306.

[37] Winzer, P.J., Fidler, F., Matthews, M.J. *et al.* (2005) 10-Gb/s upgrade of bidirectional CWDM systems using electronic equalization and FEC. *Journal of Lightwave Technology*, **23**(1), 203–210.

[38] Mazurczyk, V.J., Kimball, R.M., and Abbott, S.M. (1997) Using optical noise loading to estimate margin in optical amplifier systems. *Optical Fiber Communication Conference, OFC97, TuP5*, p. 85.

[39] Fishman, D.A. and Jackson, B.S. (1997) Erbium-doped fiber amplifiers for optical communications, In: *Optical Fiber Telecommunications IIIB*, (eds I.P. Kaminow and T.L. Koch), Academic Press, ISBN: 0123951712.

[40] Essiambre, R.-J., Raybon, G., and Mikkelsen, B. (2002) *Optical Fiber Telecommunications Systems v. IV-B*, Academic Press Inc. (London) Ltd, ISBN: 0-12-395173-9.

[41] Kim, C., Lee, Y., Ji, S. *et al.* (2004) Performance of an OSNR monitor based on the polarization-nulling technique. *Journal of Optical Networking*, **3**, 388–395.

4

Enabling Technologies

Stefano Santoni, Roberto Cigliutti, Massimo Giltrelli, Pasquale Donadio, Chris Matrakidis, Andrea Paparella, Tanya Politi, Marcello Potenza, Erwan Pincemin and Alexandros Stavdas

4.1 Introduction

The management and deployment of efficient networks, particularly for broadband services, is the result of many brilliant inventions and discoveries in the optical and telecommunication fields. In the last 15 years, optical fiber transmission systems have progressed enormously in terms of information handling capacity and link distance. Some major advances in transmitter technology, both electronic and optoelectronic, have made this possible. The single longitudinal-mode infrared emission from semiconductor distributed feedback (DFB) laser diodes (LDs) can be coupled efficiently into today's low-loss silica single-mode fibers. In addition, these lasers can be switched on and off with transition times on the order of 100 ps to provide data rates up to approximately 10 Gbit/s. The outstanding performance of these LDs has, in fact, provoked considerable effort to improve the switching speed of digital integrated circuits in order to fully utilize the capability of the optoelectronic components.

In this chapter, we discuss the overall basic point-to-point fiber optic transmission system: the optical transmitters, the optical receivers, the fiber optic cable, the optical amplifiers (OAs), the optical filters, and multiplexers. We shall introduce some information about the characteristics of LDs used in directly modulated optical transmitters and the electronic circuits to drive them. A brief description of the various components used in optical transmission, such as lasers and light-emitting diodes (LEDs), is also provided.

4.2 Transmitters

4.2.1 Introduction

The design of a state-of-the-art laser transmitter requires the combining of several different technologies. The LD is, of course, the critical component whose properties govern

Core and Metro Networks Edited by Alexandros Stavdas

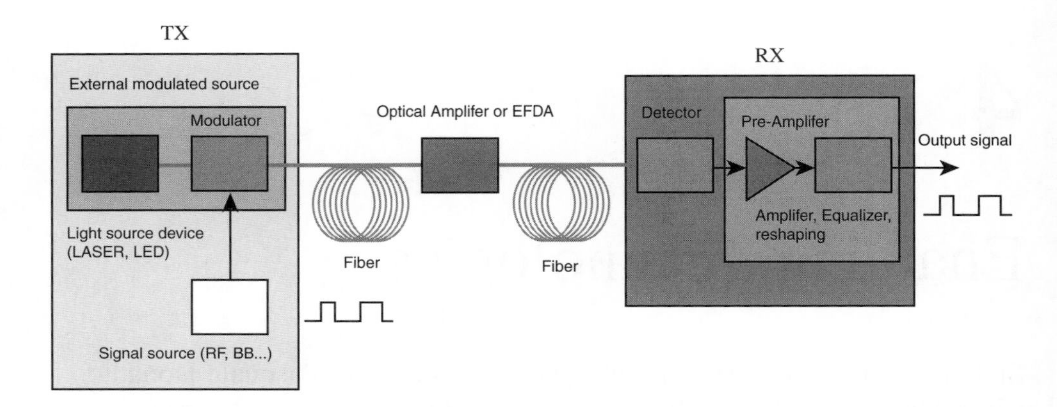

Figure 4.1 Optical transmission system

the transmitter performance. Its electrical and optical properties are both very important. The electrical driver circuit and the link between it and the laser are also critical design areas. Finally, the package design is a specialized combination of a microwave and optical enclosure.

The optical transmission system (Figure 4.1) converts digital or electrical signals into optical signals. The optical signals are first modulated and then transmitted over optical fibers.

The most common devices used as the light source in optical transmitters are the LED and the LD. In a fiber optic system, these devices are mounted in a package that enables an optical fiber to be placed in very close proximity to the light-emitting region in order to couple as much light as possible into the fiber. In some cases, the emitter is even fitted with a tiny spherical lens to collect and focus "every last drop" of light onto the fiber and in other cases, a fiber is "pigtailed" directly onto the actual surface of the emitter. The most popular wavelengths of operation for optical transmitters are 850, 1310, and 1550 nm.

4.2.1.1 Modulation Schemes

Modulation is the process of converting a signal to a form that is suitable for the transmission medium. In optical networks, the digital signal is converted to an optical signal. Depending on the nature of this signal, the resulting modulated light may be turned on and off or may be linearly varied in intensity between two predetermined levels, as depicted in the Figure 4.2.

Various schemes are employed to modulate signals. These schemes depend on the application, the availability and limitations of the components, and the medium of transmission. These schemes are discussed in the following sections.

On–Off Keying

On–off keying (OOK) is a type of modulation that turns the carrier signal on for a 1 bit and off for a 0 bit. The data signal must be coded because of the difficulty in determining the difference between a 0 bit and a 1 bit.

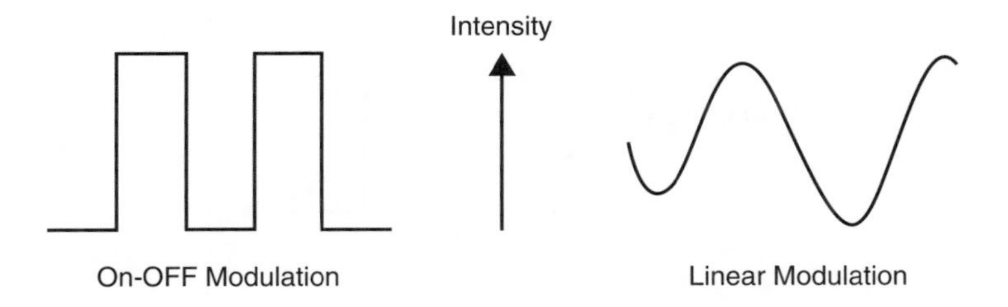

Figure 4.2 Modulation schemes

Consider the difference between sending the message 10 and the message 110 using OOK. To send the first message, you simply turn the switch on to send 1 and then turn it off to send 0. To send the second message, you again turn the switch on to send the first 1. The switch must also be on to send the second 1. If you turn the switch off between the first 1 and the second 1, this is interpreted as a 0. So the switch must be left on for the second 1 and then turned off for the final bit, 0. In both cases, the actions performed for the two messages are the same. The switch is turned on and then turned off. The only difference between the two is that the switch might be left on longer to send two 1s than to send a single 1. Therefore, to interpret signals in this system successfully, the receiver must measure the time the switch was left on to determine whether this act represented a single 1 or two 1s. For the receiver to distinguish a single 1 from a pair of 1s, the sender and receiver must agree on a precise amount of time that will be used to send all single symbols. The information transmitted using OOK is shown in Figure 4.3.

The switching rate of the light source limits the system's data rate.

Pulse Modulation Scheme

The carrier signal is used as a pulse train. Square pulses, raised cosine pulses,[1] or sine function pulses can be used. The characteristics of the pulse train that can be varied are its amplitude and width. The two types of pulse modulation scheme are pulse-amplitude modulation (PAM) and pulse-width modulation (PWM).

To increase the transmission bandwidth, other complex schemes, such as quadrature PAM (QPAM) and pulse-code modulation (PCM), are used.

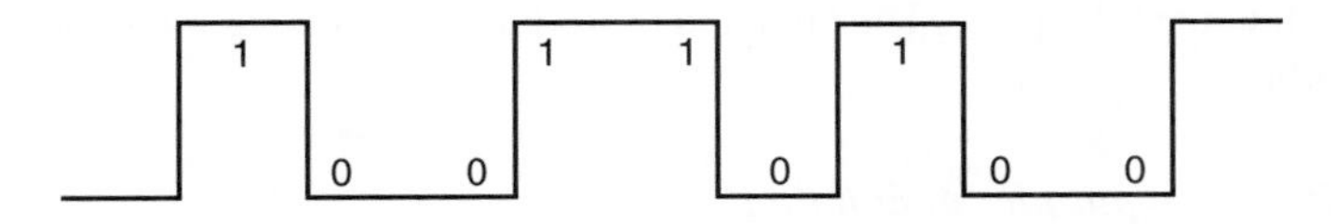

Figure 4.3 OOK, used for binary data transmission. The 1 state implies the presence of a signal and 0 indicates the absence of a signal

[1] There is no standard definition for raised cosine functions. As the name suggests, you add a constant to change the average value of the cosine function. Raised cosine functions are used in signal processing in Fourier transforms.

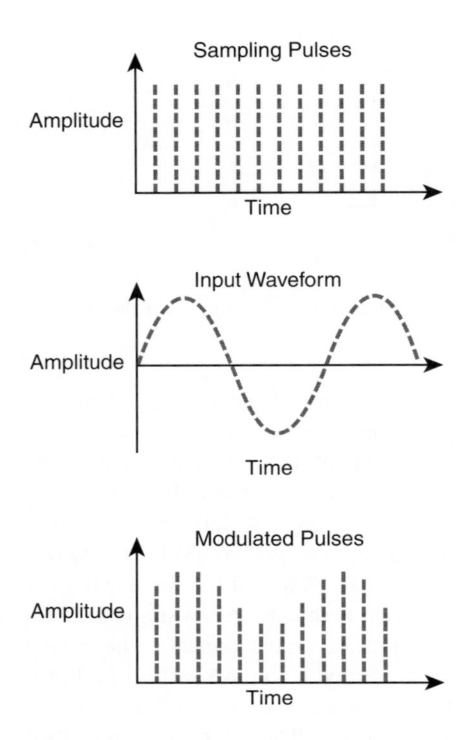

Figure 4.4 PAM, in which the amplitude of the input signal is sampled at the sampling frequency

Pulse-Amplitude Modulation

PAM, shown in Figure 4.4, is a modulation scheme that generates a sequence of pulses whose amplitude is proportional to the amplitude of the sampled analog signal at the sampling instant. The amplitude of the samples is modulated by the sampled frequency.

Pulse-Width Modulation

With PWM, a pulse alternates periodically between a high value and a low value. Figure 4.5 shows PWM. Here, the input sinusoidal signal is modulated by a sawtooth wave and the output is in the form of pulses. The pulse width is less at the positive peak and increases as the input signal reaches the negative peak. The signal is reproduced at the receiving end using a pulse-width demodulator.

Quadrature Pulse-Amplitude Modulation

With QPAM, two sinusoidal carriers, one exactly 90° out of phase with the other, are used to transmit data over a given physical channel. Each carrier can be modulated independently, transmitted over the same frequency band, and separated by demodulation at the receiver, because the carriers occupy the same frequency band and differ by a 90° phase shift. For a given bandwidth, QPAM enables data transmission at twice the rate of standard PAM without any degradation in the bit error rate (BER).

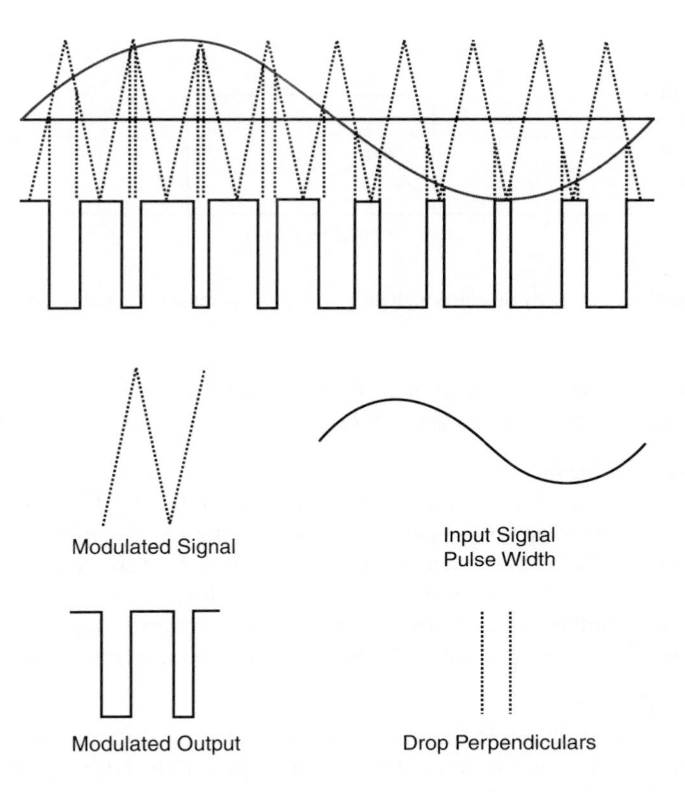

Figure 4.5 The PWM scheme, in which the amplitude of the signal is determined by the width of the pulse

Phase-Shift Keying

Phase-shift keying (PSK) is a technique for switching phases in response to the signal. Quadrature versions of PSK are called QPSK or 4-PSK.

In radio channels, PSK modulation is useful because the decoder does not need to keep track of the intensity of the received signal. For greater noise immunity in these applications, a differential modulation scheme is used, in which the changes are encoded.

Keying Scheme

In PSK, either the frequency or the phase of a carrier signal is keyed in response to patterns of 1s and 0s. Frequency-shift keying (FSK) is the process of keying between two different frequencies. In FSK, the two binary states, logic 0 (low) and logic 1 (high), are each represented by an analog waveform. Logic 0 is represented by a wave at a specific frequency and logic 1 is represented by a wave at a different frequency. A modem converts the binary data from a computer to FSK for transmission over telephone lines, cables, optical fiber, or wireless media. The modem also converts incoming FSK signals to digital low and high states that the computer can understand.

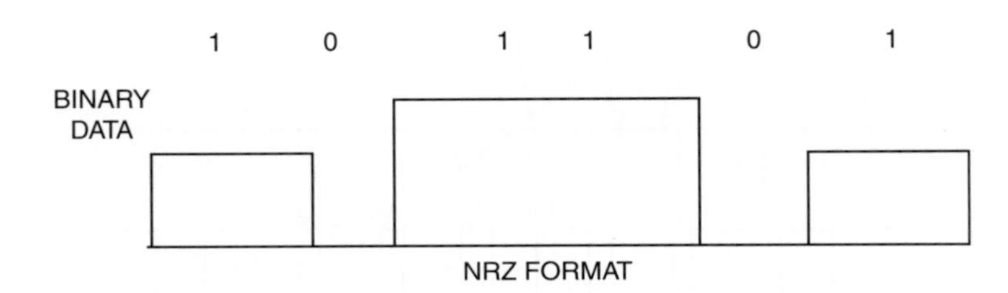

Figure 4.6 The NRZ coding format, in which the pulse does not return to zero level during the 1 bit pulse

The most important keying scheme is the OOK modulation scheme. OOK modulation uses signal formats such as nonreturn to zero (NRZ), return to zero (RZ), and short-pulse format.

Nonreturn-to-Zero Format

In the NRZ type of signal format, shown in Figure 4.6, the bandwidth used by the signal is smaller than that of other signal formats. The 0 bit does not have a pulse, and the 1 bit pulse forms a bit interval. If the bit pulses occur continuously, then there are a corresponding number of bit intervals. For example, if there are two continuous 1-bit pulses, then there will be two bit intervals. One disadvantage of using this format is that a lengthy string of 1 or 0 bits fails to make transitions. This format is generally used for high-speed communication.

Return-to-Zero Format

In the RZ signal format, shown in Figure 4.7, the 0 bit does not have a pulse and a 1 bit forms half the bit interval. The bandwidth used here is twice that of the NRZ signal format. Unlike with NRZ, lengthy strings of 1 bits produce transitions.

Short-Pulse Format

In short-pulse format, shown in Figure 4.8, the 0 bit has no pulses and the 1 bit forms only a fraction of the bit interval. Short-pulse format reduces the effects of chromatic dispersion (CD). This format is also used to reduce the dispersion properties of solitons in optical communication.

The OOK modulation is successful only if transitions occur in the signal and a DC balance[2] is maintained. This is achieved using line coding or scrambling.

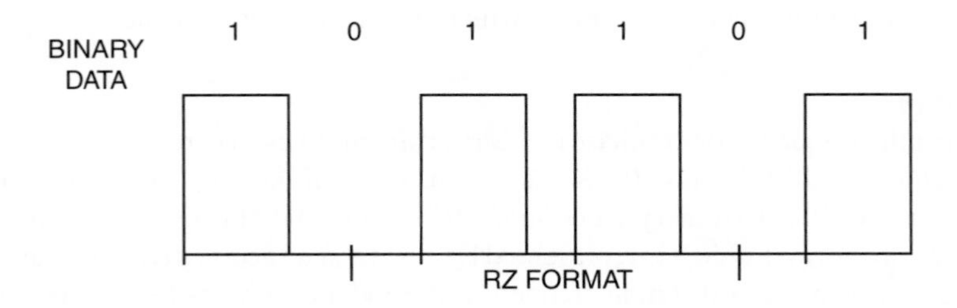

Figure 4.7 The RZ format, in which the signal returns to zero during a portion of the 1 bit pulse

[2] DC balance is achieved if the average transmission power of transmitted data is constant.

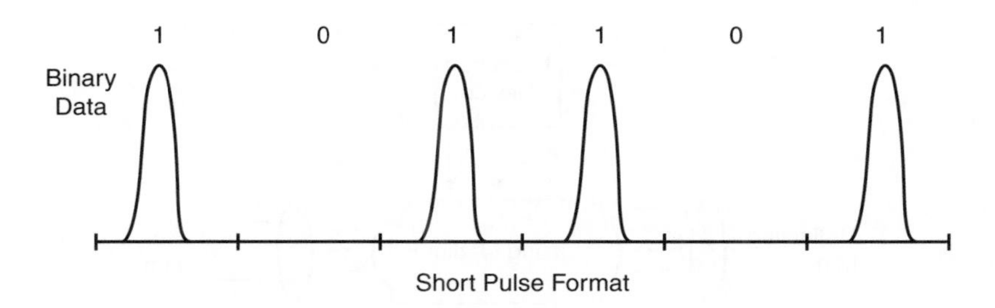

Figure 4.8 Short-pulse format, in only a fraction of the bit interval

Line Coding

Line coding is the process of encoding input data into symbols for transmission (bits). These bits undergo modulation and are then transmitted. At the receiving end, the bits are mapped to the original data. The encoding is carried out in such a way that there is DC balance and sufficient transitions occur in the signal. This type of line code is called binary block line code.

Scrambling

In scrambling, bits are translated one to one from the input data stream to the translated one. A known pseudo-random bit sequence is produced at the transmitting end by a scrambler and added on the input bit sequence. At the receiving end the descrambler subtracts the known sequence and recovers the input sequence. Scrambling has some disadvantages, such as DC imbalance and generation of long strings of 1s and 0s. However, these effects are reduced by selection of an appropriate scrambling sequence.

4.2.2 Overview of Light Sources for Optical Communications

4.2.2.1 Laser

Laser is an acronym for light amplification by stimulated emission of radiation. It works due to the interaction of light and electrons. Electrons lose energy in the form of photons. This release of energy is called photoemission. There are two types of photoemission: spontaneous and stimulated. If an electron spontaneously decays from one energy state to another, then photons are emitted. This is called spontaneous emission. If a photon interacts with an excited electron, it causes the electron to return to a lower energy level and emit a photon. The photon that induces emission of the new photon is called the stimulating photon, and the process is called stimulated emission. This results in two photons having the same energy and being in phase with each other. For stimulated emission to occur, the number of atoms in the higher energy level must be greater than the number of atoms in the lower energy state. This is called population inversion and is essential for laser action to have effect.

Some substances can exist in higher energy states for a longer time. They are said to be in the quasi-stable state. Population inversion can be achieved by energizing these substances.

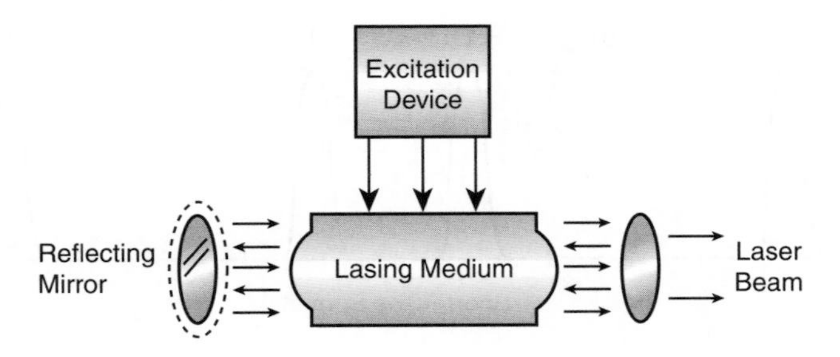

Figure 4.9 Typical structure of a laser

4.2.2.2 Structure of a Typical Laser

A typical laser consists of three important components (see Figure 4.9): gain medium, pumping source, and resonant cavity. The space between the two mirrors forms the resonant cavity. The lasing medium, which occupies the cavity, is called the gain medium, and the pumping source is the device used to excite electrons.

The lasing medium usually consists of a quasi-stable substance. The pumping source excites the electrons in the lasing medium. As a result of this, electrons start decaying to a lower energy state and emit photons. These photons further stimulate other electrons to release energy, resulting in the emission of new photons. The mirrors at both ends reflect the photons to induce further stimulated emission. This builds up to produce a high-intensity light called a laser.

The lasing medium acts like an OA, increasing the intensity of light passing through it. The gain for the amplifier is the factor by which light intensity is increased by the amplifying medium. This factor also depends on the wavelength of incoming light, the extent to which energizing the gain medium occurs, and the length of the gain medium.

The mirrors cause the light emerging from the lasing medium to reflect for more amplification. This is called positive feedback. An amplifier that works on the concept of positive feedback is called an oscillator. Of the two mirrors, one reflects the light completely and the other reflects the light partially. Light that is not reflected is transmitted through the partially reflecting mirror. This light constitutes the laser beam.

4.2.2.3 Laser Characteristics

Lasers are monochromatic and coherent sources of light; that is, the emitted light waves have the same frequency (monochromatic) and are in phase with each other (coherent). The light wavelength is related to the energy released. Based on the lasing material, absorption and emission of light of particular wavelengths are possible. The main properties of the laser are:

- **Line width.** The spectral width of a laser beam is called its line width. This affects the amount of light dispersion passing through the fiber.
- **Frequency.** Variations in laser frequency are of three types: mode hopping, mode shift, and wavelength chirp. Mode hopping is due to an unexpected rise in laser frequency because of a change in the injected current above the threshold value. A mode shift causes a change in the

laser frequency due to changes in temperature. Wavelength chirping happens due to differences in the injected current.

- **Longitudinal modes.** Lasing happens only at wavelengths that are an integral multiple of the cavity length. The set of integral multiples of the cavity length is called the cavity's set of longitudinal modes. The number of wavelengths that the laser can amplify is the number of longitudinal modes of a laser beam. Based on this, lasers can be classified as single longitudinal mode (SLM) lasers or multiple longitudinal mode (MLM) lasers. Generally, SLMs are preferred because the other modes result in dispersion of the light beam.
- **Tuning.** A laser can be tuned to different wavelengths. This phenomenon, called laser tuning, depends on the tuning time and tuning range. The tuning time is the time it takes for the laser beam to tune from one wavelength to another. The tuning range is the range of wavelengths that are tuned by the laser. A laser can be tuned continuously or tuned to selected wavelengths.

4.2.2.4 Types of Laser

The following sections discuss the different types of laser employed in optical networks.

Semiconductor Laser Diodes

In a semiconductor, electrons are present in either the valence band (they are not free from the atom) or the conduction band (in which they are free moving). Holes are created when electrons migrate from the valence band to the conduction band. During this migration, the electrons combine with the holes and produce photons.

Doping is the process of adding impurities to increase the number of holes or electrons in a semiconductor. A p-type semiconductor has a majority of holes and an n-type semiconductor has a majority of electrons. Figure 4.10 shows the structure of a semiconductor diode laser.

A semiconductor diode is similar to a normal laser, except for the addition of a pn-junction. The mirrors at both ends of the laser are perpendicular to the pn-junction. An electric voltage is applied to the pn-junction, causing electrons of the n region to interact with the holes in the p region and produce photons. These photons further stimulate the emission of more photons and produce high-resolution lasers. The frequency of the laser beam depends on the length of the

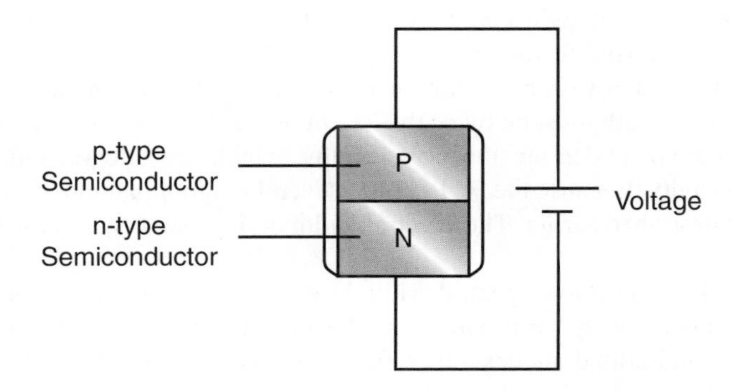

Figure 4.10 A semiconductor diode laser

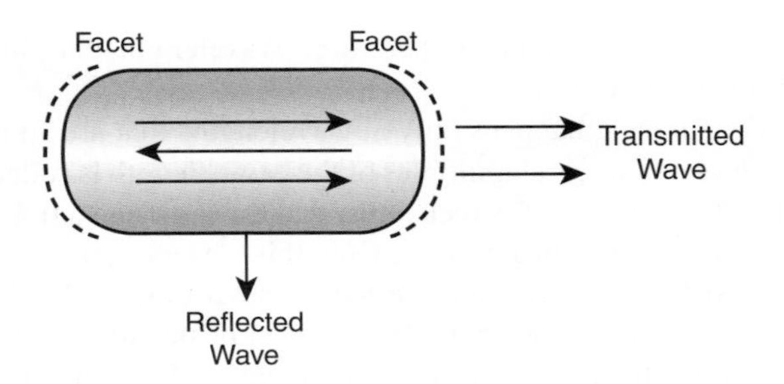

Figure 4.11 The FP laser

cavity between the mirrors. By changing the semiconductor material, light of different frequency ranges can be produced.

Semiconductor lasers differ from ordinary lasers in the following respects:

- The population inversion caused between the valence and conduction bands results in increasing the gain of the lasing material.
- The laser beam's gain spectrum is very high.
- The cavity between the mirrors is on the order of 100 μm. This increases the longitudinal mode spacing.

Fabry–Perot Lasers

The Fabry–Perot (FP) laser, shown in Figure 4.11, consists of the FP cavity as the gain medium and the two faces of the cavity, called facets. The facets are parallel to each other. One part of the light is transmitted at the right facet and the rest is reflected. The reflected wave is again reflected. The transmitted light waves for the resonant wavelength of the FP cavity are all in phase, and they add up to increase the amplitude.

Lasers oscillate if both the gain of the amplifier and the reflectivity of the mirrors are large. The point at which the laser begins to oscillate is called the lasing threshold. After the lasing threshold value is reached, the laser tends to function only as an oscillator. Spontaneous emission of electrons occurs at all wavelengths present in the amplifier's bandwidth. This results in amplification with positive feedback, which is the general characteristic of an oscillator. The feedback of light comes from the two reflecting ends of the cavity, so the feedback is called localized feedback.

The wavelength must always be within the bandwidth of the gain medium, and an integral multiple of the wavelength must be twice the length of the FP cavity for the laser to oscillate. A laser's longitudinal modes are the wavelengths, which are integral multiples of twice the length of the cavity. Because the FP laser has several longitudinal modes, it belongs to the MLM laser type described earlier. The spectral width of the laser beam in the FP laser is very large.

Optical networks require low-spectral-width lasers to work at high speeds. Thus, an SLM laser, which outputs a low-spectral-width laser beam, is beneficial. SLM can be achieved by suppressing all longitudinal modes other than the main node. The level to which this suppression is done is called the side-mode suppression ratio. The various SLM-implemented lasers are discussed in the following sections.

Distributed Feedback Lasers

A laser that uses a corrugated waveguide and that functions in an SLM is called a DFB laser. This laser is used for high-speed transmissions. It is called a DFB laser because light feedback happens in a distributed manner with the help of a set of reflectors.

This laser operates as follows. The incoming light undergoes a series of reflections. These reflected waves form the resultant transmitted wave through a process called in-phase addition. This is achieved only if Bragg's condition is satisfied. In other words, the wavelength of the cavity must be twice the period of corrugation. A number of wavelengths satisfy this condition. However, the strongest transmitted wave has a wavelength that is exactly equal to twice the corrugation period. This particular wavelength gets amplified more than the rest. All the other wavelengths are suppressed to make the laser oscillate in an SLM. If the corrugation is inside the cavity's gain region, then it is called a DFB laser. If it is outside the cavity's gain region, then it is called a distributed Bragg reflector (DBR) laser.

The disadvantage of a DFB laser is that, due to the series of reflections that occur, there are variations in wavelength and power. Using a photodetector and a thermoelectric cooler can rectify this deviation. The thermoelectric cooler avoids variations in wavelength by maintaining a constant temperature. The photodetector avoids optical power leakage in the laser.

DFB lasers are difficult to fabricate compared with FP lasers. Hence, they are more expensive. FP lasers are used only for short-distance transmissions, but DFB lasers are the main laser structures used for long-distance transmissions.

External Cavity Lasers

Lasers can be made to operate in an SLM using an external cavity. This external cavity is in addition to the primary cavity. A diffraction grating that consists of a wavelength-selective mirror is used in the external cavity. The end of the cavity that faces the grating is coated with an antireflection material. The external cavity allows only certain wavelengths to have more reflectivity and exhibit lasing. Generally, the external cavity grating is selected in such a way that only one wavelength satisfies the condition to operate in an SLM.

Filters such as the FP filter or the Bragg grating filter can also be used in external cavity lasers (ECLs). Because the cavity length of an ECL is large, it cannot be modulated at high speeds. Figure 4.12 shows the structure of an ECL.

Vertical-Cavity Surface-Emitting Lasers

These lasers work by making the length of the cavity small to increase the mode space. This ensures that only one longitudinal mode is available in the gain medium, thereby making it

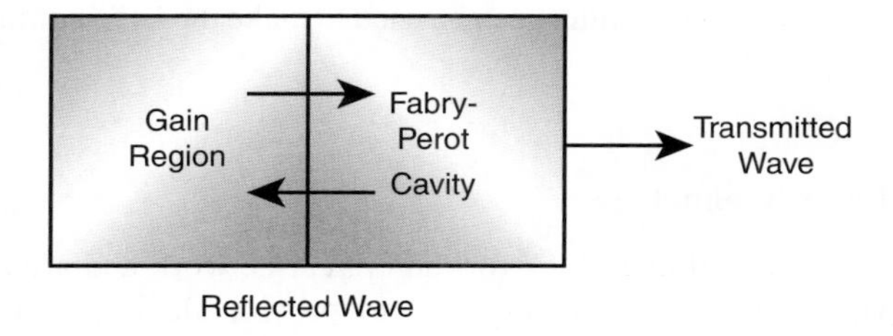

Figure 4.12 The ECL

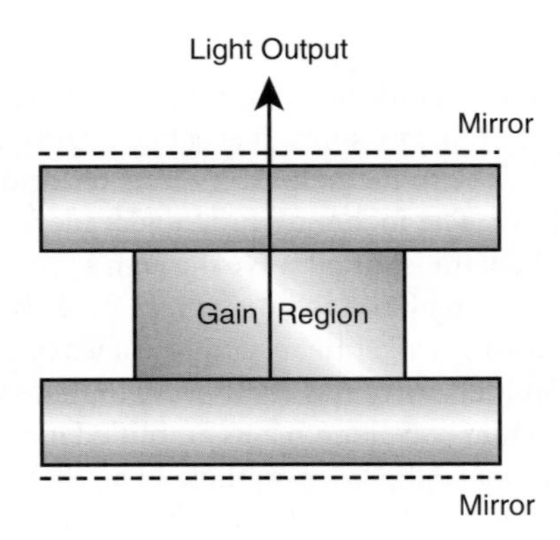

Figure 4.13 A vertical-cavity surface-emitting laser

function as an SLM laser. A vertical cavity with mirrors on the top and bottom surfaces of a semiconductor is used – hence the name vertical-cavity surface-emitting laser (VCSEL). Figure 4.13 shows the structure of a VCSEL.

Highly reflective mirrors are required in this type of laser to provide oscillation. This is because of the cavity's short length. Using alternating dielectrics with low and high refractive indexes solves this problem. This helps in high reflection as well as wavelength selection.

VCSELs have a major disadvantage. The resistance created due to injected current heats up the device, and thermal cooling is required to bring it down to room temperature.

4.2.2.5 Light-Emitting Diodes

LEDs are generally used as an alternative to the more-expensive lasers. LEDs have a pn-junction that is forward-biased. This causes the electrons in the p region to combine with the holes in the n region and produce photons. The consequence of spontaneous emission of photons is that the light coming out of the LED has a wide spectrum. Thus, LEDs cannot produce a high-intensity beam like a laser can.

In applications where a narrow spectral width is required, DFB lasers are generally used. If cost is a drawback, then LED slices can be substituted for DFB lasers. An LED slice is obtained by placing a narrow pass-band optical filter in front of the LED. This filters the spectrum that the LED emits. Many filters can be used to make the LED common and shareable for many users.

4.2.2.6 Tunable Laser Technology

Tunable lasers are a basic building block for the optical network and are adopted to reduce investments in spare equipment inventory and to provide flexibility in optical network provisioning. Future applications of optical networking will require tunable sources with very fast (sub-microsecond) reconfiguration.

Tunable Laser Characteristics

The following parameters characterize the tunable laser:

- tuning range
- tuning speed
- signal suppression during wavelength tuning
- tuning precision and stability
- output power stability
- laser relative intensity noise (RIN)
- continuous wave (CW) laser spectral width
- extinction ratio (with integrated modulator)
- side mode suppression ratio
- chirp (with integrated modulator)
- power consumption.

The critical parameters for the use in flexible and dynamic systems are the tuning speed and the tuning range.

Tunable Laser Technologies

Tunable laser technologies fall into the following main categories:

- ECLs
- VCSELs
- DFB lasers
- DBRs and multisection DBRs
- narrowly tunable lasers combination (laser array).

Some key points of widely tunable lasers are compared in Table 4.1.

All but the VCSELs are based on edge-emitting devices, which emit light at the substrate edges rather than at the surface of the LD chip. Vertical-cavity structures do the opposite.

Table 4.1 Tunable lasers: technology comparison

Tunable laser	Advantages	Disadvantages
DFB array	Well suited for integration, with established manufacturing process.	Optical performances/tuning range trade-off
	Direct modulation possible	Slow tuning time (thermal tuning – seconds)
Multisection DBR	Inherently fast switching speed	Evolving manufacturing methods, with large chip area and consequently a yield decrease
	Direct modulation possible	
	Modulator integration	Complex control
ECL	High spectral purity and output power	Medium tuning time (milliseconds–hundreds of milliseconds)
VCSEL	Low-cost technology	Low performances (optical output power)

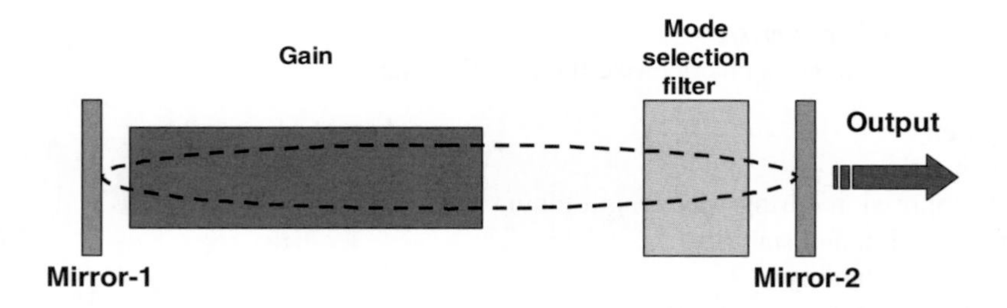

Figure 4.14 General structure of a fixed-wavelength laser

A laser's wavelength is determined by its optical cavity. Besides the characteristics of each technology, tunable lasers can emit on different wavelengths thanks to the ability to modify the parameters of the resonant cavity. The resonant wavelength is actually determined by the cavity length (mirrors distance) and by the speed of light within the medium that fills the cavity (determined by the effective refractive index).

The general structure for a fixed-wavelength laser is presented in Figure 4.14, where the resonant cavity between a pair of mirrors (with mirror-2 partially reflective) includes a gain medium and a mode selection filter (enabling resonance of just one among the possible cavity modes). The principle is applied to both edge emitters and VCSELs.

The operating wavelength of a semiconductor laser can be modified by varying the cavity length or changing the refraction index of the propagating medium. There are several methods to modify these parameters (mechanically, with a micro-electromechanical switch (MEMS), or via thermal effects or via current injection).

For semiconductor lasers, there are three general wavelength tuning methods:

- carrier injection (or free-carrier plasma effect)
- quantum confined Stark effect (QCSE)
- temperature tuning.

Carrier injection is most widely used for tunable semiconductor lasers, due to the broadest tunability. Temperature tuning has to be considered anyway, since current injection determines temperature variations that affect the tuning (towards a reduction in tuning efficiency, due to different sign of wavelength change).

Figure 4.15 presents the general structure of a tunable laser.

The following sections describe the main approaches and structures for tunable lasers.

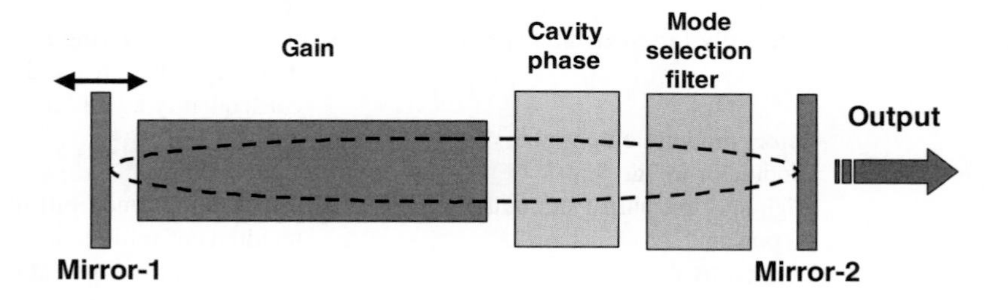

Figure 4.15 General structure of a tunable laser

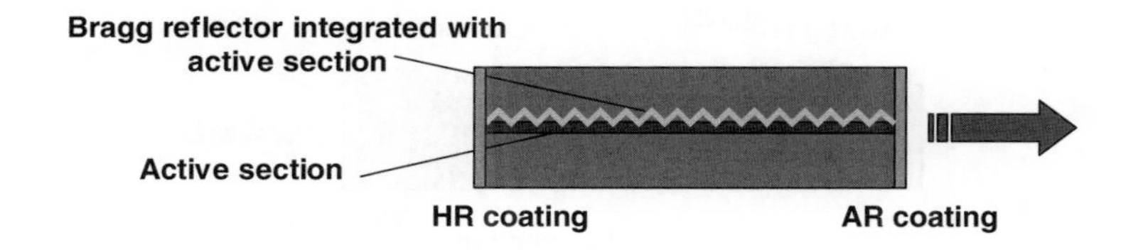

Figure 4.16 In a DFB laser, the diffraction grating is integrated into the active section

Distributed Feedback Tunable Lasers

DFB lasers incorporate a grating directly into the laser chip itself (Figure 4.16), usually along the length of the active layer: the grating (acting as a mode selection filter) reflects a single wavelength back into the cavity, forcing a single resonant mode within the laser, and producing a stable, very narrow bandwidth output.

DFB emitted optical power can be 20 mW, and the lasing action in the DFB also stabilizes the carrier density in the device, leading to small wavelength drift over time (typically a 0.1 nm shift over 25 years) and enabling operation at 25 GHz channel spacing. Other DFB typical characteristics are narrow linewidth and high optical purity (high side-mode suppression ratio).

DFB is tunable in terms of current and temperature and the tuning rate is of the order of 0.1 Å/mA and 1 Å/°C respectively. DFB lasers are tuned by controlling the temperature of the LD cavity. Because a large temperature difference is required to tune across only a few nanometers, the tuning range of a single DFB laser cavity is limited to a small range of wavelengths, typically under 5 nm. The typical tuning speed of a DFB laser is of the order of several seconds.

This laser is well suited for production in large volumes (the manufacturing process is established), but the tuning range is narrow and it can be challenging to maintain optical performance over a wide temperature range.

Distributed Bragg Reflector Tunable Lasers

A variation of the DFB laser is the DBR laser. It operates in a similar manner except that the grating, instead of being etched into the gain medium, is positioned outside the active region of the cavity (see Figure 4.17), also simplifying the epitaxial process. Lasing occurs between two grating mirrors or between a grating mirror and a cleaved facet of the semiconductor.

Tunable DBR lasers (see Figure 4.18) are made up of a gain section, a mirror (grating) section (for coarse tuning), and a phase section, the last of which creates an adjustable phase shift between the gain material and the reflector (to align cavity mode with the reflection peak, for fine tuning). Tuning is accomplished by injecting current into the phase and mirror sections, which changes the carrier density in those sections, thereby changing their refractive index

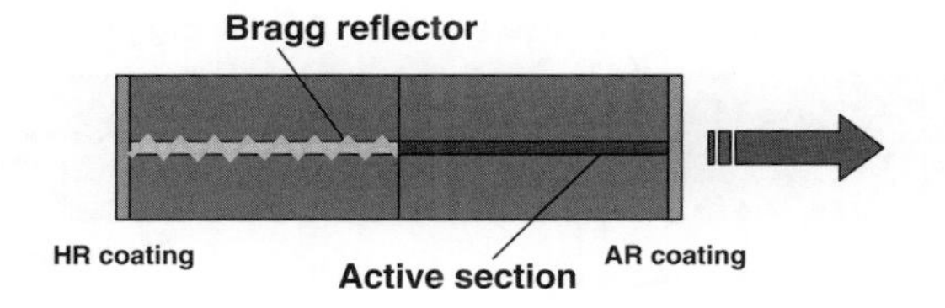

Figure 4.17 In a DBR laser, the grating is contained in a separate section

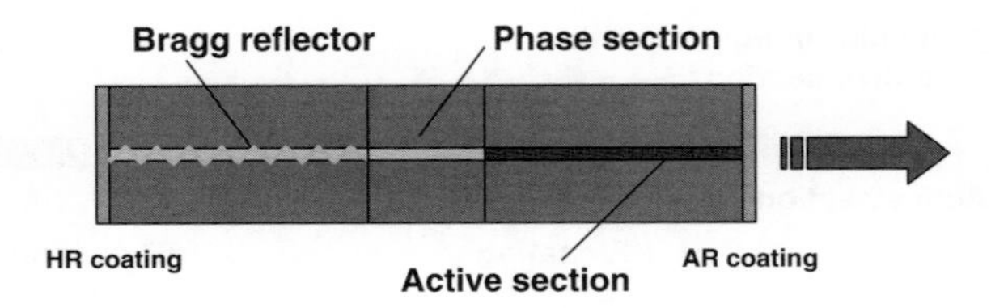

Figure 4.18 Tunable DBR laser general structure

(temperature can also be used to control refractive index changes, with lower tuning speed). Thus, at least three control parameters have to be managed, increasing the complexity of the system; moreover, the refractive index to current relation changes with time (due to p–n junction degradation, a certain current corresponds to a smaller carrier density).

The tuning range in a standard DBR laser seldom exceeds about 10 nm. Wider tuning can be achieved adding other sections besides gain and phase sections, and the various possible solutions are described below.

Being based on electrical effects, tuning speed is much faster than DFB, while optical output power of DBR lasers is generally lower than for DFB lasers.

Multisection Distributed Bragg Reflector Tunable Lasers

In order to improve tunable DBR performance, different solutions, based on the concept of incorporating additional elements (control/gain sections) to the basic tunable DBR, have been proposed.

One among them is the sampled grating DBR (SG-DBR). SG-DBR (see Figure 4.19) uses two gratings (placed at the opposite ends of the gain section) with a slightly different step, thus obtaining two wavelength combs, with a slight offset. During tuning (obtained by varying the current flowing into the front and rear gratings and phase section), the gratings are adjusted so that the resonant wavelengths of each grating are matched. The difference in blank spacings of each grating means that only a single wavelength can be tuned at any one time. Owing to this arrangement (exploiting the Vernier-like effect of reflection peaks of the two grating sections), fine tuning of the combs results in a significant change in the resonant lambda and thus in a wider tuning range.

SG-DBR lasers are a special case of a more general structure (superstructure grating DBR), where the front and rear gratings can be sampled with a modulation function (such as a linearly chirped grating), obtaining different shapes of envelope of the reflectivity peaks (the reflection envelope shape depends on Fourier components of the modulating function).

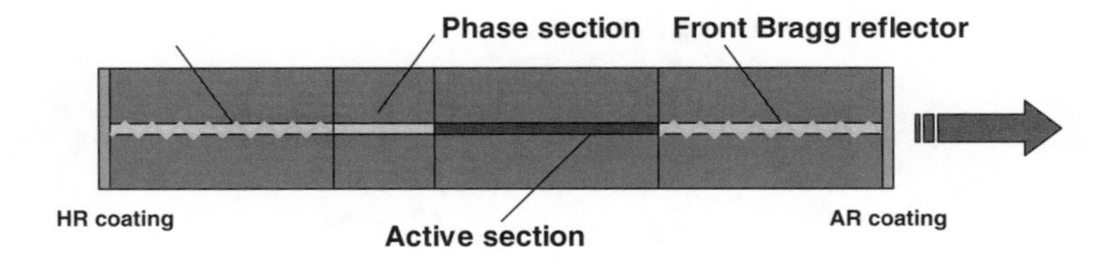

Figure 4.19 Tunable SG-DBR laser general structure

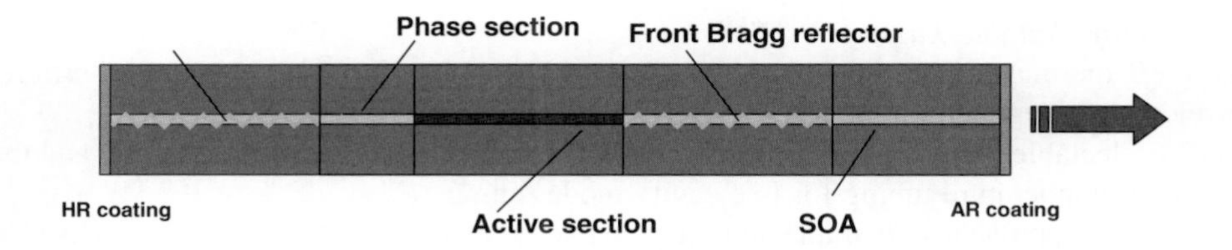

Figure 4.20 Tunable SG-DBR laser with SOA

Tuning is not continuous (the Vernier-like effect means that wavelength will jump in steps, so quasi-continuous wavelength tuning can be achieved using the gain and mirrors sections, while the phase section provides fine tuning) and multiple sections are involved, so requiring a more complex control than a standard DBR or DFB laser.

The output power is typically less than a standard DBR and is about 2 mW, due to more passive sections in the cavity. Moreover current-injection-based refractive index tuning produces an increase in absorption that results in a power variation of several decibels across the tuning range (so large variations in gain current would be needed across channels to maintain a constant channel-to-channel power). For these reasons, semiconductor OA (SOA) integration (enabling higher output power, constant power level for all channels, and blank output power during the tuning process) is a possible choice, even at the cost of increased complexity (Figure 4.20). Further integration with an electro-absorption modulator (EAM) is, in general, possible.

Switching wavelength is inherently fast (tens of nanoseconds), but the involvement of control algorithms in the tuning process and thermal drift should also be also considered.

The manufacturing process of SG-DBR lasers is similar to that of DBR lasers. This technology is also suitable for EAM and SOA integration.

Another version of a multisection DBR is the *grating-assisted coupler with sampled reflector* (GCSR) laser, which contains four sections (gain, Bragg reflector, coupling, and phase correction) and is tuned using three currents. The current-controlled waveguide coupler acts as a coarse tuner to select a narrow range of wavelengths from the modulated Bragg reflector (providing a comb of peaks and which is itself current controlled to provide a level of selection), to the phase-correction section (also current controlled), which acts as a fine tuning section (see Figure 4.21). The concept is to match the reflection peak spacing of the sampled grating to the filter width of grating-assisted coupler. GCSRs operate over a wide tuning range, on the order of 40 nm. As for other multisection lasers, power output is limited (around 2 mW) and can be increased, at the expense of tuning range, by eliminating the coarse tuning section.

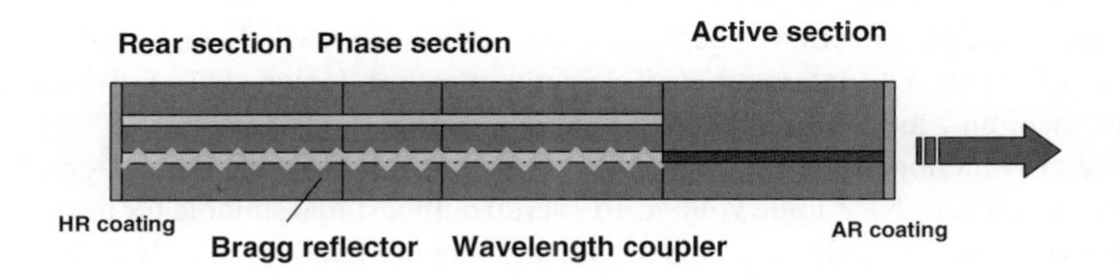

Figure 4.21 Tunable GCSR laser

Laser Array Tunable Lasers
The DFB thermal tuning range can be expanded by having an array of lasers of different wavelengths integrated on the same chip.

DFB selectable arrays operate selecting the DFB array element for coarse tuning and then exploiting temperature tuning for fine cavity mode tuning.

Common approaches to implement the coarse selection in DFB arrays are based on integrated on-chip combiners or on off-chip MEMS deflectors able to route the proper beam on the laser output.

The advantages of the on-chip combiner approach are mainly the reliability and spectral characteristics, which are basically the same as fixed-wavelength sources. Disadvantages are relevant to the trade-off between power and tuning range (sometimes an SOA is added to counterbalance the combiner losses that increase with the number of lasers), reduced yield, and large real-estate requirements.

MEMS-based devices can improve optical output power and decrease chip size, but introduce an element that could affect overall reliability.

External Cavity Tunable Lasers
Tunable ECLs are edge-emitting lasers containing a gain chip (conventional FP laser chip) and separate gratings or mirrors to reflect light back into the laser to form a cavity.

To tune the laser output, a grating or another type of narrow-band tunable mirror is adjusted in a way that produces the desired wavelength. This type of tuning usually involves physically moving the grating or the mirror, even if different mechanisms have been adopted.

ECLs can achieve wide tuning ranges (greater than 40 nm), although the tuning speed is determined by the mirror tuning (it can take tens of milliseconds to change wavelengths in the case of mechanical solutions). ECLs are widely used in optical test and measurement equipment due to the high purity of their emission together with very high output power over a broad range of wavelengths.

4.2.3 Transmitters for High Data-Rate Wavelength-Division Multiplexing Systems

4.2.3.1 Introduction

The transmission of large amounts of information along fiber-optic links requires high data rates and wavelength-division multiplexing (WDM) transmission, with or without the use of erbium-doped fiber amplifiers (EDFAs).

These systems need to be optimized both for performance and cost, in order to provide the correct transmission quality, depending on the application requirements. Transmission sources are designed with different criteria, depending on the distance reach. For a transport network, a typical classification (i.e., International Telecommunication Union (ITU)) divides interfaces for short-, medium-, long- and ultra-long-haul application.

The overall functionality of a transmitter is to convert a serial input signal, typically represented as a binary NRZ logic voltage, to a serial output signal suitable for the transmission on the fiber-optic media, typically a binary optical power modulation, namely OOK or intensity modulation (IM). The advantage of this choice is the extremely simple receiver architecture needed that can be realized with a simple photodiode (p-i-n or avalanche type photodiode), commonly indicated as direct detection (DD).

Today, technologies employed to achieve optical transmission functionality can be summarized as:

- direct modulated laser (DML) through laser current modulation;
- externally modulated laser (EML) through electro-absorption modulation (attenuation modulation) in semiconductor material (EAM);
- EML through electro-optic modulation (phase modulation) in an $LiNbO_3$ Mach–Zehnder modulator (MZM).

In some cases technology allows the possibility to integrate in the same package the external modulation function with the laser CW source: this applies to an EML with electro-absorption (EA-EML).

Cost increases with transmission complexity; therefore, the typical application areas of these transmitters are:

- short-reach interfaces – DML;
- medium-reach interfaces – EA-EML and DML;
- long- and ultra-long-reach interfaces – MZ-EML.

Typically, short-reach and some medium-reach applications are developed with laser sources in the 1.3 μm or 1.55 μm transmission windows. For WDM transmission systems, which are mainly of interest for long- and ultra-long-reach applications, modulated sources are optimized to work in the 1.55 μm transmission window, where the fiber loss is lowest and the EDFA is available, but also require precise wavelength allocation of the laser in the ITU-grid.

4.2.3.2 Direct Modulated Laser Sources

The DFB laser can be engineered to be used not only as a CW source, but also as an optical source modulated directly through its injection current. The static and dynamic behavior of a DFB laser (i.e., a laser emitting light consisting of SLM with mean optical wavelength λ_0) can be simulated by means of the solution of two coupled nonlinear rate equations [1–3]. Starting with Maxwell's equations and including some phenomenological aspects, the physical processes describing wave propagation, charge carrier injection, spontaneous and stimulated emission of photons (recombination processes) in the active laser layer can be modeled by large signal rate equations. For a single-mode laser the dynamic regime is described by the following rate equations:

$$\begin{cases} \dfrac{\mathrm{d}N(t)}{\mathrm{d}t} = \dfrac{I(t)}{eV} - \dfrac{N(t)}{\tau_\mathrm{n}} - g_0(N(t) - N_0)\dfrac{S(t)}{1+\varepsilon S(t)} \\ \dfrac{\mathrm{d}S(t)}{\mathrm{d}t} = \Gamma g_0(N(t) - N_0)\dfrac{S(t)}{1+\varepsilon S(t)} - \dfrac{S(t)}{\tau_\mathrm{p}} + \Gamma\beta_\mathrm{S}\dfrac{N(t)}{\tau_n} \end{cases} \quad (4.1)$$

where:

$N(t)$ = carrier density (electrons) [1/m^3];
$S(t)$ = photon density in the active area [1/m^3];
$I(t)$ = total injection current (modulating signal plus bias) [A];
g_0 = differential gain factor [m^3/s];
ε = gain compression factor [m^3];
N_0 = carrier density at transparency [1/m^3];
τ_n = (average) lifetime of carriers [s];
τ_p = (average) lifetime of photons [s];
Γ = optical mode confinement factor [·];
β_S = spontaneous emission coupled into lasing mode [·];
e = electron charge (=1.602×10^{-19} A s);
V = active laser layer volume (height × width × length) [m^3].

Within the first equation in Equation 4.1, the first right-hand side term describes the amount of carriers delivered by the injection current, the second describes the loss of carriers due to spontaneous emission (electron–hole recombination), and the third represents the loss due to stimulated emission.

Within the second equation in Equation 4.1, the first right-hand side term describes the increase of photons due to stimulated emission, the second represents the loss of photons in the active layer due to some nonradiative loss processes (e.g., Auger recombination), and the third describes the generation of photons due to spontaneous emission.

The parameters τ_n,τ_p,g_0,ε, N_0, and β_S are strongly device dependent and they are derived indirectly from a defined set of measurements through numerical fitting with analytical expressions derived from the above rate equations [4–6].

The output optical power $P(t)$ and frequency chirp $\Delta f(t)$, describing the optical behavior of the output signal, can be calculated from the photon density $S(t)$ and carrier (electron) density $N(t)$ by means of the following equations:

$$P(t) = \eta h \nu_0 \frac{V}{2\Gamma\tau_p} S(t) \tag{4.2}$$

$$\Delta f(t) = \frac{\alpha}{2\pi}\left\{\frac{\Gamma g_0}{2}[N(t) - \bar{N}] - \frac{1}{\tau_p}\right\} \tag{4.3}$$

where:

η = quantum efficiency [·];
h = Planck's constant (=6.626×10^{-34} J s);
ν_0 = (mean) optical frequency (single mode) [Hz];
α = linewidth enhancement factor (Henry factor) [·];
$\bar{N}$ = average carrier density (steady state) [1/m^3];

The injection current $I(t)$ is then derived from the laser input current $I_{LD}(t)$ after considering the presence of electric parasitics (junction capacitance and series contact resistance). The equation that describes this electrical filtering effect in the time domain is

$$\frac{\mathrm{d}I(t)}{\mathrm{d}t} = \frac{1}{RC}[I_{\mathrm{LD}}(t) - I(t)] \tag{4.4}$$

where R [Ω] is the series resistance, C [F] is the junction capacitance, and $I_{\mathrm{LD}}(t) = $ [A] is the input current of the LD.

Finally, the total electric field of a directly modulated single mode laser is expressed by the following relationship:

$$E_{\mathrm{out}}(t) = \sqrt{P(t)}\exp\left\{ \mathrm{j}\left[(2\pi\nu_0)t + \int_0^t \Delta f(\tau)\,\mathrm{d}\tau \right] \right\} \tag{4.5}$$

Typical values of the above listed parameters for a laser suitable for 10 Gbit/s modulation are as follows: $\lambda_0 = 1530\,\mathrm{nm}$; $V = 40\,\mu\mathrm{m}^3$; $\alpha = 5$; $\Gamma = 0.1$; $g_0 = 10.5 \times 10^{-6}\,\mathrm{cm}^3/\mathrm{s}$; $\tau_{\mathrm{n}} = 1\,\mathrm{ns}$; $\tau_{\mathrm{p}} = 2.4\,\mathrm{ps}$; $\varepsilon = 2.75 \times 10^{-17}\,\mathrm{cm}^3$; $\beta_S = 2 \times 10^{-5}$; $n_0 = 1 \times 10^{18}\,\mathrm{cm}^{-3}$; $\eta = 0.5$.

To obtain intensity modulation, the optical power of the laser should be driven from near zero (OFF-state) to the nominal power (ON-state) through a large injection current variation (large signal condition). In this operating mode the output laser behavior is described again from the rate equations as long as other effects are kept small enough by laser structure design [7,8]. An example of an optical modulated signal obtained with direct laser modulation is shown in Figure 4.22.

Some extended laser models should be considered when the intrinsic structure is realized with complex cavity design, like active quantum wells regions, coupled passive external cavities, active coupled laser cavities, external optical feedback, and so on, or when high-order carrier or photon recombination/generation phenomena take place.

There is extensive literature on how to include each specific laser design issue in the basic rate equations, to achieve a more realistic model suitable for system simulation or laser specification [9–11].

4.2.3.3 Small-Signal Amplitude-Modulation Response

The small-signal amplitude modulation (AM) frequency response, defined as the photon density variation with respect to the carrier density variation, is an important characteristic of the laser that could be used to characterize the laser bandwidth and to extract laser parameters:

$$H_{\mathrm{AM}}(\omega) = \frac{\omega_{\mathrm{R}}^2}{\omega_{\mathrm{R}}^2 - \omega^2 + \mathrm{j}\gamma_{\mathrm{e}}\omega} \tag{4.6}$$

where in the second-order transfer function the terms are the relaxation oscillation

$$\omega_{\mathrm{R}}^2 = \left(\frac{\Gamma g_0}{eV}\right)(I - I_{\mathrm{th}})$$

the damping factor

$$\gamma_{\mathrm{e}} = \frac{1}{\tau_n} + K\left(\frac{\omega_{\mathrm{R}}}{2\pi}\right)^2$$

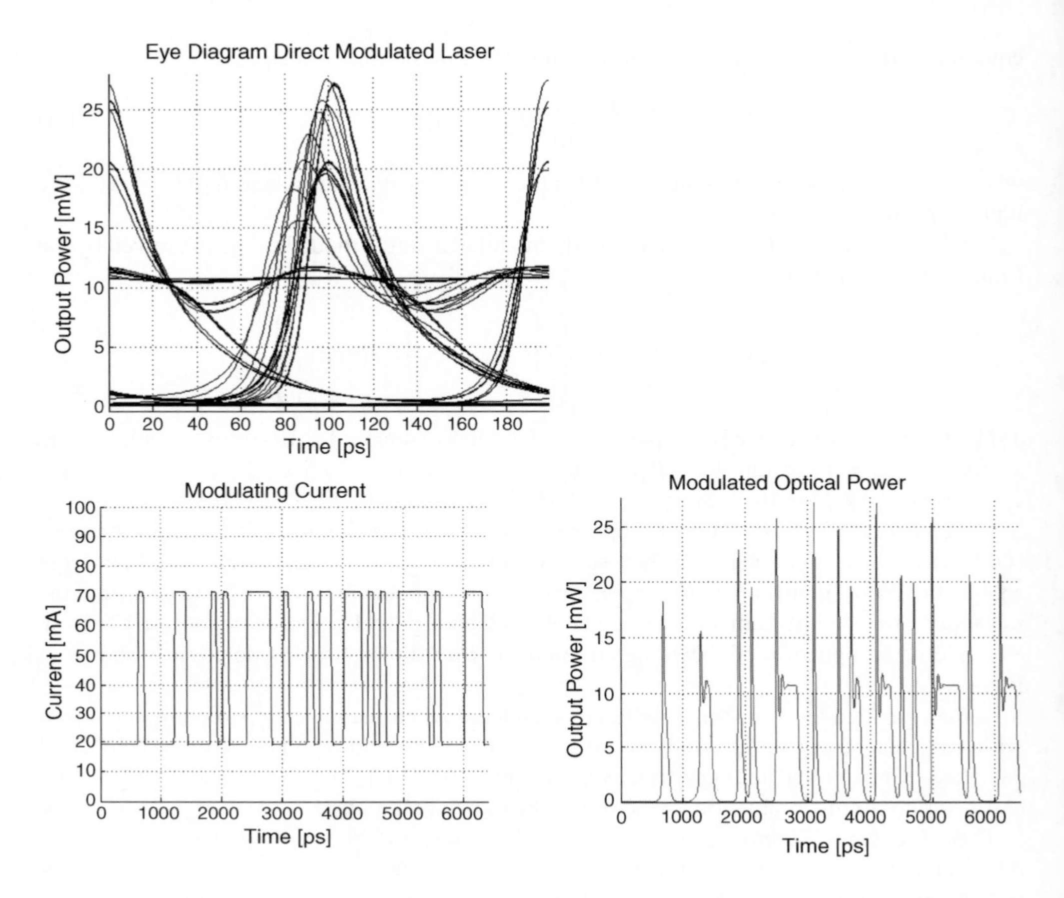

Figure 4.22 Laser direct modulation through rate equation simulation

and the so called "K-factor"

$$K = 4\pi^2\left(\tau_\mathrm{p} + \frac{\varepsilon}{g_0}\right)$$

All these terms (ω_R^2, γ_e, K) can be extracted from laser optoelectronic transfer function measurements (S_{21} optical to electrical measurements) at different bias points in the small-signal modulation regime (Figure 4.23).

The small-signal AM response also gives an idea of the bandwidth modulation capability of the laser structure itself.

Chirp and Laser Linewidth Enhancement Factor

The effect of the current injection is to introduce a phase (frequency) modulation of the output optical field in addition to the AM. The instantaneous output frequency variation of the laser under modulation can be expressed as

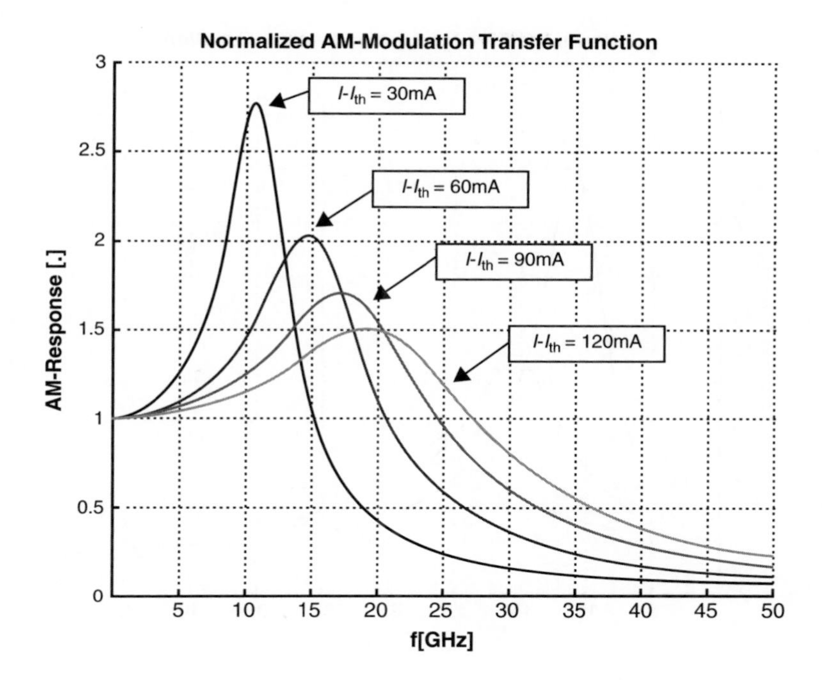

Figure 4.23 Theoretical normalized AM transfer function

$$\Delta f(t) \hat{=} \frac{1}{2\pi}\frac{\mathrm{d}\phi(t)}{\mathrm{d}t} = \frac{\alpha}{4\pi}\left[\underbrace{\frac{1}{P(t)}\frac{\mathrm{d}P(t)}{\mathrm{d}t}}_{1} + \underbrace{\kappa P(t)}_{2}\right] \tag{4.7}$$

where $P(t)$ and $\phi(t)$ are the instantaneous optical power and phase respectively.

The frequency deviation is the sum of two contributions: the first term is the *transient chirp term*, induced by variation of the optical output power as in all optical modulators, and the second term is the *adiabatic chirp term*, proportional to the optical output power by the parameter κ, which is associated with the laser nonlinear gain [12,13]. For IM-DD transmission over fiber-optic media the laser chirping can induce a penalty or be an advantage in distance reach, depending on the sign and magnitude of the chirp [14–16].

The linewidth enhancement factor α, responsible for the laser chirp, is also responsible of the laser *phase noise* and, therefore, of the *CW laser linewidth* [17].

Small-Signal Frequency-Modulation Response

The small-signal frequency-modulation (FM) response is always present in a laser modulated by injection carrier [18,19], and for the application in IM systems it is an undesired characteristic of the laser. The normalized carrier–FM transfer function is

$$H_{FM}(\omega) = \frac{\omega_{\mathrm{R}}^2 + \mathrm{j}\omega(1/\tau_{\mathrm{z}})}{\omega_{\mathrm{R}}^2 - \omega^2 + \mathrm{j}\gamma_e\omega} \tag{4.8}$$

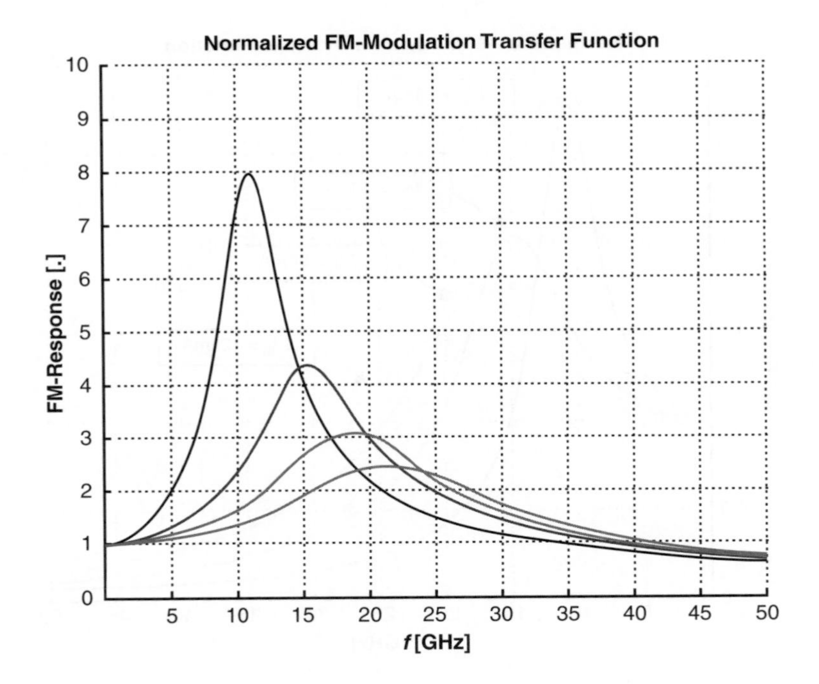

Figure 4.24 Theoretical normalized FM transfer function

where, in addition to the second-order transfer function terms, like in the AM small-signal response, there is an extra "zero" at the numerator of the transfer function characterized by the parameter

$$\tau_z = \frac{\omega_R^2}{\gamma_e} \tag{4.9}$$

The FM response measurement (Figure 4.24) is useful for intrinsic laser parameters estimation.

Phase Noise and Relative Intensity Noise

The internal spontaneous emission term is responsible for the noise generated by the laser. This term (which represents the Langevin noise) generates a random fluctuation in the number of carriers which results in both amplitude and phase random fluctuation of the laser output electric field component. The electric field of a laser biased around a bias point I_O is, therefore, expressed as

$$E(t) = [E_o + \Delta E_n(t)]e^{j([\phi_o + \phi_n(t)]} \tag{4.10}$$

where E_o and ϕ_o are the deterministic values corresponding to the bias current I_O and $E_n(t)$ and $\phi_n(t)$ are the fluctuating components. The effect of these fluctuations is the CW laser linewidth of the laser from the phase noise and the *intensity noise* superimposed to the output power (namely RIN).

The laser linewidth has a Lorentzian line-shape power spectral density:

$$S_{\phi_n}(\omega) = \left(\frac{\sigma^2_{\phi_n}}{\frac{\sigma^2_{\phi_n}}{4} + \omega^2} \right) \tag{4.11}$$

with $\sigma^2_{\phi_n}$ the phase noise variance that is related to laser physical design by

$$\sigma^2_{\phi_n} = \left[\frac{(1+\alpha^2)\sigma^2_{\phi_o}}{|E_o|^2} \right] \tag{4.12}$$

where $\sigma^2_{\phi_o}$ is the spontaneous diffusion coefficient. The full-width half-maximum (FWHM) of the power spectral density of the laser phase noise is the laser linewidth and it is equal to

$$\Delta f_L = \frac{\sigma^2_{\phi_n}}{2\pi} \tag{4.13}$$

The power spectral density of RIN and frequency noise $\dot{\phi}_n(t)$ are respectively

$$S_{RIN}(\omega) = 8\pi P^2_{co} \Delta f_{Lo} \left[\frac{\omega^2 + \gamma_e^2}{(\omega_R^2 - \omega^2)^2 + \gamma_e^2 \omega^2} \right] \tag{4.14}$$

and

$$S_{\dot{\phi}}(\omega) = \Delta f_{Lo} \left[1 + \frac{\omega_R^2}{(\omega_R^2 - \omega^2)^2 + \gamma_e^2 \omega^2} \right] \tag{4.15}$$

In the above expressions, the parameter Δf_{Lo} is the Shawlow–Townes linewidth of the laser, related to the effective laser linewidth by the relationship

$$\Delta f_L = (1+\alpha^2)\Delta f_{Lo} \tag{4.16}$$

The laser noise is extensively described in numerous fundamental works, for example [20–24], where all the relationships are derived considering the quantum nature of the noise process.

Laser noise is, in general, not taken into account in the performance of IM-DD systems, because of the dominance of other electrical and optical noise sources in the system. Optical systems where laser noise affects the performance are sub-carrier analog systems and all the systems with coherent receivers. Nevertheless, the noise analysis helps to estimate the modeling parameters of the laser subject to current modulation.

Turn-on Timing Jitter

The *turn-on delay* characteristic of semiconductor lasers is induced by the current injected into the device when the modulation current starts to switch from values below the laser threshold; therefore, the carrier density has to be built up. The time it takes for the laser regime carrier density to reach its threshold value is called the *turn-on time*. Depending on the type of LD and on the bias and operating conditions, this turn-on time may vary from several hundred picoseconds to several nanoseconds. This turn-on time effectively leads to a shortening of the optical pulse, yielding increased intersymbol interference, higher timing jitter, and ultimately a worse system performance.

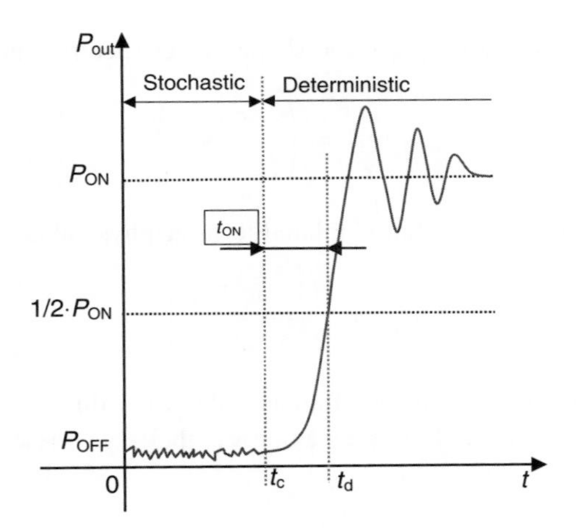

Figure 4.25 Turn-on laser characteristics

Following the model in Ref. [25], the turn-on delay t_{on} is schematically illustrated in Figure 4.25. The turn-on delay is a sum of two components: a *stochastic time-delay* t_{c}, which is related to the mean photon number in the cavity at the pulse rise instant, and a *deterministic time-delay* t_{d}, which is given by the following expression

$$t_{\mathrm{on}} = \frac{1}{2\pi f_{\mathrm{r}}}\sqrt{2\ln\left(\frac{S_{\mathrm{on}}}{\langle S_{\mathrm{c}}\rangle}\right)} \tag{4.17}$$

where:

f_{r} = relaxation oscillation frequency
S_{on} = photon number in the ON-state
$\langle S_{\mathrm{c}}\rangle$ = average photon number of the *absorbing barrier*.

The value of $\langle S_{\mathrm{c}}\rangle$ is independent of the specific turn-on event and it is a characteristic of each laser.

The *stochastic time-delay* t_{c} is a random variable with probability distribution function

$$p_{\mathrm{c}}(t_c) = \ln 2\frac{B\tau_{\mathrm{n}}}{T_{\mathrm{O}}}\left(1-\frac{t_{\mathrm{c}}}{T_{\mathrm{O}}}\right)^{\ln 2B\tau_{\mathrm{e}}-1} \tag{4.18}$$

where:

B = bit rate
τ_{n} = carrier lifetime
$T_{\mathrm{O}} = \tau_{\mathrm{n}} I_{\mathrm{th}}/I_{\mathrm{on}}$ = characteristic time-delay (corresponding to turn-on delay below threshold for low bit-rates: $B\tau_{\mathrm{n}} \ll 1$ for zero-bias and ultralow threshold laser $I_{\mathrm{on}} \gg I_{\mathrm{th}}$)
I_{th} = laser threshold current
I_{on} = ON-state laser current.

To avoid severe performance degradation it is important to determine the correct bias point of the laser in order to keep the laser OFF-state in a region that results in a compromise between two opposite transmission characteristics: the lowest OFF-state optical power to achieve the largest transmitter extinction ratio and the highest OFF-state current to have the lowest timing jitter.

Because the laser threshold current has a strong temperature dependence, it is necessary to introduce an external circuitry to keep the laser performance at its optimum for any temperature.

4.2.3.4 Electro-Absorption Modulator Laser Sources

Optical transmitters for intensity (amplitude) modulation in medium-haul systems are conventionally realized using a DFB laser in CW operation (as a source of the optical carrier) followed by an external modulator. A kind of external modulator is the EAM that is realized with a semiconductor junction used as a fast optical attenuator.

Because it is basically realized with the same materials as the laser source, it is possible to integrate the EAM in the same package with the laser, obtaining an extremely compact and cheap solution.

The Franz–Keldysh effect (FKE) and the QCSE are the physical effects available in semiconductor for electro-absorption modulation of the optical carrier. Both effects are driven by an electrical field: the FKE is obtained in a conventional bulk semiconductor and the QCSE is a phenomenon that appears in quantum well structures. These electro-absorption effects have the maximum efficiency near the bandgap of the semiconductor.

Franz–Keldysh Effect

The FKE is the change of optical transmission in semiconductor pn-junctions in the presence of an external electric field, due to the change of the wave-function overlap in the bandgap [2] (Figure 4.26). The wave-function of the electrons in the bandgap decays exponentially. As an electric field is applied to the semiconductor, the slope of the band edges is increased in the carrier depletion zone. Thus, the probability of a tunneling process due to the increased overlap of the exponential decays of the wave functions increases. The overlap of the electron and hole wave-functions allows for a photon energy less then the semiconductor bandgap to be absorbed with barrier tunneling [26].

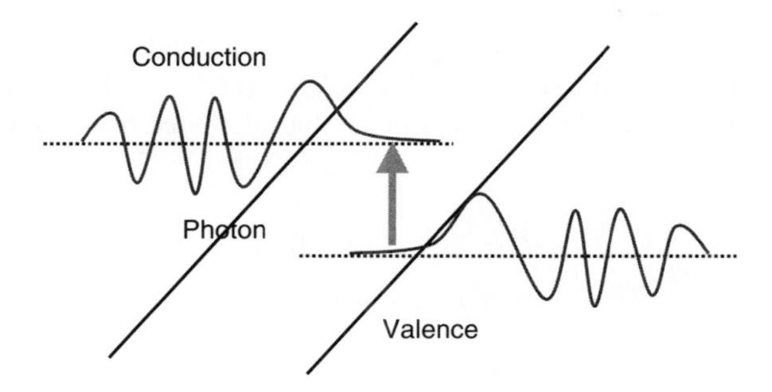

Figure 4.26 FKE in semiconductors

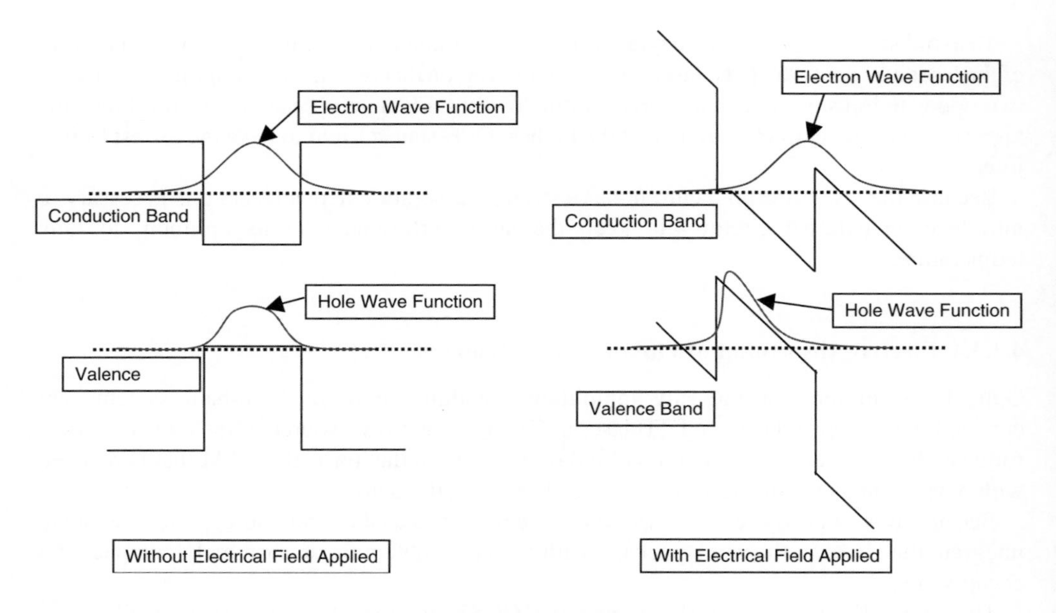

Figure 4.27 QCSE

Therefore, the absorption becomes stronger with higher field strengths. For a fixed geometry of the semiconductor the absorption depends directly on the applied voltage.

Quantum-Confined Stark Effect

The quantum well structure realized in a semiconductor provides an increase of electron and hole wave-functions. The application of an external electric field modifies the wave-function overlap and, therefore, the photon absorption. The physical mechanism of energy gap modification and the occurrence of excitonic states inside the bandgap due to the creation of electron holes in hydrogenic orbit of the atoms in the quantum well structure are at the base of the QCSE modulator [27,28] (Figure 4.27).

For an optical carrier, FKE- and QCSE-based modulators have the same behavior.

Complex Refractive Index

The modulation behavior of the EAM is due to the interpretation of the optical power absorption coefficient variation in terms of a complex refractive index [29] $n = n' - \mathrm{j}n''$. The electric field $\bar{E}_{\mathrm{out}}$ at the output of the modulator can be calculated from the input electric field $\bar{E}_{\mathrm{in}}$ by

$$\bar{E}_{\mathrm{out}} = \bar{E}_{\mathrm{in}} \exp(-\mathrm{j}\gamma L) \tag{4.19}$$

where j is the imaginary unit, L is the modulator length, and γ is the propagation constant.

Using the relationship:

$$\gamma = \frac{2\pi}{\lambda_0} n = k_0 n \tag{4.20}$$

Equation 4.19 can be rewritten as

$$\bar{E}_{\text{out}} = \bar{E}_{\text{in}} \exp(-\text{j}k_0 n' L - k_0 n'' L) \tag{4.21}$$

The two terms in the exponent describe the absorption losses and the phase shift caused by the modulator respectively:

$$|\bar{E}_{\text{out}}| = |\bar{E}_{\text{in}}| \exp(-k_0 n'' L) \tag{4.22}$$

$$\Phi = -k_0 n' L \tag{4.23}$$

Owing to the Kramers–Kronig relations, the real and imaginary parts of the refractive index are correlated. This means that changing the transmission in the modulator automatically leads to a phase shift of the optical output signal.

Describing the behavior of an electro-absorption modulator, it is common practice to use the *linewidth enhancement factor* α, that is often called *α-parameter* or chirp-parameter. The *α-parameter* is defined as

$$\alpha = \frac{\partial n'}{\partial n''} \tag{4.24}$$

Combining Equations 4.22 and 4.23, a relation between the output power that is proportional to $|\bar{E}_{\text{out}}|^2$ and the phase Φ can be obtained:

$$\frac{\text{d}\Phi}{\text{d}t} = \frac{\alpha}{2} \frac{1}{P_{\text{out}}} \frac{\text{d}P_{\text{out}}}{\text{d}t} \tag{4.25}$$

Integration of equation Equation 4.25 yields the instantaneous phase of the output signal:

$$\Phi(t) = \frac{\alpha}{2} \ln[P_{\text{out}}(t)] \tag{4.26}$$

Note that Equation 4.26 assumes small-signal conditions for the chirp parameter α [30].

Optical Output Power

The optical output power is determined describing the transmission of the modulator as a function of the modulation voltage $v_{\text{mod}}(t)$, [31]. The transmission characteristic can be approximated from the following power law:

$$P_{\text{out}}(t) = P_{\text{out},0} \exp\left\{ -\left[\frac{v_{\text{mod}}(t)}{V_{\text{o}}} \right]^{a} \right\} \tag{4.27}$$

where $P_{\text{out},0}$ is the device optical output power in CW at zero bias, V_{o} is a *characteristic voltage* at which the output power is decreased at $P_{\text{out,o}}/e$, and a is a *power coefficient* depending on the specific EAM construction (Figure 4.28).

The transmission of the modulator is controlled by the modulation voltage $v_{\text{mod}}(t) = v_{\text{mod,AC}}(t) + v_{\text{bias}}$, where v_{bias} is the voltage necessary to bias the modulator at its optimum modulation point and $v_{\text{mod,AC}}(t)$ is the modulation signal.

The modulating voltage v_{mod} applied to the EAM is limited to the values V_{min} and V_{max}, in order to keep the device in its safe working area to avoid transmission degradation effects or electrical device damage.

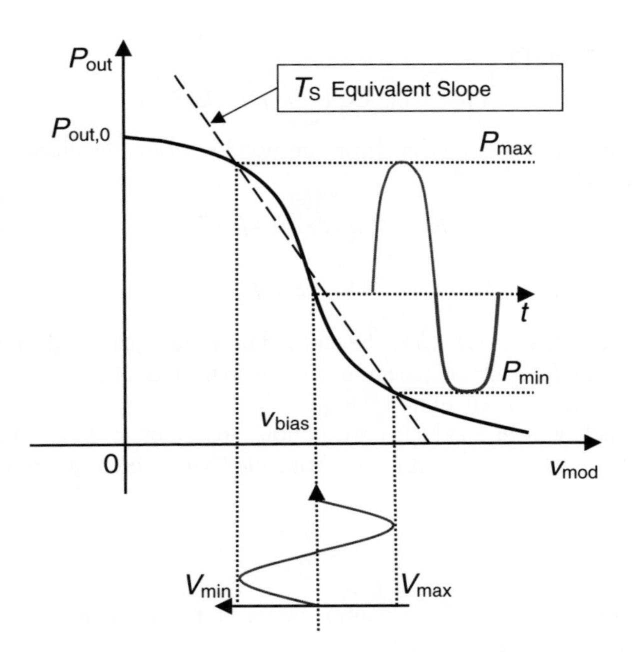

Figure 4.28 Electro-absorption modulation characteristics

Sometimes a simplified characteristic considers only the *attenuation slope* T_S between the two output powers P_{min} and P_{max} as a description of the modulation characteristics.

Static Frequency Deviation

For the cases where modulator and laser are monolithically integrated, a change in the refractive index in the modulator region results in a shift of the resonant frequency of the laser structure. Thus, a frequency deviation can occur that is often called a static chirp. The static frequency deviation can be expressed by

$$\Delta f_{out}(t) = \Delta f_{in}(t) + v_{mod}(t)\eta_{stat} \tag{4.28}$$

Thus, the static frequency deviation $v_{mod}(t)\eta_{stat}$ of the EAM adds linearly to the static frequency deviation of the signal at the input of the EAM module [32].

Dynamic Chirp

According to Equations 4.26 and 4.127, a change of the optical output power $P_{out}(t)$ leads to a change of the instantaneous phase $\Phi(t)$ and, therefore, to a frequency deviation. Typically, the α-parameter of the EAM depends on the modulation voltage v_{mod} applied. The real α-parameter should be estimated measuring it from the output instantaneous phase [30].

Typical values for EAM of parameters are as follows:

- $P_{out,o} = 2\,\text{dBm}$ (EAM transmission at zero voltage)
- $V_o = -1.3\,\text{V}$ (characteristic voltage at $P_{out,o}$/e output power)
- $a = 2$ (for modulator based on FKEs)
- $V_{min} = -2.5\,\text{V}$ (minimum allowed EAM input voltage)

- $V_{max} = 0.5$ V (mMaximum allowed EAM input voltage)
- $V_{bias} = -1.5$ V (bias voltage)
- $h_S = 0$ MHz/V (static frequency deviation: "0" for an EAM separate from laser source)
- $T_S = 5$ dB/V (transmission slope)
- $\alpha = 0.3$ (dynamic chirp parameter).

4.2.3.5 Mach–Zehnder Modulator Laser Sources

Intensity (amplitude) modulation used in optical transmitter for long-haul systems is realized using a DFB laser in CW operation as an optical carrier source followed by an MZM for data modulation. This MZM is referred to as an external optical modulator and is realized using an optical waveguide in a material with relevant electro-optic effect; for example, titanium diffused waveguides on lithium niobate ($Ti:LiNbO_3$).

A typical transmitter block diagram is shown in Figure 4.29.

Pockel's Effect

The linear electro-optic effect or Pockel's effect [33,34] is the change in the refractive index that occurs when an external electric field is applied to a crystal (e.g., $LiNbO_3$ substrate). Hereby, the refractive index variation $\Delta n(t)$ of the crystal is linearly proportional to the applied electric field $e(t)$. The magnitude of the effect is critically dependent on the orientation between the electric field and the electro-optic crystal axis. Further, an optical wave that propagates in the crystal along an axis orthogonal to the electro-optic crystal axis is subject to a change in the optical path length, because of the refractive index variation induced by the external electrical field. This variation introduces an optical phase change:

$$\Delta\Phi(t) = \int_0^L \Delta\beta(t)\,\mathrm{d}l = \frac{2\pi}{\lambda}\Delta n(t)L \tag{4.29}$$

where L is the device length and $\Delta n(t)$ is the refractive index change induced by the external electric field variation in the optical waveguide. For $LiNbO_3$, the electro-optic effect

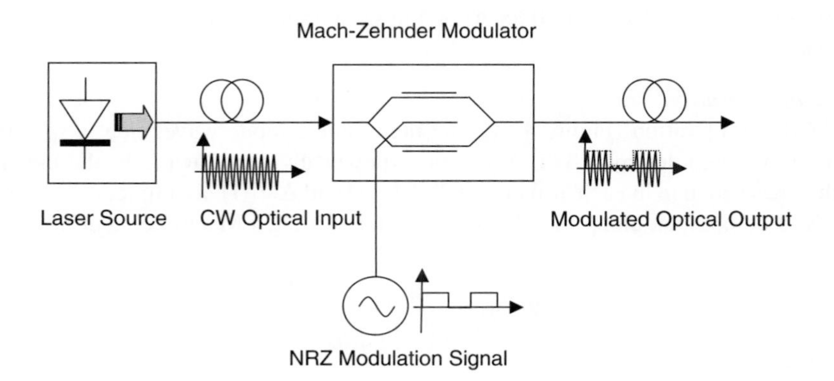

Figure 4.29 NRZ transmitter block diagram

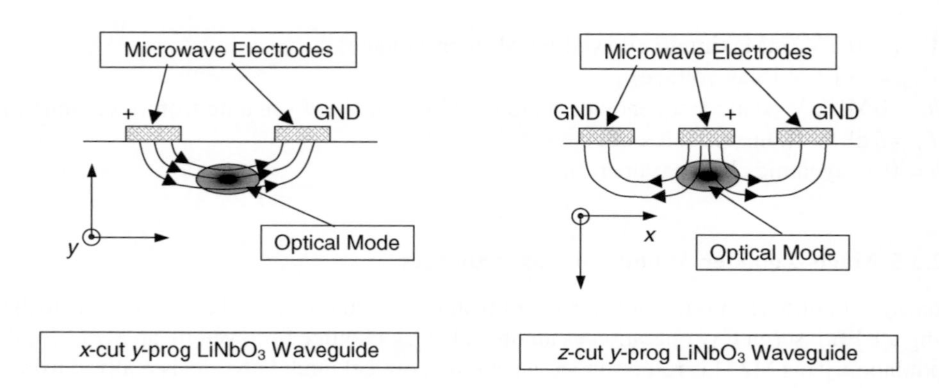

Figure 4.30 Electric field in optical waveguides of an MZM

is obtained for the optical waveguide aligned with the y-axis and the external electric field applied along the z-axis (or x-axis) as in Figure 4.30, and the refractive index variation is

$$\Delta n(t) = \Gamma\left(\frac{n_o^3}{2}\right) r_{33} e(t) \tag{4.30}$$

where Γ is the overlap integral between the electrical and the optical mode fields the transverse (x–y plane) to the propagation axis (z), defined as

$$\Gamma = d \iint_A \tilde{E}_o(x, y) \left|\tilde{E}_{MW}\right|^2 \mathrm{d}x\,\mathrm{d}y \tag{4.31}$$

where $\tilde{E}_o(x, y)$ is the normalized electrical field component of the optical wave ($\tilde{E}_o(x, y) = 1/d$ for a field uniform in the waveguide), $\tilde{E}_{MW}(x, y)$ is the normalized electrical field component of the microwave signal in the optical waveguide section; r_{33} is the coefficient of the electro-optical tensor, n_o is the ordinary refractive index at (zero electric field) of the $LiNbO_3$, and $e(t)$ is the external electric field applied along the optical waveguide.

Since the optical wavelength λ is much smaller than the electrode length L, slight changes of the refractive index result in a significant phase modulation. The MZM use the phase modulation due to the Pockel's effect to obtain an amplitude modulation of the incoming optical carrier.

Principle of Operation

In an MZM configuration (Figure 4.31), the input single-mode waveguide is split into two single-mode waveguides by a 3 dB Y-junction (power divider). Owing to the electro-optic effect, the phase change in each arm (1 and 2) $\Delta\Phi_1(t)$ and $\Delta\Phi_2(t)$, as a function of the applied electric field component along the electro-optic axis for each Mach–Zehnder arm $e_1(t)$ and $e_2(t)$ is

$$\Delta\Phi_1(t) = \frac{\pi}{\frac{\lambda}{2}\left(\frac{2}{n_o^3}\right)\frac{1}{\Gamma_1 r_{33} L_1}} e_1(t) \tag{4.32}$$

$$\Delta\Phi_2(t) = \frac{\pi}{\frac{\lambda}{2}\left(\frac{2}{n_o^3}\right)\frac{1}{\Gamma_2 r_{33} L_2}} e_2(t) \tag{4.33}$$

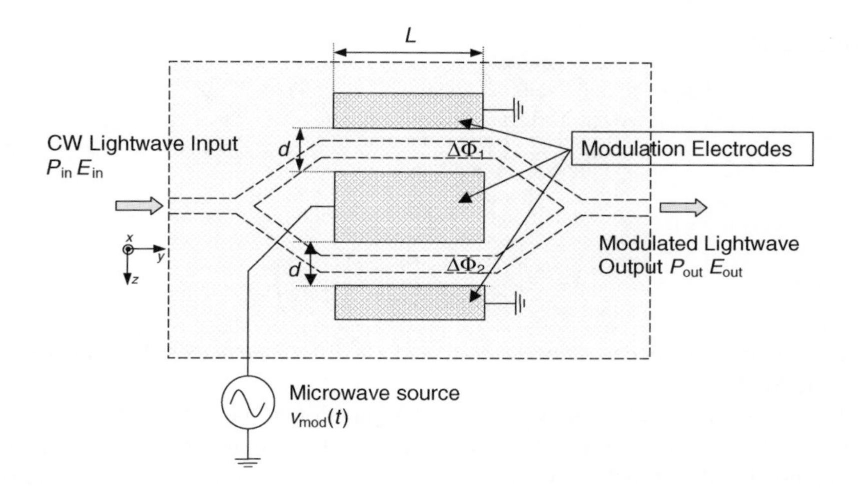

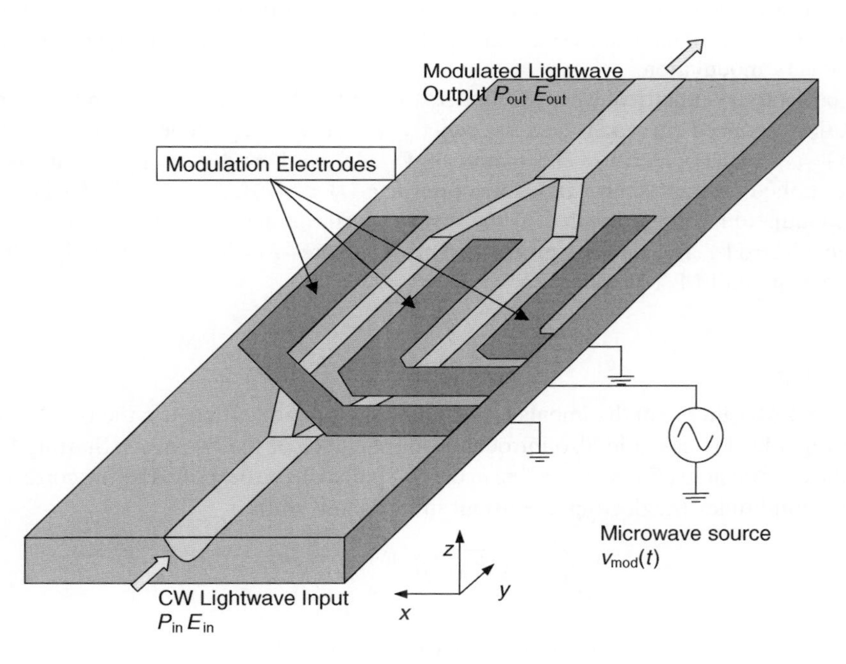

Figure 4.31 General configuration of an MZM

Considering an MZM with a push–pull-mode electrode configuration, the relative phase difference between the two beams

$$\Delta\Phi(t) \hat{=} \Delta\Phi_1(t) - \Delta\Phi_2(t) \tag{4.34}$$

is the variable that controls the intensity modulation of the MZM. The relationship between the relative phase difference and the applied modulation voltage $v_{mod}(t)$ results in the following equation:

$$\Delta\Phi(t) = \frac{\pi}{\left(\frac{\lambda}{n_o^3}\right)\frac{1}{r_{33}}\left(\frac{1}{\frac{\Gamma_1 L_1}{d_1}+\frac{\Gamma_2 L_2}{d_2}}\right)} v_{\mathrm{mod}}(t) \hat{=} \frac{\pi}{V_\pi} v_{\mathrm{mod}}(t) \tag{4.35}$$

where

$$V_\pi \hat{=} \left(\frac{\lambda}{n_o^3}\right)\frac{1}{r_{33}}\left(\frac{1}{\frac{\Gamma_1 L_1}{d_1}+\frac{\Gamma_2 L_2}{d_2}}\right) \tag{4.36}$$

is the switch voltage needed to induce a relative phase difference of exactly π between the two arms of the MZM.

The two waves then recombine via a second Y-junction at the output. If the waves are in phase ($\Delta\Phi = 0$), then they interfere constructively and combine into a fundamental mode, which is guided in the output waveguide. If the waves are out of phase ($\Delta\Phi = \pi$), then they interfere destructively and are transformed to a higher order mode and, therefore, lost into the substrate. Thus, the MZM can use interference between two waves to convert phase modulation into intensity modulation.

In a perfectly symmetrical Mach–Zehnder structure and electrode configuration, the phase modulation induced in each arm is exactly equal in magnitude with opposite sign ($\Delta\Phi_1(t) = -\Delta\Phi_2(t)$), because the match of $\Gamma_1 = \Gamma_2 = \Gamma$ and $L_1 = L_2 = L$ happens and also the applied external electric fields are opposite $e_1(t) = -e_2(t) = v_{\mathrm{mod}}(t)/d$, where $v_{\mathrm{mod}}(t)$ is the modulation voltage applied to the centre electrode and $d_1 = d_2 = d$ is the distance between electrodes (signal and ground). The V_π expression for the perfectly efficiency-matched push–pull MZM reduces to

$$V_\pi = \left(\frac{\lambda}{n_o^3}\right)\frac{1}{r_{33}}\left(\frac{d}{2\Gamma L}\right) \tag{4.37}$$

In a real MZM, other than the imbalance of phase modulation efficiency, the two Y-junctions and waveguides losses could also introduce an imbalance of the two arms. Starting from the input, the electric field ($\bar{E}_{\mathrm{in}} = |E_{\mathrm{in}}|e^{j\Phi_{\mathrm{in}}}$) in the two paths can be described by the corresponding complex amplitudes (neglecting a constant initial phase shift):

$$\bar{E}_1 = \sqrt{k_{\mathrm{in}}}\sqrt{a_1}e^{j\Delta\Phi_1}|E_{\mathrm{in}}|e^{j\Phi_{\mathrm{in}}} \tag{4.38}$$

$$\bar{E}_2 = \sqrt{1-k_{\mathrm{in}}}\sqrt{a_2}\, e^{j\Delta\Phi_2}|E_{\mathrm{in}}|e^{j\Phi_{in}} \tag{4.39}$$

The output electric field at the Y-junction output $\bar{E}_{\mathrm{out}}$ is obtained by combining the fields $\bar{E}_1$ and $\bar{E}_2$:

$$\begin{aligned}\bar{E}_{\mathrm{out}} &= \sqrt{k_{\mathrm{out}}}\bar{E}_1 + \sqrt{1-k_{\mathrm{out}}}\bar{E}_2 \\ &= (A_1 e^{j\Delta\Phi_1} + A_2 e^{j\Delta\Phi_2})|E_{\mathrm{in}}|e^{j\Phi_{\mathrm{in}}}\end{aligned} \tag{4.40}$$

The loss parameters k_{in} and k_{out} ($0 \le k_{\mathrm{in}} \le 1$ and $0 \le k_{\mathrm{out}} \le 1$) account for the nonideal 3 dB power-splitting ratio of the input and output Y-junction respectively. The loss parameters a_1 and a_2 ($0 \le a_1 \le 1$ and $0 \le a_2 \le 1$) refer to the optical propagation losses of the waveguides of the two MZM arms. As previously noted, the nonequal phase modulation of the two arms is taken into account through $\Delta\Phi_1(t)$ and $\Delta\Phi_2(t)$.

The cumulative effect of the power losses k_{in}, k_{out} and a_1, a_2 is condensed in the parameters A_1 and A_2, which represent the *total optical imbalance* on the electric field component in the two arms of the MZM:

$$A_1 = \sqrt{k_{\text{in}}}\sqrt{a_1}\sqrt{k_{\text{out}}} \tag{4.41}$$

$$A_2 = \sqrt{1-k_{\text{in}}}\sqrt{a_2}\sqrt{1-k_{\text{out}}} \tag{4.42}$$

Usually, the following parameters are used to summarize the MZM behavior:

- $\gamma \hat{=} A_2/A_1$, which represents the MZM optical unbalance
- $\Phi_m \hat{=} (\Delta\Phi_1 + \Delta\Phi_2)/2$, which represents the MZM residual phase
- $\Delta\Phi \hat{=} \Delta\Phi_1 - \Delta\Phi_2$, which represents the MZM modulation phase.

Then the output electric field is

$$\begin{aligned}\bar{E}_{\text{out}} &= A_1(\mathrm{e}^{\mathrm{j}\Delta\Phi_1} + \gamma\mathrm{e}^{\mathrm{j}\Delta\Phi_2})|E_{\text{in}}|\mathrm{e}^{\mathrm{j}\Phi_{\text{in}}} \\ &= A_1[\mathrm{e}^{\mathrm{j}(\Delta\Phi/2)} + \gamma\mathrm{e}^{-\mathrm{j}(\Delta\Phi/2)}]\mathrm{e}^{\mathrm{j}\Phi_{\text{m}}}|E_{\text{in}}|\mathrm{e}^{j\Phi_{\text{in}}}\end{aligned} \tag{4.43}$$

and the output electric field phase is

$$\Phi_{\text{out}} = \Phi_{\text{in}} + \Phi_{\text{m}} + \arctan\left[\left(\frac{1-\gamma}{1+\gamma}\right)\tan\left(\frac{\Delta\Phi}{2}\right)\right] \tag{4.44}$$

For the case $k_{\text{in}} = k_{\text{out}} = \frac{1}{2}$, $a_1 = a_2 = a$, which represents an *optically balanced loss* MZM (ideal 3 dB Y-junctions and symmetric optical losses between arms), the MZM exhibits the following loss parameters: $A_1 = A_2 = \sqrt{a}/2$ and $\gamma \hat{=} 1$; therefore, the output electric field magnitude is

$$\begin{aligned}\bar{E}_{\text{out}} &= \frac{\sqrt{a}}{2}(\mathrm{e}^{\mathrm{j}\Delta\Phi_1} + \mathrm{e}^{\mathrm{j}\Delta\Phi_2})|E_{\text{in}}|\mathrm{e}^{\mathrm{j}\Phi_{\text{in}}} \\ &= \frac{\sqrt{a}}{2}\left[\mathrm{e}^{\mathrm{j}[\Phi_{\text{m}} + (\Delta\Phi/2)]} + \mathrm{e}^{\mathrm{j}[\Phi_{\text{m}} - (\Delta\Phi/2)]}\right]|E_{\text{in}}|\mathrm{e}^{\mathrm{j}\Phi_{\text{in}}} \\ &= \frac{\sqrt{a}}{\sqrt{2}}\left[\sqrt{1+\cos(\Delta\Phi)}\right]|E_{\text{in}}|\mathrm{e}^{\mathrm{j}\Phi_{\text{out}}}\end{aligned} \tag{4.45}$$

where the associated phase is

$$\Phi_{\text{out}} = \Phi_{\text{in}} + \Phi_{\text{m}} \tag{4.46}$$

Output Power Characteristic

The optical output power is determined as follows (without considering the field impedance mismatch of the optical waveguides):

$$\begin{aligned}P_{\text{out}} &= |\bar{E}_{\text{out}}|^2 = A_1^2[1 + \gamma^2 + 2\gamma\cos(\Delta\Phi)]P_{\text{in}} \\ &= A_1^2(1+\gamma^2)\left[1 + \frac{2\gamma}{1+\gamma^2}\cos(\Delta\Phi)\right]P_{\text{in}}\end{aligned} \tag{4.47}$$

As a result, the intensity modulation transfer function of the MZM is obtained [35,36]:

$$\frac{P_{\text{out}}}{P_{\text{in}}} = A_1^2\left[1+\gamma^2+2\gamma\cos(\Delta\Phi)\right] \tag{4.48}$$

For the ideal MZM (optically balanced with no losses), $A_1 = A_2 = \sqrt{a}/2$, $\gamma \hat{=} 1$, and $a = 1$, Equation 4.48 leads to the expression

$$\frac{P_{\text{out}}}{P_{\text{in}}} = \frac{1}{2}[1+\cos(\Delta\Phi)] = \cos^2\left(\frac{\Delta\Phi}{2}\right) \tag{4.49}$$

which is the well-known sinusoidal characteristic of the MZM intensity modulation transfer function.

Extinction Ratio and Insertion Loss

Practically, to characterise the performance of a MZM, two (static) parameters are used: the "*DC extinction ratio*" (on-of-ratio) and the "*DC insertion loss*". The advantages of introducing these parameters is that they are directly measurable. The DC extinction ratio ε refers to the full swing modulation and is defined as

$$\varepsilon_{[\text{dB}]} \hat{=} 10\log(\varepsilon) = 10\log\left(\frac{P_{\text{out}_{\max}}}{P_{\text{out}_{\min}}}\right) \tag{4.50}$$

and the DC insertion loss a_{ins} is given by

$$a_{\text{ins}} \hat{=} 10\log(a_{\text{ins}}) = 10\log\left(\frac{P_{\text{out}_{\max}}}{P_{\text{in}}}\right) \tag{4.51}$$

From the above definitions it is possible to link this macroscopic parameter to the internal MZM parameters with the following relationships:

$$\varepsilon = \left(\frac{\gamma+1}{\gamma-1}\right)^2 \tag{4.52}$$

$$a_{\text{ins}} = A_1(1+\gamma)^2 \tag{4.53}$$

Inverting the above relationship, it is possible to substitute the macroscopic parameter ε (DC extinction ratio) and a_{ins} (DC insertion loss) to the internal MZM parameter as follows:

$$\gamma = \frac{\sqrt{\varepsilon}-1}{\sqrt{\varepsilon}+1} \tag{4.54}$$

$$A_1 = \frac{a_{\text{ins}}}{4\varepsilon}(\sqrt{\varepsilon}+1)^2 \tag{4.55}$$

For an optically balanced MZM ($\gamma = 1$ and $A_1 = A_2 = \sqrt{a}/2$) the device parameter values are $\varepsilon = \infty$ and $a_{\text{ins}} = a$, which corresponds to an infinite extinction ratio and no insertion loss if $a = 1$.

Intensity Modulation Transfer Function

The intensity modulation transfer function for the nonideal MZM uses Equations 4.44 and 4.48. The transfer function characteristic to respect the macroscopic parameters, derived using the definitions above, can be written as

$$\frac{P_{\text{out}}}{P_{\text{in}}} = \frac{a_{\text{ins}}(\varepsilon+1)}{2\varepsilon}\left\{1 + \frac{\varepsilon-1}{\varepsilon+1}\cos\left[\frac{\pi}{V_\pi} v_{\text{mod}}(t)\right]\right\} \tag{4.56}$$

Figure 4.32 illustrates, as an example, the power ratio versus the modulation voltage that includes the effects of nonideal parameters.

For an optically balanced MZM, the intensity modulation characteristic simplifies to Equation 4.49.

Output Phase

The instantaneous output phase Φ_{out} associated with the electric field at the nonideal MZM output results from Equation 4.44:

$$\Phi_{\text{out}}(t) = \Phi_{\text{in}}(t) + \Theta(t) \tag{4.57}$$

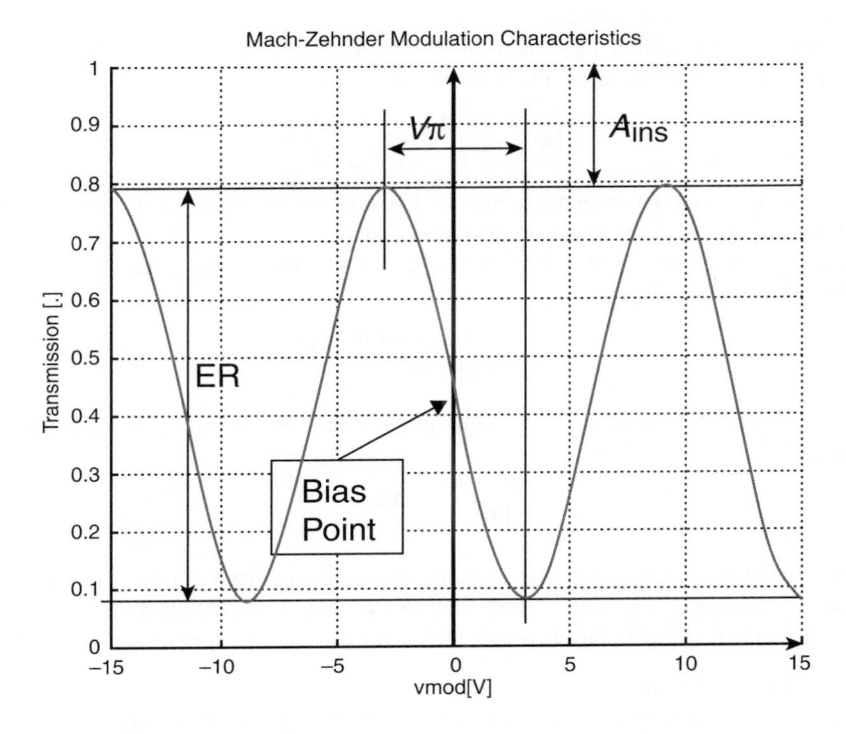

Figure 4.32 MZ transmission characteristic with ER = 10 dB and $A_{\text{ins}} = 1$ dB

where the phase introduced by the MZM is

$$\Theta(t) \hat{=} \Phi_{\rm m}(t) + \arctan\left\{\frac{1}{\sqrt{\varepsilon}}\tan\left[\frac{\Delta\Phi(t)}{2}\right]\right\} \tag{4.58}$$

and it is responsible for the MZM frequency chirp.

For an optically loss-balanced MZM ($\varepsilon = \infty$), the phase contribution simplifies to

$$\Theta(t) = \Phi_{\rm m}(t) \tag{4.59}$$

Then the modulator contribution to the output phase is due to the term

$$\Theta(t) = \frac{\Delta\Phi_1(t) + \Delta\Phi_2(t)}{2} \tag{4.60}$$

The ideal push–pull mode ($\Delta\Phi_1(t) = -\Delta\Phi_2(t)$ for any time instant t) for an optically loss-balanced MZM has modulator residual phase $\Theta(t) = 0$; therefore, there is no contribution from the device to the output electric field phase.

This condition, which verifies when MZM is both *optically* (loss) and *electrically* (phase modulation) balanced, is the desired chirp-free case.

Frequency Chirp

The frequency associated with the phase of the output electric field is the time derivative of the phase modulation induced in the MZM, $\mathrm{d}\Theta(t)/\mathrm{d}t$, and appears as a residual frequency modulation $\Delta f(t)$ of the modulator.

The *chirp* can be calculated by means of the common formula given in Ref. [29] for the frequency chirp used in the laser direct modulation case:

$$2\pi\Delta f = \frac{\mathrm{d}\Theta}{\mathrm{d}t} = \frac{\alpha}{2}\frac{1}{P_{\rm out}}\frac{\mathrm{d}P_{\rm out}}{\mathrm{d}t} \tag{4.61}$$

where the well-known α-parameter indicates the relation between chirp and (normalized) amplitude modulation of the modulated output signal.

For the case of an optically balanced MZM, the so-called chirp parameter δ is defined according to Refs. [37,38] as

$$\delta \hat{=} \frac{\Delta\Phi_1(t) + \Delta\Phi_2(t)}{\Delta\Phi_1(t) - \Delta\Phi_2(t)} \tag{4.62}$$

Thus, using the above definition, the equation for the output phase in Equation 4.60 is simplified to

$$\Theta(t) = \delta\frac{\Delta\Phi(t)}{2} \tag{4.63}$$

Therefore, an MZM is chirp free if it is characterized by a chirp parameter $\delta = 0$.

In general, MZM is considered as a low-chirp device with respect to other optical modulation methods, because it is practically possible to control the chirp parameter with proper design.

Output Frequency Deviation

The total frequency deviation of the output electric field is, in the case of the chirp-free MZM, identical to the frequency deviation of the MZM input electric field:

$$\Delta f_{\text{out}}(t) = \Delta f_{\text{in}}(t) \tag{4.64}$$

So, frequency deviation $\Delta f_{\text{in}}(t)$ is preserved from input to output if the MZM frequency chirp is equal to zero. Otherwise, a residual frequency modulation $\Delta f_m(t)$ is added to the input one (if present):

$$\Delta f_{\text{out}}(t) = \Delta f_{\text{in}}(t) + \Delta f_{\text{m}}(t) \tag{4.65}$$

with

$$\Delta f_{\text{m}}(t) \hat{=} \frac{1}{2\pi} \frac{\mathrm{d}\Theta(t)}{\mathrm{d}t} \tag{4.66}$$

Frequency Response of the Transfer Function

The MZM radio-frequency (RF) electrode is a microwave transmission line that must be optimized for both the maximum efficiency of the modulator driver and the maximum electro-optic-induced effect [39–41].

Consider (as an example) the applied modulation microwave voltage

$$v_{\text{mod}}(t) \hat{=} V_{\text{o}}\text{Re}[\mathrm{e}^{\mathrm{j}(2\pi f t)}] \tag{4.67}$$

consisting (for simplicity) of a single sinusoidal tone at frequency f; a nonuniform modulation voltage will be displaced along the electrode, as happens along a transmission line terminated by its characteristic impedance:

$$v_{\text{MW}}(z, t) \hat{=} V_{\text{o}}\text{Re}[\mathrm{e}^{-\gamma_{\text{MW}} z + \mathrm{j}(2\pi f t)}] \tag{4.68}$$

(where MW denotes microwave) so that at the modulator electrode input $(z = 0)$ $v_{\text{mod}}(t) = v_{\text{MW}}(0, t)$.

The term $\gamma_{\text{MW}} = \alpha_{\text{MW}} + \mathrm{j}\beta_{\text{MW}}$ is the propagation coefficient of the microwave transmission line formed by the modulation electrode that considers the effect of the phase coefficient $\beta_{\text{MW}} = 2\pi f n_{\text{MW}}/c$, where n_{MW} is the effective refractive index and the loss coefficient $\alpha_{\text{MW}} = \alpha_{\text{MW0}}\sqrt{f}$, which are both dependent on the signal frequency. The lightwave carrier wave traveling in the Mach–Zehnder waveguide under the electrode (starting from $z = 0$ and ending to $z = L$) is modulated by a nonuniform modulation voltage:

$$v_{\text{MW-rel}}(z, t) = V_{\text{o}}\text{Re}[\mathrm{e}^{-\alpha_{\text{MW}} z - j2\pi f(d_{12} z - t)}] \tag{4.69}$$

The mismatch between the effective ordinary refractive index (at zero electric field) of the guided optical mode n_{o} and the effective refractive index of the electrical (microwave) mode n_{MW} generates a relative drift between the electrical and the optical waves that is considered with the walk-off parameter $d_{12} = (n_{MW} - n_o)/c$.

The optical modulation is determined by the cumulative induced phase shift introduced along the electrode interaction length n_{MW}. The walk-off between the electrical and the optical waves determines a frequency-dependent reduction of the integrated phase shift, which can be calculated by integrating the phase change induced by the relative modulation voltage over the electrode length:

$$\begin{aligned}\Delta\phi(t,f) &= \int_0^L \Delta\beta(t,f,z)\,\mathrm{d}z \\ &= \int_0^L \left\{\frac{2\pi}{\lambda}\Gamma\left(\frac{n_o^3}{2}\right)r_{33}\left[\frac{v_{\text{MW-rel}}(z,t)}{d}\right]\right\}\mathrm{d}z\end{aligned} \tag{4.70}$$

Introducing the following variables:

$$\xi \mathrel{\hat{=}} 2\pi f d_{12} \tag{4.71}$$

$$\Delta\phi_o \mathrel{\hat{=}} \left[\frac{2\pi}{\lambda}\left(\frac{n_o^3}{2}\right)r_{33}\frac{\Gamma V_o L}{d}\right] = \pi\left[\left(\frac{n_o^3}{\lambda}\right)r_{33}\frac{\Gamma L}{d}\right]V_o = \frac{\pi}{V_\pi}V_o \tag{4.72}$$

where

$$V_\pi \mathrel{\hat{=}} \left[\frac{1}{r_{33}}\left(\frac{\lambda}{n_o^3}\right)\frac{d}{\Gamma L}\right]$$

is the DC V_π, n_o and is the effective refractive index (at zero electric field) of the guided optical mode for the ordinary ray n_e, r_{33} is the electro-optic coefficient, λ is the optical wavelength (in vacuum), L is the RF electrode length, d is the gap between the RF electrode and the ground, and Γ is the overlap integral (Equation 4.31). Thus, phase modulation results:

$$\begin{aligned}\Delta\phi(t,f) &= \frac{\pi}{V_\pi}V_o\left(\frac{1}{L}\right)\left[\int_0^L \{\mathrm{Re}[\mathrm{e}^{-\alpha_{\mathrm{MW}}z-\mathrm{j}(\xi z-2\pi f t)}]\}\,\mathrm{d}z\right] \\ &= \frac{\pi}{V_\pi}\mathrm{Re}\left((V_o\mathrm{e}^{\mathrm{j}2\pi f t})\mathrm{e}^{-[(\alpha_{\mathrm{MW}}+\mathrm{j}\xi)L]/2}\frac{\sinh\{[(\alpha_{\mathrm{MW}}+\mathrm{j}\xi)L]/2\}}{[(\alpha_{\mathrm{MW}}+\mathrm{j}\xi)L]/2}\right)\end{aligned} \tag{4.73}$$

A simplified frequency model of the MZM could be extracted from the above considerations introducing in the *conventional* model a *transform function* $H(f)$ that consider the *frequency-dependent* characteristics of the modulation efficiency of the microwave electrode:

$$\Delta\phi(t,f) = \frac{\pi}{V_\pi}\mathrm{Re}\left[(V_o\mathrm{e}^{\mathrm{j}2\pi f t})H(f)\right] \tag{4.74}$$

This *transform function* $H(f)$ introduces a *frequency-dependent attenuation* (expressed by the modulus value $|H(f)|$) and a *frequency-dependent phase* (expressed by the argument value $\angle H(f)$). The amplitude of the microwave modulating signal V_o is therefore modified by the term (Figure 4.33)

$$|H(f)| = \mathrm{e}^{-\alpha_{\mathrm{MW}}L/2}\sqrt{\frac{\sinh^2(\alpha_{\mathrm{MW}}L/2)+\sin^2(\xi L)/2}{(\alpha_{MW}L/2)^2+(\xi L/2)^2}} \tag{4.75}$$

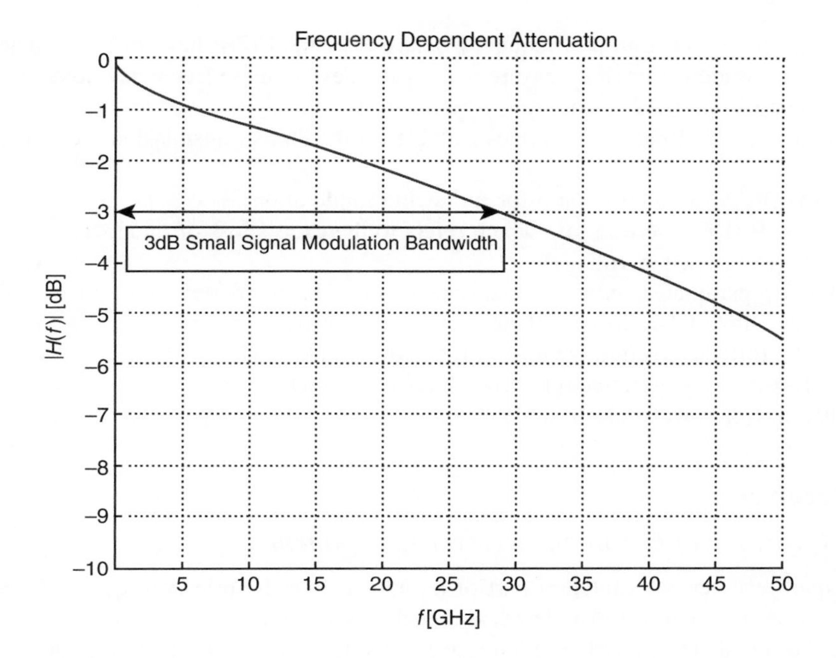

Figure 4.33 Frequency-dependent attenuation introduced by microwave electrodes with $\alpha_{\mathrm{MW,o}} = 0.2$ dB/(cm $\sqrt{}$GHz), $n_{\mathrm{MW}} = 2.14$, $n_o = 2.12$, $L = 20\,\mathrm{mm}$

This term is set an upper limit to MZM on the attainable frequency which in turn is restricting higher-speed data rate modulation. Nevertheless, microwave matching in a frequency range of the modulation signal due to impedance mismatch between microwave termination load and modulation driver is also important (Figure 4.34).

In the case of a perfect match between the modulation electrode transmission line and the termination load, $Z_{\mathrm{L}}(f) = Z_{\mathrm{MW}}(f)$, the modulation voltage applied is

$$v_{\mathrm{mod}}(t) = \mathrm{Re}\left\{\frac{Z_{\mathrm{L}}(f)}{Z_S(f) + Z_{\mathrm{L}}(f)}\left[V_S\,\mathrm{e}^{\mathrm{j}(2\pi f)t}\right]\right\} \tag{4.76}$$

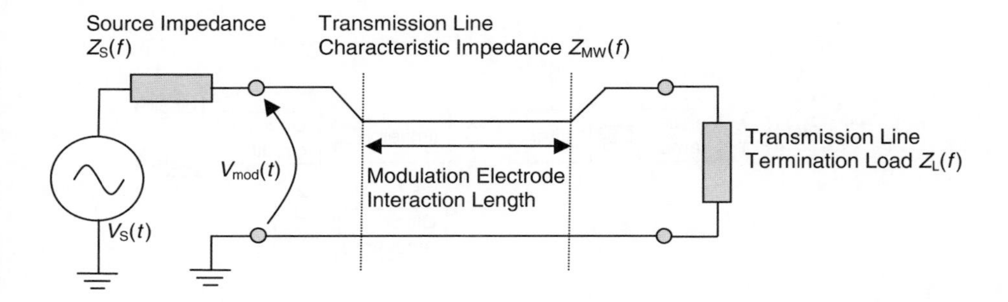

Figure 4.34 Microwave equivalent circuit

The parameters involved in the frequency behavior of the MZM have to be extracted from hybrid electro-optic parameter measurement of the device in the frequency domain (i.e., the measurement of microwave scattering parameters S_{11} and S_{21}).

Examples of MZM parameter values suitable for 40 Gbit/s transmission are as follows:

- ER = 15 dB (DC extinction ratio for full swing modulation)
- $A_{ins} = 1.5$ dB (DC insertion loss which refers to the excess loss of the device)
- $V_{\pi} = 3$ V (DC π-switch voltage)
- $\delta = 0$ (chirp parameter, only for ideal transmission characteristic: range -1 to $+1$)
- $\alpha_{MW,o} = 0.2\,\mathrm{dB}/\sqrt{\mathrm{GHz}}$ (loss coefficient of the microwave electrode)
- $n_{MW} = 2.2$ (effective refractive index of microwave electrode)
- $n_o = 2.15$ (ordinary effective refractive index of $LiNbO_3$)
- $L = 40$ mm (microwave electrode length).

4.3 Receiver

4.3.1 Overview of Common Receiver Components

The optical receiver is an integral part of an optical link. Its role is to detect the received optical power, translate it to an electrical signal, amplify the signal, retime, regenerate, and reshape the signal. Then it either remains as an electrical signal or it is used to drive the next optical transmitter of the concatenated fiber link. The main components of an optical receiver are the photodetector and a front-end amplifier, and we will look into those in this sectio. An optical receiving system typically comprises a clock and data recovery circuit. An OA can optionally be placed before the photodetector. Figure 4.35 illustrates an optical receiving system.

An optical receiver performs well when it can achieve the desired system performance with the minimum amount of received optical power at the required speed (bit rate). System performance criteria include Bit Error Rate (BER) and eye margin/extinction ratio for digital systems, signal-to-noise ratio (SNR) for analog systems, and dynamic range (ratio of largest to smallest signal levels for which a specific value of BER or SNR can be achieved). It is important to note that the speed of the optical receiving system is very important.

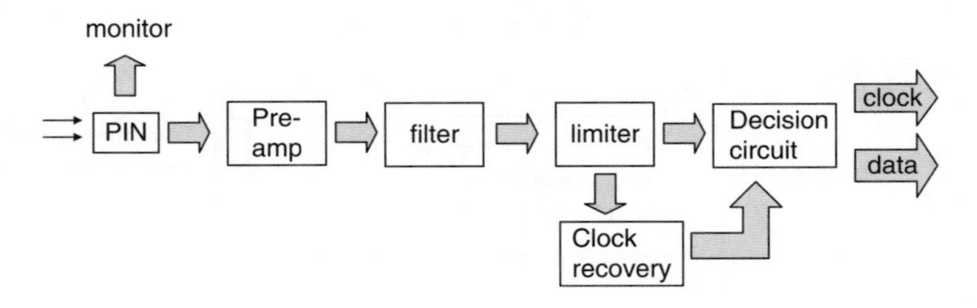

Figure 4.35 A typical optical receiving system; an optical pre-amp may be used before the PIN, for example, and different variants exist

4.3.1.1 Photodetectors

Photodetection is a procedure that is based on the quantum-mechanical interaction among electrons and photons in a semiconductor [55] and specifically on absorption. Electrons in the valence band absorb the photons which are incident on the semiconductor. These electrons get excited and return to the conduction band, leaving a vacancy or hole in the valence band. If an appropriate voltage is applied to the semiconductor, then the generated electron–hole pairs give rise to photocurrent. Ideally, the energy of the incident photon equals the order of the generated photocurrent. Mathematically, the energy of the incident photon is represented as follows:

$$E = hf_{\mathrm{c}} = \frac{hc}{\lambda} \tag{4.77}$$

where:

f_{c} = frequency of the photon
h = Planck's constant
c = velocity of light
λ = wavelength.

The value of λ that corresponds to the energy of the incident photon is at least equal to the cutoff wavelength; that is, the wavelength that corresponds to the bandgap energy of the semiconductor.[3]

The ratio of the energy of the optical signal absorbed to the photocurrent generated is called the efficiency of the photodetector η. A highly efficient photodetector is constructed in such a way that its efficiency is close to unity. Indeed, this efficiency is attained using a semiconductor slab of suitable thickness in the photodetector. This is because the power absorbed by the semiconductor is denoted as

$$P_{\mathrm{abs}} = l – \mathrm{e}^{-\alpha l} \tag{4.78}$$

where l is the thickness of the semiconductor slab, α is the absorption coefficient of the material and e is Exp.

Hence, the efficiency of the photodetector η is calculated as follows:

$$\eta = \frac{P_{\mathrm{abs}}}{P_{\mathrm{in}}} = l – \mathrm{e}^{-\alpha\lambda} \tag{4.79}$$

where P_{in} is the power of the incident optical signal.

Note that the value of the absorption coefficient depends on the wavelength of the incident light and, evidently, the semiconductor photodetector is transparent to wavelengths greater than the cutoff wavelength.

We define the responsivity of the photodetector as

$$R = \frac{I_{\mathrm{P}}}{P_{\mathrm{in}}} \tag{4.80}$$

where I_{P} is the average current generated by the photodetector and P_{in} is the power of the incident optical signal.

---

[3] Bandgap energy is the energy difference between the valence and conduction bands.

If a fraction n of the incident photons is absorbed and the receiver generates photocurrent, the responsivity of the photodetector becomes

$$R = \frac{en}{hf_c} \tag{4.81}$$

where e is the electronic charge and n is the the refractive index.

Substituting the value of hf_c from Equation 4.77 results in

$$R = \frac{en\lambda}{hc} = \frac{n\lambda}{1.24}[\mathrm{eV}/\mu\mathrm{m}] \tag{4.82}$$

Therefore, the responsivity is expressed with respect to the wavelength. If the photodetector is designed to achieve efficiency close to unity, then the responsivity of the photodetector is also close to unity.

A semiconductor slab has some disadvantages. The electrons in the conduction band start recombining with holes in the valence band before they enter the external circuit to create the photocurrent. This compromises the efficiency of the photodetector. Therefore, the electrons must be forced out of the semiconductor. This is ensured using a photodiode, which consists of a pn-junction with a reverse-bias voltage applied to it. Figure 4.36 shows the principle of operation of the photodiode.

The pn-junction creates an electric field. By applying a reverse-bias voltage, electrons near the depletion region move into the n-type region. This prevents recombination of electrons with holes in the p region. The holes near the depletion region drift to the p region and the photocurrent is generated.

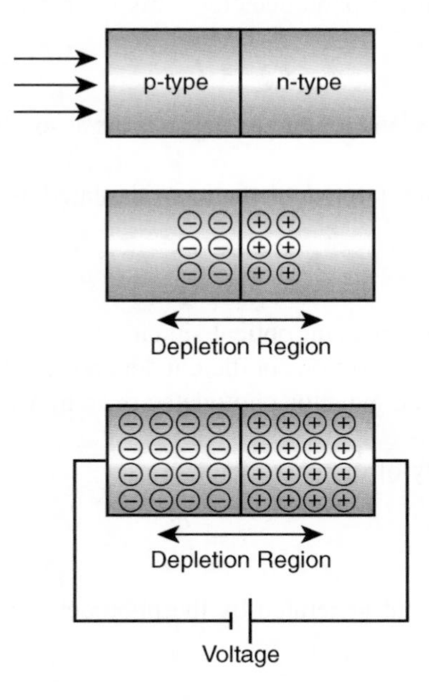

Figure 4.36 The operating principle of a photodiode

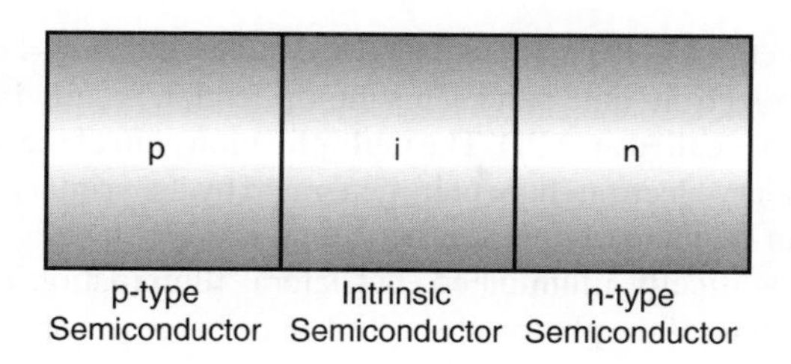

Figure 4.37 A pin photodiode

4.3.1.2 Pin Photodiode

A pin photodiode, shown in Figure 4.37, consists of an intrinsic semiconductor sandwiched between p-type and n-type semiconductors. The intrinsic semiconductor covers the depletion region and its width is greater than in p-type and n-type semiconductors. The majority of light absorption occurs in the depletion region. A semiconductor material can be placed in the region to provide the required wavelength. This wavelength is larger than the cutoff wavelength of the semiconductor and there is no light absorption in this region. A pin photodiode can also be constructed with two junctions of different semiconductor materials. This is called a heterostructure or double heterojunction. In p-i-n photodiodes, the photocurrent originates in the drift of electrons and holes absorbed in the intrinsic region. In fast photodiodes, the short transit times through a thin space-charge region (intrinsic i-layer) ensure high bandwidth. If the recombination times are much larger than the transit times, then a large part of the photogenerated carriers is collected in the contact regions. However, there is a trade-off between bandwidth and external quantum efficiency, because less incident photons are absorbed when the thickness of the intrinsic region is decreased. Decreasing the thickness of the i-region is then favorable to speed, but this increases the device capacitance C. As a result, the device area must be reduced if the charging time of the capacitance (RC with $R = 50\,\Omega$) becomes larger than the transit times.

The trade-off between bandwidth and responsivity disappears in optical waveguide structures, when light is injected laterally and propagates in a direction perpendicular to charge transport. The length and the thickness of the i-region can be optimized independently, so that all photons coupled to the device are absorbed and transit times are short. However, owing to the small dimensions of the optical waveguide, it is hard to obtain low coupling losses. Evanescent coupling can be used to improve the coupling efficiency in laterally illuminated photodiodes. Light is injected in an input waveguide and is progressively coupled to the absorption layer by energy transfer between different modes. Light injection, light absorption, and charge transport can then be optimized independently with careful design of the layer stacking and the lateral dimensions. The design of the input waveguide may have a strong impact on device packaging and cost.

4.3.1.3 Avalanche Photodiode

In photodetectors, a photon can produce only one electron. If an intense electric field is generated, then more electrons can be made to excite from the valence band to the conduction band. The resulting electron–hole pairs are called secondary electron–hole pairs. These pairs, in turn, generate more electron–hole pairs. This multiplication is called avalanche multiplication, and the photodiode is called an APD. The multiplication gain of the APD is the mean value of the number of secondary electron–hole pairs generated by the primary electron. The value of multiplication gain can be increased to a large extent.

Usually, APDs are vertically illuminated, but lateral illumination is interesting, since it allows independent designs of the photon absorption and charge transport functions, as in p-i-n photodiodes.

4.3.1.4 Other Types of Photodiode [56]

In metal–semiconductor–metal (MSM) photodiodes [57], the electric field is applied between interdigitated electrodes which form Schottky contacts on the semiconductor layers. The photo-induced carriers then drift in the lateral direction imposed by the field, which in turn are collected very rapidly if the interelectrode spacing is small. Decreasing the interelectrode spacing results in increased capacity; hence, it is pointless to reduce the interelectrode spacing beyond the limit in which the charging time of the device is larger than the transit time between the electrodes. Owing to their planar technology, MSM photodiodes can be easily integrated with field-effect transistors.

Uni-traveling-carrier (UTC) photodiodes are back-illuminated and light is absorbed in a rather thin ($\sim$0.25 μm) p-type contact region. The holes recombine at the p + contact layer, whereas the electrons move towards large-gap n-type layers by diffusion. The electrons then go through a p-n heterojunction, are injected into an intrinsic (i) drift region, and collected at an n-type contact layer. As a result, there are only electrons in the drift region. When saturation occurs, electrons accumulate in the drift region in UTC photodiodes [4], whereas in p-i-n photodiodes the holes accumulate. It turns out that the maximum linear photocurrent density is lower in p-i-n structures because the hole velocity is smaller, and UTC photodiodes can handle higher carrier and photon densities than p-i-n photodiodes.

In traveling-wave photodetectors (TWPDs), a waveguide photodetector is embedded in a transmission line. Very high bandwidth can be obtained when the optical and electrical group velocities are matched, and also when impedance matching on a resistive load is achieved [59]. In TWPDs, it is hard to obtain exact velocity matching. This can be achieved most effectively by periodically loading the transmission line with localized capacitances.

In the velocity-matched distributed photodetector, a cplanar strip transmission line is periodically loaded by the capacitances of discrete evanescently coupled photodiodes [60]. This allows almost independent optimization of the optical waveguide, the transmission line, and the photodetector. A small fraction of the optical power is absorbed in each photodiode, and the power handling capability is excellent if the waveguide losses are weak.

4.3.1.5 Front-End Amplifiers

There are two types of front-end amplifier: high-impedance and transimpedance. Figures 4.38 and 4.39 show the circuits for these two types of front-end amplifier.

There are two considerations for designing the amplifier. The thermal noise that gets added to the photocurrent is inversely proportional to the load resistance R_L. Therefore, R_L must be made

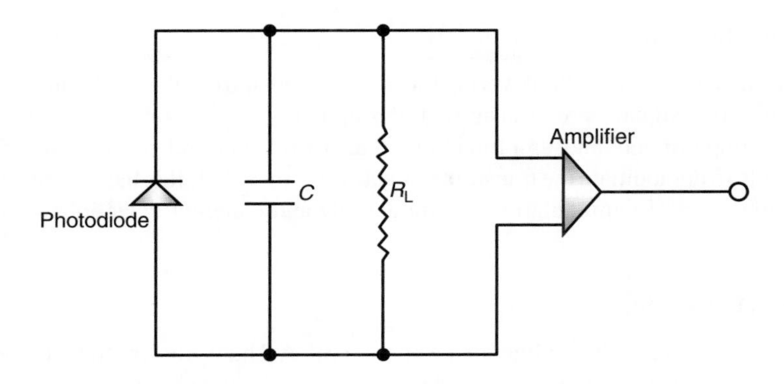

Figure 4.38 A circuit for a high-impedance front-end amplifier

small to reduce the thermal noise current. On the other hand, the bandwidth of the photodetector is also inversely proportional to the output load resistance R_P. Therefore, the front-end amplifier is selected based on a trade-off between photodiode bandwidth and the noise current.

In the case of a high-impedance front-end amplifier (Figure 4.38):

$$R_P = R_L \tag{4.83}$$

where R_L is the load resistance and R_P is the output load resistance.

In the case of a transimpedance front-end amplifier (Figure 4.39):

$$R_P = \frac{R_L}{A+1} \tag{4.84}$$

where A is the amplifier gain.

The photodiode bandwidth is increased by a factor of $A + 1$. The thermal noise is higher than in the case of a high-impedance front-end amplifier. Nevertheless, it is used in most optical networks because the increase in thermal noise current is quite low.

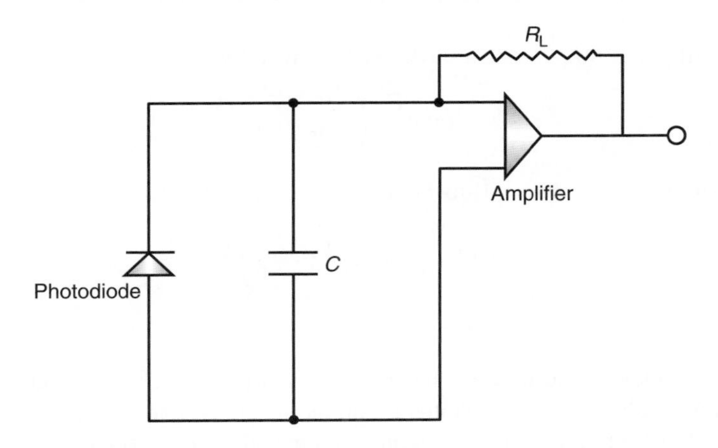

Figure 4.39 A circuit for a transimpedance front-end amplifier

4.3.1.6 Demodulation

Demodulation is the process of recovering the original signal from the modulated input signal. After modulation, signals are transmitted through an optical link and undergo several transmission impairments and amplifications that distort and add noise to the signal. The receiver is able to demodulate the transmitted data correctly only if this has an acceptable BER. There are two types of demodulation technique: DD and coherent detection.

4.3.1.7 Direct Detection

DD is based on the availability of light in the bit interval. A 0 bit is transmitted if there is no light and a 1 bit is transmitted if light is available. Let us consider detection in two cases: the ideal system and a practical system. In an ideal system, it is assumed that no noise gets added during transmission. We will now derive the value of the BER for ideal and practical DD receivers.

Ideal Receiver

The stream of photons arriving at the receiver is considered a Poisson process. Usually, no errors are related to 0-bit transmission, because it is assumed that no photons are received during 0-bit transmission. Photons are received only when the 1 bit is transmitted.

The rate at which photons arrive at the receiver is given by

$$\frac{P}{hf_c} \tag{4.85}$$

where P is the signal power, h is Planck's constant, and f_c is the carrier frequency.

If B is the bit rate, then the number of photons received during the 1 bit is given as P/hf_cB.

The probability that n photons are received at a bit interval of $1/B$ is given

$$\exp\left[-\left(\frac{P}{hf_cB}\right)\right]\frac{(P/hf_cB)^n}{n!} \tag{4.86}$$

where P is the signal power, h is Planck's constant, f_c is the carrier frequency, and B is the bit rate.

The probability that no photons are received is given by

$$\exp\left[-\left(\frac{P}{hf_cB}\right)\right] \tag{4.87}$$

If an equal number of 1 bits and 0 bits are considered, then the BER is given by

$$\text{BER} = \frac{1}{2}\exp\left[-\left(\frac{P}{hf_cB}\right)\right] \tag{4.88}$$

Practical Receiver

Receivers are not ideal because, when they receive the transmitted bits, various noise currents are created along with the resulting photocurrent: thermal noise and shot noise. In particular, if optical preamplifications is used,then noise due to spontaneous emission in OAs and so on must be considered.

Shot Noise Current

The photodetector at the receiving end generates electrons randomly even if the intensity of input light remains constant. This current is called the shot noise current. It is a component of the resulting photocurrent. The photocurrent is given by

$$I = I + i_s \tag{4.89}$$

where I is the constant current and i_s is the value of the current due to the arrival of photons.

If the electrical bandwidth of the receiver is B_e, then the variance of shot noise current for a pin receiver is given by

$$\sigma^2_{shot} = 2eIB_e \tag{4.90}$$

where $I = RP$, R is the responsivity of the photodetector , e is electron charge and B_e is electric filter bandwidth.

Thermal Noise Current

This current is created due to random movement of electrons at a finite temperature. It can be considered a Gaussian random process. If the load resistance of the photodetector is R_L, then the photocurrent is given by

$$I = I + i_s + i_t \tag{4.91}$$

where i_t is the thermal current.

The thermal current i_t has a variance given by

$$\sigma^2_{thermal} = \frac{4k_B T}{R(B_e)} \tag{4.92}$$

where R is resistance, T is temperature, k_B is Boltzmann's constant, and B_e is the receiver's electrical bandwidth.

The receiver's electrical bandwidth must be at least half the value of the optical bandwidth to avoid signal distortion. At the receiving end, transistors, at the front-end amplifier, create the thermal noise current. The noise is defined by the noise figure, which is the factor by which the thermal noise present at the amplifier input is enhanced at the amplifier output. The variance of thermal noise current using a front-end amplifier is given by

$$\sigma^2_{thermal} = \frac{4k_B T F_n B_e}{R_L} \tag{4.93}$$

where F_n is the noise figure.

The variance of the thermal noise current is always larger than that of the shot noise current. Shot noise current also results from an APD. This current is caused by the avalanche multiplication gain G_m.

The photocurrent is given by

$$I = R_{APD} P \tag{4.94}$$

where R_{APD} is the responsivity of the APD.

The variance of current due to an APD is given

$$\sigma^2_{shot} = 2eG^2_m F_A(G_m) RPB_e \tag{4.95}$$

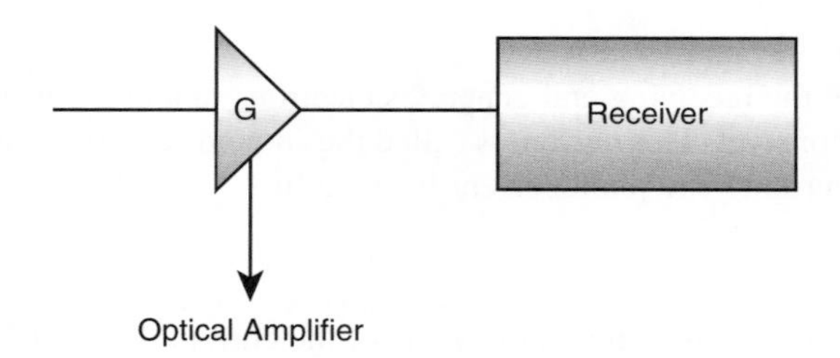

Figure 4.40 An OA is placed before a receiver to amplify the received signal

where $F_A(G_m)$ is the excess noise factor of the APD, e is electric charge and B_e is electric field bandwidth. It increases as the gain increases.

$$\sigma^2_{\text{thermal}} = \frac{4k_B T B_e}{R_L} \tag{4.96}$$

The photocurrent can now be considered a Gaussian random process with variance given by

$$\sigma^2 = \sigma^2_{\text{thermal}} + \sigma^2_{\text{shot}} \tag{4.97}$$

Noise Due to Spontaneous Emission

The OA in the receiving end gives rise to noise current due to spontaneous emission. Figure 4.40 shows an OA.

The noise power due to spontaneous emission for each polarization mode is given by

$$P_N = n_{sp} hf_c (G-1) B_o \tag{4.98}$$

where P_N is the noise power, G is the amplifier gain, B_o is the optical bandwidth, and n_{sp} is the spontaneous emission factor based on population inversion occurring in the amplifier. There are two polarization modes in a single-mode fiber, so the noise power becomes twice P_N.

At the receiving end, the optical preamplifier is generally placed before a pin photodiode.

The current generated by the photodetector is given by

$$I = RGP \tag{4.99}$$

where P is the optical power received, $G =$ is the gain of the preamplifier, and R is the responsivity of the photodetector.

The current produced by the photodetector is proportional to the optical power. The electric field generates noise that beats against the signal (signal–spontaneous beat noise) and noise that beats against itself (spontaneous–spontaneous beat noise).

The variance due to the different noise currents is given by the following equations:

$$\begin{aligned} \sigma^2_{\text{thermal}} &= I_t^2 B_e \\ \sigma^2_{\text{shot}} &= 2eR[GP + P_n(G-1)B_o]B_e \\ \sigma^2_{\text{sig-spont}} &= 4R^2 GPP_n(G-1)B_e \\ \sigma^2_{\text{spont-spont}} &= 2R^2[P_n(G-1)]^2(2B_o - B_e)B_e \end{aligned} \tag{4.100}$$

where e is electric charge and B_e is electric field bandwidth.

If the amplifier gain is large, then the thermal and shot noise values are very low. By filtering the noise before it reaches the pin photodiode, the optical bandwidth can be decreased. This reduces the spontaneous–spontaneous beat noise. This makes the signal–spontaneous beat noise the major noise component. The noise figure of the amplifier, which is the ratio of input SNR (SNR_i) to output SNR (SNR_o), is given by the following equations:

$$\begin{aligned} SNR_i &= \frac{(RP)^2}{2R_e P B_e} \\ SNR_o &= \frac{(RGP)^2}{4R^2 PG(G-1) n_{sp} h f_c B_e} \end{aligned} \tag{4.101}$$

The noise figure is given

$$F_n = \frac{SNR_i}{SNR_o} = 2n_{sp} \tag{4.102}$$

4.3.1.8 Coherent Detection

A DD receiver has disadvantages, such as thermal and noise currents. Coherent detection is a technique that improves the receiver's sensitivity. This type of detection increases signal gain by mixing it with a light signal from a "local oscillator" laser. The major noise component here is shot noise due to the local oscillator. Figure 4.41 depicts a coherent receiver.

The incoming light signal is mixed with a local oscillator and is sent to the photodetector. Assuming that the phase and polarization of the two waves are the same, the power of the photodetector is given by

$$\begin{aligned} P_r(t) &= \left[\sqrt{2aP}\cos(2\pi f_c t) + \sqrt{2P_{LO}}\cos(2\pi f_{LO} t)\right]^2 \\ P_r(t) &= aP + P_{LO} + 2\sqrt{aPP_{LO}}\cos[2\pi(f_c - f_{LO})t] \end{aligned} \tag{4.103}$$

where P is the power of the input signal, P_{LO} is the power of the local oscillator, f_c is the carrier frequency of the input signal, and f_{LO} is the frequency of the local oscillator waves

If the local oscillator power P_{LO} is made large, then the shot noise component dominates the other noise components in the receiver.

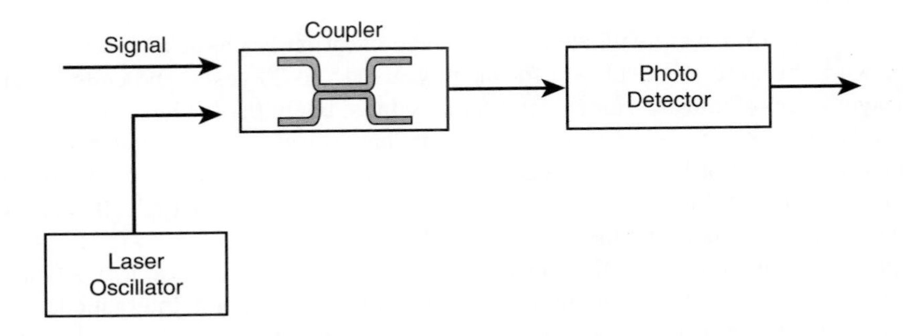

Figure 4.41 Coherent detection to improve the receiver's sensitivity

Noise variance is given by the following equations:

$$\begin{aligned} \sigma_1^2 &= 2eI_1B_e \\ \sigma_0^2 &= 2eI_0B_e \end{aligned} \tag{4.104}$$

You can neglect P when compared with P_{LO}, because it has a value less than -20 dBm, and P_{LO} is 0 dBm.

Therefore, the value of BER is given by

$$\text{BER} = Q\sqrt{\frac{RP}{2eB_e}} \tag{4.105}$$

where e is electric charge and B_e is electric field bandwidth.

If $B_e = B/2$, then BER $= Q\sqrt{M}$, where M is the number of photons per bit.The sensitivity of a coherent receiver is better than that of a DD receiver.

4.3.1.9 Clock Recovery

In nearly all light-wave systems it is necessary to recover the clock in the process of regenerating the signal. Several commonly used methods of clock recovery make use of phase-locked loops or surface acoustic wave filters, with the latter being the most widely used for high-frequency systems, which have strict requirements on the phase transfer function. Because the commonly used NRZ format does not contain energy at the clock frequency, it is necessary to use some form of nonlinear device, such as a rectifier or exclusive OR gate, to produce frequency components at or near the clock frequency. This signal is then passed through a narrowband filter such as a surface acoustic wave filter or dielectric resonator to extract the clock. The extracted clock, suitably phased, is used to clock the decision circuit and may be used elsewhere in the terminal or regenerator as needed.

4.3.1.10 Decision Circuit

The decision circuit receives the signal (plus noise) from the postamplifier along with the recovered clock. The decision circuit produces an output which is reshaped and retimed.

4.4 The Optical Fiber

The optical fiber infrastructure of a telecommunication carrier is a basic but essential element of its network, because it is deployed for many years (15 to 20 years) and called to support several generations of transmission systems (at 2.5 Gbit/s, 10 Gbit/s, 40 Gb/s, and in a mid-term future at 100 Gbit/s). A bad choice of fiber can have disastrous consequence over both performance and cost of long-haul transmission networks. Optical cables used today by the last generation of WDM transmission systems at 10 and 40 Gbit/s were typically deployed by the incumbent operators in the first part of the 1990s.

The physical characteristics of the optical fibers (for long-haul transmission) available on the market have been standardized by international organizations like the International Electrotechnical Commission (IEC) or the ITU. The different fiber categories which exist at the IEC and at the ITU are summarized in Table 4.2.

Table 4.2 Optical fiber normalization at the IEC and ITU

Fiber type	IEC	ITU
NDSF	B1.1	G.652
Pure silica core fiber	B1.2	G.654
DSF	B2	G.653
Dispersion-flattened fiber	B3	—
NZDSF	B4	G.655
NZDSF for large bandwidth transport (on the S-, C- and L-bands)	—	G.656

Today, terrestrial transport networks of incumbent operators are principally equipped with standard single-mode fiber (SSMF), compliant with the G.652 standard; but, until recently, stiff competition between nonzero dispersion-shifted fibers (NZDSFs), compliant with the G.655 standard, and nondispersion-shifted fibers (NDSFs) or SSMF has occurred. The most well-known G.655 fibers were LEAF™, Truewave™, and Teralight™ fibers manufactured by Corning, OFS Fitel, and Draka Comteq respectively. They have been deployed since the beginning of the 2000s typically by the new carriers in the USA and Europe and by the Chinese operators that did not have modern optical transport networks. The incumbent carriers have globally kept their legacy NDSF-based transport networks and are still resistant to replacing their G.652 fibers by G.655 fibers. Indeed, G652 fibers have shown their superiority to support 10 Gbit/s dense WDM (DWDM) transmission systems with 50 GHz channel spacing (owing to their high local CD ensuring a good decorrelation between WDM channels in the presence of nonlinear effects) and a good level of performance at 40 Gbit/s as well when the right modulation formats and dispersion map are chosen.

4.4.1 Short Introduction to the Waveguide Principle

The CD D of an optical fiber is the sum of two terms: a first term related to the media dispersion and a second term related to the waveguide dispersion. The media and waveguide dispersions are respectively represented in dark and grey in the plots of Figure 4.42. By playing on the manufacturing parameters of the waveguide (media and profile of the waveguide), the spectral dispersion profile of the fiber and its null-dispersion wavelength λ_0 or its third-order CD D' can be adjusted precisely.

SSMFs (plot 1 in Figure 4.42) for which the null-dispersion wavelength λ_0 is near 1300 nm, dispersion-shifted fibers (DSFs) (plot 2 in Figure 4.42) for which λ_0 is near 1550 nm, or dispersion-flattened fibers (DFFs) (plot 3 in Figure 4.42) whose dispersion does not vary in the bandwidth of the EDFAs can thus be designed and manufactured. Some simple rules, useful for designing single-mode optical fibers, are presented here. The first two parameters to control are (i) the loss level in the useful bandwidth of the EDFAs and (ii) the null dispersion wavelength λ_0, which is shifted towards the high wavelengths when germanium is added in silica (which increases its refractive index) or, in contrast, towards the low wavelengths when fluorine is introduced (which decreases its refractive index). However, adding doped materials in the pure silica increases the fiber losses. It is thus essential to find a trade-off between these parameters. Moreover, the dispersion of germanium-doped silica is not sufficient to shift the null dispersion wavelength up to 1550 nm. It is then necessary to increase the waveguide dispersion and to play

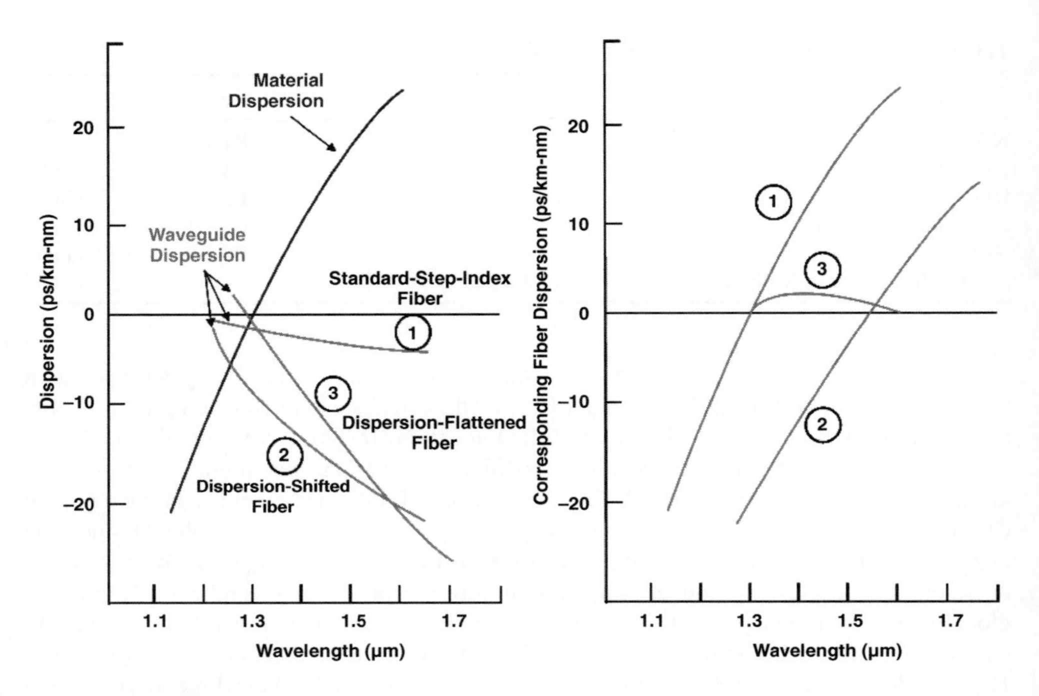

Figure 4.42 CD of the various fibers available on the market, with, on the right-hand side the dispersion of the material (SiO_2) and of the waveguide for three fiber types (standard step index, dispersion-shifted, and dispersion flattened fibers), and on the left-hand side the resulting fiber dispersion

with its refractive index profile. Dispersion of a waveguide with a normalized mode frequency V and a propagation constant b is given by the expression

$$D_W(\lambda) = -\frac{N_2^2}{n_2 c\lambda}\left(\frac{n_1^2-n_2^2}{2n_1^2}\right)V\frac{\partial^2(bV)}{\partial V^2} \quad \text{with} \quad \begin{cases} V = k_0^2 a(n_1^2-n_2^2)^{1/2} \\ b = \dfrac{(\beta/k_0)-n_2}{n_1-n_2} \\ N_2 = n_2-\lambda\dfrac{\partial n_2}{\partial\lambda} \end{cases} \tag{4.106}$$

where c is the light speed in vacuum, λ the light wavelength, n_1 the fiber core refractive index, n_2 the fiber cladding refractive index, a the core radius, β the propagation constant, and k_0 the wave number in the vacuum. Increasing the absolute value of D_W is then equivalent to increasing $(n_1^2-n_2^2)$,which, for constant V, is equivalent to reducing the core radius a. A DSF, whose null dispersion wavelength λ_0 has been fixed at 1550 nm, has to exhibit a mode effective area lower than that of an SSMF (50 μm^2 for a DSF against 80 μm^2 for an SSMF). Identically, dispersion-compensating fibers (DCFs) used to compensate the SSMF dispersion, whose dispersion coefficient (largely negative at 1550 nm: -80 ps/(nm km)) is obtained owing to a substantial increase of the absolute value of D_W, have an extremely low mode effective area (around 20 μm^2). The corollary of the rise of the absolute value of D_W, and consequently of the core–cladding refractive index difference (namely $n_1^2-n_2^2$), is an increase of the losses of the fiber (0.25 dB/km for DSF and 0.6 dB/km for DCF compared with 0.2 dB/km for SSMF).

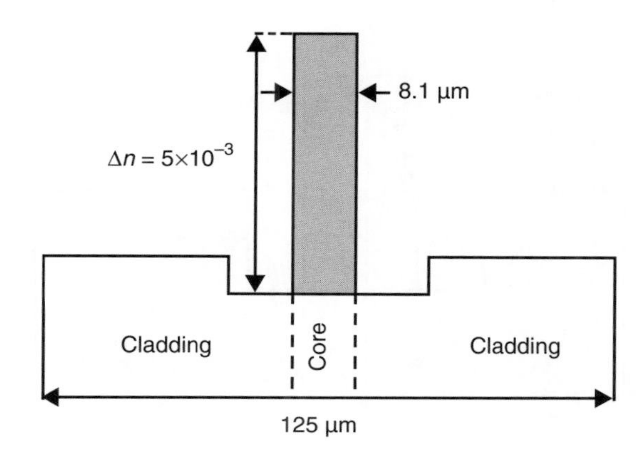

Figure 4.43 Index profile of a depressed-cladding fiber

The massive presence of doped materials (and some impurities) in the fiber structure combined with a stronger confinement of light energy in the fiber core increase the loss level due to Rayleigh scattering. To increase the core-cladding refractive index difference without increasing the losses (in particular due to Rayleigh scattering), a depressed-cladding refractive index profile technique is used. After this technique, the refractive index of the cladding, located near the fiber core, is reduced due to the addition of fluorine leading to a lower value compared to the external cladding refractive index (Figure 4.43).

With a particular type of index profile, it is possible to impose on the fiber a flattened CD on an extended bandwidth. These fibers, called DFFs, have an index profile with a W shape. More generally, the requirement of the transmission system designers to have a fiber with a dispersion profile that is as flat as possible is driven by the need to have the most homogeneous transmission, independent of the wavelength or the amplification band (S-, C- or L-band) considered (see Figure 4.47 in the next section). This is the reason why the fiber suppliers try to reduce the third-order dispersion of their most recent products (mainly their latest G.655 fibers). A W profile divides the fiber refractive index into three distinct regions: the core with refractive index n_1 and radius a_1, the internal cladding with refractive index n_2 and external radius a_2, and the external cladding with refractive index n_3 (Figure 4.44).

The W profile is in fact a particular type of depressed cladding profile, whose refractive index differences $\Delta n_+ = n_1 - n_3$ and $\Delta n_- = n_3 - n_2$ are almost equal, the external radius of the internal cladding a_2 being around twice the core radius a_1. In these conditions, the external cladding has a major influence on the waveguide properties, because a non-negligible portion of the mode energy propagates inside it. With this technique, the effective area of the LP_{01} mode in the modern fibers (in particular the fibers belonging to the G.655 ITU recommendation) can be substantially increased. To understand how a W index profile flattens the fiber dispersion, it is important to recall that the effective area of the LP_{01} mode increases with the wavelength. Therefore, it reinforces the guiding effect of the external cladding. However, when the wavelength decreases, the mode is more confined and its effective area is lower; the core influence on the guidance is then predominant. In other words, when the wavelength increases, the mode initially guided by the core migrates progressively towards the external cladding. The

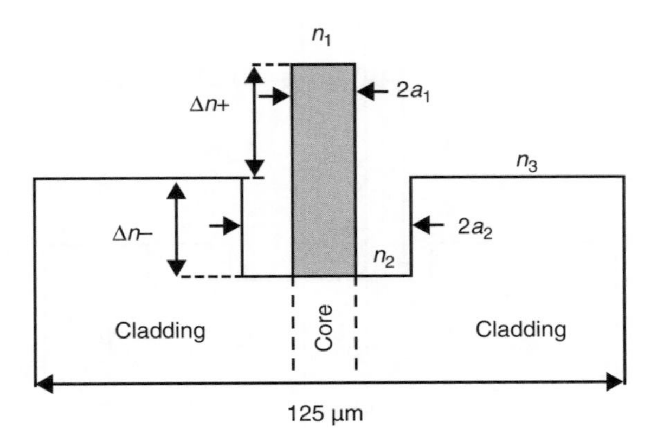

Figure 4.44 W index profile

mode which saw primarily the group index of the fiber core ends by being essentially sensitive to the group index of the external cladding.

As the properties in terms of CD between the core and the external cladding are opposite, the mode migration from the core to the external cladding as the wavelength increases generates a flattening of the CD (seen by the mode). However, the presence in the external cladding of a non-negligible part of the mode energy results in an enhancement of the micro-bending loss sensitivity. This limitation can be partially alleviated by incorporating an extra cladding of refractive index n_4 (Figure 4.45) to the index profile described in Figure 4.44: n_4 is lower than n_3 in order to increase the mode confinement and, thus, to reduce the macro-bending losses.

4.4.2 Description of Optical Single-Mode Fibers

As explained earlier, the optical fiber infrastructure of a telecommunication carrier is an essential element of its network, like the transport equipment that is installed over it. Among all the available fibers (see Table 4.2), the NDSF or G.652 fiber, also called SSMF, is the most

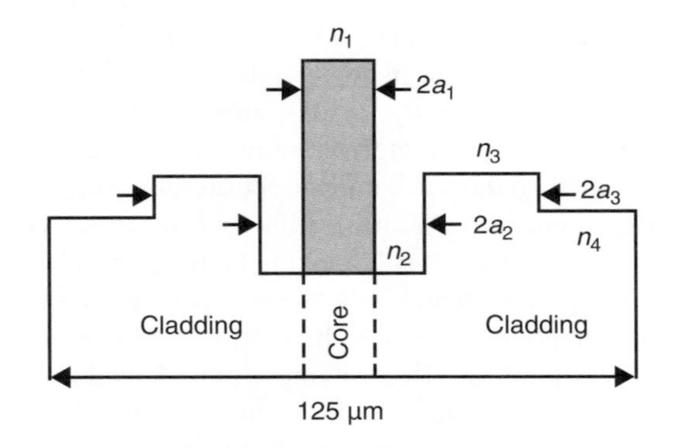

Figure 4.45 Index profile with four claddings

commonly deployed in the world (90% of fiber installed), followed by the G.655 fibers, particularly used by the new operators and deployed especially in the USA and China. The G.653 fibers (implemented, for example, by NTT, Telecom Italia, Telmex, and Telecom Argentina) have almost totally disappeared from the market today because of their poor level of performance in WDM configuration, and concerned operators have launched a campaign to replace them progressively. Avoiding dispersion in single-channel transmission at 1550 nm resulted unfortunately in exacerbation of cross-nonlinearities (both cross-phase modulation (XPM) and four-wave mixing (FWM)), and exotic strategies like unequal channel spacing or use of L-band have to be used to cope with nonlinear effects. The last, but not the least limiting, factor of the DSF resided finally in its polarization mode dispersion (PMD). Because of the core–cladding index difference being two times higher than that of G.652 fibers, the PMD of DSF is globally higher than that of G.652 fibers.

G.652 fibers are especially used by incumbent operators on their legacy fiber infrastructure. Manufactured and commercialized for more than 20 years, the G.652 fiber is a product in constant evolution, in particular when one considers its spectral attenuation profile and its PMD level. Its cut-off wavelength is around 1260 nm, which means that SSMF is single mode from 1260 nm up to the upper amplification bands used in modern optical communications. Its CD is null at 1310 nm and in the range 16–18 ps/(nm km) in the C-band. Its loss level is ~0.20 dB/km, but recently Corning proposed a G.652 fiber with an ultra-low level of losses at ~0.17 dB/km at 1550 nm. Its fundamental mode effective area is equal to 80 μm^2 at 1550 nm. The high CD value of SSMF at 1550 nm, which is detrimental in single-channel transmission, provides an interesting protection against interchannel nonlinear effects in WDM transmission. G.652 fibers are thus particularly indicated when high spectral efficiency transmissions are implemented, in particular when 10 Gbit/s channels spaced by 50 GHz are transmitted. Its high CD enables a high decorrelation of transmitted channels and a low sensitivity to cross-nonlinearities. Owing to the implementation of DCF modules, the maximum reach of transmission systems propagating on SSMF is compatible with ultra-long-haul application, even if dispersion compensation impacts substantially on the cost of optical links. This is the reason why the fiber suppliers tried at the end of the 1990s to develop a "fiber matched to WDM". The idea was to reduce the CD value in order to decrease the cost of compensation (or even suppress it totally at 10 Gbit/s) while keeping a high enough value to ensure a good protection against cross-nonlinearities. In other terms, the objective was to find the best trade-off between CD and nonlinear effect exacerbation. Different fibers of this family, called NZDSF, were developed along the years by different manufacturers. Nonetheless, it can be noted that all the equipment vendors announce significantly longer transmission distances when their 10 Gbit/s WDM systems are installed over SSMF than when they use NZDSFs. Another non-negligible advantage of G.652 over G.655 fiber is its price, which is 2.5 times cheaper. Today, the price per kilometer of G.652 fiber is ~ €10, while for G.655 fiber it is rather ~ €25. Since 2002, the fiber price has been divided by a factor of 2. While in 2000, G.655 fibers represented ~15% of the fiber sales, today they represent less than 5%.

At 40 Gbit/s and higher, SSMF shows some limitations when used with conventional OOK modulation formats. Its high CD at 1550 nm stimulates action of intra-channel nonlinear effects. The intra-channel FWM (IFWM) and intra-channel XPM (IXPM) are exacerbated more in SSMF than in other fiber types with lower CD (in particular, G.655 fibers), leading to the creation of pulse timing jitter, amplitude fluctuations over the "1" of the binary sequence and "ghost" pulses apparition in the "0". In order to ensure that these nonlinear interactions do

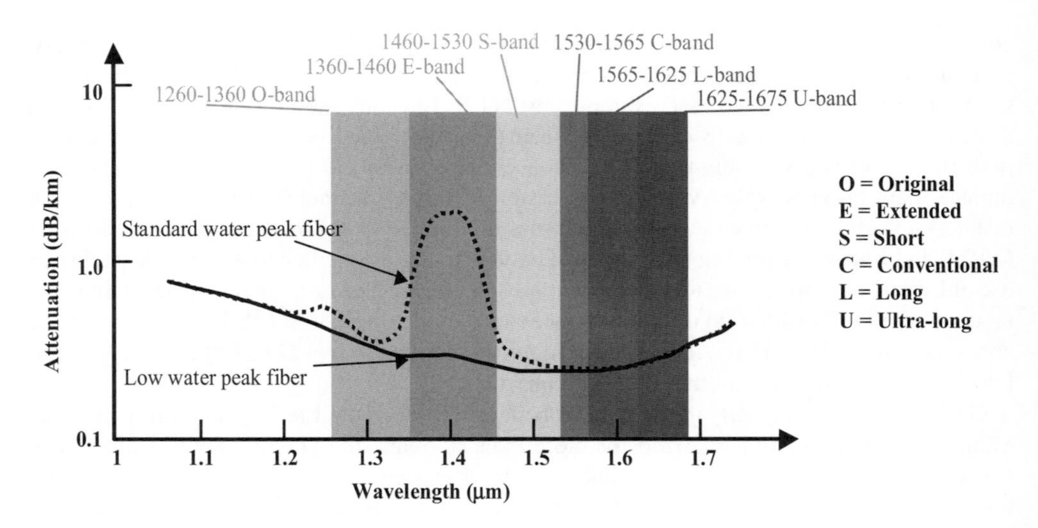

Figure 4.46 Spectral attenuation profile of G.652A and G.652C fibers

not perturb too much transmission, particular care has to be brought to the design of the dispersion map, as well as to the choice of the modulation format. Implementing, for example, differential phase-shift keying (DPSK) modulation permits one to push farther the nonlinear threshold of the transmission while gaining ~3 dB of optical SNR (OSNR) sensitivity when compared with OOK formats.

In 2003, a new version of the G.652 fiber was proposed by fiber manufacturers. This fiber, corresponding to the G.652.C standard, is characterized by a low attenuation level in the band 1380–1400 nm, obtained due to the suppression of the water peak (see Figure 4.46 and Table 4.3). This fiber has progressively replaced all other types of G.652 fibers, and today all G.652 fibers sold on the market have no water peak. G.652.C fibers are particularly recommended for metropolitan networks, where a large bandwidth starting at 1400 nm and finishing at 1600 nm is used. They can be useful as well for future Raman-based WDM

Table 4.3 G.652 fiber standardization at the ITU

Physical features	ITU G.652.A	ITU G.652.C (G.652.D)
Cable cut-off wavelength (nm)	$\leq$1260	$\leq$1260
Attenuation in cable (dB/km)	$\leq$0.40 (at 1310 nm) $\leq$0.35 (at 1550 nm) $\leq$0.40 (at 1625 nm)	$\leq$0.40 (1310–1625 nm) $\leq$0.30 (1550 nm) $\leq$ measured at 1310 nm (at 1383 $\pm$ 3 nm)
Mode field diameter at 1310 nm (μm)	(8.6–9.5) $\pm$ 0.7	(8.6–9.5) $\pm$ 0.7
CD slope S_o (ps/(nm^2 km))	$\leq$0.093	$\leq$0.093
Null dispersion wavelength (nm)	Between 1300 and 1324	Between 1300 and 1324
PMD (ps/km$^{1/2}$)	$\leq$0.20	$\leq$0.50 (0.20)

long-haul transmission systems at 40 and 100 Gbit/s, for which the Raman pumps are advantageously located in the area where the loss level is decreased (see also Section 4.5.7). The implementation of Raman amplification on such low water-peak SSMFs could permit one to merge the E, S, C, and L bands into a single one, opening a transmission window from 1260 nm up to 1680 nm, able to carry 30 or 40 Tbit/s of traffic. Note that this improvement of the SSMF features has not resulted in an increase of its cost.

Recently as well, fiber manufacturers have realized many efforts to adapt their costly G.655 fiber-based fabrication process to SSMF. These methods, based on the implementation of the high mode coupling manufacturing process, have enabled a decrease in the PMD of G.652 fibers from 0.1 ps/km$^{1/2}$ up to 0.06 ps/km$^{1/2}$, and even 0.04 ps/km$^{1/2}$ (under the condition of choosing the best fiber samples). Intrinsically, the core of G.652 fibers, being less doped in germanium (which can be considered from the PMD point of view as impurities) and consequently purer than that of G.655 fibers, means the PMD of G.652 fibers has to be the best.

The G.653 fiber, also termed DSF, is characterized by a very low CD at 1550 nm. This fiber type appeared at the beginning of the 1990s to enable transmission at 2.5 and 10 Gbit/s over transoceanic distances. Indeed, on G.652 fibers, in the absence of periodic in-line dispersion compensation, the maximum reach was limited to several hundreds of kilometers at 2.5 Gbit/s and only 80–100 km at 10 Gbit/s. The main interest of this new fiber type consisted principally in its robustness to the bit rate increase. The G.653 fiber has been largely deployed in Japan, in Italy, and in Central and South America. The low level of CD at 1550 nm of the DSF, which constituted its main advantage at its creation, is rapidly becoming its main drawback. Indeed, in WDM transmission, a low CD at 1550 nm exacerbates the impact of cross-channel nonlinearities (FWM and XPM) and decreases significantly the performances of the DSF when high spectral efficiency transmissions are implemented. A way to overcome this problem is to use WDM systems over the L-band, where the CD is comparable to that of G.655 fibers in the C-band. Nonetheless, with such a fiber type, it will be impossible to open a bandwidth of 400 nm able to carry 30 or 40 Tbit/s (as mentioned above for the G.652 fiber). Moreover, DSF was three times more expensive than G.652 fiber.

Another important limiting factor of DSF is its PMD. Because of a core–cladding refractive index difference two times higher than that of G.652 fibers, the PMD of the DSF is higher. However, the high mode coupling manufacturing process can also be applied to the fabrication of the DSF. Even if G.653 fibers are effectively not being deployed today, some work is being done at the ITU and a new recommendation (detailed in Table 4.4) was written at the end of 2003, in particular to define DSF with a low PMD level.

Table 4.4 G.653 fibers standardization at the ITU

Physical features	ITU G.653.A	ITU G.653.B
Cable cut-off wavelength (nm)	$\leq$1270	$\leq$1270
Attenuation in cable at 1550 nm (dB/km)	$\leq$0.35	$\leq$0.35
Mode field diameter at 1550 nm (μm)	(7.8–8.5) $\pm$ 0.8	(7.8–8.5) $\pm$ 0.8
CD on the range 1525–1575 nm, $\lvert D(\lambda)\rvert$ (ps/(nm km))	$\leq$3.5	$\leq$3.5
CD slope S_o (ps/(nm^2 km))	$\leq$0.085	$\leq$0.085
Null dispersion wavelength (nm)	Between 1500 and 1600	Between 1500 and 1600
PMD (ps/km$^{1/2}$)	$\leq$0.50	$\leq$0.20

G.655 fibers, also termed NZDSF, were designed at the end of the 1990s in order to reconcile the advantages of SSMF and DSF. Originally, G.655 fibers were presented as the best trade-off between G.652 and G.653 fibers. In particular, their level of CD is sufficiently high to limit the impact of cross-channel nonlinearities (FWM and XPM) and sufficiently low to enable the transport of 10 Gbit/s channels with a moderate need for dispersion compensation units. In addition, G.655 fibers were intended to be the best compromise for 40 Gbit/s WDM transmission.

In reality, G.655 fibers have never demonstrated their capability to support 10 Gbit/s DWDM transmission systems with performance superior to that of G.652 fibers. SSMF has even shown higher capacity than all G.655 fiber types to transport very high spectral efficiency DWDM systems, especially when 25 GHz channel spacing is implemented (even if rarely proposed by equipment suppliers). The continuous increase of the CD at 1550 nm of the most recent G.655 fibers (for example, the Truewave fiber family, with the Truewave, Truwave-RS™, and Truewave-Reach™ fibers) has shown that the optimal trade-off between their different physical features (namely CD, CD slope, effective area) has not been found yet. Figure 4.47 defines the CD range where the NZDSF fibers can be designed. Practically, the NZDSF+ family (whose dispersion at 1550 nm is positive) cannot have on the C-band a dispersion lower than 1 ps/(nm km), while the NZDSF− family (whose dispersion at 1550 nm is negative) cannot have on the C-band a dispersion higher than −1 ps/(nm km).

Today, different products are commercially available. These last ones can be arranged in three categories: G.655.A, G.655.B, and G.655.C. Table 4.5 details the differences existing between these various classes of G.655 fibers.

The G.655 fiber deployed most is without contest the LEAF fiber (manufactured by Corning), followed by the Truewave-RS™ fiber from OFS Fitel (previously Lucent) and the Teralight fiber from Draka Comteq (previously Alcatel). To motivate interest for the operators to equip their long-distance networks with G.655 fibers, fiber suppliers have insisted on two arguments: first, the accumulated CD and the corresponding compensation amount and cost are

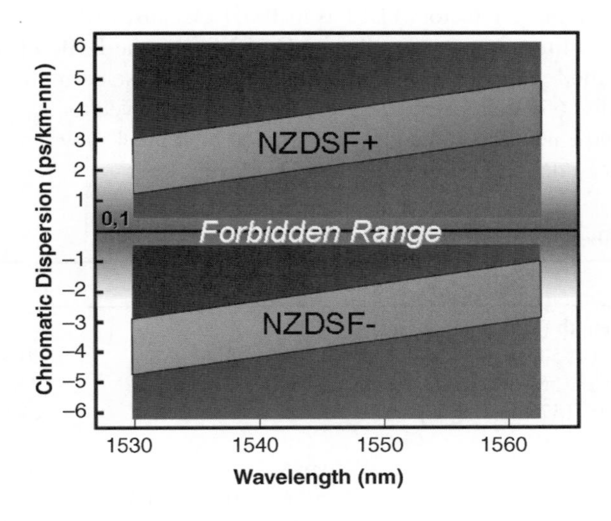

Figure 4.47 CD range where are defined NZDSF fibers

Table 4.5 G.655 fibers standardization at the ITU

Physical features	ITU G.655.A	ITU G.655.B (ITU G.655.C)
Cable cut-off wavelength (nm)	≤1450	≤1450
Attenuation in cable (dB/km)		
at 1550 nm	≤0.35	≤0.35
at 1625 nm		≤0.4
Mode field diameter at 1550 nm (μm)	(8–11) ± 0.7	(8–11) ± 0.7
CD over the range 1530–1565 nm (ps/(nm km))		
D_{min}	≥0.1	≥1
D_{max}	≤6	≤10
Dispersion sign	Positive or negative	Positive or negative
$D_{max} - D_{min}$ (ps/(nm km))		≤5.0
CD over the range 1565–1625 nm	—	?
Dispersion sign	—	positive or negative
PMD (ps/km$^{1/2}$)	≤0.50	≤0.50 (0.20)

lower after one span of G.655 fiber rather than after one span of G.652 fiber; second, G.655 fibers have intrinsically a better PMD than G.652 fibers. As mentioned previously, G.655 fibers have a more complex index profile than G.652 fibers, with in particular a core largely more doped in germanium than G.652 fibers. Consequently, there is no intrinsic reason for G.655 fibers to exhibit a lower PMD than G.652 fibers if the same fabrication process is applied to both G.655 and G.652 fibers.

Figure 4.48 presents the CD profile as a function of the wavelength of the different G.655 fibers, as well as G.652 and G.653 fibers (for comparison). E-LEAF (E for enhanced) is the most recent version of the LEAF.

The G.656 ITU recommendation was published in 2004 in order to define an NZDSF adapted for DWDM transmission in the S-, C-, and L-bands. Indeed, some of the fibers compliant with the G.655 standard have a very low level of CD (~0 ps/nm) in the S-band, making them

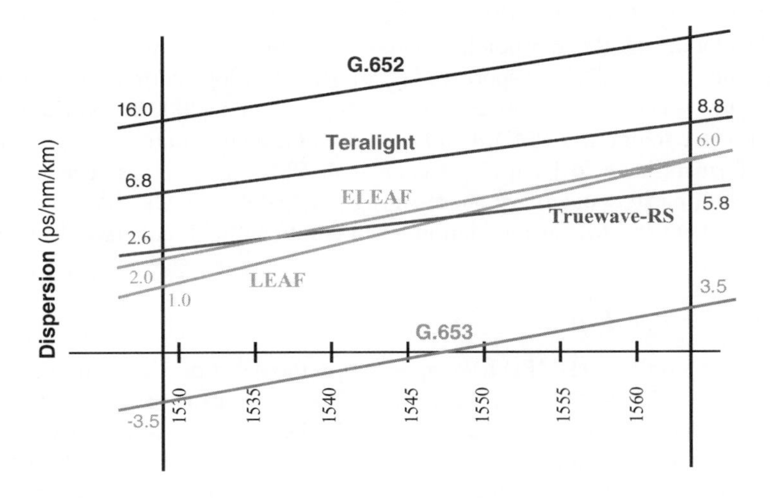

Figure 4.48 CD of the G.652, G.653 and G.655 fibers

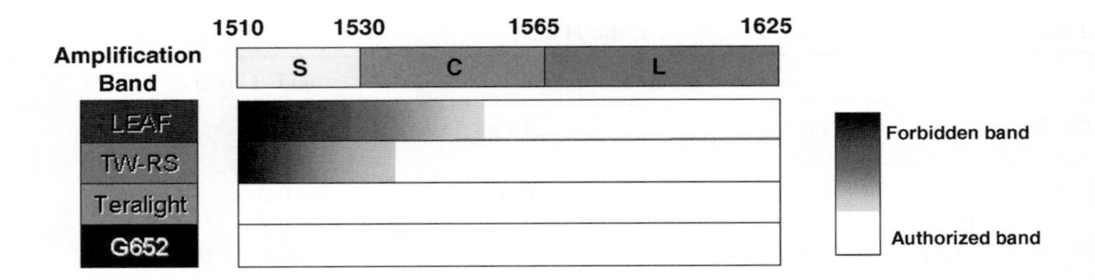

Figure 4.49 Fiber ability to support some traffic over the different amplication band

inappropriate for efficient DWDM long-reach transport. The G.656 ITU standard defines the features of the NZDSFs devoted to large bandwidth DWDM transport. The CD level has to be in the range 2–14 ps/nm on the 1460–1625 nm wavelength range. Some of the existing G.655 fibers, like Teralight or Truewave-Reach fibers, were compliant with the new standard, which is not the case for LEAF and Truewave-RS fibers (Figure 4.49).

It is thus not easy for a carrier to choose a fiber for its network when so many fiber types are available on the market. As a fiber is installed for a minimum duration of 15–20 years, a bad choice can be disastrous. Often, the operators adopted a conservative approach and continue to deploy the last versions of G.652 fiber. Nonetheless, to explain this choice, one can affirm that the SSMF has a satisfactory performance at 10 Gbit/s, particularly when high spectral efficiency systems (25 and 50 GHz channel spacing) are deployed. At 40 Gbit/s, some limitations (intra-channel nonlinearities like IFWM and IXPM) appear on SSMF, especially for ultra-long-haul applications. It can then be advantageous to use G.655 fibers to support such high bit-rate transmission systems: indeed, at 40 Gbit/s, the intra-channel nonlinear effects are limited on G.655 fibers because of their lower level of CD, while the cross-nonlinearities impact is significantly decreased because of the larger channel spacing used (100 GHz) in the 40 Gbit/s transmission systems. Recently, equipment suppliers (faced with operators' refusals to replace their "old" G.652 fiber-based cables by the new G.655 fiber-based cables) have looked at developing Tx/Rx transponder technologies able to cope with transmission limitations exacerbated by 40 Gbit/s transport. Today, DPSK technology permits one, for example, to reach maximum transmission distances at 40 Gbit/s greater than 2000 km on G.652 fibers. But DPSK is not more tolerant to PMD than standard NRZ modulation format, and alternative modulation formats have to be used to cope with PMD. Nortel has recently introduced polarization-multiplexed quaternary phase-shift keying (Pol-Mux QPSK) technology, which can be very resilient to PMD accumulation when combined with coherent detection.

4.4.3 Special Fiber Types

Dispersion-managed fibers (DMFs) have appeared in the last 4 or 5 years in the world of the optical transmission. These fibers consist of two fiber types: a fiber with a large mode effective area ($\sim$110 μm^2) and positive CD (+ 20ps/(nm km)) called super-large effective area fiber, and a fiber with a low mode effective area ($\sim$30 μm^2) and negative CD ($-$40 ps/(nm km)) called inverse dispersion fiber. Consequently, it is unnecessary to periodically insert dispersion-compensating modules in the transmission line to compensate the dispersion accumulated in the fiber spans. The total losses of a dispersion map are thus considerably reduced, from 30 dB to about 22 dB. In terms of OSNR, it is highly valuable because each decibel of OSNR gained by this reduction of dispersion map losses permits one to extend the reach of the transmission system and/or to increase the granularity of the WDM channels. Moreover, as a 100 km DMF span consists of one-third super-large effective area fiber, the overall nonlinear effects are reduced in the DMF spans due to the large effective area of first span section. Moreover, DMF is particularly useful when implementing distributed Raman amplification (with both forward and backward pumping, or with only backward pumping). When compared with a transmission line employing SMF, the reach of the system can be nearly doubled by using DMF. Unfortunately, because of its significant cost, it is mainly used in ultra-long-haul or "express" networks (for example, to link the West coast to the East coast in the USA) or submarine networks.

4.5 Optical Amplifiers

The transmission of near-infrared signals in optical fibers is affected by two basic limitations: attenuation and dispersion (Figure 4.50). *Attenuation* arises from the fact that the glass fiber is not perfectly transparent; actually, it causes a gradual decrease of the signal intensity during propagation.

Dispersion causes distortion and broadening of pulses traveling through the fiber, because signal components at different wavelengths have different speeds, on account of the dispersive nature of glass and of the laws of guided propagation.

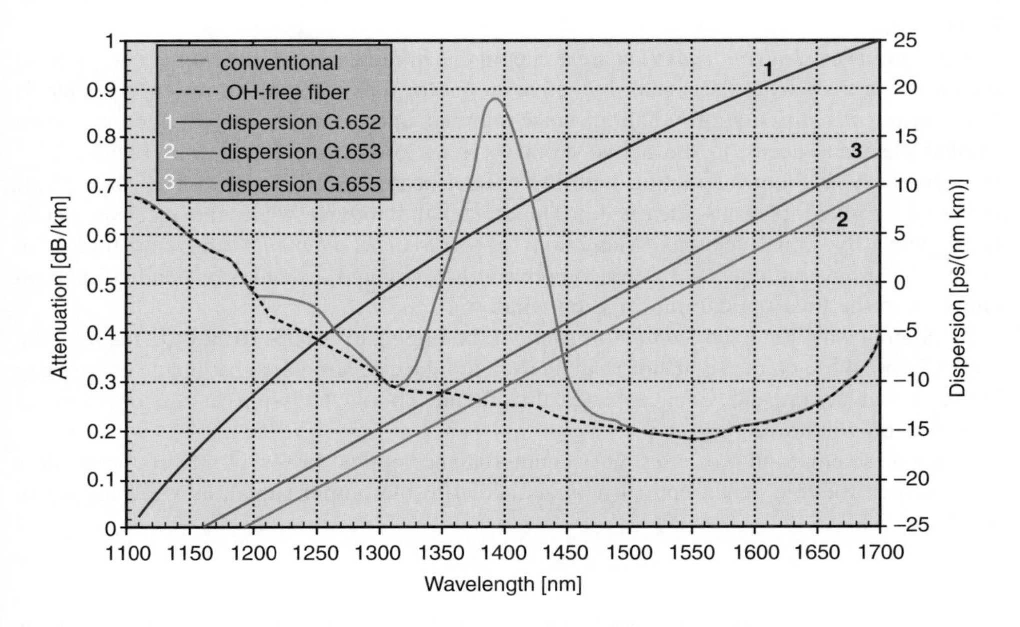

Figure 4.50 Attenuation and dispersion of transmission fibers

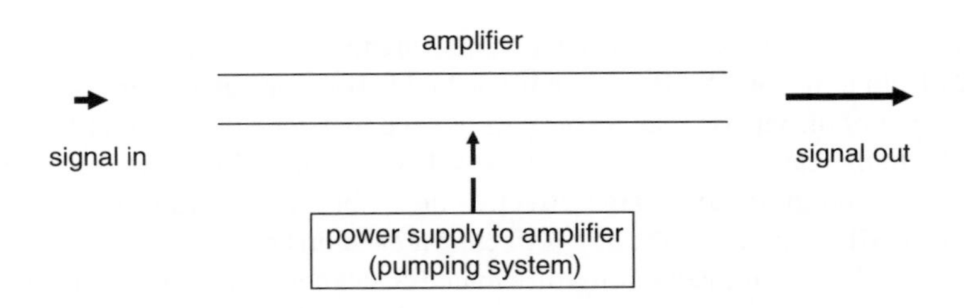

Figure 4.51 Schematics of an OA

Attenuation and dispersion contribute to set a maximum transmission distance beyond which the optical signal – the coding of an analogue waveform or a sequence of digital pulses – cannot be detected any longer nor decoded with sufficient precision.

The use of OAs has promoted an important evolution in telecommunication systems since the end of the 1980s. These devices overcome the problem of losses in any optical link, and make it possible to speak about a *virtually lossless optical network*.

The OA is an optical device (basically, in glass fiber form or as a microcrystal of semiconductor material) which accepts a weak signal at the input and resends it out with a much higher output power level. The gain in signal level[4] is obtained at the expense of a (optical or electrical) power supply from the outside, made by a suitable subsystem called the *pumping system* (Figure 4.51).

This power supply is used to excite the atoms (or ions) of the glass dopant or of the semiconductor that constitute the laser active medium, making their outer electrons jump to a higher energy state. In the presence of the input optical signal, this energy can be released under the form of further photons "stimulated" by those of the signal, as sketched in Figure 4.52.

Three kinds of *electron transition* are acting in the amplifier: (i) *absorption* of energy from the pump, or even from the signal, by active atoms (or ions) on their lower energy state 1; (ii) *spontaneous emission* of radiation (noise photons, uncorrelated to signal photons) when excited electrons decay to the active atom (or ion) lower state (level 1 in Figure 4.52) spontaneously, without interaction with the optical beam; and (iii) *stimulated emission* of radiation (emitted photons identical to input signal photons) when the electron decay is triggered by signal photons. Many ions of *rare-earths elements* (the lanthanides, in Figure 4.53) have an energy difference between levels 1 and 2 that corresponds to photon emission in the fiber-optical transmission windows.

In a similar way, some *semiconductor elements* belonging to Groups III and V of the periodic table (Figure 4.53, on the right side) can be used to fabricate structures having an energy gap between conduction and valence bands that favors photon emission in the fiber-optical transmission windows.

Using these emission processes, OAs counterbalance optical losses. However, the result is not obtained for free, since optical noise is added to the output signal, as we shall see in Section 4.5.4.

[4] The optical gain can reach very high values (in excess of 40 dB) at low input signals ($\tilde{<}-40$ dBm), over a gain bandwidth of the order of 4–9 THz.

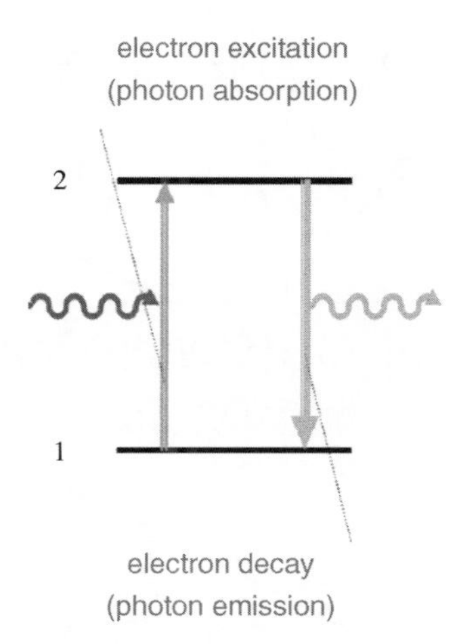

Figure 4.52 Electron transitions and photon processes in a laser active medium

4.5.1 Introduction to Optical Amplifiers

Most efficient OAs are made with special optical fibers, with their core containing some kind of doping substance – in most cases trivalent ions of rare-earth elements – such as erbium, praseodymium, neodymium, thulium, and so on that show radiative transitions fitting in the

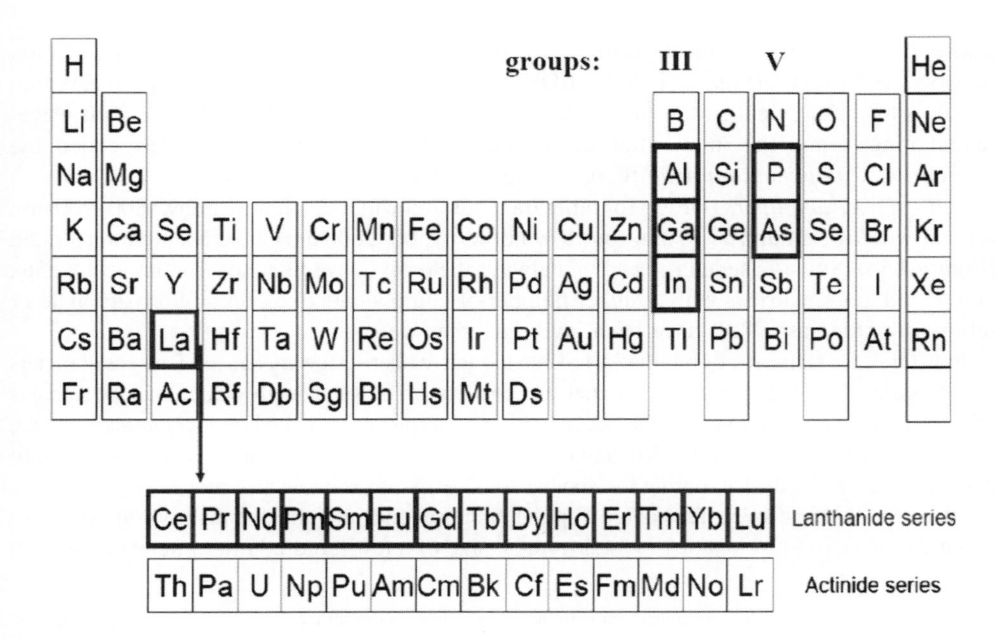

Figure 4.53 Mendeleev's table: lower cells indicate the lanthanide series; right cells indicate elements of the Groups III and V

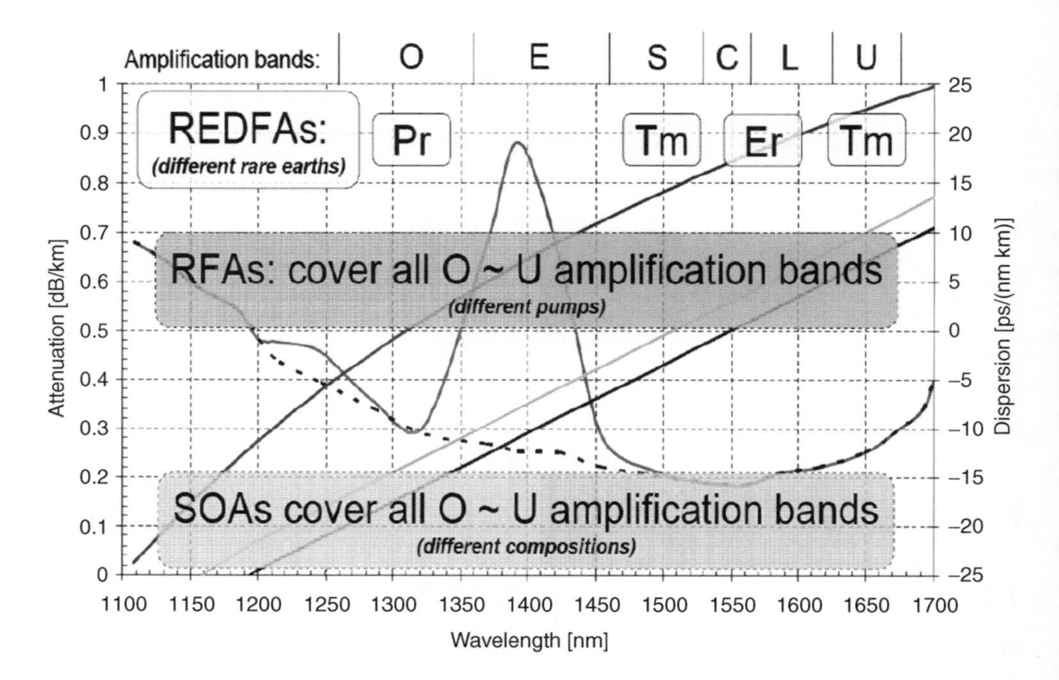

Figure 4.54 Amplification bands defined by ITU-T and available amplifier types (REDFA: rare earth-doped fiber amplifier)

transmission windows. Currently, the way to obtain optical gain in telecommunication networks is through the use of EDFAs. EDFA performance is excellent in the spectral region which is historically called the *third telecommunication window* (the 1530–1565 nm wavelength range), and that now, after the advent of EDFAs themselves, is just called the *conventional amplification band* (C-band, Figure 4.54).

Optical fiber amplifiers (OFAs) for other transmission windows (first window: 800–900 nm; second window: around 1310 nm, fourth window:[5] 2500–2600 nm) have been demonstrated (Figure 4.54). Silica-based glasses are not always the most convenient host for active dopants; compound glass matrices with a higher number of components (e.g., on a fluorozirconate or tellurium oxide basis [88]) have been investigated as well.

In an OFA, the signal to be amplified is fed into the active fiber – a few meters to a few tens of meters long (Figure 4.55). Together with the signal, an optical beam of wavelength shorter than the signal is injected in the amplifier, to excite the active medium; this process is called *optical pumping*. At the output end of the fiber the amplified signal is emitted; in band with it, there is some optical noise generated inside the device, during the spontaneous decay of excited active atoms (or ions).

Also, SOAs have been fabricated for the first, second and third transmission windows (Figure 4.54). An SOA is a tiny crystal (typical size of a few hundred micrometers as seen in

[5] Anticipated in non-silica fibres for transmission in the mid-infrared, but never used in practice. On the contrary, the second and third windows have been covered by the amplification bands shown in Figure 5.54; O: original band; E: extended band; S: short-wavelength band; C: conventional band; L: long-wavelength band; U: ultra-long-wavelength band [89].

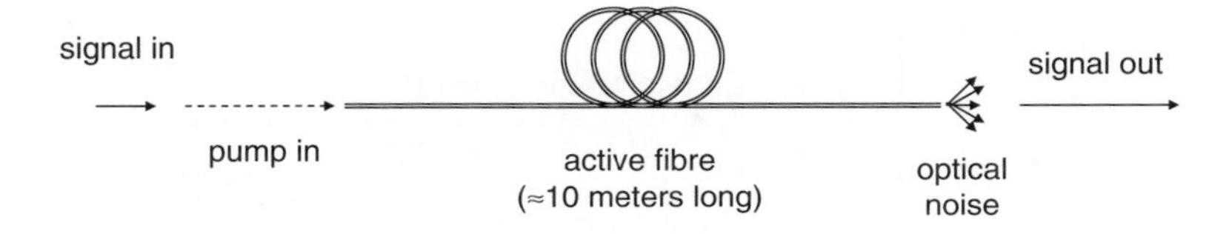

Figure 4.55 Schematics of an OFA

Figure 4.56) made with semiconductor elements from Groups III and V of the periodic table (right-side cells in Figure 4.53). SOA performance in terms of gain and noise is good, but high coupling losses to the transmission line – because of the marked difference between the circular geometry of the transmission fiber and the Cartesian symmetry of the SOA; see Figure 4.56 – and intrinsic nonlinearity (Section 4.5.5) prevent the wide use of SOAs to compensate for propagation, branching, and insertion losses. Their rapid gain dynamics are a better fit for various signal processing and photonic logic operations (e.g., wavelength converters, nonintrusive detectors, bi-stable devices).

These are the reasons why application in point-to-point and point-to-multipoint optical links is preferentially left to EDFAs. However, SOAs modified for weak nonlinear behavior are occasionally also in this sector, because of their small size and low cost.

Apart from specific differences (pumping is electrical in SOAs and optical in OFAs), the behavior of SOAs is analogous to OFAs; optical transitions now involve charge carriers (electrons and holes) between the valence and the conduction bands of the semiconductor (Figure 4.57), which can be approximated as a two "effective energy-level system". Photons are emitted when carriers relax across the crystal bandgap, the width of which is suitable for emission in the transmission windows.

As seen above, SOAs and OFAs (EDFAs and other rare-earth-doped fiber amplifiers) are devices where optical amplification is a consequence of stimulated emission induced by radiation, which travels through an excited laser medium. Optical amplification can also be due to a process of *stimulated scattering* in a waveguide. This is what happens in Raman fiber

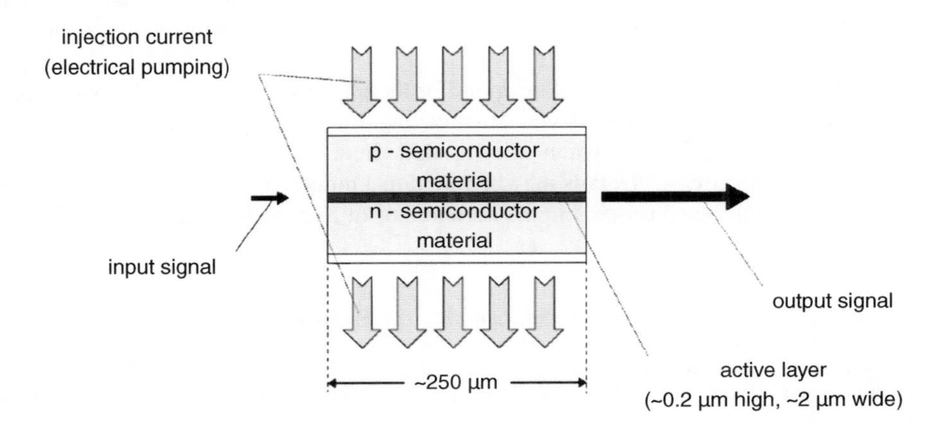

Figure 4.56 Illustration of an SOA

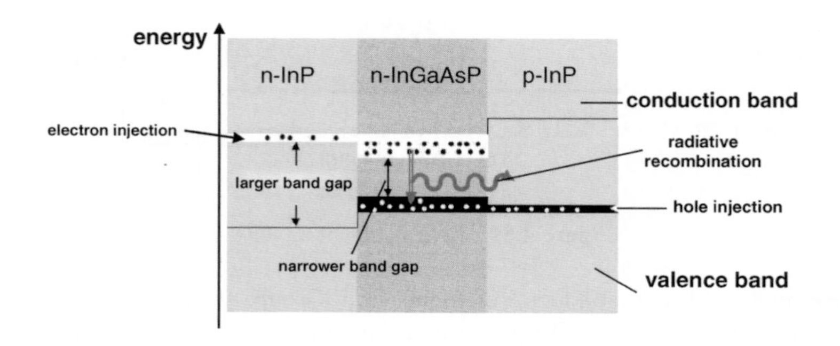

Figure 4.57 Band structure across a double-heterostructure[6] semiconductor optoelectronic source, made with the elements indium, gallium, arsenic, and phosphorus

amplifiers (RFAs) and Brillouin fiber amplifiers – the so-called *nonlinear* amplifiers – where light scattering is induced by optical or acoustic phonons (glass vibrations) respectively.

RFAs offer optical gain by exploiting a coherent scattering of photons inside an *ad hoc fiber* (lumped RFAs) or *directly along the transmission fiber* (distributed RFAs). Raman scattering is pumped by optical radiation of a power level strong enough to produce the coherent diffusion process (Figure 4.58). Usually, RFAs are used to boost the optical performance over very long spans (over 40–70 dB of attenuation without intermediate access points), for which EDFAs are not adequate. Distributed RFAs are also attractive because they produce little noise with respect to all kinds of lumped amplifiers.[7] Indeed, while a lumped RFA may be a few hundred meters long, a distributed RFA shares the length of the transmission fiber, producing an amplification process distributed all along the physical span. On the contrary, with lumped devices, optical amplification is provided abruptly where the amplifier is sited. Analogous considerations are valid for the noise generation process.

The drawback of the use of RFAs is that, due to the high pump power required for operation, joints and connectors all along the transmission line must be replaced with fusion splices, to avoid lumped attenuation points that could cause thermal damage of the fiber.

Brillouin fiber amplifiers work analogously to RFAs, but will not be considered here, because they have a very narrow gain bandwidth that is not applicable to high bit-rate-modulated signals.

In the following section, a concise description of the physics relevant for OA operation is given. Specific consideration will be devoted to traveling-wave[8] (TW) devices. A TW amplifier exploits the single-pass gain, which means that in Figure 4.51 the optical signal travels once along the device.[9] Details may be found in a number of classical works [91–93]. The case of SOAs will be considered in the second half of Section 4.5.3 and that of RFAs in Section 4.5.7.

[6] A heterostructure is a device realized with alloys of semiconductor elements having different bandgaps, as shown in Figure 4.57. See, for instance, Ref. [90].

[7] This characteristic is often referred to as the "negative noise figure" of distributed Raman amplifiers.

[8] OFAs are traveling-wave amplifiers. SOAs may be realized as both TW amplifiers and FP amplifiers, the latter having an appreciable reflectivity at the facets and exploiting multipass gain.

[9] On the contrary, in an FP amplifier the signal bounces back and forth within the device, which is surmounted by partially reflecting facets; see Section 4.5.8.

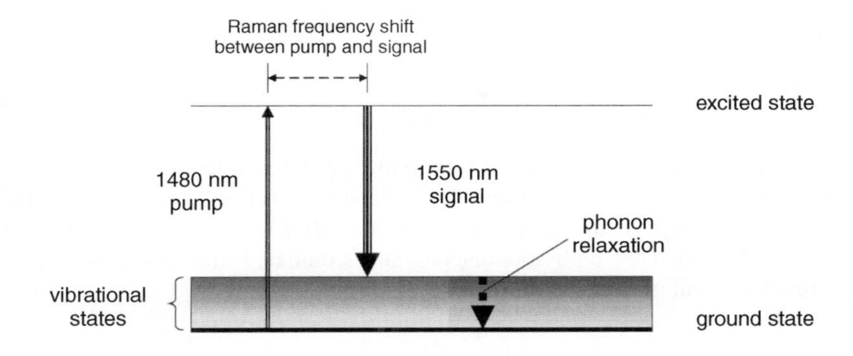

Figure 4.58 Energy states involved in Raman scattering

4.5.2 Principle of Operation

This section will discuss the basic relations for the operation of a TW amplifier, working on a (basic) two-level spectroscopic scheme (see Figure 4.52): both OFAs and SOAs are population-inversion OAs. Unlike OFAs and SOAs, the amplification effect in RFAs is achieved by the nonlinear interaction between the signal and vibrations of the glass structure, but we will be not concerned with its details, here; the interested reader can consult [53], Chapter 5, and Section 4.5.7.

Consider a set of atoms with two energy levels 1 and 2 of energies E_1 and E_2, and electronic population densities N_1 and N_2 (the number of electrons per unit volume belonging to those energy states) respectively (Figure 4.52). An atom excited at the upper level E_2 is not in an equilibrium condition and, sooner or later, must return to the ground state through emission of a photon of frequency ν such that $E_2 = E_1 + h\nu$, h being Planck's constant. In the absence of any radiation field at this frequency, this process takes place with a certain probability, dependent on the active medium properties only, and is called *spontaneous emission*. If at a given time t there are N_2 atoms per unit volume on state 2, then the number of atoms that will drop to the ground state in unit time is given by the expression

$$\frac{dN_2}{dt} = -AN_2 \equiv -\frac{N_2}{\tau_{sp}} \tag{4.107}$$

The coefficient A [1/s] is the *spontaneous emission rate* (compare with Section 5.2.2.1); the reciprocal quantity $\tau_{sp} = A^{-1}$ gives the spontaneous emission lifetime (in seconds) of the upper laser level, which is an average measure of how long the excited state "lasts" when it is isolated from other physical systems (i.e., when no optical signal is present).

When electromagnetic radiation resonating with the energy level separation is present in the active medium, induced – or stimulated – processes also take place. The transition of an atom from level 1 to level 2 with the absorption of a photon of frequency ν is the well-known *absorption* process. An inverse process also exists in which an excited atom drops from level 2 to level 1, emitting one such photon *in a way induced* by the presence of other photons; this is called *stimulated emission*. The probability $B(\nu)$ [1/s] of both induced processes is the same and is proportional to the intensity $J(\nu)$ [W/m^2] of the optical radiation at frequency ν and to the

A coefficient:

$$B(\nu) = \frac{c^2 J(\nu)}{8\pi n_r^2 h\nu^3 \tau_{sp}} g(\nu) =: \sigma(\nu)\frac{J(\nu)}{h\nu} \tag{4.108}$$

The preceding expression contains the vacuum light speed c and the laser medium refractive index n_r; the function $g(\nu)$ is the *line-shape*, normalized to unit area. The stronger the radiation intensity J is, the stronger is the stimulated emission rate B – and the shorter the effective lifetime (i.e., determined by both spontaneous and stimulated emission processes) of the excited level. The combination of parameters

$$\sigma(\nu) = \frac{c^2}{8\pi n_r^2 \nu^2 \tau_{sp}} g(\nu) \tag{4.109}$$

is called the *transition cross-section* [m^2]. According to Equation 4.108, the induced transition probability at frequency ν is the cross-section σ times the number of photons $J(\nu)/h\nu$ at the right frequency, traveling through the lasing medium.

The introduction of the optical frequency ν probably deserves a comment. The laser line corresponding to the energy level difference E_2-E_1 is not monochromatic, but presents a certain spread B_o since the energy levels involved are broadened by the interaction among the different atoms in the solid state of matter (e.g., Stark splitting [94]; see also Section 4.5.6). Optical frequencies ν belong to such a linewidth B_o.

When the device is in thermodynamical equilibrium with its surroundings, most atoms of the laser medium will be at the ground state E_1 and only a small number will be on the excited state E_2; the ratio of the electronic population densities is expressed through the *Boltzmann probability factor*:

$$\frac{N_2}{N_1} = \exp\left(-\frac{E_2-E_1}{kT}\right) = \exp\left(-\frac{h\nu}{kT}\right) \tag{4.110}$$

where k is Boltzmann's constant and T the absolute temperature. In this condition, $N_1 > N_2$ and absorption prevails over stimulated emission. By pumping energy from the outside[10] to excite the laser medium atoms, the electronic populations of the active material can be inverted (*population inversion*: $N_1 < N_2$) with respect to thermodynamical equilibrium; stimulated emission can then overcome absorption and the active medium can supply *optical gain.* In fact, the power generated per unit volume [W/m^3] along the axial coordinate z of the amplifier can be calculated from Equations 4.108 and 4.109 as

$$P(\nu) = (N_2-N_1)B(\nu)h\nu = (N_2-N_1)\sigma(\nu)J(\nu) \tag{4.111}$$

This locally generated power $P(\nu)$ contributes to the rate of growth of the light intensity $dJ(\nu)/dz$ in the fiber or amplifier waveguide, along which the absorption and emission processes take place[11] simultaneously. Taking fiber or waveguide nonresonant attenuation (e.g., scattering loss, absorption by impurities) into account with a coefficient α, the *evolution*

[10] Pumping is done optically with OFA and RFAs, and electrically with SOAs.

[11] See Section 5.2.2.1 for a corresponding analysis in the time domain.

equation for the light along the device cavity (fiber or waveguide) is obtained in the form

$$\frac{dJ(\nu)}{dz} = (N_2 - N_1)\sigma(\nu)\Gamma J(\nu) - \alpha J(\nu) \quad (4.112)$$

In the preceding equation, Γ is an *overlap factor* between the regions where population inversion occurs and where the light distribution is different from zero over the fiber (or waveguide) transverse cross-section at coordinate z.

In general, Equation 4.112 must be solved numerically. However, a *small signal approximation* can give useful insight on the device operation. For that, suppose that the population inversion term $N_2 - N_1$ is uniform along the device and set

$$\gamma(\nu) = (N_2 - N_1)\sigma(\nu)\ [1/\mathrm{m}] \quad (4.113)$$

for the *gain coefficient* (at frequency ν). Then Equation 4.112 has the solution

$$J(\nu, z) = J(\nu, 0)\exp\{[\Gamma\gamma(\nu) - \alpha]z\} \quad (4.114)$$

As long as the gain coefficient overcomes background losses, the light intensity grows exponentially along the TW-OA, starting from the input intensity $J(\nu, 0)$. The ratio

$$G(\nu, L) := \frac{J(\nu, L)}{J(\nu, 0)} = \exp\{[\Gamma\gamma(\nu) - \alpha]L\} \quad (4.115)$$

at the amplifier output (axial coordinate $z = L$) is called the *amplifier gain* at frequency ν. Equation 4.114 is only an approximation because it does not take into account the fact that, as the amplified signal[12] $J(\nu, z)$ grows stronger and stronger, it depletes the population inversion $N_2 - N_1$, which becomes z-dependent itself. Indeed, as more and more photons induce stimulated emission, the density of atoms on the upper level 2 decreases while the ground-state population increases; this phenomenon is called *gain saturation*.

As observed in Ref. [92], the nature of the overlap factor Γ differs significantly for various kinds of amplifiers. In SOAs,[13] the waveguide cross-section is asymmetrical and determines transverse electric (TE)-polarized and transverse magnetic (TM)-polarized optical modes. Since $\Gamma_{TE} > \Gamma_{TM}$, as long as the gain coefficient γ is isotropic, the TE-polarization will substantially prevail. On the contrary, in OFAs, both Γ and γ correspond to a circular symmetry and the gain coefficient is independent of field polarization. In scattering amplifiers involving nonlinear light interaction in a fiber, gain is only different from zero along the polarization of the pump beam, unless this dependence is balanced by optical pumping with multiplexed orthogonal polarization modes.

4.5.3 Gain Saturation

As already noted, Equation 4.114 is only an approximation for the small-signal condition into the amplifier. In general, things are not so simple and the gain is saturated by the amplified

[12] Upon integrating the light intensity $J(\nu, z)$ over the device transverse cross-section, the signal power $P(\nu, z)$ is obtained. The gain coefficient is generally defined with respect to power levels rather than intensity levels. Within the present approach, we assume no difference between the two definitions.

[13] In SOAs, Γ is known as the *optical confinement factor*.

signal, as described by the expression

$$\gamma(\nu) = \frac{\gamma_0(\nu)}{1 + J(\nu)/J_{\text{sat}}} \tag{4.116}$$

where γ_0 is the *unsaturated gain coefficient* (the one in the absence of any optical signal); the *saturation intensity* J_{sat} describes the strength of the phenomenon: when the radiation intensity reaches the value J_{sat}, the gain coefficient is halved with respect to its unsaturated value. For a general solution, Equation 4.116 should be put into Equation 4.114 for (numerical) integration. But the device modeling is even more complicated; Equation 4.113 shows that even the gain coefficient γ_0 depends on the lasing medium excitation.

4.5.3.1 Optically Pumped Amplifiers

This means that in optically pumped amplifiers (such as EDFAs) a second differential *equation for the pump beam* with intensity J_{P}

$$\frac{\text{d}J_{\text{P}}(\nu)}{\text{d}z} = N_1\sigma_{\text{P}}(\nu)\Gamma_{\text{P}} J_{\text{P}}(\nu) - \alpha_{\text{P}} J_{\text{P}}(\nu) \tag{4.117}$$

should be considered (and integrated) in parallel with Equation 4.112. For other kinds of pumping, Equation 4.117 would be replaced by an analogous expression for the pump source (see Equation 4.127 for the SOA case).

The model is completed by a set of *rate equations for the population densities*, in particular for the inversion population density:

$$\frac{\text{d}(N_2 - N_1)}{\text{d}t} = -\left[K_{\text{P}} + \frac{r\sigma(\nu)J(\nu)}{h\nu}\right](N_2 - N_1) + H_{\text{P}}N_{\text{tot}} \tag{4.118}$$

where K_{P} and H_{P} are quantities related to pump strength, N_{tot} is the total density of active atoms (or ions), and r is a numerical factor depending on the *spectroscopy* of the laser system.

Indeed, as far as optically pumped systems are concerned, it is easy to see that the active atoms *cannot* have two energy levels only involved in the laser action. Stimulated emission from level 2 to level 1 is directly competing with pumping from 1 to 2 (pump photons and signal photons would be the same!) and population inversion between levels 1 and 2 would not be reached. For optically pumped systems, one or two energy levels must be involved in the laser mechanism besides the two laser levels; these are the basic types of the *three-level laser* and *four-level laser* systems (Figure 4.59). The parameter r in Equation 4.118 is $r = 2$ for the three-level laser and is unity for the four-level laser. Also, the values of the pump parameters K_{P} and H_{P} depend on the spectroscopic scheme: K_{P} is substantially the lifetime of the upper laser level, while H_{P} is related to the pump strength.

In a three-level laser, electronic transitions involve the ground state 1, the upper laser level 2 and a pump level 3 (left side of Figure 4.59). Pump photons[14] are absorbed by electrons at the ground state 1, which jump to level 3. From there, electrons relax to the upper laser level 2 by nonradiative decay: the amplifier works on the transition between levels 2 and 1.

[14] Pump photons are at a shorter wavelength than signal photons, now, and are no longer in direct competition with stimulated emission.

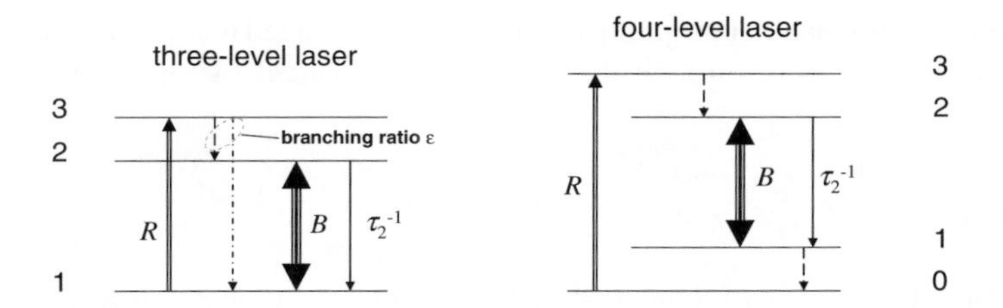

Figure 4.59 Three- and four-level laser systems

In a four-level system, the lower laser level does not coincide with the ground state 0 anymore. Electrons are pumped from level 0 to level 3,[15] then rapidly decay to the upper laser level 2. The amplifier again works in the channel between levels 2 and 1. Finally, the lower laser level 1 is depopulated by decay to the ground state 0 and this last process makes the *four-level system more efficient in reaching population inversion than the three-level scheme* (right side of Figure 4.59). In that figure, R [1/s] denotes the pumping rate, B [1/s] the induced process rate of Equation 4.108, τ_2 [s] the lifetime of the upper laser level,[16] and ε the branching ratio of the pump level 3.[16] Then coefficients K_P and H_P can be described by

$$K_P = \varepsilon R + \frac{1}{\tau_2}, \quad H_P = \varepsilon R - \frac{1}{\tau_2} \qquad \text{(three-level system)} \tag{4.119}$$

$$K_P = \frac{1}{\tau_2}, \quad H_P = R \qquad \text{(four-level system)} \tag{4.120}$$

With the amplifier in steady-state conditions, Equations 4.116 and 4.118 give

$$\frac{\gamma}{\sigma} = N_2 - N_1 = \frac{H_P N_{tot}}{K_P + \frac{r\sigma(v)J(v)}{hv}} = \frac{(N_2 - N_1)_0}{1 + \frac{r\sigma(v)J(v)}{K_P hv}} = \frac{\gamma_0}{\sigma\left[1 + \frac{r\sigma(v)J(v)}{K_P hv}\right]} \tag{4.121}$$

where the saturation intensity can be written

$$J_{sat,3} = \frac{K_{P,3} hv}{2\sigma(v)} \quad \text{and} \quad J_{sat,4} = \frac{K_{P,4} hv}{\sigma(v)} \tag{4.122}$$

for three-level and four-level systems respectively. For the four-level laser:

$$K_{P,4} = \tau_2^{-1} \quad \text{(four-level system)} \tag{4.123}$$

is the inverse of the total lifetime of the upper laser level. In a three-level system:

$$K_{P,3} = \tau_2^{-1}\left(1 + \varsigma \frac{J_P}{J_P|_{th}}\right), \quad \text{(three-level system)} \tag{4.124}$$

[15] In a three-level system time τ_2 is the same as the spontaneous lifetime τ_{sp} in Equation 4.1. This does not hold in a four-level system, where $\tau_2 < \tau_{sp}$, since the upper laser level may also decay to level 0 directly.

[16] In the three-level system, ε denotes the fraction of electrons that decay from level 3 to level 2, with respect to the total number of electrons decaying from level 3. This branching ratio does not influence the rate equations for the four-level laser.

also depends on the intensity of the pump radiation J_P, normalized to the transparency threshold $J_P|_{th}$, at which $N_2 - N_1 = 0$, through the Boltzmann statistical factor (compare with Equation 4.110):

$$\varsigma = \frac{1+\exp(-h\nu/kT)}{1-\exp(-h\nu/kT)}, \quad \text{where} \quad \exp\left(-\frac{h\nu}{kT}\right) = \left.\frac{N_2}{N_1}\right|_{\text{no pump}} \tag{4.125}$$

At weak pumping, both systems tend to work in the same way (since $\varsigma J_P/J_P|_{th} \sim 1$). The main difference between them appears as the pump level becomes stronger and stronger. *In a four-level system the depletion of population inversion is uniquely due to stimulated emission.* On the contrary, *in a three-level system, population inversion is a balance between stimulated emission and pumping.* In fact, three-level system atoms that decay to the lower laser level through stimulated emission deplete the inversion, but this latter rises again when the population of level 1 is in turn excited by the pump beam.

4.5.3.2 Electrically Pumped Amplifiers

Up to now, we have mainly considered optically pumped amplifiers. For an electrically pumped SOA, Equation 4.112 is conveniently rewritten as

$$\frac{dJ(\nu)}{dz} = (n-n_0)A_g\Gamma J(\nu) - \alpha J(\nu) \tag{4.126}$$

while a rate equation for the charge carrier density n in the semiconductor is

$$\frac{dn}{dt} = -\frac{n}{\tau_{sp}} - A_g(n-n_0)\frac{J(\nu)}{h\nu} + \frac{j}{ed} \tag{4.127}$$

In Equations 4.126 and 4.127, t is time, τ_{sp} is the (density-dependent) carrier relaxation time, A_g is the (linear) gain coefficient, n_0 is the carrier density at transparency (i.e., when the semiconductor optical waveguide is transparent to radiation), j is the injection current density, e is the electric charge, and d is the thickness of the region in the SOA, where stimulated emission appears (active layer in Figure 4.56).

Equations 4.112 and 4.118 (as well as Equations 4.126 and 4.127) are coupled together via the stimulated emission rate $(N_2-N_1)\sigma(\nu)J(\nu)$ (rewritten as $(n-n_0)A_gJ(\nu)$ for the SOA case).

4.5.4 Noise

In a laser medium under population inversion conditions, stimulated emission is induced by the presence of the signal beam and the signal is amplified at the expense of the pumping energy. The stimulated photons are identical to signal photons and add to them coherently. When the active atoms are excited, they will also decay in a spontaneous way, emitting photons in the signal band, but in a totally uncorrelated way to the signal itself. This *local spontaneous emission* adds to the coherent signal anywhere in the amplifier, and Equation 4.112 suggests that all spontaneous emission generated locally is amplified itself via the gain coefficient (amplified spontaneous emission, ASE). ASE is the device noise that was mentioned above. It can be shown (see Section 5.2.2.1) that the noise power emitted as ASE out of the amplifier is given by

$$P_{\mathrm{ASE}} = m_{\mathrm{t}}(G-1)n_{\mathrm{sp}}h\nu B_{\mathrm{o}} \tag{4.128}$$

where m_{t} is the number of noise modes ($m_{\mathrm{t}} = 1$ for SOAs, $m_{\mathrm{t}} = 2$ for OFAs), $n_{sp} \geq 1$ is the so-called *spontaneous emission factor*, the number of photons emitted spontaneously that couple to the device optical field, and B_{o} is the optical transition bandwidth, since the two laser levels are not perfectly sharp in energy, but have a finite energy extension. Equation 4.128 tells simply that P_{ASE} is the local spontaneous emission $m_{\mathrm{t}}n_{\mathrm{sp}}$ times the photon energy $h\nu$, integrated over the optical bandwidth B_{o}, and amplified by the active medium by $(G-1)$, with G the optical gain (Equation 4.115). The spontaneous emission is amplified by $G-1$ (not G) since it is generated *inside* the amplifier and cannot exploit its gain fully. Though for $G \gg 1$ there is little difference between factors $G-1$ and G, it is easy to see that the first is the correct one, since in the limit of a transparent active medium $(G \rightarrow 1)$ no net spontaneous emission occurs. From Equation 4.128 it can be seen that, depending on the gain coefficient G and the bandwidth B_{o}, ASE noise can reach appreciable values. For $G \sim 32$ dB, one spontaneous photon in each ASE mode ($m_{\mathrm{t}}n_{\mathrm{sp}} = 2$) and $B_{\mathrm{o}} \sim 4.2$ THz (the EDFA gain bandwidth), P_{ASE} can be as high as 1 mW. Since the signal linewidth at very high bit-rates B will have a band $\sim B$, the amount of noise overlap to the signal can be controlled by filtering the amplifier output around the signal. For example, a noise reduction by a factor $B/B_{\mathrm{o}} \sim 100$ would be possible even for 40 Gbit/s optical flows.

The noise properties of an OA are often described by means of the so-called amplified noise figure:

$$F \cong \frac{1}{G}\left(1 + \frac{2P_{\mathrm{ASE}}}{m_{\mathrm{t}}B_{o}h\nu}\right) = \frac{1+2N}{G} \tag{4.129}$$

with N being the number of noise photons. Section 5.2.2.1 gives a more detailed treatment of noise in optical (fiber) amplifiers.

4.5.5 Gain Dynamics

From Equation 4.118 it is seen that population inversion has a relaxation time[17]

$$\left(\tau_{2,\mathrm{eff}}\right)^{-1} = K_{\mathrm{P}} + \frac{r\sigma(\nu)J(\nu)}{h\nu} = K_{\mathrm{P}}\left[1 + \frac{J(\nu)}{J_{\mathrm{sat}}}\right] \tag{4.130}$$

Then, Equations (4.120–123) give

$$\tau_{2,\mathrm{eff}} = \frac{\tau_2}{1 + J(\nu)/J_{\mathrm{sat}} + (1-\delta)J_{\mathrm{P}}/J_{\mathrm{P}}|_{\mathrm{th}}}, \quad \text{with} \quad \delta = \begin{cases} 0 & \text{for a three-level system} \\ 1 & \text{for a four-level system} \end{cases} \tag{4.131}$$

For OFAs: $\tau_2 \sim 10$ ms, $\tau_{2,\mathrm{eff}} = 0.05$–$10$ ms; for SOAs: $\tau_2 \sim 1$ ns, $\tau_{2,\mathrm{eff}} = 0.01$–$0.2$ ns.

[17] If more than one signal is simultaneously present, then the saturation term $J(\nu)/J_{\mathrm{sat}}$ must be replaced by the corresponding sum $\sum_i J(\nu_i)/J_{\mathrm{sat}}$ at the various signal frequencies $\nu_1, \nu_2, \ldots$.

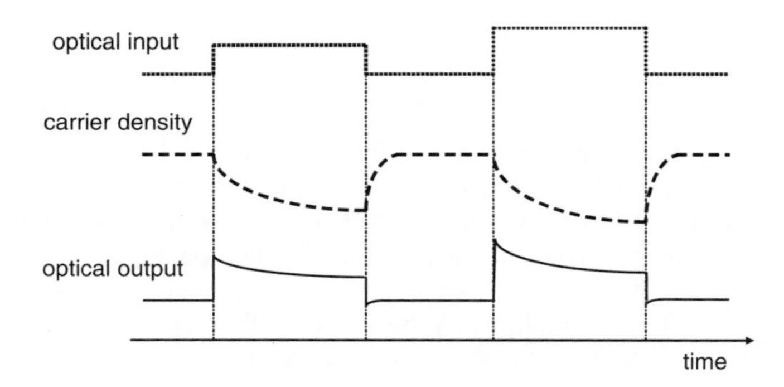

Figure 4.60 SOA response to modulation of the optical input

As a consequence, *OFAs and SOAs exhibit totally different dynamical properties*. At the modulation rates used in optical communications, OFAs are virtually independent of patterning effects (the differential saturation between low- and high-frequency components during intensity modulation) and are independent of waveform distortion and of nonlinear crosstalk among different channels: the signal is amplified by a saturated gain determined by the average power level. On the contrary, the gain dynamics of SOAs will follow the modulation rate (up to ~100 GHz!) and produce a set of transmission distortions, in digital as well as in analog transmission (Figure 4.60).

4.5.6 Optical Fiber and Semiconductor Optical Amplifiers

Rare-earths are used as laser active materials in OFAs because of their peculiar *electronic structure*. Hosted in a glass matrix (or in a crystal), rare-earth elements M (those of the lanthanide series in Figure 4.53) behave as trivalent M^{3+} ions. The electronic structure is such that these M^{3+} ions have a lot of radiative transitions[18] at optical frequencies of interest for fiber-optic telecommunications. These transitions are quite sharp, atomic-like, even if they occur in a solid. Indeed, the spectroscopic characteristics of M^{3+} ions are scarcely influenced by external conditions, because these ions have a completely filled outer electronic orbit, which efficiently screens the inner ionic structure from external perturbations [94–96]. In conclusion, rare-earth ions have optical transitions around 1310 and 1550 nm, with definite and attractive spectral characteristics. Moreover, particular aspects of these features can be tailored most conveniently, at least in principle, by modifying the composition of the glass host matrix (e.g., silica doped with germanium, silica co-doped with germanium and aluminum, fluoro-zirconate ZBLAN glass, tellurium oxide glass, and so on).

The first step in OFA modeling is a sufficiently detailed knowledge of the amplification scheme on which the device performance is based. These schemes involve a number of energy levels (not just two; see Figures 4.59 and 4.62) of the active ion M^{3+} in the glass host. Each level has a fine structure in terms of energy states due to the Stark splitting caused by the local electric field in the host (Figure 4.61). In this way, groups of states dependent on the value of the

[18] In an energy level diagram, a radiative transition is a transition supported by the emission/absorption of photons.

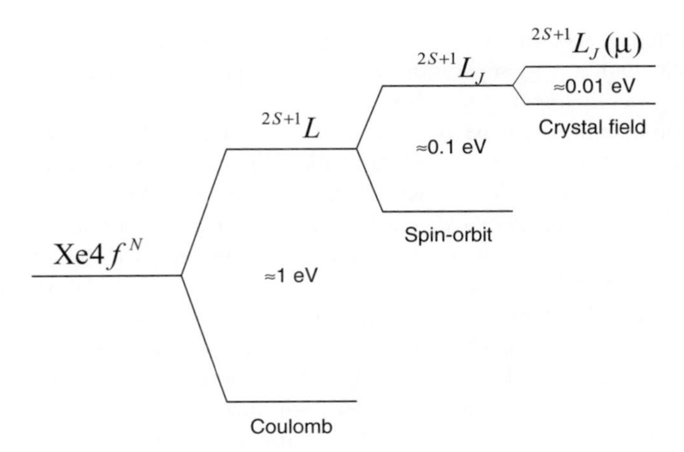

Figure 4.61 Schematics of rare earth ion energy levels, as determined by electrostatic, spin-orbit, and crystal field interactions

ion total angular momentum J arise (called J-manifolds) having a width of some hundredths of an electron-volt.

Varying the characteristics of the glass matrix, the centre of every J-manifold is little influenced, but the distribution of states within it can change considerably [94–98], influencing the width and shape of the gain band. In glasses, local variations of symmetry and electric charge around the active ions generate a statistical distribution of Stark splitting; thence the spectrum cannot generally be resolved in single lines. The effects of this kind of *inhomogeneous spectral broadening* are independent of the temperature and affect the transitions between different J-manifolds.

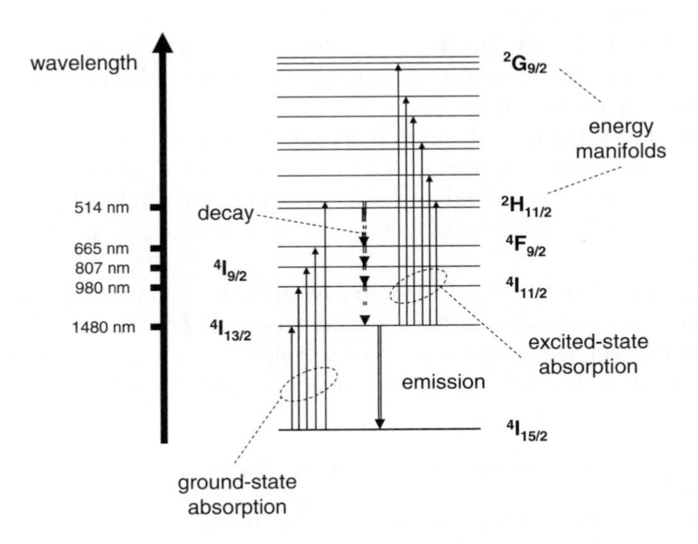

Figure 4.62 Energy levels of Er^{3+} ions in glass

On the other hand, relaxation mechanisms between states in one *J*-manifold are extremely rapid due to their small energy separation, which favors phonon-assisted nonradiative decays, causing a *homogeneous line broadening*.

This second broadening effect becomes weaker and weaker as the temperature lowers, since thermal conditions influence the number of available phonons (phonons are the elementary excitations of glass vibrations). This is actually the picture for OFAs [97,98], when one usually speaks of an *inhomogeneously broadened homogeneous transition line*.

In SOAs, the main difficulty is to relate the optical properties to the electrical injection current distribution. An SOA is a small crystalline structure (size ~100 μm) made by a pile of semiconductor layers of suitable composition, grown on an *n*-type substrate. The structure core is the active layer (Figure 4.56), where lasing takes place: its dimensions are 2–5 μm in width, 0.1–0.2 μm in height, and 250–400 μm in length. Another type of SOA, called a multi-quantum-well structure, has the active region partitioned into a set of thinner layers (the quantum wells), the thickness of which is 0.01–0.02 μm. For SOAs, the pumping mechanism is realized by the bias current in the p–n junction structure of the semiconductor crystal, generally composed by elements from Groups III and V of the periodic table (Figure 4.53).

Optical transitions take place between filled electron states at the conduction band edge and free electron states (i.e., holes), at the valence band edge. Pumping efficiency is increased by adopting a *double heterostructure* (Figure 4.57), which provides the *electrical confinement* of charge carriers around the active region. Besides this, the behavior of the refractive index on the double heterostructure layers is such that it also favors the optical confinement for radiation around the active region itself. Effectiveness and stability of operation are assured using the so-called *index-guided* structures, where the spatial distribution of voltage barriers internal to the crystal and/or a suitable geometry of metal contacts to the device guarantee a better confinement to the injection current.

The band structure of the semiconductor device, characterized by its bandgap (Table 4.6[19]), can be approximated as a discrete two-level laser system. Indeed, intra-band relaxation times are very short compared with the inter-band dynamics of charge carriers, which is featured by a lifetime of the order of 200 ps–4 ns. However, this simple model does not take into account some effects related to the detailed band structure, such as band filling, hot carrier, carrier leakage, and bimolecular recombination.

The actual band structure is much more complicated than a two-level system, as Figure 4.63 suggests. A *direct bandgap*[20] is present between the conduction band on one side and heavy-hole and light-hole valence bands – originated by the complex dynamics of electron states in the crystal – and a split-off valence band due to spin-orbit interaction for the electrons in the semiconductor.

From this scheme, it emerges that, to a reasonable approximation, the semiconductor crystal can be considered as a gain material subject to *homogeneous broadening*. This means that all carriers in an SOA react in the same way at any wavelength in the gain bandwidth. Moreover, since two effective laser levels represent the conduction and valence bands, the laser emission

[19] For example, according to the alloy composition, InGaAsP has a bandgap of 0.73–1.35 V, corresponding to photon emission at a wavelength of 900–1700 nm.

[20] A semiconductor is called direct bandgap when the extrema of the conduction and valence bands are located at the same value of the state wavenumber in the crystal; see Figure 4.63. In this case only, radiative recombinations across the bandgap can actually take place.

Table 4.6 Bandgap of some semiconductor compounds

Material	Composition	Bandgap (eV)	Wavelength (nm)
Indium phosphide	InP	1.35	920
Gallium arsenide	GaAs	1.42	870
Aluminum gallium arsenide	AlGaAs	1.40–1.55	800–900
Indium gallium arsenide	InGaAs	0.95–1.24	1000–1300
Indium gallium arsenide phosphide	InGaAsP	0.73–1.35	900–1700

mechanism follows a four-level scheme to a good approximation, because pumping does not influence the characteristics of the lower laser level, in contrast to what happens in EDFAs (see Figure 4.59).

4.5.7 Raman Amplifiers

Raman amplifiers are important for telecommunications; for that reason this section is devoted to a detailed discussion of their properties.

4.5.7.1 Distributed Raman Amplification

In recent years, interest in distributed Raman amplification has constantly increased in the scientific community due to some features that make this kind of amplification technology competitive when compared with other techniques (such as EDFAs). These features include (but are not limited to) the capability of achieving amplification at any wavelength of interest (by changing the wavelength of the Raman pump), a low level of noise introduced into the

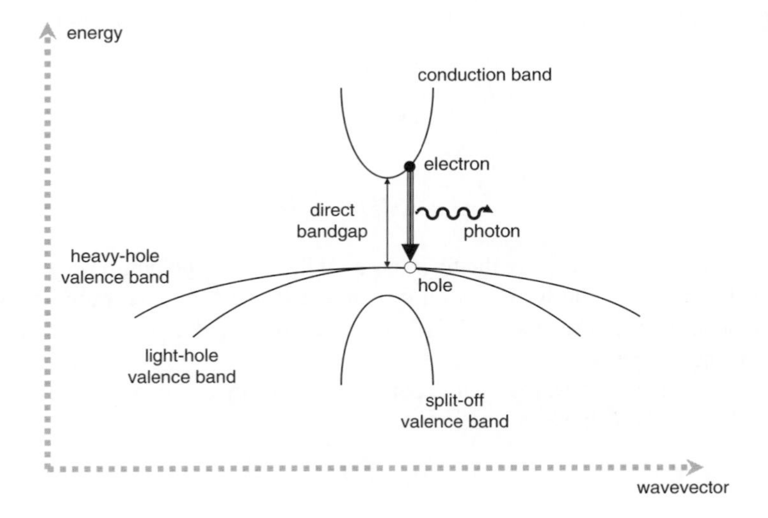

Figure 4.63 Schematic band structure of a III–V semiconductor

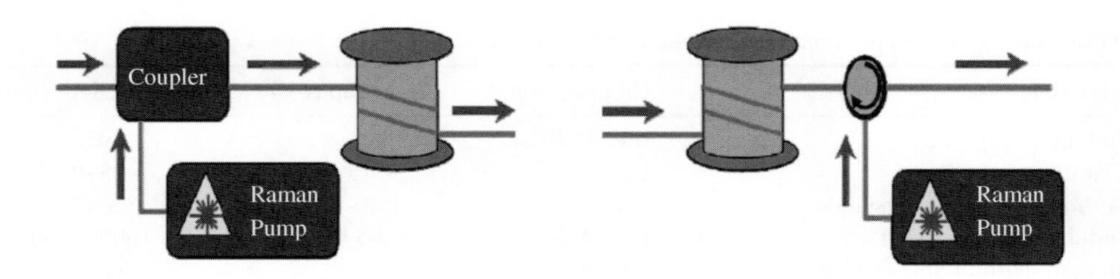

Figure 4.64 Raman amplification: co-propagating configuration (left); counter-propagating configuration (right)

system, and *no need to use a different medium to achieve amplification apart from the transmission fiber itself.*

Figure 4.64 shows the simplest configurations involving a single signal (propagating from the left to the right in both pictures) and a single Raman pump, propagating in the same direction as the signal in the left picture (co-propagating configuration), and in the opposite direction in the right picture (counter-propagating configuration).

More complicated configurations, including more Raman pumps both co- and counter-propagating, can also be used to improve the behavior of the distributed Raman amplification (to flatten the amplifier gain on a certain band or to compensate the fiber losses locally, to cite a few examples).

4.5.7.2 System Modeling

To evaluate the performances of a Raman system, we need a mathematical model of the pump(s) and signal(s) interaction; this model, in the CW and quasi-CW conditions – and for single pump/single signal interaction – can be found in Ref. [53]. It can be introduced here with a suitable modification of Equations 4.112 and 4.117 for the intensities J_S and J_P of the signal and the pump:

$$\varepsilon\frac{dJ_P}{dz} = -g_R\frac{\lambda_S}{\lambda_P}J_PJ_S - \alpha_PJ_P \tag{4.132}$$

$$\frac{dJ_S}{dz} = g_RJ_PJ_S - \alpha_SJ_S \tag{4.133}$$

In the above relations, α_S and α_P are the fiber losses at the signal and pump wavelengths λ_S and λ_P respectively and g_R is the Raman gain coefficient (for a reference pump at wavelength $\lambda_0 = 1\ \mu m$); the numerical factor $\varepsilon = \pm 1$ (+1 for the co-propagating configuration, −1 for the counter-propagating one).

It should be noted that the Raman gain coefficient scales inversely to the wavelength of the pump, according to the following relation:

$$\frac{g_{R1}}{g_{R2}} = \frac{\lambda_2}{\lambda_1} \tag{4.134}$$

where g_{Rx} is the Raman gain coefficient related to a pump at wavelength λ_x.

Integrating the intensity over the fiber cross-section,[12] one gets the optical power:

$$J_i(z)A_{\mathrm{eff}} = P_i(z), \quad (i = S, P) \tag{4.135}$$

with A_{eff} the fiber effective area. With this substitution, Equations 4.132 and 4.133 become

$$\varepsilon\frac{\mathrm{d}P_{\mathrm{P}}}{\mathrm{d}z} = -\frac{g_{\mathrm{R}}}{A_{\mathrm{eff}}}\frac{\lambda_{\mathrm{S}}}{\lambda_{\mathrm{P}}}\mathrm{P}_{\mathrm{P}}P_{\mathrm{S}} - \alpha_{\mathrm{P}}P_{\mathrm{P}} \tag{4.136}$$

$$\frac{\mathrm{d}P_{\mathrm{S}}}{\mathrm{d}z} = \frac{g_{\mathrm{R}}}{A_{\mathrm{eff}}}P_{\mathrm{P}}P_{\mathrm{S}} - \alpha_{\mathrm{S}}P_{\mathrm{S}} \tag{4.137}$$

Power equations will be used in the following to discuss in some detail amplifier noise (Section 4.5.7.4).

4.5.7.3 On/Off Raman Gain

In this section, the definition of on/off Raman gain is given, which will be useful in the following. The on/off Raman gain at a specified position z along the fiber is the ratio between the signal powers at z when the Raman pump is turned on and off respectively (hence the name).

As an example, Equations 4.136 and 4.137 for the counter-propagating case model the system when the pump is turned on; to solve them, the *undepleted pump approximation* is used;[21] that is, Equation 4.136 is simply

$$-\frac{\mathrm{d}P_P}{\mathrm{d}z} = -\alpha_{\mathrm{P}}P_{\mathrm{P}} \tag{4.138}$$

which can be immediately solved to obtain

$$\mathrm{P}_{\mathrm{P}}(z) = P_{\mathrm{P}}(0)\exp(\alpha_{\mathrm{P}}z) \tag{4.139}$$

In the counter-propagating case, $P_{\mathrm{P}}(0)$ is unknown, but $P_{\mathrm{P}}(L)$ is assigned, L being the fiber length; using Equation 4.139, we get

$$P_{\mathrm{P}}(z) = P_{\mathrm{P}}(L)\exp[-\alpha_{\mathrm{P}}(L-z)] \tag{4.140}$$

Using this expression in the signal Equation 4.137, after few mathematical steps we easily obtain

$$P_{\mathrm{S}}(z) = P_{\mathrm{S}}(0)\exp\left[-\alpha_{\mathrm{S}}z - g_{\mathrm{R}}P_{\mathrm{P}}(L)\mathrm{e}^{-\alpha_P L}\frac{\mathrm{e}^{\alpha_{\mathrm{P}}z}-1}{\alpha_{\mathrm{P}}}\right] \tag{4.141}$$

When the Raman pump is turned off (i.e., $P_{\mathrm{P}}(z) \equiv 0$), the following relation holds:

$$\frac{\mathrm{d}P_{\mathrm{S}}}{\mathrm{d}z} = -\alpha_{\mathrm{S}}P_{\mathrm{S}} \tag{4.142}$$

[21] This approximation is similar to the small-signal approximation discussed in Section 4.5.2.

so that

$$P_S(z) = P_S(0)\exp(-\alpha_S z) \tag{4.143}$$

Finally, the on/off Raman gain is obtained from the ratio of Equations 4.141–4.143

$$G_{\text{on/off}}(z) = \exp\left[-g_R P_P(L) e^{-\alpha_P L} \frac{e^{\alpha_P z} - 1}{\alpha_P}\right] \tag{4.144}$$

It is particularly interesting to evaluate the on/off Raman gain at the end of the fiber; in this specific case, we obtain

$$G_{\text{on/off}}(L) = \exp\left[-g_R P_P(L) \frac{1 - e^{-\alpha_P L}}{\alpha_P}\right] \tag{4.145}$$

4.5.7.4 The Amplified Spontaneous Emission Noise

As already noted in Section 4.5.4, to evaluate the performance of an amplifier, the noise introduced by the device into the system should be taken into account. RFAs are not an exception to this general rule; the predominant noise component introduced by the Raman effect is ASE noise, due to a spontaneous conversion of a fraction of the pump power on all of the Raman band and to the subsequent amplification of the converted power.

To include ASE in the model, the simplest way is to add a single (spontaneously emitted) photon to the signal photons; this photon is then amplified just as the coherent signal. To include this photon, we need to rewrite the power equations (Equations 4.136 and 4.137) in terms of the photon numbers. This can be more easily accomplished by the following change of variables [52]:

$$P_i(z) = \frac{E_i}{t_g} = n_i(z) h\nu_i \frac{v_g}{L} \tag{4.146}$$

where $i = S, P, P_i$ and n_i are the power and the number of photons of the radiation component at frequency ν_i (having energy $E_i = n_i h\nu_i$), v_g is the group velocity of the light into the fiber, and L is the length of the fiber span, so that $t_g = L/v_g$ is the time relating power P_i to energy E_i. Applying those substitutions, one obtains

$$\varepsilon \frac{dn_P}{dz} = -\gamma_0 n_P n_S - \alpha_P n_P \tag{4.147}$$

$$\frac{dn_S}{dz} = \gamma_0 n_P n_S - \alpha_S n_S \tag{4.148}$$

where n_P and n_S are the number of pump and signal photons respectively and γ_0 is the Raman gain coefficient for the photon equations that is defined as follows:

$$\gamma_0 = \frac{g_0}{A_{\text{eff}}} \frac{\nu_P^2}{\nu_0} \frac{h v_g}{L} \tag{4.149}$$

g_0 being the Raman gain coefficient related to a reference pump at wavelength $\lambda_0 = 1$ μm (with ν_0 the corresponding frequency).

Before discussing the noise contribution, a simplifying consideration can be done. If one chooses the co-propagating case[22] ($\varepsilon = 1$) and neglects fiber losses, adding Equations 4.147 and 4.148 to each other gives

$$\frac{\mathrm{d}(n_P + n_S)}{\mathrm{d}z} = 0 \tag{4.150}$$

This condition means that (neglecting the losses) the total number of photons is a constant; so, for every photon that is lost from the pump, there is a new photon acquired by the signal. This confirms that the Raman effect is a *single-photon* process.

As said before, to take ASE into account, one spontaneously emitted (signal) photon is introduced in the photon equations; then, Equations 4.147 and 4.148 should be modified accordingly:

$$\varepsilon \frac{\mathrm{d}n_P}{\mathrm{d}z} = -\gamma_0 n_P (n_S + 1) - \alpha_P n_P \tag{4.151}$$

$$\frac{dn_S}{dz} = \gamma_0 n_P (n_S + 1) - \alpha_S n_S \tag{4.152}$$

The solutions of these equations give the evolution of signal and noise.

4.5.7.5 Noise Figure of Raman Amplifiers

To evaluate the performance of the system, and to make a comparison with different amplifying solutions, it is necessary to quantify the noise of distributed Raman amplification. To do so, one introduces the noise figure F of an RFA accordingly to the definition given by Refs [53,99]; that is:

$$F(z) = \frac{1 + 2N(z)}{G_{\mathrm{on/off}}(z)} \tag{4.153}$$

where $N(z)$ is the number of photons due to the ASE and $G_{\mathrm{on/off}}(z)$ is the on/off Raman gain, as defined in Section 4.5.7.3.

So, with RFAs we also come to a formally identical expression for the amplifier noise as for OFAs (Equation 4.129).

4.5.8 Lasers and Amplifiers

The distinction between laser *oscillators* and *amplifiers* is that lasers are amplifiers equipped with optical feedback (from terminal mirrors). In the case of semiconductor devices, LDs are simpler to manufacture than SOAs, since the standard fabrication process of the devices provides them with cleaved end facets, acting as mirrors (Figure 4.65).

[22] Equation 4.150 can be shown to hold also for the counter-propagating case.

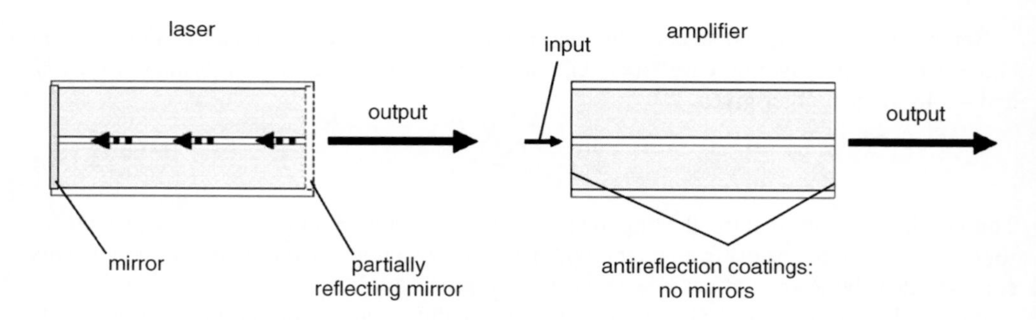

Figure 4.65 LD (left) and SOA (right)

This approach directly yields FP-LDs with multimode emission spectra, since their resonance wavelengths are given by the condition

$$\lambda = \frac{2n_g L}{m}, \quad m \text{ an integer} \tag{4.154}$$

In Equation 4.154, L is the device cavity length (250–500 μm) and n_g is the effective refractive-index[23] of the active layer (4–4.5); an integer number of half-wavelengths should fall in the cavity for constructive interference conditions. Then the typical separation between FP resonances (called the free spectral range (FSR)) is given by

$$\mathrm{FSR} = \frac{\lambda^2}{2n_g L} \tag{4.155}$$

LD sources suitable for transmission in single-mode optical fibers need a single-frequency emission spectrum. The suppression of FP resonances[24] is also an issue in TW-SOAs, where the single-pass gain of the material is efficiently exploited, with peak gain values of 20–25 dB at injection currents of 100–150 mA; the 3 dB amplification band is ~10 THz. This 3 dB band $\Delta\nu_g$ is related to the reciprocal coherence time of the optical transition T_r by $\Delta\nu_g = (\pi T_r)^{-1}$; for semiconductors used in SOAs, $T_r \sim 0.03$ ps.

On the other hand, the gain for an FP-SOA can be expressed by the formula [92]

$$G_{\mathrm{FP}} = \frac{(1-R_1)(1-R_2)G}{[(1-\sqrt{R_1 R_2}G)^2 + 4G\sqrt{R_1 R_2}\sin^2\phi]} \tag{4.156}$$

where G is the single-pass gain as given by Equation 4.115 and R_1 and R_2 are the reflectivities of the two facets. The gain spectrum has rather sharp maxima when the optical phase

$$\phi = \frac{2\pi n_g L}{\lambda} \tag{4.157}$$

[23] That is, the group refractive index of the optical mode in the device cavity.

[24] Mirror reflectivity can be reduced or eliminated by different techniques: *antireflection coating* of end facets, *windowed* geometries, *tilted* cavities [53]. One speaks of TW-SOAs when the residual reflectivity R at end facets is lower than 0.1–1%. For higher R values, devices with different operation come into existence: injection-locked SOAs or FP SOAs, whether they are operated above or below laser threshold.

is an integer number (compare with Equation 4.154). According to Equation 4.156, a rather high gain level can be obtained even with a small single-pass gain G, since mirror reflectivity can easily reach 30%; this means instability of the FP amplifier with respect to the TW device. Moreover, the periodic resonances in the G_{FP} spectrum are only ~1 GHz wide and are not suitable for high bit-rate signal amplification. Furthermore, these resonances depend upon temperature through cavity characteristics.

4.6 Optical Filters and Multiplexers

4.6.1 Introduction

Optical multipexing is a technique used in optical-fiber networks for enhancing the capacity of point-to-point links as well as for simplifying the routing process within the optical layer. In today's core network the main multiplexing technique is the optical domain equivalent of frequency-division multiplexing, namely WDM, for reasons to be shown clearly in the following paragraphs. In WDM, several wavelengths, each one modulated by a separate data pattern, are launched into (multiplexed) or decoupled from (demultiplexed) an optical fiber (Figure 4.66).

On the component level there is a fundamental difference in the nature of the devices used for WDM and for other multiplexing techniques. It is the passive nature of these devices that gives WDM all these desirable characteristics generally identified as "transparency"; namely, bit-rate independence, as well as modulation format and protocol insensitivity. Because of its dominant role in real telecommunication systems, we will only consider wavelength multiplexing for the remaining part of this chapter.

In this section, a specific set of optical components that are required for the implementation of WDM systems is addressed, namely optical (de-)multiplexers and filters both with similar functionality: to facilitate channel separation from the rest of the WDM comb [55]. Optical multiplexers are devices that combine the output of several receivers and launch it as a set into

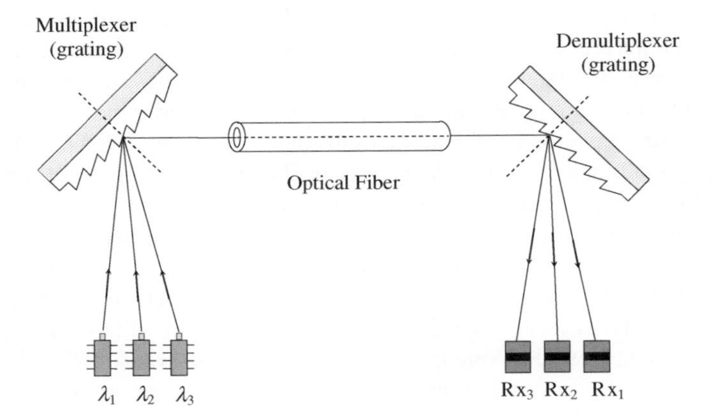

Figure 4.66 The conceptual description of WDM: three wavelength channels are multiplexed, transmitted through an optical fiber, demultiplexed at the end, and detected from the corresponding receivers

an optical fiber and the demultiplexers split the received multichannel comb into individual channels and, hence, wavelengths. Filters, on the other hand, have a similar functionality; but, as the name implies, they filter a specific wavelength out of a band or ASE noise, while tunable filters filter out specific wavelengths that can be altered by tuning the passband of the optical device, usually selected at the end of the WDM link or at a switch. In this section, the functionality and the physics of these devices, which are very similar, will be discussed.

When seen from the point of view of technical applications, the most important phenomena of light are *interference* and *diffraction*. Hence, the techniques used for optical (de-)multiplexing or filtering (regardless of the form in which they appear) are primarily based on one of them. There is no satisfactory explanation of the difference between these two terms [62]. However, for any practical reason, when two optical sources interfere the result is called *interference*, whilst the term *diffraction* is more appropriate when there are a large number of sources. For optical (de-)multiplexing purposes, the exploitation of two-beam interference is made through devices based on division of the amplitude of the incident beam before they are superimposed again. Under this category are devices like the Mach–Zehnder, the Michelson, and the Sagnac interferometers. An important family of (de-)multiplexing devices are based on arrangements involving multiple divisions of the amplitude or multiple divisions of the wavefront of the incoming wave and they are called either *interference filters* (FP interferometers, multilayer thin-film filters, and fiber Bragg gratings (FBGs)) or *diffraction gratings* (integrated optic, free-space, or acousto-optic devices) respectively.

4.6.2 Optical (De-)Multiplexing Devices

4.6.2.1 Functionality

The choice of the technology to be used depends strongly on the type of application under consideration. Hence, for low- to medium-capacity networks – that is, for up to eight-wavelength-channel WDM systems (with bit rates ranging from 644 Mbit/s to 10 Gbit/s per wavelength) – all the aforementioned devices could be used indistinguishably (Figure 4.67). When the total number of wavelength channels N is the predominant consideration for the choice of a technology (in particular when $N \geq 32$), the diffraction gratings are the primary candidates. Nevertheless, regardless of the technological platform, higher wavelength channel count can be obtained by adding up groups of band-optimized devices. For example, (de-)multiplexing devices with up to 60 channels are commercially available using interleaving of band-optimized interference filters in a parallel or cascadeable configuration (Figure 4.67c and d), whilst several hundred to thousand channels could be produced with band-optimized diffraction gratings.

In Figure 4.67, the four different arrangements produce the same final result from a systems point of view. In Figure 4.67b, the star coupler facilitates in distributing the same multiwavelength signal to all its N ports (each one will collect $1/N$ of the original optical power). Then a filter will select the requested wavelength. From a functionality point of view, the final outcome is the same as if a diffraction grating is used. Nevertheless, the performance in terms of crosstalk and losses might be different. In any case, the diffraction-grating-based devices are expected to dominate in the high-capacity systems and, therefore, will be dealt with in more detail here. In the sections below, a more detailed comparison between the technological platforms will be provided.

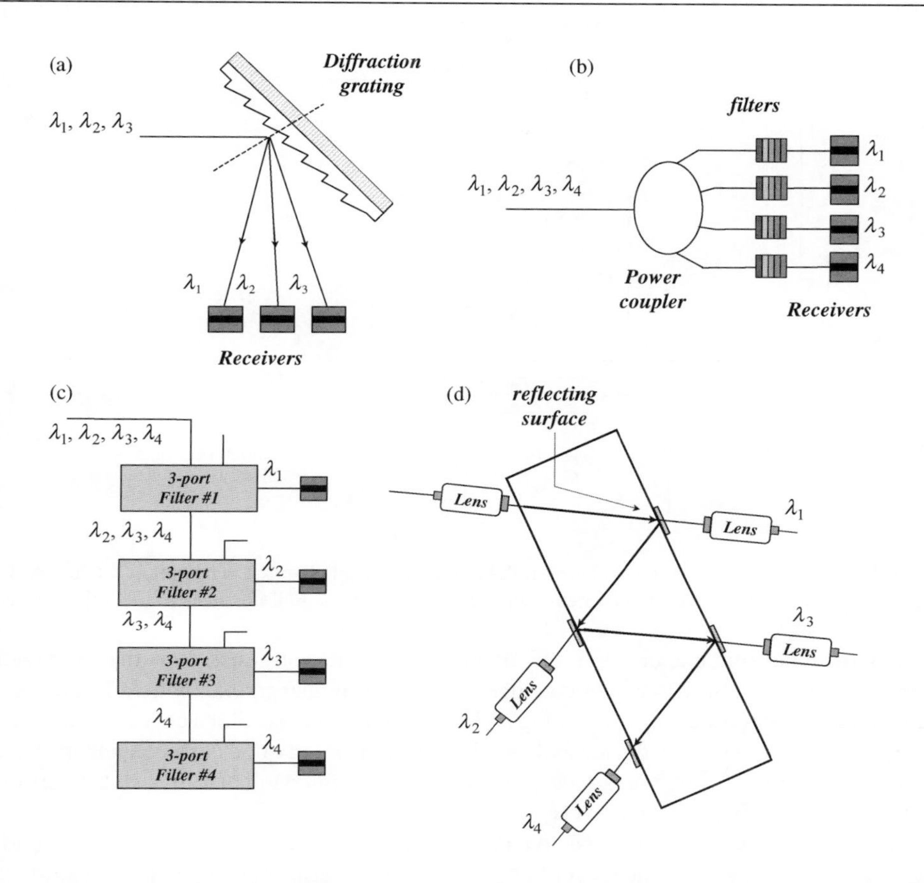

Figure 4.67 Four equivalent demultiplexing arrangements: (a) diffraction grating; (b) star coupler (for broadcasting) and fixed-wavelength filters; (c) a cascade (bus) of three-port devices; (d) modified "cascade" configuration for two-port devices based on interference (thin-film) filters

4.6.2.2 (De-)Multiplexer Performance Considerations

For assessing the quality of any (de-)multiplexing device, a number of interrelated issues have to be considered. These include the *spectral spacing between adjacent channels*, the *total number of wavelength channels*, the *passband flatness*, the *coupling losses*, and the *level of outband crosstalk*. The best device is the one that allows the largest number of channels with the flattest bandpass, the smallest coupling losses per channel, and the largest optical isolation between adjacent channels with the minimum spectral separation between them.

From Equation 4.160, it is concluded that the larger the size of the grating W is (that is, $W = Nd$), the sharper the intensity distribution is and the wider the spatial separation between two wavelength channels can be. Given that sufficiently large optical isolation is available between two adjacent channels, the number of grooves/slits should be considerably larger than from what is required for fulfilling the Rayleigh criterion.

The *coupling loss* of any (de-)multiplexer for a given wavelength channel is defined as the ratio of the incoming power to the outgoing power. This could vary with wavelength and

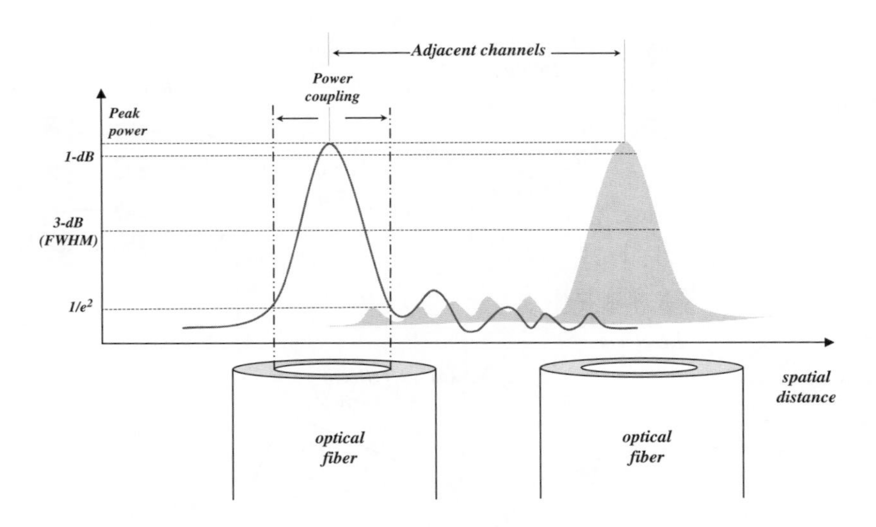

Figure 4.68 Coupling efficiency and crosstalk between adjacent channels. The part of the intensity not coupled to the destined output fiber smears with the wavelength signal destined to the adjacent fibers

depends on many parameters. For diffraction grating devices, these are the propagation material (free-space, Si, III–V semiconductor), the optical aberrations (which depend on the size of auxiliary optics for a planar grating and the clear aperture size for a concave grating), the mode mismatch between the device and the fiber (for integrated optic devices), and the type of the final receptor (e.g., detector, single-mode fiber, or multimode fiber with a clear aperture of 20 μm, 10 μm and 50 μm respectively).

A fraction of the optical power of a given wavelength that is not coupled to the corresponding outgoing receptor (detector, fiber) could be coupled to the adjacent channels' receptors, thus generating *crosstalk*. In this way, crosstalk is the unintended coupling of signals from adjacent channels due to device imperfections. These phenomena can be better understood by considering the impulse response of any (diffraction grating) demultiplexer (Figure 4.68). The impulse response has an intensity distribution with a Gaussian-like central part (due to the Gaussian intensity distribution emitted from a single-mode fiber) and a sinc-squared function ($\sin^2 x/x^2$) distribution at the outer parts. As a practical rule of thumb, good optical isolation (low crosstalk) is achieved when the ratio of the optical signal bandwidth (measured at the $1/e^2$ point from its peak value) over the channel spacing is less than 0.25. The fact that the main part of the impulse response has a Gaussian intensity distribution profile leads to passband narrowing when many of these devices are cascaded. For this reason, optical techniques are necessary in order to flatten the passband.

4.6.2.3 Diffraction Gratings

Principle of Operation

A diffraction grating is any physical arrangement that is able to alter the phase (optical length) between two of its successive elements by a fixed amount. The impact of this progressive phase alteration becomes evident at the far-field intensity distribution. Consider the case of a plane

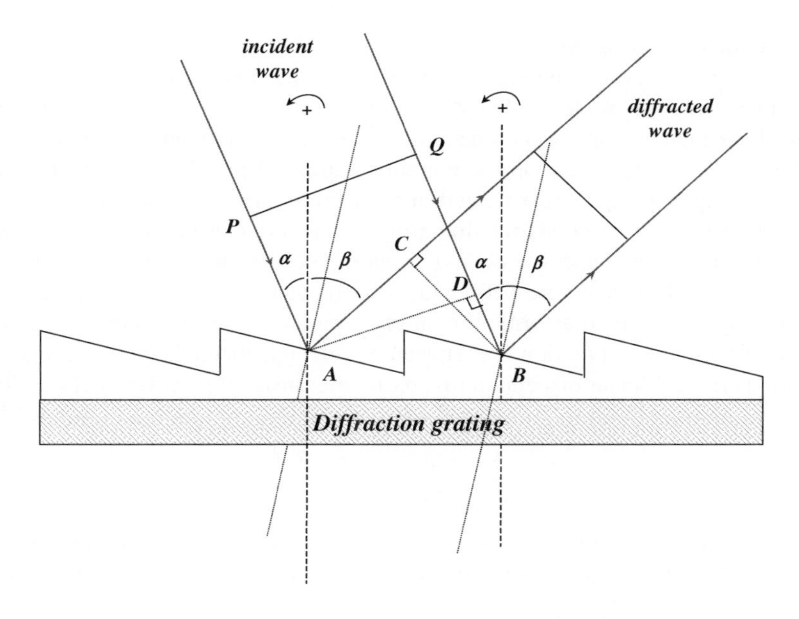

Figure 4.69 The principle of operation of a planar diffraction grating

reflection grating (Figure 4.69). The incident plane wavefront PQ first reaches the point A, which then becomes a source of secondary wavelets, and hence it advances point B. Finally, the incident wavefront reaches B, which then becomes a source of secondary wavelets. These wavelets are exceeding those originating from A at the same time. Hence, the path difference from the corresponding points of the two neighboring grooves (spaced by d), as measured at a distant point of observation, is

$$\mathrm{AD} - \mathrm{BC} = d(\sin\alpha - \sin\beta) \tag{4.158}$$

If the incident beam is on the same side of the normal as the diffraction beam, then the minus sign in Equation 4.158 should be replaced with a *plus*. For a more detailed presentation the reader is referred to classical textbooks like [63,64]. In any case it can be shown that the far-field intensity distribution of a planar diffraction grating is the same as that of N rectangular slits modulated by the diffraction envelope of a large slit. Constructive interference occurs when

$$d(\sin\alpha - \sin\beta) = m\lambda \tag{4.159}$$

where m is a constant called the diffraction order and λ is the wavelength (at free space) of the channel. The number N of the grooves/slits determines the sharpness of the principal maximum of the intensity distribution. For example, according to the Rayleigh criterion for the resolution limit, two equal-intensity wavelengths spaced by $\Delta\lambda$ are just resolved if the spatial distance between them is such that the two intensity distributions are crossing each other at 0.8 of their maximum value. The theoretical resolution limit for any diffraction grating is defined as

$$R = \frac{\lambda}{\Delta\lambda} = mN \tag{4.160}$$

Diffraction Grating Classification

The diffraction gratings could be either concave or planar and they can operate either in a reflection or transmission mode. A planar diffraction grating – regardless of the technological platform and the mode of operation – cannot be used as a standalone component when it is employed as a (de-)multiplexer in optical communications. A practical (de-)multiplexer based on a planar grating is always implemented in a spectrographic configuration employing two auxiliary optical components: one for collimating the incoming beam (that is, for transforming the spherical wave to a plane wave) and a telescopic system for focusing the outgoing beam. The wavelength channels are diffracted at different angles determined by Equation 4.160 and the telescopic system transforms this angle separation into a spatial separation at the image plane. This spatial separation Δx between wavelength channels $\Delta\lambda$ in the image plane is given by the reciprocal linear dispersion ($\mathrm{d}\lambda/\mathrm{d}x$ in nm/mm). Differentiating (Equation 4.159) gives

$$\frac{\mathrm{d}\lambda}{\mathrm{d}x} = \frac{\mathrm{d}\lambda}{f\,\mathrm{d}\beta} = \frac{\mathrm{d}\cos\beta}{mf} \tag{4.161}$$

where f is the focal length of the focusing part of the spectroscopic system. In practice, when the spectrograph is used as a (de-)multiplexer, Δx is dictated by the minimum distance between the output waveguide/fiber cores.

Spectrograph Overview

The most important configurations for planar grating spectrographs are the *Ebert–Fastie* and the *Czerny–Turner* (Figure 4.70). In the former case, the spectrograph is constructed from a planar grating and a large concave mirror (or lens). The Czerny–Turner configuration offers the alternative of using two smaller concave mirrors instead of a single large one. The main drawback of these configurations is the use of the auxiliary optics off-axis, which is something that generates large aberrations. As a result, a point source is imaged as a geometrical extended entity that degrades the performance of the optical system.

A concave grating does not need auxiliary optics, since it is a complete spectrograph. The corrugated surface provides the necessary diffraction for wavelength separation or recombination, whilst the geometrical properties of the concave surface allow focusing of the diffracted wavelengths. Since concave gratings operate off-axis they also suffer from large geometric aberrations. However, there are specific geometric arrangements, called focal curves, which minimize the adverse effect of these aberrations. The best-known focal curve of concave gratings is the *Rowland circle* (Figure 4.70). For a concave substrate with radius of curvature R, the Rowland circle has a diameter R and it is tangent to the apex of the substrate. The important characteristic of this geometric locus is that when a point source A is placed upon it ($r_\mathrm{A} = \mathrm{OA} = R\cos\alpha$) an image free from second- and third-order, as well as reduced fourth-order, Seidel meridional aberrations is produced at a location B on the Rowland circle ($r_\mathrm{B} = \mathrm{OB} = R\cos\beta$).

4.6.2.4 Practical Diffraction Grating Devices

Arrayed Waveguide Grating

The most widely deployed (and studied) type of a grating-based (de-)multiplexer is the arrayed waveguide grating (AWG). This is a two-dimensional integrated optic device (see Refs [65,66]

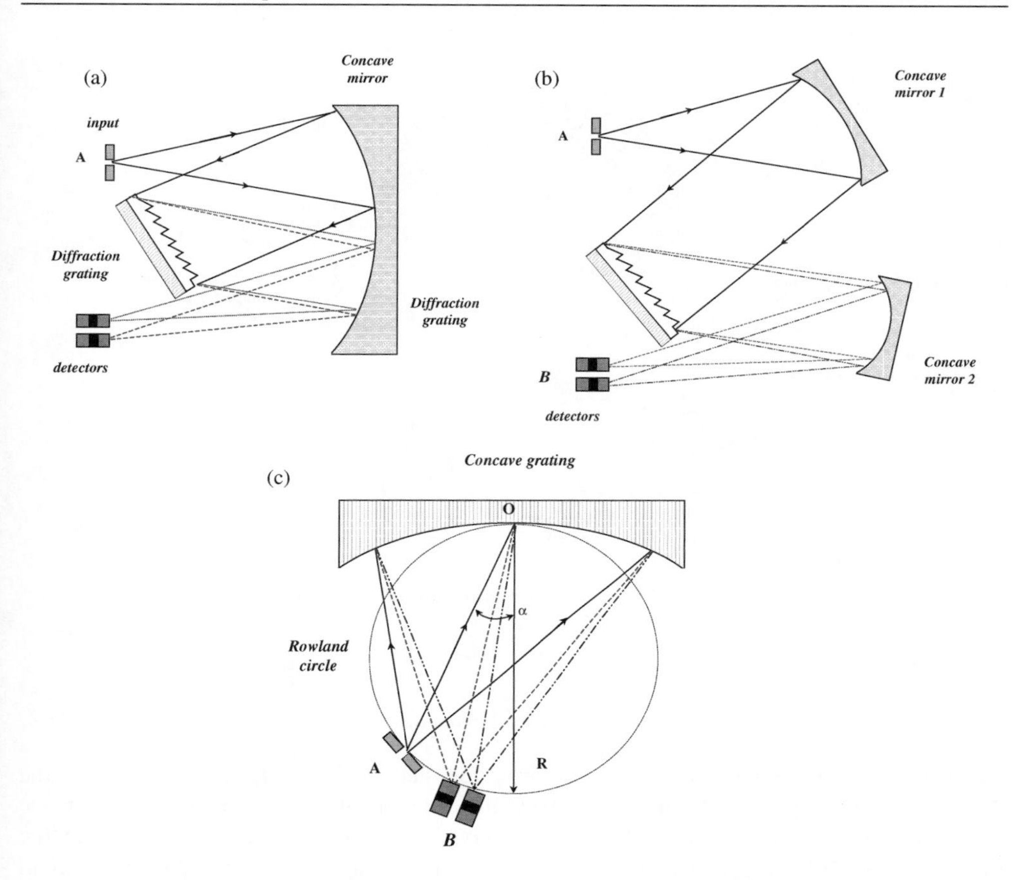

Figure 4.70 Spectrographs: (a) Ebert–Fastie; (b) Czerny–Turner; (c) Rowland circle

and references cited within). A special geometric arrangement of two slab waveguides and an array of single-mode waveguides forms a spectrographic set-up based on a transmission grating. The principle of operation, when the device is used as a demultiplexer, is as follows. The multiwavelength channel signal enters the device from the input slab waveguide (Figure 4.71a), where it freely propagates. The input and output slab waveguides in most cases are constructed using the Rowland circle (Figures 4.70c and 4.71b). In principle, other geometric arrangements (generalized focal curves) are also possible. In any case, aberration-free focal curves are mandatory, since the array of the single-mode waveguides is placed at the circumference of a spherical arc. The signal is coupled to the array of the waveguides probably via tapering for best coupling conditions. The array of waveguides plays the role of the grooves/slits in classical gratings. Despite the use of the slabs on a Rowland circle, the entire set-up has a planar grating configuration (a diffraction grating and two auxiliary optical systems). The length of the array waveguides is chosen such that the optical path length difference ΔL between adjacent waveguides is equal to an integer multiple of the central wavelength of the (de-)multiplexer.

Owing to the additional phase change introduced by the arrayed waveguide length difference (which results in "hardwiring" all phases), the corresponding grating equation (Equation 4.158)

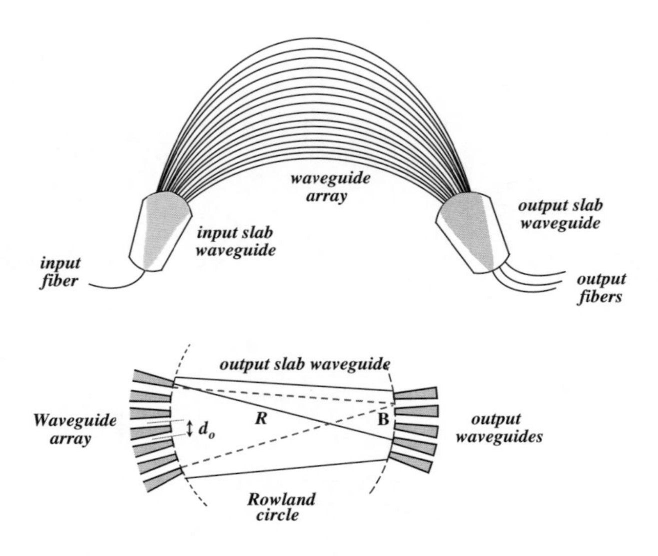

Figure 4.71 Schematic illustration of the AWG

is modified to

$$n_s d_o(\sin\alpha + \sin\beta) + n_c \Delta L = m\lambda \tag{4.162}$$

where n_s is the refractive index of the waveguide slab, n_c is the refractive index of the waveguides in the array (in the most general case, they are not the same), and d_o is the distance between two successive waveguides in the array. To understand the reasons behind the implementation of the AWG using this additional path length difference ΔL, one should consider the following.

Advances in integrated optic fabrication techniques based on lithographic etching made possible the demonstration of AWGs on InP, Si/SiO_2, and $LiNbO_3$. Despite these advances, there are limitations on the size of the wafers that could be practically used. Owing to these size restrictions (resulting in small focal lengths), the required linear dispersion, Equation 4.161, is obtained by operating the AWG in high diffraction orders. From Equation 4.162, the diffraction order equals $\Delta L = m^{\mathrm{AWG}}\lambda/n_c$ and $m^{\mathrm{AWG}} + 1 = m$. However, the free-spectral-range (FSR) of the AWG is FSR $= \Delta\lambda = \lambda/m$; and given that $50 < m < 100$, then FSR < 30 nm. This is one of the main limitations of this approach; that is, the AWG can only be used in the context of a limited spectral range. From an engineering point of view this problem can be solved, as mentioned earlier, by cascading coarse AWG demultiplexers followed by band-optimized fine-granularity AWGs. In this way, a 480 WDM channel 10 GHz-spaced (de-)multiplexer has been reported comprising a WDM coupler, two 100 GHz-spaced AWGs followed by 64 10 GHz-spaced AWGs [67]. The largest reported number of channels with a single AWG is 64 channels spaced by 50 GHz (0.4 nm) [67] or 128 channels spaced by 25 GHz [68], both developed on silica.

To improve the performance of the AWGs in terms of diffraction efficiency, modifications of known spectroscopic techniques have been applied [69,70]. The techniques require varying the optical path length (and hence the "hardwired" phase difference) between the waveguides in a nonuniform way, resulting in redistribution of the energy at the image plane. The effect is further improved after a suitable defocusing of the input/output waveguides. The former method is the integrated-optic equivalent of techniques used in aberration-corrected holographic concave gratings, where the intensity distribution in the image plane is altered when the corresponding grooves are not equidistant parabolas.

Practical constraints in the fabrication of the AWGs are also attributed to the difficulty of the lithographic system to truly simulate a focal curve like the Rowland circle. In addition, this mount requires the axis of the input/output waveguides to point towards the pole of the slab, implying that the waveguides have to be tilted with respect to each other. When this condition is not met, the consequent *vignetting* degrades the performance of the outer wavelength channels in the spectrum. Overall, the AWG is a very good candidate device for a demultiplexer operating within a single transmission band (like the C band of the EDFA), but it is problematic or impractical for wider optical bandwidth applications. Another issue, common to all integrated optic devices, is the inherent birefringence of the materials used that leads to TE/TM PMD-like problems. Also, temperature controllers are needed for thermal stabilization of the operational conditions of the devices.

Other Integrated Optic Spectrographs

Other practical integrated optic spectrographs used as (de-)multiplexers include a *two-dimensional concave grating* [71–74] and a modified Czerny–Turner configuration [75]. The former type has been implemented on both silica and III–V semiconductor compounds. A Rowland circle or generalized focal curves have been used for producing aberration-free images [72]. The fabrication of this device, in contrast to the AWG, requires deeply etched grating facets, which is achieved using ion-beam etching. A subsequent problem is the attainable degree of verticality of the grating wall. Another important consideration is associated with the rounding errors of the diffraction grating facets due to lithographic inaccuracies [73]. Again, owing to the limited size of the wafers used, the requested linear dispersion (which, in the current case, is expressed as $\mathrm{d}\lambda/\mathrm{d}x = n_s d \cos\beta/mf$, where n_s is the refractive index of the slab) is achieved by operating the grating at high orders.

The difference between the two-dimensional concave gratings and the AWGs is that in the arrayed waveguide case the grating constant (pitch) equals the distance between two successive waveguides. Given that the waveguide length is of the order of a millimeter, the waveguides need to be sufficiently apart to avoid exchange of energy between them (in Si-based devices the distance is at the order of 20 μm). This restriction does not apply to two-dimensional concave gratings. As a result, a grating pitch of few micrometers is feasible and the requested linear dispersion is attained by operating the device at a lower order than the AWG (typically at 10th to 30th order). So, these devices may operate in a wider spectral range compared with their AWG counterparts. With this technology, a device with 120 channels and 0.29 nm channel spacing has been reported [74]. A simple rule for identifying the trade-off between grating pitch and diffraction order for the integrated optic concave gratings can be obtained by solving the equation for the linear dispersion and the grating equation, which leads to

$$\frac{m}{d} = \frac{2\lambda n_s \sin\alpha + 2n_s\sqrt{(\mathrm{d}\lambda/\mathrm{d}x)^2 f^2 + \lambda^2 - (\mathrm{d}\lambda/\mathrm{d}x)^2 f^2 \sin^2\alpha}}{2[(\mathrm{d}\lambda/\mathrm{d}x)^2 f^2 + \lambda^2]} \tag{4.163}$$

On the other hand, the Czerny–Turner mount consisting of a transmission grating and two parabolic mirrors used off-axis has been reported [75]. A paraboloid compensates the spherical aberration when it is operated off-axis. However, it introduces a significant amount of meridional coma. A Czerny–Turner spectrograph should be deployed using two spherical mirrors with different radii of curvature in order to compensate for meridional coma. In any case, both (de-)multiplexer types requiring temperature control and compensation due to material birefringence e.g. AWGs exhibit polarization mode dispersion.

Free-Space Gratings

Practical free-space optical (de-)multiplexers can be found either in the form of planar grating mounts or in the form of a holographic concave grating. Free-space grating multiplexers were the first to be tested in conjunction with WDM system experiments. The most established planar grating (de-)multiplexer is implemented based on a modification of the Ebert–Fastie configuration, where the source and the image are almost collocated at the optical axis of a parabolic mirror. This is now a commercial product called STIMAX [77]. Operating an optical system on-axis results in an aberration-free image from second- and third-order Seidel aberrations. Fourth-order aberrations (spherical aberrations) do exist and they are compensated by means of the parabolic mirror.

The STIMAX (de-)multiplexer has a very high wavelength channel count (Table 4.7). The main limitation of this configuration is the rapid increase of third-order Seidel aberrations (coma) for the outer wavelength channels of the spectrum due to the use of the parabolic mirror off-axis. Mechanical and thermal stability issues have been successfully addressed. Owing to the free-space operation, PMD-like problems are inherently absent, whilst the polarization dependence of the diffraction efficiency is an issue especially for wider bandwidth applications.

Another configuration uses a planar grating together with retro-reflectors at the image plane that facilitate in producing a zero-dispersion focused light, reducing bandwidth narrowing due to cascaded (de-)multiplexers [77]. In principle, this approach provides a bandwidth flattening technique that is still applicable even when the device is operated as a demultiplexer, something that is not the case with other solutions used in AWGs [78]. A variant of this technique is used to eliminate the polarization dependence of the diffraction efficiency [79].

Concave diffraction gratings are single-element optical systems that simultaneously provide dispersive and focusing properties. A single optical element is very attractive due to easier optical alignment and reduced packaging problems. For this reason, concave gratings were used as (de-)multiplexers even in the very early stages of WDM transmission [80]. However, it has been recognized that the main disadvantage of this optical system is the large inherent aberrations associated with the spherical concave substrate; thus, aberration-corrected gratings need to be deployed [81].

The holographic concave gratings are the primary candidates for providing aberration-corrected (de-)multiplexers [82]. The principle of operation of the holographic concave grating is the following (Figure 4.72). The concave substrate is covered with a suitable photoresist material sensitive to a specific wavelength (e.g., Ar^+ laser). A laser beam is split and focused in two pinholes such that two coherent point sources (C and D) are created. In general, the resulting spherical waves interfere on the substrate, generating fringes that are neither equidistant nor straight. This is the recording phase. When the device is used as a demultiplexer, the single-mode fiber carrying a multiwavelength signal is placed at point A. Then, a particular wavelength is imaged at point B, which, however, is not a point source due to large optical aberrations. The aberrations can be eliminated by placing the two point sources C and D in a

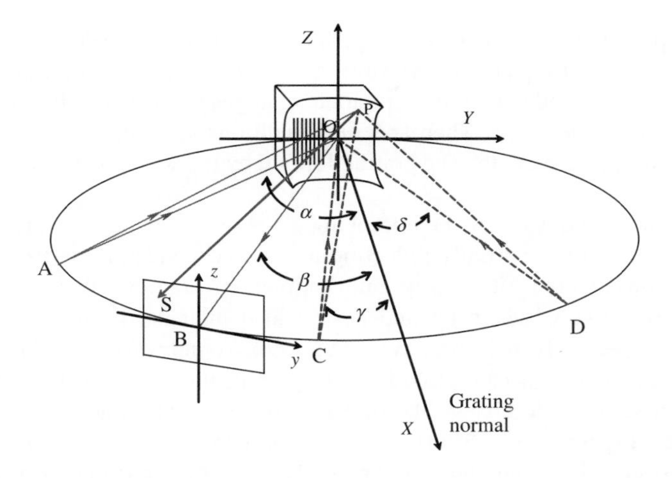

Figure 4.72 Schematic illustration of a holographic concave grating

position such that the fringes generated introduce a phase shift that cancels out the aberrations introduced by the spherical substrate. It has been demonstrated that up to a thousand wavelength channels can be (de-)multiplexed using these devices, covering a spectral range of 200 nm [83]. Nevertheless, these devices are still available only in laboratories. As with all free-space devices, mechanical stability and polarization-dependent diffraction efficiency are issues; in particular, when the number of wavelength channels is more than 100, the problem of an efficient fiber mount has to be tackled. The performance of all diffraction grating devices is summarized in Table 4.7.

Acousto-Optic Grating

This is an *active* device. The principle of operation of this device is based on the interaction of light with sound resulting in a transmission grating. A sinusoidal sound wave traveling at the

Table 4.7 Performance of diffraction grating devices

	Channel spacing (nm)	No. of channels	Losses (dB)	Crosstalk (dB)	Comments
AWG[a]	0.8	40	<6	<−20	Si/SiO_2
AWG[a]	0.8	Up to 40	<8	<−25	1 dB band, ∼0.16 nm
AWG[a] flat passband	0.8	Up to 40	<9	<−24	1 dB band, ∼0.32 nm
AWG [68][b]	0.2	128	3.5–5.9	<−16	Si/SiO_2
Free-space planar[a]	0.8	Up to 40	< 5.5	<−30	1 dB band, >28 GHz
STIMAX[a]	0.8	Up to 64	< 5.5	<−30	—
Minilat[a]	0.8	Up to 92	< 8	<−33	1 dB band, ∼0.2 nm
Holographic concave[b] [83]	0.4	64	< 8	<−30	—
2D concave[b] [71]	1	50	16	<−19	InP
2D concave[b] [74]	0.29	120	20–40	<−44	Si based

[a]Commercially available products.
[b]Laboratory results.

surface of an appropriate material generates periodic variations of the density (or strain) of the material according to the frequency of the wave. As a result, the macroscopic effect is a periodic change of the refractive index and these periodic changes act as partially reflecting mirrors. Hence, an incident plane wave, when specific conditions are met, will be diffracted at an angle according to its wavelength. The grating formed by the sound wave is a dynamic (time-varying) one.

The effect of the sound wave on the impinging plane wave can be understood in two ways. The distance between two "partially reflecting mirrors" depends on the frequency Ω of the sound wave. Conservation of energy and momentum require that $\omega_\mathrm{i} = \omega_\mathrm{d} + \Omega$ and $\mathbf{K}_\mathrm{i} = \mathbf{K}_\mathrm{d} + \mathbf{K}$ respectively, where the indexes i and d indicate the incident wave and the diffracted wave respectively. $\mathbf{K}$ is the wave vector of the sound wave; and since $\Omega \ll \omega_i$, then $\omega_\mathrm{i} \cong \omega_\mathrm{d}$. Thus, apart from a negligible frequency shift, the effect of the sound wave on a multiwavelength signal is to change the direction of propagation according to wavelength (demultiplexing) (Figure 4.73a). Alternatively, it could be argued that an optical path length (phase) delay occurs between the two "partially reflecting mirrors" similar to what is produced by the grooves of a diffraction grating. When the distance between them satisfies the grating equation (which now is termed the *Bragg condition*), the elementary planes add constructively. The intensity distribution of the impulse response has the known sinc-squared form and its sharpness will depend on the number of reflecting mirrors; that is, the total interaction length L (the sharpness will depend on the ratio L over the wavenumber determined by Ω).

4.6.3 Overall Assessment of (De-)Multiplexing Techniques

Having presented the main technological platforms currently used for optical (de-)multiplexing, it would be interesting to highlight their pros and cons. As pointed out earlier, the main question when a technology is assessed is the type of application in mind. In general, the configurations illustrated in Figure 4.67b and c have different drawbacks.

A layout like the one in Figure 4.67 based on a power coupler and optical filters is a flexible solution up to approximately eight wavelength channels. Beyond this point, splitting losses tend to be high so that the grating solution is advised. Optical filters with tailor-made spectral characteristics (e.g., through wavelength) can be used in this configuration, leading to a

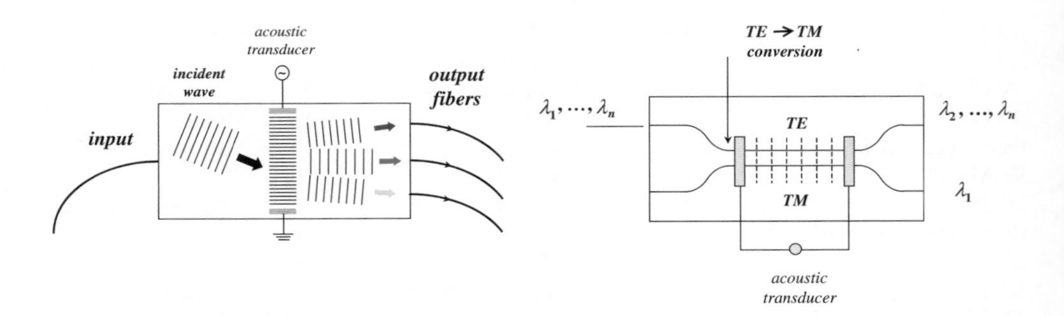

Figure 4.73 (a) The principle of operation of an acousto-optic grating; (b) a four-port AOF

(de-)multiplexer construction offering a wavelength comb with unequal channel spacing allocation (to combat, for example, fiber nonlinearities like FWM). Nevertheless, when systems with a large number of wavelength channels are desired, the cost of the system scales proportionally to channel count.

The "bus" architecture of Figure 4.67c is implemented only via three-port or four-port devices. Thus, the loss performance is not uniform across the spectrum of interest: the first channel has the lowest losses, whilst the final channel suffers from the worst losses. It is this loss figure that determines the maximum number of channels to be used per band. In general, the performance of the optical filters is good in terms of optical isolation. The loss performance of the device itself is good, but for practical applications a cascade of other optical components, like band (de-)multiplexers, is required.

With diffraction gratings, the cost is not proportional to channel count; indeed, devices with a very large number of channels have been demonstrated. AWG suffers from polarization-induced phenomena, whilst free-space gratings do not. Free-space gratings offer perhaps the best optical isolation from all (de-)multiplexing devices and they do not have any restrictions with respect to the total optical bandwidth they can handle. In principle, the free-space devices can explore the parallelism of optics to generate many (de-)multiplexers in parallel or to be used in conjunction with other free-space devices such as MEMSs in optical crossconnects.

4.6.4 Optical Filters

4.6.4.1 Acousto-Optic Filters

When an acoustic wave is applied on an acoustically active material, the induced birefringence (i.e., a dissimilar change of the refractive index for the ordinary and the extraordinary rays) of the medium alters the state of polarization of the incident wave. This principle is used for constructing acousto-optic filters (AOFs) (Figure 4.73b). A multiwavelength signal enters a four-port device like a single coupler design. At the input stage, the directional coupler splits by 50% (−3 dB) the incoming signal. Also, the light at the lower branch undergoes a phase delay of $\pi/2$. That is, at the lower branch, a polarization rotation is observed from the TE to the TM mode. When no acoustic wave is applied, another $\pi/2$ phase delay occurs at the output part and all channels exiting from the symmetrical to the input port (e.g., upper-in, upper-out). When an acoustic wave is applied, the exact matching conditions are altered via the acousto-optic effect and the requested channel is selected from the lower output port. An interesting feature of the AOF is that many acoustic frequencies could co-propagate, thus allowing a simultaneous selection of more than one channel. Hence, the AOF can be used as a band-selecting filter in hierarchical (coarse/fine) WDM (de-)multiplexing since it has a tunability of hundreds of nanometers. The crosstalk figure of the AOF is not as good as that of the other commercial diffraction gratings.

4.6.4.2 Interference Filters

These filters appear in the literature under many different names, such as dielectric thin films, multilayer interferometric filters, multistack thin-layer filters, and so on. Further, they are

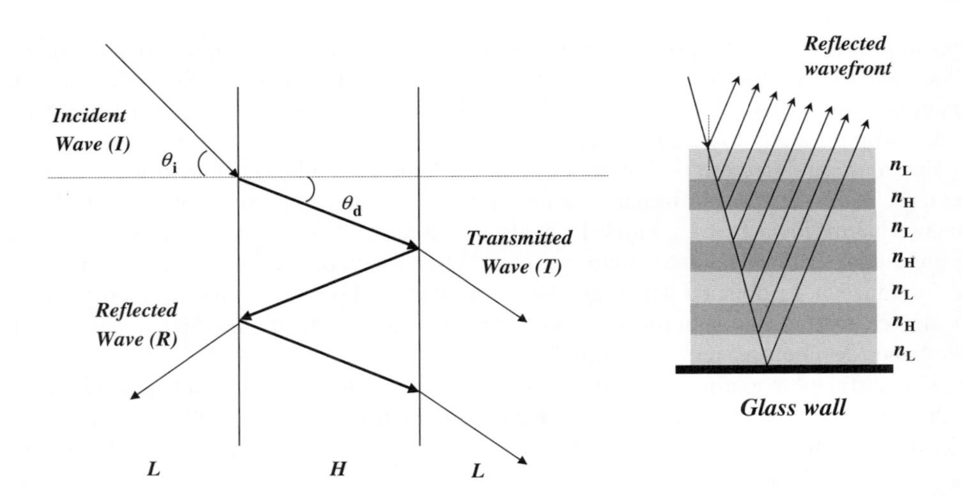

Figure 4.74 (a) An FP etalon; (b) a multilayer stack operated in a reflection mode

constructed from many different compounds ranging from liquid crystals to various oxides (SiO_2 or TiO_2) to III–V semiconductors. Nevertheless, the principle of operation for all these structures is easily understood by considering a FP etalon [84]; that is, an interferometer based on multiple divisions of the amplitude. Note that collimating optical devices (like bulk or GRID rod lenses) are mandatory at the input/output.

Let us assume that a material A with higher refractive index (n_H) than a material B (n_L) forms a cavity as shown in (Figure 4.74a). The reflected and transmitted intensities of the FP etalon, I_R and I_T respectively, normalized to the incident intensity, are given by

$$\frac{I_R}{I_i} = \frac{4\,R\sin^2(\delta/2)}{(1-R)^2 + 4\,R\sin^2(\delta/2)}, \qquad \frac{I_T}{I_i} = \frac{(1-R)^2}{(1-R)^2 + 4\,R\sin^2(\delta/2)} \tag{4.164}$$

where $\delta = 4\pi n_H \cos\theta_d L/\lambda$ and R is the fraction of the intensity reflected at each interface like n_H/n_L and n_L/n_H.

Maximum power transfer to the reflected beam occurs when the thickness of the high-refractive region is equal to a quarter-wavelength ($\delta/2 = m\pi/2$), whilst maximum transmission occurs when the thickness is one half-wavelength thick ($\delta/2 = m\pi$), assuming that $R \cong 1$. The former case is depicted in Figure 4.74b. Many different configurations could emerge by adding up such multistacks; for example, two structures, as shown in Figure 4.74b, with a space layer between them form an additional FP cavity of the type HLH(2L)HLH, assuming that each layer has a quarter-wavelength thickness. It can be shown that sharper cut-off characteristics are obtained by increasing the number of layers or cavities. However, when all cavity lengths are the same, the overall structure produces a narrower passband. Hence, layers of unequal thickness are used, something that requires the addition of further layers for phase matching, leading to a complex optimization problem. Also, the passband increases with increasing values of n_H/n_L [85]. Overall, practical constructions lead to filters offering optical isolation up to –30 dB. The losses range between 1 and 5 dB, depending on the material, number of channels, and so on.

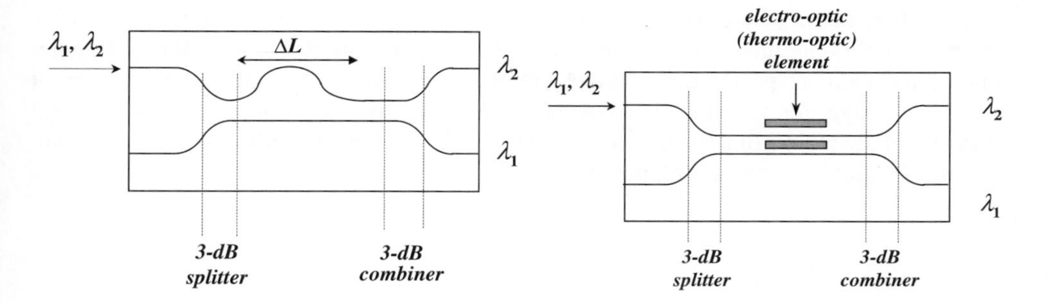

Figure 4.75 (a) The asymmetric Mach–Zehnder and (b) the symmetric configuration

4.6.4.3 Mach–Zehnder Filters

These filters are four-port devices that can be either *passive* or *active.* The integrated optic version of the Mach–Zehnder interferometer is illustrated in Figure 4.75 and it consists of three parts. The input and the output parts are 3 dB couplers, whilst the central section has two waveguide arms (upper and lower) with different path lengths (Figure 4.75a). The configuration is called *asymmetrical.* For a signal entering the upper port, the overall phase difference due to the asymmetric length is

$$\Delta\phi = \frac{2\pi n_{\mathrm{u}}}{\lambda}(L+\Delta L) - \frac{2\pi n_{\mathrm{l}}}{\lambda}L = \frac{2\pi n}{\lambda}\Delta L \tag{4.165}$$

where n_{u} and n_{l} are the refractive indexes of the upper and lower waveguides respectively. In this case they are assumed to be equal to n. The upper ports are labeled as 1 and 1′ at the input and output respectively, and likewise the lower ports as 2 and 2′ respectively. The transmittance from port 1 to 1′ is given by T_{u} and from 1 to 2′ by T_{l}.

$$T_{\mathrm{u}} = \sin^2(\Delta\phi), \qquad T_{\mathrm{l}} = \cos^2(\Delta\phi) \tag{4.166}$$

when $(2\pi n/\lambda_1)\Delta L = (m+1)\pi/2$ and $(2\pi n/\lambda_2)\Delta L = m\pi$, with m an integer, then T_{u} is unity for λ_1 and zero for λ_2 and vice versa (T_{l} is unity for λ_2 and zero for λ_1). In this way, the wavelengths λ_1 and λ_2 are collected from different ports. When the device is carefully designed, it can operate as a wavelength channel (de-)interleaver. The *symmetric* configuration (Figure 4.75b) has waveguide arms of equal length, so the phase difference occurs due to electro-optic- or thermo-optic-induced [86] change of the refractive index, resulting in different n_{u} and n_{l} when the control signal is on.

The loss figure depends on the number of wavelengths that can be demultiplexed from a single module, as well as the host material (Si or III–IV semiconductor). In general, the optical isolation between adjacent channels is not better than −30 dB.

Fiber Bragg Gratings

This technology allows the realization of very good fixed filters, but tunability is not easily and effectively achievable. The FBG is a mature and versatile technology with advantages such as being an all-fiber technology (low insertion loss, high power) and very selective in wavelength. An FBG can be obtained by writing periodic index changes in the fiber core by means of ultrviolet exposure through a phase mask, generating an interference pattern, so achieving a wavelength-selective, reflective filter. Volume Bragg grating technology is based on recording the grating on a bulk medium, obtaining an optical function similar to an FBG. This three-dimensional approach provides, in principle, more functionalities than with a standard fiber or waveguide Bragg grating, since the volume the holography allows integrating of different filtering and switching functions in a small space, due to the possibility of using the third dimension with the so-called volume gratings.

4.6.5 Tunable Filters

Tunable filters have various applications in the networks. They are key enablers for ASE suppression in tunable receivers, for signal rerouting applications in several reconfigurable optical add–drop multiplexer (ROADM) architectures, and for channel selectors in optical performance monitoring applications, with different technical requirements in the various applications.

Tunable filters should ideally exhibit fast and accurate tuning performances, with a wide tuning range (ideally covering C-band or L-band). Insertion loss is an important parameter, and optical spectrum characteristics (in terms of bandwidth, steepness, channels isolation) depend strongly on the application. For tunable receivers or network rerouting, the required wavelength selection is based on the ITU-T grid (50 or 100 GHz), but for some network monitoring applications it could be preferable to scan the whole spectrum with sufficient resolution to observe several performance parameters, such as channel power, stability of the transmitting wavelengths, and out-of-channel noise. While for monitoring applications the scan speed is also important, the tuning speed is not always a critical parameter; for routing applications a target requirement really depends on the network scenario (a tuning speed of a few milliseconds is acceptable for current applications, while future optical burst switching, for example, may require switching fabric with switching time in the order of microseconds). Tunable filter technology to be applied in signal rerouting applications such as ROADMs needs to be able to support hitless wavelength selection; for example, the wavelengths not filtered must not be disrupted during filter tuning. Requirements on filter spectrum characteristics (amplitude and phase) for applications in a dynamic all-optical network are also dependent on acceptable Q factor penalties due to filtering effects in transit nodes.

One solution is the FP configuration (Figure 4.74a), which basically is a resonant cavity which is tuned by changing the length of the cavity or the refractive index of the material filling the cavity. In the first case, wavelength tuning is achieved by changing the distance between the parallel plates (micro-optics and piezo control or MEMS, C-MEMS are mostly used). In the second case (fixed parallel plates), the filter is tuned by changing the refractive index of a material that fills the cavity between the plates. This is mostly performed by exploiting the electro-optic effect, using materials such as lithium niobate or liquid crystals or by means of multilayer structures.

The main characteristic of FP technology is the high selectivity, while a major drawback is the inherent trade-off between wide tuning and narrow filter bandwidth (achievable with two high reflectivity dielectric mirrors), due to the FSR of the FP structure. Another point is that several implementations show a nonuniform insertion loss within the tuning range.

Diffraction grating physical implementations can be on single-mode fiber and planar lightwave circuit waveguides with thermo-optic or piezo-electric tuning mechanisms, as well as based on acousto-optic crystal or diffraction MEMS. Diffraction gratings generally show a limited selectivity.

However, tunable filters can be characterized by the tuning technology used.

Opto-mechanical technology employs macroscopic electromechanical actuators to redirect the optical beam; these can be used to switch the different wavelengths after demultiplexing, so achieving rerouting functionality. The opto-mechanical-based devices are bulky and not scalable; this limits the interest in these components.

A versatile approach to tunable filters uses MEMS technology to alter the position of a movable mirror. MEMS technology is similar to the opto-mechanical one, but it uses microscopic elements (e.g., mirrors and electromechanical actuators) to redirect beams and it offers a good scalability. Thanks to micromachining technology, the MEMS-based filter structure fits in small form-factor packages.

Several approaches to tunable filters are possible with MEMS; for example, tunable cavity (i.e., FP), diffraction grating, rotating etalon and "linear sliding" filters. While diffraction gratings shows interesting characteristics and have been proposed as a basic building block for other optical components, such as tunable lasers, FP filters are attractive thanks to wide tuning capability, low polarization, and reduced processing steps.

The FP filter figure of merit is the finesse. A high finesse value corresponds to sharper transmission peaks and narrow bandwidth. The mirror spacing determines a single peak within a given wavelength band. Applying electrostatic forces to the movable mirror alters the spacing, allowing the filter to tune across its free spectral range.

MEMS evolved from integrated-circuit silicon processing technology, so the materials involved are silicon oxide, silicon nitride, aluminum, and even nickel. The limited range of mechanical properties can introduce design constraints for silicon-based MEMS devices; moreover, the rigidity of silicon MEMS requires higher voltage to provide the force for mechanical deflection (electrostatic or piezo forces are used to bend or deflect one of the MEMS layers). As the voltage is increased, it reaches a critical level beyond which the head will be pulled onto the substrate, causing electric discharge and damage. This introduces a limit to the tuning range achievable by a standard MEMS filter. A class of MEMS devices uses as the main element highly compliant polymeric materials. Compliant MEMS (CMEMS) technology uses a soft material called an elastomer that is an order of magnitude less stiff than silicon, so requiring lower voltages to achieve a given mechanical deflection and showing a mechanical range of motion that is larger than with silicon MEMS for equivalent voltages.

In general, diffraction grating MEMSs show suitable spectral characteristics to support tunable receiver applications, while CMEMSs have spectral characteristics more suitable for monitoring applications. The tuning speed is anyway in the range of tens/hundred milliseconds.

Liquid-crystal molecules' orientation affects the optical properties of the liquid crystal itself, which in turn affects the state of polarization of the light, so achieving switching/tuning of the optical beams. Liquid crystals can be electrically controlled.

One approach to implementing a tunable filter is to exploit the electro-optic properties of a material (i.e., the possibility of changing the material's refractive index by applying an electrical field) filling a resonant cavity or being part of a Bragg grating. In order to achieve a large tuning range with a low voltage and without bulky devices the material shall have a large electro-optic coefficient. One of the materials showing good electro-optic properties is liquid crystals, which are organic materials consisting of geometrically anisotropic molecules, leading to long-range orientational order and a number of mesophases, which are thermodynamic phases with physical properties intermediate between those of pure liquids and pure solids. Liquid crystals have a strong anisotropy with an extraordinary refractive index along the molecule axis that is much higher than the ordinary refractive index. They also tend to align with an external electric field.

In case of liquid crystals in a Bragg grating, a liquid crystal layer is applied on top of a planar waveguide to create an "active cladding" above the core. By varying the voltage applied to the liquid crystals, individual gratings can be turned on or off (liquid-crystal molecules aligned parallel to the waveguide axis matching the refractive index of the cladding material or molecules rotated off axis with refractive index changes), or varied in intensity to interact with the evanescent field of the light passing through the glass core of the waveguide below.

Since liquid crystals can be sensitive to temperature variation, a thermo-electric cooler should be used to avoid drifts (and to fine tune the center wavelength of the filter). Various tuning ranges (>100 nm have been demonstrated) can be achieved by adopting different liquid crystal types, depending on their dielectric anisotropy characteristics. Key performances of liquid-crystal-based filters are fast switching speed, narrow linewidth, good passband shape, and low insertion loss.

In order to tune FBG filters, both mechanical methods, such as stretching of the FBG, and thermal methods are used (with inherently low tuning speed, due to the fact that the thermo-optic effect takes place in a relatively large zone), since the period size of the index perturbations and the effective refractive index will be changed, resulting in a shift of the center wavelength.

Commonly adopted tunability methods are thermo-mechanical, piezo-electric, and actuator/stepper motor. The thermo-mechanical method is realized by adopting a bi-metal structure (to tighten the fiber) with a thermo-electric cooler. The key parameter is the temperature control accuracy, which will turn into filter resolution. The tuning speed is typically low, since the thermo-optic effect take place in a relatively large zone and the tuning range is normally in the order of 10 nm.

The piezo-electric tuning method has a good resolution and improves tuning speed, but requires high voltage and generally has a limited tuning range. Moreover, to avoid environmental effects, the FBG needs to be maintained at high temperatures and inherent hysteresis of piezo-electric devices can require the adoption of an external reference.

The actuator tuning method via a stepper motor provides a tunability range and speed similar to the piezo-electric tuning method, but with a more reliable solution.

Tuning the FBG center wavelength with strain results in a wider tuning range (tunability for axial compressive stress is wider than that due to tensile-stress mode), achieving a tuning range of several tens of nanometers.

Thin-film interference filters (TFFs) are the most widely deployed type of static fixed WDM filters. The tunable counterparts of TFFs still have limited diffusion with mechanically rotated or translated filters. Attempts to obtain a tunable TFF are normally based on index control, achieved with electro-optic or piezo-electric effects. In general, the results do not provide wide tunability, acceptable insertion loss, and low-voltage operation, while some advances in tunable TFFs appeared, proposing a different, thermo-optic, tuning mechanism exploiting the large thermo-optic coefficients in thin films of amorphous Si.

Single-cavity FP filters with up to tens of nanometers tuning range are obtained with temperature changes of hundreds of degrees generated in microscopic volumes (to obtain tuning time in the order of milliseconds) using internal heater films. Thin-film interference coatings are not limited to single-cavity designs, but multicavity filters can be designed as well. The advantage of the thin-film technology is the possibility of designing multicavity filters with steep skirts and flattened passbands for other applications, such as tunable add–drop filters and dynamic gain equalizers.

Gel/oil: this approach is similar to the thermo-optical one; it consist in heating a gel/oil element, changing its refractive index; in this way it is able to redirect the beam to the desired waveguide. The gel/oil-based devices suffer from very poor reliability and high insertion losses.

Magneto-optic: based on the Faraday effect, this allows a very fast switching (femtoseconds). The magneto-optic-based switches normally have modest optical performances, due to the weakness of the effect involved.

Acousto-optic tunable filters (AOTFs) are tuned by utilizing the surface acoustic-optic effects. AOTFs require acoustic-optical materials as a substrate; for instance, $LiNbO_3$. AOTFs are often large in size due to the acoustic-optical interaction requirements and require high power consumption to operate.

References

[1] Agrawal, G.P. and Dutta, N.K. (1986) *Long-Wavelength Semiconductor Lasers*, Van Nostrand Reinhold, New York.

[2] Yariv, A. and Yeh, P. (2007) *Photonics, Optical Electronics in Modern Communications*, 6th edn, Oxford University Press.

[3] Petermann, K. (1991) *Laser Diode Modulation and Noise*, Kluwer Academic, Dordrecht, The Netherlands.

[4] Cartledge, J.C. and Srinivasan, R.C. (1997) Extraction of DFB laser rate equation parameters for system simulation purposes. *IEEE Journal of Lightwave Technology*, **15** (5), 852–860.

[5] Bjerkan, L., Royset, A., Hafskjaer, L., and Myhre, D. (1996) Measurement of laser parameters for simulation of high-speed fiberoptic system. *IEEE Journal of Lightwave Technology*, **14** (5), 839–850.

[6] Gao, J., Li, X., Flucke, J., and Boeck, G. (2004) Direct parameter-extraction method for laser diode rate-equation model. *IEEE Journal of Lightwave Technology*, **22** (6), 1604–1609.

[7] Sukhoivanov, I. (1996) Modeling of semiconductor lasers with large signal modulation. *Vth International Conference on Mathematical Methods in Electromagnetic Theory, MMET 1996 Proceedings*, pp. 416–419.

[8] Habibullah, F. and Huang, W.P. (2006) A self-consistent analysis of semiconductor laser rate equations for system simulation purpouse. *Optics Communications*, **258**, 230–242.

[9] McDonald, D. and O'Dowd, R.F. (1995) Comparison of two- and three-level rate equations in the modeling of quantum-well lasers. *IEEE Journal of Quantum Electronics*, **31** (11), 1927–1934.

[10] Hui, R.-Q. and Tao, S.-P. (1989) Improved rate equations for external cavity semiconductor lasers. *IEEE Journal of Quantum Electronics*, **25** (6), 1580–1584.

[11] Morthier, G. (1997) An accurate rate-equation description for DFB lasers and some interesting solutions. *IEEE Journal of Quantum Electronics*, **33** (2), 231–237.

[12] Koch, T.L. and Linke, R.A. (1986) Effect of nonlinear gain reduction on semiconductor laser wavelength chirping. *Applied Physics Letters*, **43** (10), 613–615.

[13] Harder, C., Vahala, K., and Yariv, A. (1983) Measurement of the linewidth enhancement factor-alpha of semiconductor laser. *Applied Physics Letters*, **42** (4), 328–330.

[14] Linke, R.A. (1985) Modulation induced transient chirping in single frequency lasers. *IEEE Journal of Quantum Electronic*, **QE-21** (6), 593–597.

[15] Cartledge, J.C. and Burley, G.S. (1989) The effect of laser chirping on lightwave system performance. *IEEE Journal of Lightwave Technology*, **7** (3), 568–573.

[16] Koch, T.L. and Bowers, J.E. (1984) Nature of wavelength chirping in directly modulated semiconductor lasers. *IEE Electronic Letters*, **20** (25/26), 1038–1040.

[17] Henry, C.H. (1982) Theory of the linewidth of semiconductor lasers. *IEEE Journal of Quantum Electronic*, **QE-18** (2), 259–264.

[18] Vanwikelberge, P., Buytaert, F., Franchois, A. *et al.* (1989) Analysis of the carrier-induced FM response of DFB lasers: theoretical and experimental case studies. *IEEE Journal of Quantum Electronic*, **QE-25** (11), 2239–2254.
[19] Tucker, R.S. (1985) High-speed modulation of semiconductor lasers. *IEEE Journal of Lightwave Technology*, **LT-3** (6), 1180–1192.
[20] Vahala, K. and Yariv, A. (1983) Semiclassical theory of noise in semiconductor lasers – part I. *IEEE Journal of Quantum Electronic*, **QE-19** (6), 1096–1101.
[21] Vahala, K. and Yariv, A. (1983) Semiclassical theory of noise in semiconductor lasers – part II. *IEEE Journal of Quantum Electronic*, **QE-19** (6), 1102–1109.
[22] Henry, C.H. (1986) Phase noise in semiconductor lasers. *IEEE Journal of Lightwave Technology*, **LT-4** (3), 298–311.
[23] Yamamoto, Y. (1983) AM and FM quantum noise in semiconductor lasers – part I: theoretical analysis. *IEEE Journal of Quantum Electronic*, **QE-19** (1), 24–46.
[24] Yamamoto, Y. (1983) AM and FM quantum noise in semiconductor lasers – part II: comparison of theoretical and experimental results of AlGaAs lasers. *IEEE Journal of Quantum Electronic*, **QE-19** (1), 47–58.
[25] Obermann, K., Kindt, S., and Petermann, K. (1996) Turn-on jitter in zero-biased single-mode semiconductor lasers. *IEEE Photonic Technology Letters*, **8** (1), 31–33.
[26] Miller, D.A.B., Chemla, D.S., and Schmitt-Rink, S. (1986) Relation between electroabsorption in bulk semiconductor and in quantum wells: the quantum-confined Franz–Keldish effect. *Physical Review B*, **23** (10), 6976–6982.
[27] Miller, D.A.B., Chemla, D.S., Damen, T.C. *et al.* (1985) Electric field dependence of quantum absorption near the band gap of quantum-well structures. *Physical Review B*, **32** (2), 1043–1060.
[28] Harrison, P. (2005) *Quantum Wells Wires and Dots*, 2nd edn, John Wiley & Sons, Ltd, Chichester.
[29] Koyama, F. and Iga, K. (1988) Frequency chirping in external modulators. *IEEE Journal of Lightwave Technology*, **6** (1), 87–93.
[30] Devaux, F., Sorel, Y., and Kerdiles, J.F. (1993) Simple measurement of fiber dispersion and of chirp parameter of intensity modulated light emitter. *IEEE Journal of Lightwave Technology*, **11**, 1937–1940.
[31] Mitomi, O., Wakita, K., and Kotaka, I. (1994) Chirping characteristic of electroabsorption-type optical-intensity modulator. *IEEE Photonic Tehcnology Letters*, **6** (2), 205–207.
[32] Kim, Y., Kim, S.K., Lee, J. *et al.* (2001) Characteristics of 10 Gb/s electroabsorption modulator integrated distributed feedback lasers for long-haul optical transmission systems. *Optical Fiber Technology*, **7**, 84–100.
[33] Ebeling, K.J. (1992) *Integrierte Optoelektronik*, 2nd Auflage, Springer-Verlag, Berlin.
[34] Pollock, C.R. (1995) *Fundamentals of Optoelectronics*, Irwin, Chicago.
[35] Kim, H. and Gnauck, A.H. (2002) Chirp characteristics of dual-drive Mach–Zehnder modulator with a finite DC extinction ratio. *IEEE Photonics Technology Letters*, **14** (3), 298–300.
[36] Bravetti, P., Balsamo, S., Villa Montoya, J.A. *et al.* (2005) Unbroadened-spectrum chirped modulation: effects of the chirp-inducing mechanism on the spectral broadening in $LiNbO_3$ modulators. *IEEE Photonics Technology Letters*, **17** (3), 564–566.
[37] Wenke, G. and Klimmek, M. (1996) Considerations on the $\boldsymbol{\alpha}$-factor of nonideal, external optical Mach–Zehnder modulators. *Journal of Optical Communications*, **17**, 42–48.
[38] Djupsjöbacka, A. (1992) Residual chirp in integrated-optic modulators. *IEEE Photonics Technology Letters*, **4** (1), 41–43.
[39] Chung, H., Chang, W.S.C., and Adler, E.L. (1991) Modeling and optimization of traveling-wave $LiNbO_3$ interferometric modulators. *IEEE Journal of Quantum Electronics*, **27** (3), 608–617.
[40] Alferness, R.C. (1982) Waveguide electrooptic modulators. *IEEE Transaction on Microwave Theory and Technique*, **MTT-30** (8), 1121–1137.
[41] Doi, M., Sugiyama, M., Tanaka, K., and Kawai, M. (2006) Advanced $LiNbO_3$ optical modulators for broadband optical communications. *IEEE Journal of Selected Topics in Quantum Electronics*, **12** (4), 745–750.
[42] Kondo, J., Kondo, A., Aoki, K. *et al.* (2002) 40-Gb/s X-cut $LiNbO_3$ optical modulator with two-step back-slot structure. *IEEE Journal of Lightwave Technology*, **20** (12), 2110–2114.
[43] Koyama, F. and Iga, K. (1988) Frequency chirping in external modulators. *IEEE Journal of Lightwave Technology*, **6** (1), 87–93.
[44] Arecchi, F.T. and Schulz-Dubois, E.O. (1988) *Laser Handbook*, North Holland, Amsterdam.
[45] Agrawal, G.P. (1989) *Nonlinear Fiber Optics*, Academic Press, Inc., San Diego, CA.

[46] Siegman, A.E. (1986) *Lasers*, University Science Books, Mill Valley, CA.
[47] Kazovsky, L., Benedetto, S., and Willner, A. (1996) *Optical Fiber Communication Systems*, Artech House, Chapter 5.
[48] Jeruchim, M., Balaban, P., and Shanmugan, S. (1992) *Simulation of Communication Systems*, Plenum Press, New York, Chapter 4.
[49] Chua, L.O. and Lin, P.-M. (1975) *Computer-Aided Analysis of Electronic Circuits*, Prentice-Hall, Englewood Cliffs, NJ.
[50] Santagiustina, M. and Giltrelli, M. (2004) Semianalytical approach to the gain ripple minimization in multiple pump fiber Raman amplifiers. *IEEE Photonic Technology Letters*, **16**, 2454–2456.
[51] Agrawal, G.P. (2001) *Non Linear Fiber Optics*, 3rd edn, Academic Press.
[52] Auyeung, J. and Yariv, A. (1978) Spontaneous and stimulated Raman scattering in long low-loss fibers. *IEEE Journal of Quantum Electronics*, **14**, 347–351.
[53] Desurvire, E. (1994) *Erbium Doped Fiber Amplifiers: Principles and Applications*, John Wiley & Sons, Inc., New York.
[54] Kobyakov, A., Vasilyev, M., Tsuda, S. *et al.* (2003) Analytical model for Raman noise figure in dispersion managed fibers. *IEEE Photonic Technology Letters*, **15**, 30–32.
[55] Agrawal, G.P. *Fiber-optic Communication Systems*, 3rd edn, John Wiley & Sons, Inc., New York.
[56] OPTMIST Project: Technology Trend Documents http://www.ist-optimist.unibo.it/tech.asp.
[57] Chou, S.Y. and Liu, M.Y. (1991) Nanoscale tera-hertz metal–semiconductor–metal photodetectors. *IEEE Journal of Quantum Electronics*, **27**, 737.
[58] Shimizu, N., Miyamoto, Y., Hirano, A. *et al.* (2000) RF saturation mechanism of InP/InGaAs uni-travelling-carrier photodiode. *Electronics Letters*, **36**, 750.
[59] Chiu, Y.J., Fleischer, S.B., and Bowers, J.E. (1998) High-speed low-temperature-grown GaAs p-i-n travelling-wave photodetector. *IEEE Photonic Technology Letters*, **10**, 1012.
[60] Islam, M.S., Murthy, S., Itoh, T. *et al.* (2001) Velocity-matched distributed photodetectors and balanced photodetectors with p-i-n photodiodes. *IEEE Transactions on Microwave Theory and Technique*, **49**, 1914.
[61] Koester, C. (1968) Wavelength multiplexing in fiber optics. *Journal of the Optical Society of America*, **58** (1), 63–67.
[62] Feynman, R. (1983) *Lectures in Physics*, Addison-Wesley.
[63] Born, M. and Wolf, E. (1980) *Principles of Optics*, 6th edn, Pergamon Press.
[64] Longhurst, R. (1986) *Geometrical and Physical Optics*, 3rd edn, Longman.
[65] Takahasi, H., Suzuki, S., Kato, K., and Nishi, I. (1990) Arrayed waveguide grating for wavelength division multi/demultiplexing with nanometer resolution. *Electronics Letters*, **26** (2), 87–88.
[66] Smit, M. and van Dam, C. (1996) Phasar-based WDM-devices: principles design and applications. *IEEE Journal on Selected Topics of Quantum Electronics*, **2** (2), 236–250. Also: Smit, M. (1988) *Electronics Letters*, **24**, (7), 385–386.
[67] Takada, K., Yamada, H., and Okamoto, K. (1999) 480 channel 10 GHz spaced multi/demultiplexer. *Electronics Letters*, **35** (22), 1964–1966.
[68] Okamoto, K., Moriwaki, K., and Ohmori, Y. (1995) Fabrication of a 64 × 64 arrayed-waveguide grating multiplexer on Si. *Electronic Letters*, **31** (3), 184–186.
[69] Okamoto, K., Syuto, K., Takahashi, H., and Ohmori, Y. (1996) Fabrication of 128-channel arrayed-waveguide grating multiplexer with 25 GHz channel spacing. *Electronic Letters*, **32** (16), 1474–1476.
[70] Doerr, C. and Joyner, C.H. (1997) Double-chirping of the waveguide grating router. *IEEE Photonics Technology Letters*, **9** (6), 776–778.
[71] Doerr, C., Shirasaki, M., and Joyner, C.H. (1997) Chromatic focal plane displacement in the waveguide grating router. *IEEE Photonics Technology Letters*, **9** (6), 776–778.
[72] Soole, J., Scherer, A., Leblanc, H. *et al.* (1991) Monolithic InP-based grating spectrometer for wavelength-division multiplexed systems at 1.5 μm. *Electronic Letters*, **27** (2), 132–134.
[73] McGreer, K. (1995) A flat-field broadband spectrograph design. *IEEE Photonics Technology Letters*, **7** (4), 397–399.
[74] Deri, R., Kallman, J., and Dijaili, S. (1994) Quantitative analysis of integrated optic waveguide spectrometers. *IEEE Photonics Technology Letters*, **6** (2), 242–244.
[75] Sun, Z., McGreer, K., and Broughton, J. (1998) Demultiplexing with 120 channels and 0.29-nm channel spacing. *IEEE Photonics Technology Letters*, **10** (1), 90–92.

[76] Gibbon, M., Thompson, G., Clements, S. *et al.* (1989) Optical performance of integrated 1.5 μm grating wavelength multiplexer on InP-based waveguide. *Electronic Letters*, **25** (16), 1441–1442.
[77] Highwave-tech.com. http://www.highwave-tech.com/products/.
[78] Nishi, I., Oguchi, T., and Kato, K. (1987) Broad passband multi/demultiplexer for multimode fibres using a diffraction grating with retroreflectors. *IEEE Journal of Lightwave Technology*, **LT-5** (12), 1695–1700.
[79] Soole, J.B., Amersfoort, M.R., Leblanc, H.P. *et al.* (1996) Use of multimode interference couplers to broaden the passband of wavelength dispersive integrated WDM filters. *IEEE Photonics Technology Letters*, **8** (10), 1340–1342.
[80] http://www.photonetics.com.
[81] Watanabe, R., Nosu, K., Harada, T., and Kita, T. (1980) Optical demultiplexer using concave grating in 0.7–0.9 μm wavelength region. *Electronic Letters*, **16** (3), 106–108.
[82] Kita, T. and Harada, T. (1983) Use of aberration corrected concave gratings in optical multiplexers. *Applied Optics*, **22** (6), 819–825.
[83] Stavdas, A., Bayvel, P., and Midwinter, J.E. (1996) Design and performance of concave holographic gratings for applications as multiplexers/demultiplexers for wavelength routed optical networks. *Optical Engineering (SPIE)*, **35**, 2816–2823.
[84] Stavdas, A., Manousakis, M., Droulias, S. *et al.* (2001) The design of a free-space multi/demultiplexers for ultra-wideband WDM networks. *IEEE Journal of Lightwave Technology*, **19** (11), 1777–1784.Also: Stavdas, A. (1995) Design of multiplexer/demultiplexer for dense WDM wavelength routed optical networks. PhD thesis, University of London.
[85] Yariv, A. (1985) *Optical Electronics*, HRW International Edition, 3rd edn.
[86] Hecht, E. (1987) *Optics*, 2nd edn, Addison-Wesley.
[87] Offrein, B.J., Bona, G.L., Horst, F. *et al.* (1999) Wavelength tunable optical add-after-drop filter with flat passband for WDM networks. *IEEE Photonics Technology Letters*, **11** (2), 239–241.
[88] Yamada, M. and Shimizu, M. (2003) Ultra-wideband amplification technologies for optical fiber amplifiers. *NTT Technical Review Letters*, **1** (3), 80–84.
[89] ITU-T Supplement no. 39. Optical system design and engineering considerations, October 2003.
[90] Thompson, G.H.B. (1980) *Physics of Semiconductor Lasers*, John Wiley & Sons, Inc., New York.
[91] Bjarklev, A. (1993) *Optical Fiber Amplifiers: Design and System Applications*, Artech House, Boston, MA.
[92] Shimada, S. and Ishio, H. (eds) (1994) *Optical Amplifiers and their Applications*, John Wiley & Sons, Ltd, Chichester.
[93] Sudo, S. (ed.) (1997) *Optical Fiber Amplifiers – Materials, Devices, and Applications*, Artech House, Boston, MA.
[94] Condon, E.U. and Shortley, G.H. (1977) *The Theory of Atomic Spectra*, 9th corr. repr., Cambridge University Press, Cambridge, UK.
[95] Dieke, G.H. (1968) *Spectra and Energy Levels of Rare Earth Ions in Crystals*, Interscience, New York, NY.
[96] Reisfeld, R. and Jørgensen, C.K. (1977) *Lasers and Excited States of Rare Earths*, Springer-Verlag, Berlin.
[97] Zemon, S., Lambert, G., Miniscalco, W.J. *et al.* (1989) Characterization of Er^{3+}-doped glasses by fluorescence line narrowing. *Proceedings of SPIE Conference on Fiber Laser Sources Amplifiers*, vol. **1171**, p. 219.
[98] Desurvire, E. and Simpson, J.R. (1990) Evaluation of 4I15/2 and 4I13/2 Stark level energies in erbium-doped aluminosilicate glass fibers. *Optics Letters*, **15** (10), 547.
[99] Kobyakov, A. *et al.* Analytical model for Raman noise figure in dispersion managed fibers. *IEEE Photonic Technology Letters*, **15**, 30–32, 20.

5

Assessing Physical Layer Degradations

Andrew Lord, Marcello Potenza, Marco Forzati and Erwan Pincemin

5.1 Introduction and Scope

One reason for the enormous body of research and study into the physical layer impairments arising from high bit rate, high channel count optical transmission is to understand better the fundamental capacity/distance limits for transmission. Another reason is to enable improved, more robust system designs to give sufficient operating margin, but with a good cost balance.

If a new system build could rely on a new fiber installation, then the design would be simpler. However, the fiber infrastructure belonging to many operators comprises fiber installed anywhere from the mid 1980s to the mid 2000s – that is, a 20-year period. During this time, optical fiber specifications have improved and new fiber types have been introduced. Some operators have retained a single type of fiber, to simplify infrastructure management, whilst others have bought existing fiber networks with a range of fiber types. Even within a single type of fiber, such as step-index fiber, the wide range of polarization-mode dispersion (PMD) values in a single network make system design, particularly for bit rates of 10 Git/s and beyond, challenging.

So, physical layer modeling has been developed to deal with the twin challenges of carrying more capacity over legacy infrastructure. However, in the future, it is possible that a third challenge will make even more demands on our understanding of the physical layer performance of optical systems: namely the concept of automatically setting up optical circuits quickly rather than the steady, traditional approach of commissioning and testing. If the inevitable growth in bandwidth requirements also brings with it more dynamic traffic, then networks which can rapidly track these dynamics may be more efficient and cost effective than traditional alternatives. Automatic wavelength service provisioning requires a strong knowledge of physical layer performance, because there is not enough time to do a full commissioning; instead, we trust our models to be robust enough to handle live traffic.

Core and Metro Networks Edited by Alexandros Stavdas

This chapter discusses in depth the modeling of all the major physical layer parameters that have an impact on optical transmission at high bit rates and with high spectral efficiencies. The first section examines the end-to-end system level, in which the performance can be described in terms of an optical power budget. This approach seeks to allocate penalties to the various physical layer effects, such that they can be subtracted from an overall margin. After considering all effects, there should still be sufficient margin for acceptable operation.

Optical amplifiers (OAs) counter fiber loss and, thereby, offer the potential to extend optical system distances to many thousands of kilometers. The power budget first needs to account for the noise introduced by these amplifiers, and this is dealt with in Section 5.2.

One simple solution to noise is to increase signal power, and this can be done in optical systems until nonlinear effects give rise to sufficient penalty. The nonlinear effects arise from the glass properties, including both the nonlinear Kerr effect and also the fiber chromatic dispersion (CD); they can be severely exacerbated for multichannel systems. The crosstalk arising from nonlinear effects is dealt with in Sections 5.2 and 5.3. The two supported transverse propagation modes in optical fiber lead to a range of polarization-related effects in the optical fiber, amplifiers, and passive components, and these are also dealt with in Sections 5.2 and 5.3.

In dynamically switched optical transmission, the potentially rapid changes in optical signal power, together with the sensitivity of erbium-doped fiber amplifiers (EDFAs) to their operating point, can lead to transient effects which tend to worsen as they propagate through chains of amplifiers. These effects, together with solutions, are described in Section 5.3.

The final sections of the chapter discuss compensation techniques both for chromatic and PMD and finally some application to chains of optical add–drop multiplexers (OADMs), different modulation formats, and the notion of hybrid networks in which there is some, infrequent regeneration.

5.2 Optical Power Budgets, Part I

The scope of this section is to recall the basic theory from Chapter 3, to discuss the Gaussian model of photodetection, and to apply it to estimate light path performance in the presence of several transmission impairments along an optical transport network. A system point of view will be adopted: a detailed modeling of photodetection and of amplifier noise is outside the scope of this section. Rather, the aim of this part is to introduce a convenient setting to discuss light path performance and we shall content ourselves with giving simple heuristic justifications of basic formulae. Readers willing to deepen theoretical foundations of the basic relations will be referred to Chapter 3 and to the literature.

5.2.1 Optical Signal-to-Noise Ratio and Q Factor

The present scope is to recall the basic theory of Chapter 3 to treat optical connections (*light paths*) in an optical transport network. Use will be made of parameters such as optical signal-to-noise ratio (OSNR) and Personick's Q-factor to assess a suitable performance metric for light paths in terms of the accumulation of noise (Section 5.2.2) and distortion (Section 5.2.3) along an optical route. This analysis will help in drawing a set of light path evaluation rules; as a consequence, simple power budget rules will be derived in Section 5.2.4.

Building and operating a transparent network are two separate activities, but they are closely linked. The design phase can take several weeks or months, whereas the operation of the network, including rerouting of wavelengths, is in real time.

In the *design phase*, one has to decide where to put components such as optical cross-connects (OXCs), OADMs, and OAs. To do the network design at a given bit rate, it is necessary to determine:

1. The number of *nodes*[1] and their range of attenuation.
2. The number of *amplifiers* and the range of *span*[2] lengths; for attenuation levels below, say, 20 dB the optical noise due to OAs is relatively low. If either node or span attenuation exceed these values then the optical noise will increase, thus requiring higher signal power, which in turn will increase the impact of nonlinear effects; so these aspects require the determination of:
   a. the range of optimum signal powers;
   b. the modulation format;
   c. the inline dispersion compensation strategy – that is full compensation or undercompensation together with equalization at line terminals, or precompensation, postcompensation, and so on;
   d. the need for (and type of) forward error correction (FEC);
   e. the load of wavelengths and their spacing;
   f. the optical performance monitoring (OPM) strategy to adopt.

Answers to these mutually related questions will affect the *performance of network paths*. At this stage in the building of an actual network, detailed calculations and numerical simulations can be made, as a lot of time is available. *Numerical simulations* are based on the detailed modeling of physical processes occurring on the transport plane and allow the network designer to calculate the performance of the simulated light path in terms of transmission parameters.

However, it may not be possible to go out and measure or commission every single possible path that a wavelength channel might take in the network. Rather, it could be possible to measure and characterize each *physical link*; that is to say, each physical connection through adjacent nodes. Also, given the set of nodes and the physical connectivity pattern in the network, a few *critical paths* can be individuated that are affected by the heaviest transmission impairments in terms of number of crossed nodes, of transport distance, and of required capacity. This partial investigation can still be at the basis of a *metric to measure transport quality*.

The information gathered in one of these ways allows network designers to ascertain the feasibility of a subset of important light paths in the structure, which are basic or critical for network operation; longer paths, less likely to be used, need not be feasible in a totally transparent way (may require regeneration).

The reader should remark that a detailed solution of the *whole* network problem, determining the features in (1) and (2) listed above, would mean the collection of *network information* and *traffic demands* to give *routing and wavelength assignment* criteria as a result of a suitable *linear optimization procedure* of constrained network variables. This procedure is usually

[1] Network sites of traffic generation/flexibility/termination.

[2] Distance between adjacent amplifier sites.

beyond any reasonable reach for large networks having as many as 40–80 nodes. The complexity of impairment-constrained-based routing algorithms is polynomial with respect to the number N of nodes in the network, reaching the order N^4 [1] for a fully meshed architecture.

Since the requirement of getting the optimal solution to the network problem may not be affordable for large networks, one should limit to testing the workability of *one design* of the network.[3] This more modest point of analysis would require numerical simulation of all possible connections in a fully meshed infrastructure – that is, to simulate $\frac{1}{2}N(N-1)$ links – and it may still be too onerous, even in the design phase.

However, the link connectivity of the network (i.e., the number of direct links between nodes) will be a fraction of the connectivity of a completely meshed structure. For cost reasons it is likely that, in a real network, the effective number of links will be further reduced to the order of N. With the time available in the design phase, it may be a reasonable endeavor to perform a detailed simulation of N links (resilience and load balancing issues will only multiply this number by a small integer $k \lesssim 5$). Storing the simulation results of these links, a look-up table approach could be considered where the performance of *any* route is evaluated "summing-up" the transmission penalties incurred along any link belonging to that route.

This methodology requires a *sum rule*; this is one basic issue to consider in setting a metric for light path evaluation rules based on numerical simulations of physical links. From a path simulation one gets a reliable binary answer: the path is either feasible or not feasible. However, in the case of an unfeasible path, a single simulation gives no information about the reason why this happens (which are the stronger impairments) and how the path performance could be improved (if a pseudo-linear transmission regime may be preferable to a linear one, or if relaxing from a 50 GHz to a 100 GHz channel spacing may help, or if a link not working at 40 Gbit/s may operate at the lower bit rate of 10 Gbit/s).

Therefore, an alternative or complement to the simulation approach is sought, and that is the subject of this section. It is important to collect *a set of light-path evaluation rules* describing network characteristics in an analytical or semi-analytical manner, or according to a model that, though not as accurate as a simulation may be, can provide network designers with information on transmission impairments on a whole bundle of related links. Light-path evaluation rules can have an important role both in the planning and deployment phase, at least to get a rough idea of the global design, and in the *operational phase*, supporting real-time light path (re-) routing.

The set of light-path evaluation rules should be simple and accurate to give a global view of the network. Simple, because it adds burden to the control plane (or whatever subsystem is devoted to path routing, see Chapter 2) and should be performed in real time (1 ms–1 s scale) and be accurate, because it has to work. A suitable trade-off between simplicity and accuracy has to be employed, maybe with some iterative procedure or layered approach about the detail and approximation of network analysis.

This section aims at collecting such simple rules for light-path evaluation. Starting from the definition of quality parameters for light-path performance, we propose simple (but reasonably accurate) models for transmission penalties on quality parameters and define a performance

[3] Which may result from legacy infrastructure, sector planning, heuristic solutions, or any other simplified approach to the network problem.

metric for light-path evaluation in terms of a sum rule of effects. As we saw in Chapter 3, frequently used *quality parameters* for light path evaluation are:

1. optical power level at receiver
2. Personick's Q-factor
3. OSNR
4. bit-error rate (BER)
5. eye-diagram opening (EO).

In the simplest networks without OAs, performance can be determined by power budgeting. Path attenuation A, receiver sensitivity P_{sens}, and operational margins M are used to set the optical channel power level P_0 at launch:

$$P_0 = P_{\text{sens}} + A + M \quad [\text{dBm}] \tag{5.1}$$

Moreover, the system bandwidth is compared with signal distortions to quantify transmission impairments due to possible dispersion effects and (linear) crosstalk mechanisms.

Translating these impairments into an equivalent *power penalty*[4] P_P at the receiver, the *power budget* (Equation 5.1) can be generalized to

$$P_0 = P_{\text{sens}} + A + M + P_P \quad [\text{dBm}] \tag{5.2}$$

Today, networks make a massive use of OAs (Section 4.5) that overcome path attenuation, but produce *optical noise* overlapping the signals (Section 5.2.2.1): the ratio between coherent signal power and noise power in the signal band just defines the OSNR. It is an important quality parameter since its value at detection affects the receiver sensitivity, reducing the performance with respect to the ideal case of 'infinite' OSNR[5] (i.e. the condition when actual OSNR values do not degrade performance). Low OSNR values cause a transmission penalty: the lower the OSNR, the higher the penalty, as shown in Figure 5.1.

The OSNR is a basic parameter for optical networks and can be used to assess light path *baseline* performance in the absence of any transmission penalty other than amplifier noise accumulation (Sections 4.5 and 5.2.2.1).

Personick's *Q-factor* [2,3] is a measure of the electrical signal-to-noise ratio (SNR) for a digital binary system. It is a fundamental parameter to assess end-to-end light-path performance, taking into account its optical baseline and every transmission penalty present on the transport layer (see Chapter 3).

The BER (the ratio of the number of wrongly detected bits to the total number of transmitted bits) is an operative quantity measured at signal reception to assess path end-to-end performance; see Figure 5.1. At present, the performance of a transport network is still referred to a 10^{-12} BER level, even if a 10^{-15} level is occasionally used.

Finally, the EO is used to assess channel performance in the presence of specific transmission impairments, such as CD, PMD, polarization-dependent gain (PDG), four-wave mixing (FWM), and so on, which degrade the eye pattern (Figure 5.2).

---

[4] The power penalty P_P is defined as a *positive* quantity; it means a higher power level $P_{\text{sens}} + P_P$ is required at the receiver to maintain the original performance in the presence of the penalty itself.

[5] Even in absence of any OA, OSNR is still limited by the spontaneous emission of sources: OSNR values as high as 50–60 dB can be attained in nonamplified optical networks. To give an idea of noise accumulation impact, at 10 Gbit/s bit rate, 'infinite' OSNR values are those higher than 30–35 dB in a 0.1 nm linewidth; Figure 5.1.

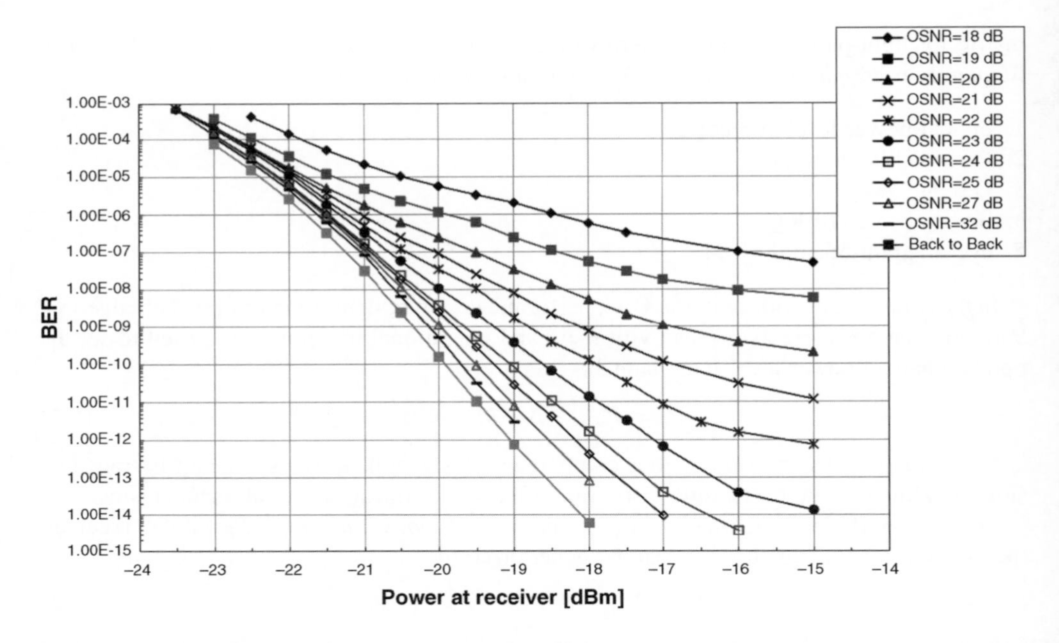

Figure 5.1 (Short-haul) receiver sensitivity for a 10 Gbit/s non return-to-zero signal. OSNR values refer to 0.1 nm linewidth. (Courtesy of Telecom Italia Lab)

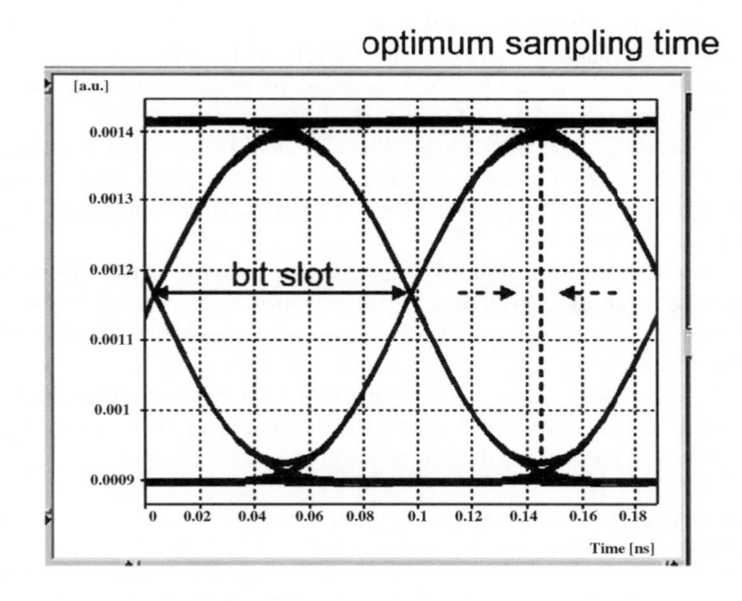

Figure 5.2 Eye-pattern

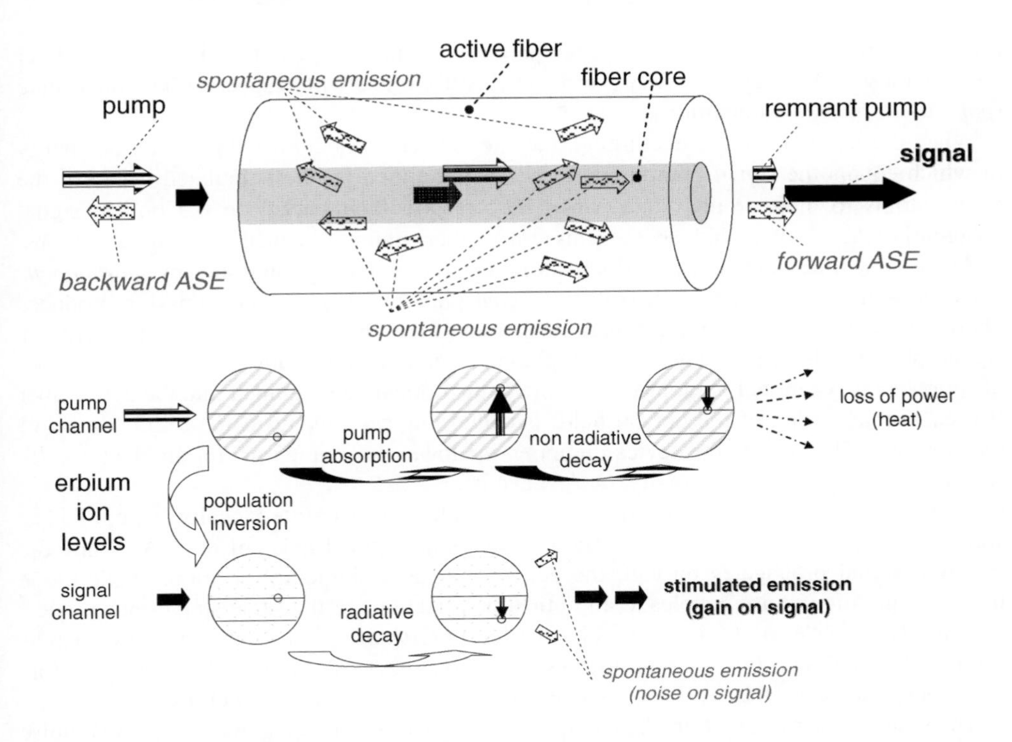

Figure 5.3 Schematic of EDFA operation. The active fiber medium is brought into population inversion by pump absorption, so emission overtakes attenuation. Stimulated emission supplies signal gain; the fraction of spontaneous emission guided inside the active fiber is amplification noise

To introduce these parameters in a quantitative way, the next subsection preliminarily discusses noise in OAs and SNR of the transmission channel.

5.2.2 Noise

This section discusses the topics of OA noise and of receiver noise. The reader is referred to Section 4.5 for the discussion of OAs and to Section 3.2 for detection theory.

5.2.2.1 Optical Noise of Fiber Amplifiers

The previous chapter covered the physics of optical fiber amplifiers (OFAs)[6] and of optical fibers in general, so only a very concise discussion will be given on OFA operation.

Light guided in the active fiber is amplified by *stimulated emission* of radiation into the active medium (Figure 5.3). The amplifier stores energy from the *pumping system* (laser diode pumps) into the active medium (erbium) atoms or ions, which jump onto an excited energy

[6] Optical transport networks may also use other types of device, such as semiconductor optical amplifiers (SOAs). Even if SOA physics are different from those of OFAs in the details (SOAs are electrically pumped and have a continuum of energy states) the formulae to be discussed in the following for OFAs also apply to SOAs with minor differences, which will be explained when needed.

level. After rapid relaxation processes that partially dissipate the excitation energy into fiber heating, these ions decay to a lower energy level, still excited with respect to the ground state (*population inversion* condition).

When the input signal travels through the excited active medium – the excitation energy of which equals the signal photon energy – it is amplified by the stimulated decay of the active atoms to their ground state, when they release their energy to the optical signal coherently, since the latter induces the emission of other photons identical to the signal photons.

At the same time, some excited erbium atoms can return to the ground state by *spontaneous emission* without any interaction with the signal photons. Spontaneous emission produces photons that superimpose on the signal incoherently: they have the same average energy as signal photons, but have uncorrelated phase, direction, and polarization. The number of spontaneously emitted photons that couple to the optical mode guided into the active fiber (for each polarization of the optical field) is called the *spontaneous emission factor* n_{sp} of the amplifier. The heavier the device pumping, the lower this number is: for an ideal (totally inverted) OA $n_{sp} = 1$; more usually, OFAs have $n_{sp} = 1.5$–2.

As long as it travels through the device, light spontaneously emitted locally is amplified as well: this *amplified spontaneous emission* (ASE) is the optical noise of the OA. While the coherent signal propagates through the active fiber as a single-polarization, single-mode field configuration, noise couples to all optical polarizations of the same single-mode field.[7] The *number of ASE modes* is denoted by m_t: for an OFA $m_t = 2$; for other devices, such as semiconductor OAs (SOAs), or multi-transverse-mode devices, different degeneracy factors may apply. However, the local spontaneous emission is always $m_t n_{sp}$ photons.

The system description of an OA can be given in terms of three parameters: gain G, noise power P_N and amplification bandwidth Λ_{amp}; G defines the ratio between input signal power P_{in} and output (coherent) signal power P_{coh}, P_N is the noise power in band with the signal and Λ_{amp} is the wavelength region where $G > 1$. For EDFAs, both G and P_N values depend on the specific signal wavelength $\lambda \in \Lambda_{amp}$ (see Figure 5.4).

Consider an optical beam or signal ϕ_{in} launched into an OFA. The beam consists of a photon flux I passing through the device cross-section in unit time. The photon flux can undergo deterministic variations in time (signal modulation), but for the moment it is supposed that the source emits under continuous wave (CW) conditions. If the signal is modulated, then the discussion below applies to the instant value $I(t)$ of flux at time t.

Even in the case of a CW signal, I fluctuates in time around an average value. This phenomenon is due to quantum mechanisms (*Heisenberg's uncertainty principle* between photon number and phase), as well as to quantum statistical effects (phase and amplitude noise of the source) and to classical statistical processes (excess intensity noise). The statistics of the optical beam are given to a good extent by the *flux mean value* (denoted by the use of angle brackets, like $\langle I \rangle$) and *flux variance* of I in time or over a statistical ensemble:

$$\langle I \rangle \equiv I_{in}, \quad \langle (I - \langle I \rangle)^2 \rangle = \langle I^2 \rangle - \langle I \rangle^2 \equiv \Sigma_{in}^2$$

[7] This is because optical transition cross-sections are largely independent of optical polarization. However, a small anisotropy is present in the polarization plane and is responsible for the PDG [4]. This effect is generally negligible (smaller than 0.1 dB) in a single OFA, but can produce appreciable effects on very long OFA chains, degrading the eye-pattern. To combat PDG and improve the signal-to-noise ratio, polarization-scrambling techniques can be adopted [5].

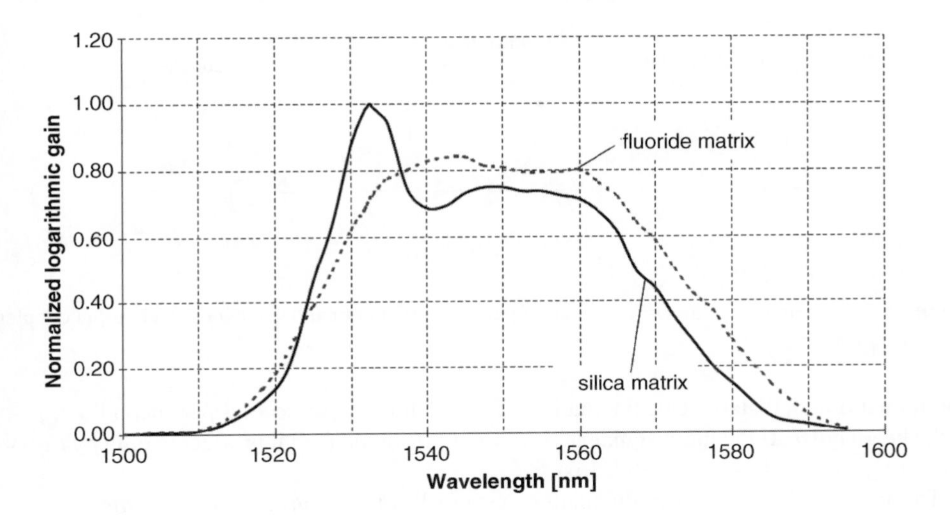

Figure 5.4 Spectral gain profile for EDFAs

The flux mean value I_{in} (in photons per second, i.e. [s^{-1}]) is the average value of the optical power entering the device, expressed in photon units:

$$P_{\text{in}} = I_{\text{in}} h \nu_S \tag{5.3}$$

In this equation, ν_S is the *peak signal frequency* in the spectral region $\Delta\nu_{\text{in}}$ occupied by the signal itself (input *signal bandwidth*) (Figure 5.5) and h is Planck's constant. The variance of

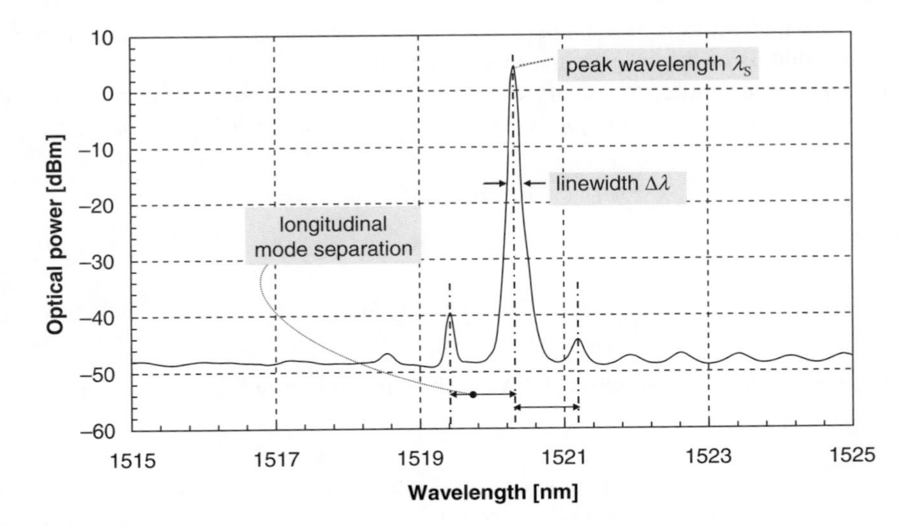

Figure 5.5 Signal spectrum showing peak wavelength λ_S and linewidth $\Delta\lambda$. Calling u the light speed in the medium where wavelengths are measured, in the frequency domain the peak frequency is $\nu_S = u/\lambda_S$ and the signal bandwidth $\Delta\nu = (u/\lambda_S^2)\Delta\lambda$. (Courtesy of Telecom Italia Lab)

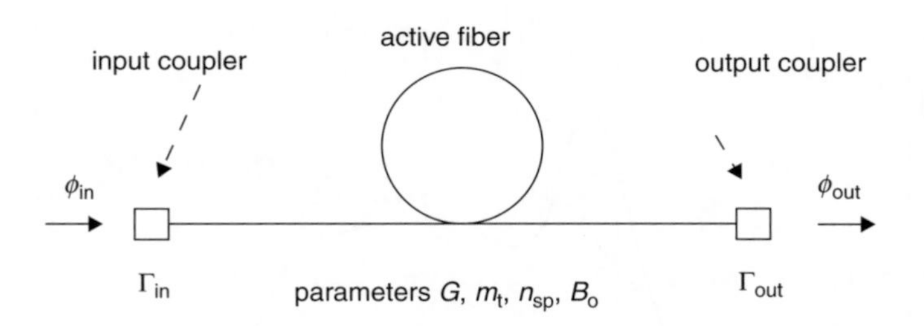

Figure 5.6 Schematics of an OFA showing input and output optical beams, input and output couplers, and the active fiber

(both quantum and statistical) fluctuations Σ_{in}^2 (in photons per second per second, i.e. $[s^{-2}]$) is the beam noise. If the input signal ϕ_{in} is a *coherent signal*, then it is affected only by *Poisson noise* (e.g. see Section 6.8 in Ref. [6]).

The variance Σ_q^2 of Poisson fluctuations establishes the *quantum noise regime*:[8]

$$\Sigma_q^2 = I_{in}\Delta\nu_{in} \tag{5.4}$$

For a single-mode optical field configuration such as the one generated by a single longitudinal-mode laser diode (Section 4.2), the input bandwidth $\Delta\nu_{in}$ can be identified with the longitudinal mode spacing of the source (Figure 5.5); this is the bandwidth that – centered around the peak frequency ν_S of the spectrum – can be associated with the signal (see Sections 5.2 and 6.5 in Ref. [6]), emitted by a laser source.

The bandwidth $\Delta\nu_{in}$ incorporates all spectral details of the signal (Figure 5.5). For a distributed-feedBack (DFB) source, typically employed in long-haul optical telecommunication systems, the CW linewidth can be as narrow as 10 MHz. However, when the source is modulated at a high bit-rate, its dynamical spectrum is dominated by the modulation spectrum and the linewidth is of the order of the modulation frequency itself.

If the input signal does not satisfy the quantum limit (Equation 5.4), then the global fluctuations Σ_{in}^2 account for both *quantum noise* Σ_q^2 and *excess statistical noise* Σ_{exc}^2,

$$\Sigma_{in}^2 = \Sigma_q^2 + (\Sigma_{in}^2 - \Sigma_q^2) \equiv \Sigma_q^2 + \Sigma_{exc}^2 \tag{5.5}$$

where

$$\frac{\Sigma_{exc}^2}{\Delta\nu_{in}} = \frac{\Sigma_{in}^2}{\Delta\nu_{in}} - \frac{\Sigma_q^2}{\Delta\nu_{in}} = \frac{\Sigma_{in}^2}{\Delta\nu_{in}} - I_{in} \tag{5.6}$$

Consider the beam ϕ_{in} fed into an OFA, as sketched in Figure 5.6.

[8] Owing to this dynamical picture, the photon flux is preferred to the photon number to describe the beam behavior. Expressing Equation 5.4 in terms of the number n_ϕ of photons passing through the beam cross-section in time $1/\Delta\nu_{in}$, one finds that the average value of n_ϕ equals its variance, a typical property of the Poisson distribution:

$$\langle n_\phi^2\rangle - \langle n_\phi\rangle^2 \equiv \frac{\Sigma_q^2}{(\Delta\nu_{in})^2} = \frac{I_{in}}{\Delta\nu_{in}} \equiv \langle n_\phi\rangle.$$

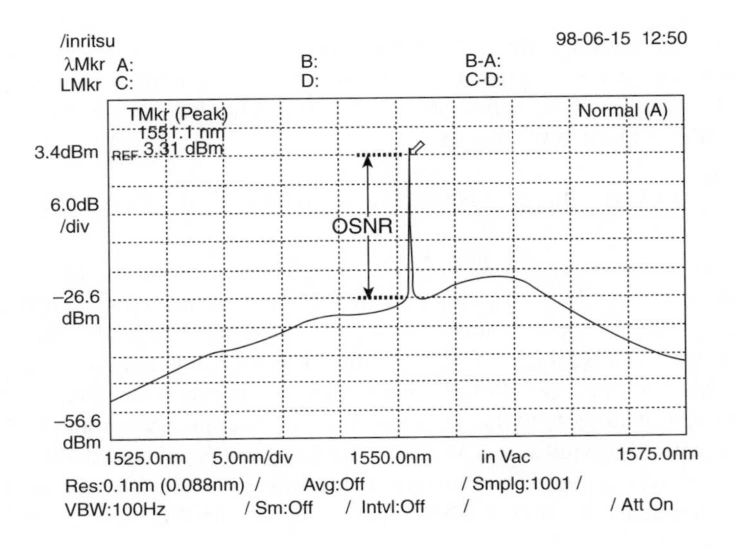

Figure 5.7 Signal and unfiltered noise background at the output of an EDFA

From left to right, the signal crosses:

- an input coupler with attenuation coefficient Γ_{in};
- the active fiber with gain G, number m_t of field configurations that contribute to signal fluctuations equivalently (transverse modes, ASE polarizations), effective spontaneous emission factor n_{sp}, and width B_o of the output spectrum;
- an output coupler with attenuation coefficient Γ_{out}.

The output bandwidth is set by the transition linewidth of the active ions, denoted by $\Delta\nu_{out}$, which is ~50 nm wide (Figure 5.7). However, this spectral region is delimited by the OFA noise, while the coherent output signal bandwidth is of the same order[9] as the input one $\Delta\nu_{in}$. In practical applications, wideband noise is intolerable and filters are used to limit the detection bandwidth and improve the in-band OSNR.[10]

The value R of the *OSNR* in linear units (the acronym OSNR itself is reserved for the corresponding logarithmic value) in an optical band $B_o \leq \Delta\nu_{out}$ is

$$R(B_o) = \frac{P_S}{P_N} = \frac{P_S}{\rho_N B_o} \quad \text{[linear units]}; \quad \text{OSNR}(B_o) = 10\log R(B_o) \quad \text{[dB]} \tag{5.7}$$

where ρ_N is the (effective) noise spectral density, assumed uniform over the optical band B_o.

A conventional bandwidth B_{ref} of 12.5 GHz, corresponding[11] to a 0.1 nm bandwidth, is generally used to calculate a *reference OSNR* and compare different configurations.

[9] If the amplifier gain is uniform over the signal frequency spectrum.

[10] If the receiver has a too wide frequency band, the electric signal-to-noise ratio would be unnecessarily degraded by wideband noise. On the other hand, the receiver bandwidth cannot be too narrow, otherwise the detector would lose efficiency, also rejecting part of the coherent signal power.

[11] In the 1500 nm wavelength region. More generally, the relation $\Delta\nu = (c/\lambda_S^2)\Delta\lambda$ holds between bandwidth $\Delta\nu$ and linewidth $\Delta\lambda$; c is the light speed in the medium where wavelengths are measured.

Indeed, the reader should realize that *OSNR values depend upon the optical bandwidth through which they are measured.* For a given signal configuration, consider two optical bandwidths $B_o = \Delta\nu_1$ and $B_o = \Delta\nu_2$ both wider than $\Delta\nu_{in}$, but narrower than $\Delta\nu_{out}$; using Equation 5.7, the respective OSNRs are

$$R(\Delta\nu_1) = \frac{\Delta\nu_2}{\Delta\nu_1} R(\Delta\nu_2) \quad [\text{l.u.}]; \quad \text{OSNR}(\Delta\nu_1) = 10\log\left(\frac{\Delta\nu_2}{\Delta\nu_1}\right) + \text{OSNR}(\Delta\nu_2) \quad [\text{dB}] \tag{5.8}$$

When $\Delta\nu_1 < \Delta\nu_2$, then $R(\Delta\nu_1) > R(\Delta\nu_2)$. For every application, the narrowest output bandwidth possible is preferable, but B_o cannot be reduced at will: all signal components must be passed to the receiver. For the same reason, the reference bandwidth B_{ref} is physically meaningful as long as it is larger than the modulated signal bandwidth, for signals modulated up to ~10 Gbit/s. For higher bit-rates, B_{ref} may still have a conventional meaning for reference OSNR, but actually it also affects the signal bandwidth. Equation 5.8 as it stands cannot be used in connection with bandwidths narrower than the coherent signal bandwidth.

Because of the ASE noise, the amplifier output is always noisier than its input. This fact is even clearer when stated in terms of SNR degradation. The SNR itself is generally defined as (Chapter 3)

$$\text{SNR} = \frac{\text{electrical signal power}}{\text{electrical power of fluctuations}} = \frac{\text{squared coherent optical power}}{\text{variance of optical fluctuations}}. \tag{5.9}$$

The first definition applies to the electrical domain; the second holds for the optical field.[12] A major difference exists between the definition in Equation 5.7 for OSNR and Equation 5.9: the OSNR is a ratio between two optical power levels. On the other hand, the SNR (Equation 5.9), even when referred to the optical domain, implies the process of photodetection: optical fluctuations manifest themselves only at the receiver, imprinting the statistical properties of radiation onto the photodetected electrical quantities.[13]

Starting from first principles, the SNR calculation is usually performed quantum mechanically, solving the master equation for the density matrix of the matter–radiation interacting system, with limitation to the matrix diagonal elements [7]. This gives the photon number probability density in the beam and allows a simplified solution, performed in the so-called *semiclassical approach* to the problem, where matter is considered quantum mechanically and radiation is treated as a microscopically and macroscopically self-consistent, albeit classical, field. In the regime of linear amplification, where saturation effects are negligible, the semiclassical approach gives the same results as the fully quantum mechanical Fokker–Planck [9,10] or Langevin equations (see Chapter 2 in Ref. [8], Section 2.2 in Ref. [11], and chapter VII in Ref. [12]).

Consider a wave packet amplified by a two-level laser active medium; consider further the probability $P_n(t)$ of finding the wave packet at time t in a photon number state $|n\rangle$ with n photons. In this picture, where the photon flux is associated with a wave packet, the photon number n corresponds to the photon flux I multiplied by the time $1/\Delta\nu$ associated with the wave packet having bandwidth $\Delta\nu$ (compare with note 8). Transitions occur between photon states

[12] An optical power P produces a photocurrent $i \propto P$. Since the signal power is measured as $R_l i^2$ on a load resistance R_l, the electrical signal power is proportional to P^2; similar considerations hold for noise terms.

[13] For ideal photodetectors, the photoelectron statistics are identical to incident photon statistics [8].

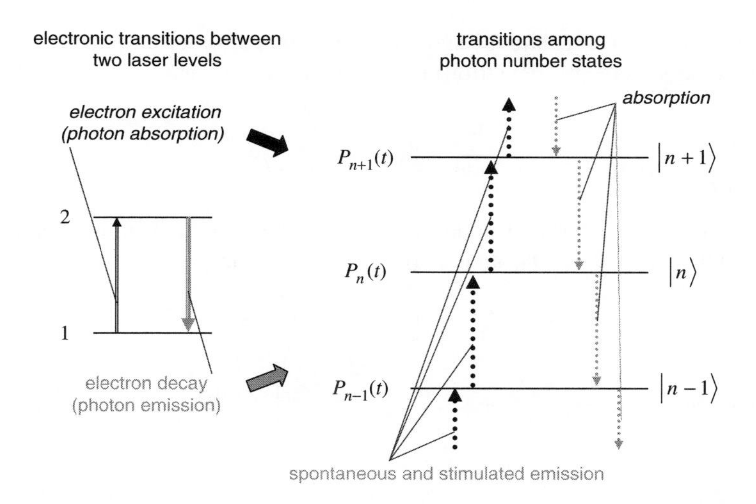

Figure 5.8 Electron transitions in the laser active medium (left) and photon number transitions in the amplified wave packet (right)

$|n-1\rangle$, $|n\rangle$ and $|n+1\rangle$ because of photon absorption and emission phenomena taking place on the electronic population densities N_1 and N_2 of the two laser levels (Figure 5.8).

The active medium couples *electron transitions* between its atomic levels (left side of Figure 5.8) with *transitions amongst photon number states* of the wave packet (right side of Figure 5.8). Electron excitation is associated with photon absorption, while electron decay is related to (spontaneous and stimulated) emission (compare with Figure 5.3). Let a [s^{-1}] be the transition probability of stimulated emission per photon; the theory of the Einstein coefficient A for laser actions assures this is also the transition probability for spontaneous emission (which is independent of the presence of photons) and that $a = A\Gamma N_2$, with Γ being an optical confinement factor.[14] Further, let b be the probability of absorption per photon, which is $b = A\Gamma N_1$. Then, the transition probability for $|n\rangle \rightarrow |n+1\rangle$ can be written as $T_{n,n+1} = a + an$: spontaneous emission is unaffected by the presence of other photons, while the stimulated emission rate is proportional to the number of photons present in the packet. Similarly, the transition probability for $|n\rangle \rightarrow |n-1\rangle$ is $T_{n,n-1} = bn$. In much the same way, $T_{n-1,n} = a + a(n-1)$ and $T_{n+1,n} = b(n+1)$. With these expressions for $T_{m,n}$s, the time evolution of the probability $P_n(t)$ is ruled by the *master equation*

$$\frac{dP_n(t)}{dt} = -[a(n+1)+bn]P_n(t) + anP_{n-1}(t) + b(n+1)P_{n+1}(t) \tag{5.10}$$

Introducing the statistical moment of the k-order for the photon number n; that is, the value of n^k averaged with probability P_n:

$$\langle n^k \rangle = \sum_{n=1}^{\infty} n^k P_n, \quad k = 1, 2, 3, \ldots$$

[14] It takes into account the superposition between optical wave packet and active medium ([11, Section 2.2.1.1]).

Equation 5.10 is used to calculate the statistical moments of order one (mean value) and two (variance or squared standard deviation) of the photon flux. For $k = 1$ and 2, Equation 5.10 gives

$$\frac{\mathrm{d}\langle n\rangle}{\mathrm{d}t} = (a-b)\langle n\rangle + a \quad \text{and} \quad \frac{\mathrm{d}\langle n^2\rangle}{\mathrm{d}t} = 2(a-b)\langle n^2\rangle + (3a+b)\langle n\rangle + a$$

respectively. These two differential equations must be solved with the initial conditions for the input wave packet: $\langle n(0)\rangle = \langle n_0\rangle$ and $\langle n^2(0)\rangle = \langle n_0{}^2\rangle$. One gets

$$\langle n(t)\rangle = \langle n_0\rangle \mathrm{e}^{(a-b)t} + \frac{a}{a-b}\left[\mathrm{e}^{(a-b)t} - 1\right]$$

$$\langle n^2(t)\rangle - \langle n(t)\rangle^2 = \langle n_0\rangle \mathrm{e}^{(a-b)t} + \frac{a}{a-b}\left[\mathrm{e}^{(a-b)t} - 1\right] + \frac{2a}{a-b}\left[\mathrm{e}^{(a-b)t} - 1\right]\langle n_0\rangle \mathrm{e}^{(a-b)t}$$

$$+ \left\{\frac{a}{a-b}[\mathrm{e}^{(a-b)t} - 1]\right\}^2 + (\langle n_0{}^2\rangle - \langle n_0\rangle^2 - \langle n_0\rangle)\mathrm{e}^{2(a-b)t}$$

If the wave packet, launched at $t = 0$ into the amplifier, emerges from the device at time t, it is immediate to conclude

$$\langle n_0\rangle \equiv \langle n_{\mathrm{in}}\rangle, \quad \langle n(t)\rangle \equiv \langle n_{\mathrm{out}}\rangle, \quad \mathrm{e}^{(a-b)t} = \mathrm{e}^{A\Gamma(N_2-N_1)} = G, \quad \frac{a}{a-b} = \frac{N_2}{N_2-N_1} \equiv n_{\mathrm{sp}}$$

where G is the internal single-pass gain and n_{sp} the spontaneous emission factor[15] of the amplifier. Then, the average photon number and its variance in the wave packet are

$$\langle n_{\mathrm{out}}\rangle = \langle n_{\mathrm{in}}\rangle G + n_{\mathrm{sp}}(G-1) \tag{5.11}$$

$$\langle n_{\mathrm{out}}{}^2\rangle - \langle n_{\mathrm{out}}\rangle^2 = \langle n_{\mathrm{in}}\rangle G + n_{\mathrm{sp}}(G-1) + 2n_{\mathrm{sp}}(G-1)\langle n_{\mathrm{in}}\rangle G$$

$$+ [n_{\mathrm{sp}}(G-1)]^2 + (\langle n_{\mathrm{in}}{}^2\rangle - \langle n_{\mathrm{in}}\rangle^2 - \langle n_{\mathrm{in}}\rangle)G^2 \tag{5.12}$$

It is convenient to recast Equations 5.11 and 5.12 of the semiclassical approach from photon number statistics into optical flux statistics. The photon number n has already been related to the flux I multiplied by the equivalent transit time $1/\Delta\nu$ of the wave packet (having bandwidth $\Delta\nu$) along the amplifier.[8] Conversely, I is associated with the integral of the number of photons at a given frequency per unit frequency interval, over the system optical band.

Upon a frequency integration, with consideration of input and output attenuations and of the number m_{t} of noise modes, Equation 5.11 gives for the average flux

$$I_{\mathrm{out}} = I_{\mathrm{coh}} + I_{\mathrm{incoh}} = G\Gamma_{\mathrm{out}}\Gamma_{\mathrm{in}}I_{\mathrm{in}} + m_{\mathrm{t}}\Gamma_{\mathrm{out}}(G-1)n_{\mathrm{sp}}B_{\mathrm{o}} \tag{5.13}$$

I_{coh} and I_{incoh} are the coherent and noise components of the optical flux respectively. This result is clear from a physical point of view. By definition, the coherent term is expressed through the device *net* gain:

$$I_{\mathrm{coh}} = G\Gamma_{\mathrm{out}}\Gamma_{\mathrm{in}}I_{\mathrm{in}}$$

[15] Also called the amplifier *inversion parameter*.

The ASE term I_{incoh}

$$I_{\text{incoh}} = m_t n_{\text{sp}} \Gamma_{\text{out}} (G-1) B_o \tag{5.14}$$

is the local spontaneous emission $m_t n_{\text{sp}}$, integrated over the optical bandwidth B_o, amplified by the active medium $(G-1)$ and attenuated by the output coupler Γ_{out}. The local spontaneous emission is amplified by $G-1$ (not G) and this corresponds to the minimum noise contribution of the active fiber that makes both input and output signal fluctuations satisfy Heisenberg's uncertainty principle [13]. Even though for $G \gg 1$ there is little difference between factors $G-1$ and G, the first is the correct one, since in the limit of a transparent active medium $(G = 1)$ no net spontaneous emission exists (see also Section 4.5).

For the flux variance, Equation 5.12 gives

$$\begin{aligned}
\frac{\Sigma^2_{\text{out}}}{\Delta\nu_{\text{out}}} &= \left(\frac{\Sigma^2_{\text{out}}}{\Delta\nu_{\text{out}}}\right)_{\text{shot}} + \left(\frac{\Sigma^2_{\text{out}}}{\Delta\nu_{\text{out}}}\right)_{\text{beat}} + \left(\frac{\Sigma^2_{\text{out}}}{\Delta\nu_{\text{out}}}\right)_{\text{exc}} \\
&= \underbrace{(I_{\text{coh}} + I_{\text{incoh}})}_{\text{shot}} + \underbrace{\left(\frac{2}{m_t B_o} I_{\text{coh}} I_{\text{incoh}} + \frac{1}{m_t B_o} I^2_{\text{incoh}}\right)}_{\text{beat}} + \underbrace{(G\Gamma_{\text{out}}\Gamma_{\text{in}})^2 \frac{\Sigma^2_{\text{exc}}}{\Delta\nu_{\text{in}}}}_{\text{excess}} \\
&= G\Gamma_{\text{out}}\Gamma_{\text{in}} I_{\text{in}} + m_t \Gamma_{\text{out}} (G-1) n_{\text{sp}} B_o + 2G(G-1)\Gamma^2_{\text{out}} \Gamma_{\text{in}} I_{\text{in}} n_{\text{sp}} \\
&\quad + m_t \Gamma^2_{\text{out}} (G-1)^2 n^2_{\text{sp}} B_o + (G\Gamma_{\text{out}}\Gamma_{\text{in}})^2 \frac{\Sigma^2_{\text{exc}}}{\Delta\nu_{\text{in}}}
\end{aligned} \tag{5.15}$$

Equation 5.15 puts semiclassical contributions to noise into evidence. The first two terms are the *quantum noise* on the amplified signal and on ASE respectively; both of them satisfy the Poisson condition (Equation 5.4) at the amplifier output. Their presence is to be expected, as a quantum limit in the expression for the variance, with respect to the mean value (Equation 5.13). The third and fourth terms are called *beat noise*, between signal and ASE and of ASE with itself respectively. Beat noise reflects the quantum statistical nature of the ASE process (see Section 2.3 in Ref. [8]). It is due to the simultaneous presence of phase and amplitude noise in the output beam (see Section 10.13 in Ref. [14]), which are statistically independent, ergodic, and steady-state processes. Remarkably enough, a stochastic description of noise in terms of classical fields with random phase and noise amplitude predicts a beating effect between signal and ASE at detection, which has exactly the same amount [15], though, in principle, no beating effects could arise between uncorrelated light beams.[16] This description is frequently referred to, since it supplies accurate spectral form factors for beat noise terms.

The last term in Equation 5.15 is the *excess statistical noise*. It can have different origins, for example the relative intensity noise (RIN) of the laser source or the effect of spurious reflections on the optical path, as for the multipath interference effect.

Equations 5.13 and 5.15 give a statistical description of the optical beam that is sufficient for SNR evaluation in the approach called *Gaussian approximation to the amplifier noise* (Section 3.2). The name is due to the fact that for a Gaussian random process the whole statistics can be reconstructed from the mean value and variance: all centered odd moments are vanishing while all centered even moments are permutations of products of the second-order moment

[16] Beat effects between independent signals can exist and have been experimentally detected, but limited to a space–time region where the corresponding phases are mutually correlated [16].

(see Section 2.2.3 in Ref. [10]). More detailed treatments of amplifier noise statistics require the determination of higher order moments of the photon flux,[17] but for a system analysis the Gaussian approximation is accurate enough [17,18].

As stressed in Section 3.2, cases where higher order moments depart from Gaussian behavior are the determination of optimum threshold for the receiver decision circuit, the analysis of situations where ASE–ASE beat noise is important, the exact statistics of nonlinear effects and the analysis of 2R-regenerators where the Gaussian model usually overestimates transmission penalties.

On the contrary, Gaussian approximations are more accurate when BER is low ($P_{\mathrm{BER}} \leq 10^{-9}$) and signal–ASE beat noise is predominant over ASE–ASE beat noise. This happens when a sufficiently high number of photons, say, $n_{\phi} \gtrsim 100\,000$ is present in a 10 m long active fiber, corresponding to an input power $P_{\mathrm{in}} \gtrsim 0.15\ \mu\mathrm{W}$. Besides this, even when noise is not normally distributed, an effective Gaussian model can often be formulated. Therefore, it is reasonable to use the Gaussian approximation for an order of magnitude BER estimate [9,11] under usual operating conditions for an optical transport system. For more accurate (though cumbersome) calculations, the reader is referred to Chapter 3.

Equation 5.9 will be used first to calculate the SNR[18] of the optical input signal for direct detection by photon counting, $L_{\mathrm{in}}^{(\mathrm{o})}$, as

$$L_{\mathrm{in}}^{(\mathrm{o})} = \frac{I_{\mathrm{in}}^2}{2B \frac{\Sigma_{\mathrm{in}}^2}{\Delta\nu_{\mathrm{in}}}} = \frac{P_{\mathrm{in}}^2}{2B(h\nu_{\mathrm{S}})^2 \frac{\Sigma_{\mathrm{in}}^2}{\Delta\nu_{\mathrm{in}}}} \tag{5.16}$$

in linear units, where B is the receiver bandwidth. The numerator in Equation 5.16 is the squared coherent signal power. The denominator is the squared power fluctuations $(h\nu_{\mathrm{S}})^2(\Sigma_{\mathrm{in}}^2/\Delta\nu_{\mathrm{in}})$ falling into the receiver bandwidth B. The *factor of two* stems from the definition of (single-sided) optical bandwidth. When a power density is integrated over its optical spectrum, the Fourier components of the optical flux $I(t)$ or of the photocurrent $i(t)$ are considered. The Fourier transform mathematically implies the presence of both positive and *negative* optical frequencies; though physically nonobservable, negative frequencies *are* in the Fourier spectrum and their contribution, though not supplying any new physical information, doubles the weight of the corresponding positive counterparts (see Section 10.3 in Ref. [14]).

In the case of a quantum limited signal, Equation 5.16 assumes the form

$$L_{\mathrm{in}}^{(\mathrm{o})}\Big|_{\mathrm{QL}} = \frac{I_{\mathrm{in}}^2}{2B \frac{\Sigma_{\mathrm{q}}^2}{\Delta\nu_{\mathrm{in}}}} = \frac{I_{\mathrm{in}}^2}{2BI_{\mathrm{in}}} = \frac{I_{\mathrm{in}}}{2B} = \frac{P_{\mathrm{in}}}{2Bh\nu_{\mathrm{S}}} \tag{5.17}$$

giving the ratio of the coherent signal power to the noise power coupled into the receiver bandwidth. This latter corresponds to twice the photon energy[19] $h\nu_{\mathrm{S}}$ in the single-sided bandwidth B; h is Planck's constant.

---

[17] The master equation can be used to obtain the time evolution for all optical moments, besides the evolution of the probability distribution function P_n itself.

[18] The acronym SNR will be reserved for the logarithmic value of the signal-to-noise ratio; the value in linear units will be denoted by L. Because of (5.9), the logarithmic value is $\mathrm{SNR} = 20 \log L$.

[19] It is well known that direct detection by photon counting has a lower SNR than both heterodyne and homodyne detection [6].

Beat and excess noise are statistical contributions to the amplifier output fluctuations. In Equation 5.14, fluctuations have been referred to an output bandwidth $\Delta\nu_{out}$, which differs from the input band $\Delta\nu_{in}$: $\Delta\nu_{in}$ is the spectral region occupied by the launch signal; in this case it can be the bandwidth of a source longitudinal mode. After the amplifier, the beam ϕ_{out} is spread over a wider spectrum due to the wideband ASE noise (see Figure 5.7). To use 'practical' notation, a set of directly measurable quantities can be defined. The *net gain* G_a of the amplifier between its input and output connectors is

$$G_a = G\Gamma_{out}\Gamma_{in} \tag{5.18}$$

It is the internal gain G of the active fiber net of coupling losses. The device *output* (coherent) *signal power* is (see Equation 5.13)

$$P_{coh} = G_a P_{in} = G_a I_{in} h\nu_S \tag{5.19}$$

Eventually, the *output* (incoherent) *ASE power* P_{ASE} superimposed to the output signal is considered. It is about[20] one-half the total ASE power generated by the device that is released both in a forward and backward direction with respect to the signal (Figure 5.3). According to Equation 5.13, P_{ASE} is given by

$$P_{ASE} = m_t\Gamma_{out}(G-1)n_{sp}h\nu_S B_o = \frac{m_t}{\Gamma_{in}}(G_a - \Gamma_{out}\Gamma_{in})n_{sp}h\nu_S B_o \tag{5.20}$$

Two other useful expressions for P_{ASE} can be given as well. The first is written in terms of the *number* N_a *of ASE photons* emitted by the amplifier in each mode:

$$N_a = \Gamma_{out}n_{sp}(G-1) = \frac{n_{sp}}{\Gamma_{in}}(G_a - \Gamma_{out}\Gamma_{in}) \tag{5.21}$$

giving

$$P_{ASE} = m_t N_a h\nu_S B_o \tag{5.22}$$

The second equation introduces the *single-polarization ASE spectral density* ρ_{ASE}:

$$\rho_{ASE} = \frac{P_{ASE}}{m_t B_o} = N_a h\nu_S \tag{5.23}$$

in a way that

$$P_{ASE} = m_t\rho_{ASE}B_o \tag{5.24}$$

Therefore, Equations 5.13 and 5.15 can be written in terms of measurable optical powers:

$$P_{out} = G_a P_{in} + P_{ASE} = P_{coh} + P_{ASE} \tag{5.25}$$

where P_{out} is the total output power of the amplifier and

$$\frac{\Sigma_{out}^2}{\Delta\nu_{out}} = \frac{P_{coh}}{h\nu_S} + \frac{P_{ASE}}{h\nu_S} + \frac{2P_{coh}P_{ASE}}{m_t B_o(h\nu_S)^2} + \frac{P_{ASE}^2}{m_t B_o(h\nu_S)^2} + G_a^2\frac{\Sigma_{exc}^2}{\Delta\nu_{in}} \tag{5.26}$$

[20] Asymmetries between forward and backward ASE actually exist, especially in weakly inverted amplifiers.

Equations 5.25 and 5.26 can be expressed through an affine transformation on signal and noise terms [8]. Introducing a 2×1 column vector for the optical flux:[21]

$$\begin{pmatrix} I \\ \sigma_{\text{opt}}^2 \end{pmatrix}, \quad \begin{cases} I & \text{mean photon flux } [\text{s}^{-1}] \\ \sigma_{\text{opt}}^2 \equiv 2B\dfrac{\Sigma^2}{\Delta\nu} & \text{variance of photon flux } [\text{s}^{-2}] \end{cases} \tag{5.27}$$

the amplifier is represented by the 2×2 *transfer matrix* and the 2×1 *noise vector*:

$$\mathbf{M}_{\text{a}} = G_{\text{a}}\begin{pmatrix} 1 & 0 \\ 1-G_{\text{a}}+2N_{\text{a}} & G_a \end{pmatrix}, \qquad \mathbf{N}_{\text{a}} = m_{\text{t}}N_{\text{a}}B_0\begin{pmatrix} 1 \\ 1+N_{\text{a}} \end{pmatrix} \tag{5.28}$$

Then, Equations 5.25 and 5.26 can be rewritten in a compact way:

$$\begin{pmatrix} I_{\text{out}} \\ \sigma_{\text{out,opt}}^2 \end{pmatrix} = \mathbf{M}_{\text{a}}\begin{pmatrix} I_{\text{in}} \\ \sigma_{\text{in,opt}}^2 \end{pmatrix} + \mathbf{N}_{\text{a}} \tag{5.29}$$

Equation 5.29 is convenient for determining the SNR gradual degradation along a chain of OAs and of attenuating sections; for attenuating sections, the noise vector is zero, $N_{\text{a}} = 0$ and gain $G_{\text{a}} > 1$ is replaced by attenuation $A < 1$:

$$\mathbf{A} = A\begin{pmatrix} 1 & 0 \\ 1-A & A \end{pmatrix} \tag{5.30}$$

It is customary to define single-sided spectral densities $\rho \equiv \Sigma^2/\Delta\nu = \sigma_{\text{opt}}^2/2B$ of optical fluctuations: they can be obtained from Equation 5.26. Considering the identification of shot, beat, and excess contributions already made in Equation 5.15, it is convenient to distinguish between a spectral density $\rho_{\text{shot,opt}}$ due to quantum (shot) noise

$$\rho_{\text{shot,opt}} \begin{cases} \rho_{\text{S-shot,opt}} = \dfrac{P_{\text{coh}}}{h\nu_{\text{S}}} = \dfrac{G_a P_{\text{in}}}{h\nu_{\text{S}}} \\ \rho_{\text{ASE-shot,opt}} = \dfrac{P_{\text{ASE}}}{h\nu_{\text{S}}} = m_{\text{t}}N_{\text{a}}B_{\text{o}} = \dfrac{m_{\text{t}}}{\Gamma_{\text{in}}}\Gamma_{\text{out}}(G_{\text{a}}-\Gamma_{\text{out}}\Gamma_{\text{in}})n_{\text{sp}}B_{\text{o}} \end{cases} \tag{5.30}$$

and a spectral density $\rho_{\text{stat,opt}}$ due to statistical (beat and excess) noise:

$$\rho_{\text{stat,opt}} \begin{cases} \rho_{\text{S-ASE,opt}} = \dfrac{2P_{\text{coh}}P_{\text{ASE}}}{m_{\text{t}}B_{\text{o}}(h\nu_{\text{S}})^2} = \dfrac{2}{m_{\text{t}}B_{\text{o}}}\dfrac{G_{\text{a}}P_{\text{in}}}{h\nu_{\text{S}}}\dfrac{P_{\text{ASE}}}{h\nu_{\text{S}}} \\ \rho_{\text{ASE-ASE,opt}} = \dfrac{P_{\text{ASE}}^2}{m_{\text{t}}B_{\text{o}}(h\nu_{\text{S}})^2} = \dfrac{1}{m_{\text{t}}B_{\text{o}}}\left(\dfrac{P_{\text{ASE}}}{h\nu_{\text{S}}}\right)^2 \\ \rho_{\text{exc,opt}} = G_{\text{a}}^2\dfrac{\Sigma_{\text{exc}}^2}{\Delta\nu_{\text{in}}} \end{cases} \tag{5.31}$$

All spectral densities (in $[\text{s}^{-1}]$) have to be multiplied by $2B$ to get the corresponding variance of the optical flux in $[\text{s}^{-2}]$.

[21] In Equation 5.27 the suffix 'opt' has been attached to σ^2 for future convenience, to distinguish from fluctuations in the electrical domain, to be introduced later.

Using Equation 5.31 it is possible to determine the output SNR:

$$L_{\text{out}}^{(\text{o})} = \frac{I_{\text{out}}^2}{\sigma_{\text{out,opt}}^2} = \frac{P_{\text{coh}}^2}{2B(h\nu_\text{S})^2 \frac{\Sigma_{\text{out}}^2}{\Delta\nu_{\text{out}}}} \equiv \frac{P_{\text{coh}}^2}{2B(h\nu_\text{S})^2 \rho_{\text{out,opt}}} \quad [\text{adim.}] \tag{5.32}$$

Since $L_{\text{out}}^{(\text{o})} \leq L_{\text{in}}^{(\text{o})}$ always, the ratio of input to output SNRs expresses the SNR degradation; in the optical domain this reads

$$D^{(\text{o})} = \frac{L_{\text{in}}^{(\text{o})}}{L_{\text{out}}^{(\text{o})}} \tag{5.33}$$

For a coherent input signal, Equation 5.33 defines the *amplifier noise factor* $F^{(\text{o})} = L_{\text{in}}^{(\text{o})}\Big|_{\text{QL}} / L_{\text{out}}^{(\text{o})}$. From Equations 5.30 and 5.31 its complete expression is

$$\begin{aligned} F^{(\text{o})} &= \frac{L_{\text{in}}^{(\text{o})}\Big|_{\text{QL}}}{L_{\text{out}}^{(\text{o})}} = \frac{P_{\text{in}}}{2Bh\nu_\text{S}} \frac{2B(h\nu_\text{S})^2 \frac{\Sigma_{\text{out}}^2}{\Delta\nu_{\text{out}}}}{P_{\text{coh}}^2} = \frac{1}{G_\text{a} P_{\text{coh}}} \frac{\Sigma_{\text{out}}^2}{\Delta\nu_{\text{out}}} h\nu_\text{S} \\ &= \frac{1}{G_\text{a} P_{\text{coh}}} \left(P_{\text{coh}} + P_{\text{ASE}} + \frac{2P_{\text{coh}} P_{\text{ASE}}}{m_\text{t} B_\text{o} h\nu_\text{S}} + \frac{P_{\text{ASE}}^2}{m_\text{t} B_\text{o} h\nu_\text{S}} + G_\text{a}^2 \frac{\Sigma_{\text{exc}}^2}{\Delta\nu_{\text{in}}} h\nu_\text{S} \right) \\ &= \frac{1}{G_\text{a}} \left(1 + \frac{P_{\text{ASE}}}{G_\text{a} P_{\text{in}}} \right) + \frac{2P_{\text{ASE}}}{G_\text{a} m_\text{t} B_\text{o} h\nu_\text{S}} \left(1 + \frac{P_{\text{ASE}}}{2G_\text{a} P_{\text{in}}} \right) + \frac{1}{P_{\text{in}}} \frac{\Sigma_{\text{exc}}^2}{\Delta\nu_{\text{in}}} h\nu_\text{S} \end{aligned} \tag{5.34}$$

In ideal conditions ($G_\text{a} \gg 1$, $n_{\text{sp}} = 1$, $\Gamma_{\text{in}} = \Gamma_{\text{out}} = 1$, $\Sigma_{\text{exc}} = 0$), the noise factor is dominated by signal–ASE beat noise:

$$F^{(\text{o})} \cong \frac{2P_{\text{ASE}}}{G_\text{a} m_\text{t} B_\text{o} h\nu_\text{S}} = \frac{2n_{\text{sp}}}{\Gamma_{\text{in}}} \frac{G_\text{a} - \Gamma_{\text{out}}\Gamma_{\text{in}}}{G_\text{a}} \cong 2$$

which defines the amplifier *quantum limit*: in ideal conditions, an OA halves the input SNR. The logarithmic value $10 \log F^{(\text{o})}$ is the *amplifier noise figure*, measured in decibels: the amplifier quantum limit is $10 \log F^{(\text{o})} = 3$ dB. Preamplifiers should operate near this limit: the preceding expression shows that (i) the device should be strongly pumped ($n_{\text{sp}} \sim 1$), (ii) input coupling losses should be extremely low ($\Gamma_{\text{in}} \sim 1$) and (iii) the gain should be fairly high ($\Gamma_{\text{out}}\Gamma_{\text{in}}/G_\text{a} \ll 1$).

Straightforward calculations show that

$$\frac{\rho_{\text{ASE-shot,opt}}}{\rho_{\text{S-shot,opt}}} = \frac{F^{(\text{o})} B_\text{o} h\nu_\text{S}}{P_{\text{in}}} \quad [\text{l.u.}]; \quad \rho_{\text{ASE-shot,opt}} - \rho_{\text{S-shot,opt}} = F^{(\text{o})} - 58 - P_{\text{in}} \quad [\text{dB}]$$

$$\frac{\rho_{\text{ASE-ASE,opt}}}{\rho_{\text{S-ASE,opt}}} = \frac{F^{(\text{o})} B_\text{o} h\nu_\text{S}}{2P_{\text{in}}} \quad [\text{l.u.}]; \quad \rho_{\text{ASE-ASE,opt}} - \rho_{\text{S-ASE,opt}} = F^{(\text{o})} - 61 - P_{\text{in}} \quad [\text{dB}]$$

$$\frac{\rho_{\text{exc,opt}}}{\rho_{\text{S-ASE,opt}}} = \frac{\Sigma_{\text{exc}}^2/\Delta\nu_{\text{in}}}{\Sigma_\text{q}^2/\Delta\nu_{\text{in}}} \frac{1}{F^{(\text{o})}} \quad [\text{l.u.}]; \quad \rho_{\text{exc,opt}} - \rho_{\text{S-ASE,opt}} = \frac{\Sigma_{\text{exc}}^2}{\Delta\nu_{\text{in}}} - \frac{\Sigma_\text{q}^2}{\Delta\nu_{\text{in}}} - F^{(\text{o})} \quad [\text{dB}]$$

$$\frac{\rho_{\text{S-shot,opt}}}{\rho_{\text{S-ASE,opt}}} = \frac{1}{G_\text{a} F^{(\text{o})}} \quad [\text{l.u.}]; \quad \rho_{\text{S-shot,opt}} - \rho_{\text{S-ASE,opt}} = -G_\text{a} - F^{(\text{o})} \quad [\text{dB}]$$

These relationships show clearly that the dominant noise term is signal–ASE beat noise. For operative conditions of an optical transmission system, excess noise fluctuations at the input would be kept as low as possible, and excess noise is usually negligible. Similarly, all noise terms other than heterodyne beating (i.e., signal–ASE beating) would be at least 15 dB below it. This is the case for very low input signal levels (say, $P_{\text{in}} \sim -40$ dBm) when ASE–ASE beat noise is $\sim$15 dB lower than signal–ASE beat noise. For intermediate input levels ($P_{\text{in}} \sim -15$ dBm) signal shot noise would be -20 dBm under heterodyne beating; at deep saturation, signal shot noise could even increase around -13 dB below heterodyne beat noise.

5.2.2.2 Electrical Noise

The scope of this section is to cast the definitions of optical noise in the electrical domain; this will allow us to embed the (optical) behavior of the transport plane into an end-to-end performance evaluation that incorporates signal detection too.

In passing from optical to electrical quantities, two remarks are in order. The first concerns the 'translation' of the optical moments into corresponding electrical quantities. The second has to do with the interplay of optical and electrical bandwidth at the receiver.

Let us first consider the expression of electrical moments. The photon flux I is simple to manage; I produces a photocurrent i in amperes:

$$i = e\eta I \quad [\text{A}] \tag{5.35}$$

where e is the elementary electrical charge and η is the quantum efficiency of the detector. Equation 5.35 simply states that a stream of I photons per second, impinging on the photodetector, generates a stream of ηI electrons per second, which gives rise to the photocurrent (Section 3.2). The first two statistical moments of the photocurrent are

$$\mu := \langle i \rangle, \quad \text{mean photocurrent [A];} \quad \sigma_{\text{el}}^2, \quad \text{photocurrent variance [A}^2\text{]} \tag{5.36}$$

In treating electrical fluctuations σ_{el}^2 corresponding to optical ones σ_{opt}^2, one should distinguish between *quantum* effects and *statistical* effects. As seen with Equations 5.30 and 5.31, σ_{opt}^2 has a quantum contribution $\sigma_{\text{shot,opt}}^2 = 2B\rho_{\text{shot,opt}}$ and a statistical contribution $\sigma_{\text{stat,opt}}^2 = \sigma_{\text{beat,opt}}^2 + \sigma_{\text{exc,opt}}^2 = 2B(\rho_{\text{beat,opt}} + \rho_{\text{exc,opt}})$: they are differently influenced by the quantum efficiency η.

This aspect is described by *Burgess's variance theorem*[22] [16]; here we will content ourselves with a heuristic approach (see Section 3.2 for details). Quantum fluctuations – which we previously called Poisson (or shot) noise, obeying the Poisson condition (Equation 5.4) – behave like an average flux. On the other hand, statistical fluctuations always result from the interaction of *two* fluxes. As a consequence, $\sigma_{\text{shop,opt}}^2$ depends *linearly* on η, while $\sigma_{\text{beat,opt}}^2$ and $\sigma_{\text{exc,opt}}^2$ both depend in a *quadratic* way on it. On the other hand, both quantum and statistical fluctuations must be multiplied by the elementary charge *squared*. So the fluctuations of the photocurrent (Equation 5.35) read

$$\sigma_{\text{shot,el}}^2 = \eta e^2 \sigma_{\text{shot,opt}}^2 \quad \text{and} \quad \sigma_{\text{stat,el}}^2 = \eta^2 e^2 \sigma_{\text{stat,opt}}^2, \text{ both in [A}^2\text{]} \tag{5.37}$$

[22] In the theory of photodetection, Burgess's variance theorem is a consequence of the basic Mandel quantum photocounting formula.

Electrical fluctuations calculated this way, and related to the optical transport plane, should be added to the effects of thermal and circuit noise. *Thermal noise* can be approximated as a Johnson–Nyquist effect in the receiver load resistance R_l (Section 3.2 or see section 6.2 in Ref. [6]):

$$\text{thermal noise spectral density} \quad \rho_T = \frac{2kT}{R_l} \quad [A^2\,Hz^{-1}] \tag{5.38}$$

at absolute temperature T, in terms of Boltzmann's constant k.

Circuit noise at the receiver is modeled as a dark current i_d that adds to the photocurrent and produces a noise density $\rho_d = ei_d$ in the receiver bandwidth (Section 3.2):

$$\text{detection circuit dark current } i_d \quad [A] \tag{5.39}$$

$$\text{circuit noise spectral density } \rho_d = ei_d \quad [A^2\,Hz^{-1}] \tag{5.40}$$

Up to this point, the operation of photodetection can be described as another affine transformation by means of a photodetection matrix $\mathbf{D}$ and a noise vector $\mathbf{V}_d$. The 'optical' vector (Equation 5.27) is mapped onto an 'electrical' vector

$$\begin{pmatrix} i \\ \sigma_{el}^2 \end{pmatrix}, \quad \begin{cases} i & \text{mean photocurrent } [A] \\ \sigma_{el}^2 & \text{electrical noise variance } [A^2] \end{cases} \tag{5.41}$$

by the transformation (see Chapter 3 in Ref. [8])

$$\begin{pmatrix} i \\ \sigma_{el}^2 \end{pmatrix} = \mathbf{D}\begin{pmatrix} I \\ \sigma_{opt}^2 \end{pmatrix} + \mathbf{V}_d \tag{5.42}$$

where for a PIN photodiode

$$\mathbf{D}_{PIN} = \eta e \begin{pmatrix} 1 & 0 \\ (1-\eta)e & \eta e \end{pmatrix} \tag{5.43}$$

For an avalanche photodiode (APD) receiver, one has the detection matrix

$$\mathbf{D}_{APD} = \eta e \langle M \rangle \begin{pmatrix} 1 & 0 \\ e\langle M \rangle(\langle M \rangle^x - \eta) & \eta e \langle M \rangle \end{pmatrix} \tag{5.44}$$

where $\langle M \rangle$ is the average APD gain and x the excess noise exponent.

For both receivers the noise vector is given by

$$\mathbf{V} = \begin{pmatrix} i_d \\ ei_d + \dfrac{2kT}{R} \end{pmatrix} \tag{5.45}$$

The second question to be discussed concerns the combined effect of optical and electrical bandwidths B_o and B_e on electrical noise, and this slightly modifies the result of Equation 5.37. While the optical spectral densities $\rho_{shot,opt}$ and $\rho_{exc,opt}$ are flat on the receiver electrical bandwidth, the ASE profile tailors beat noise form factors, accounting for the noise power effectively falling within the detector bandwidth. This means that, in passing from

Equation 5.37 to electrical noise spectral densities, measured in $[\mathrm{A}^2\,\mathrm{Hz}^{-1}]$, one has

$$\begin{aligned}
\rho_{\text{S-shot,opt}} \mapsto \rho_{\text{S-shot,el}} &= e^2\eta\rho_{\text{S-shot,opt}} \\
\rho_{\text{ASE-shot,opt}} \mapsto \rho_{\text{ASE-shot,el}} &= e^2\eta\rho_{\text{ASE-shot,opt}} \\
\rho_{\text{S-ASE,opt}} \mapsto \rho_{\text{S-ASE,el}} &= (e\eta)^2\rho_{\text{S-ASE,opt}} f_{\text{S-ASE}}(B_e, B_o) \\
\rho_{\text{ASE-ASE,opt}} \mapsto \rho_{\text{ASE-ASE,el}} &= (e\eta)^2\rho_{\text{ASE-ASE,opt}} f_{\text{ASE-ASE}}(B_e, B_o) \\
\rho_{\text{exc,opt}} \mapsto \rho_{\text{exc,el}} &= (e\eta)^2\rho_{\text{exc,opt}}
\end{aligned} \tag{5.46}$$

The ASE-dependent terms are spectrally shaped by the optical and electrical bandwidths. The form factors $f_{\text{S-ASE}}(B_e, B_o)$ and $f_{\text{ASE-ASE}}(B_e, B_o)$ are calculated with a statistical, albeit essentially classical, field description of beat noise. The reader will find detailed calculations in Ref. [15]; we shall only give an intuitive deduction. Suppose ASE is approximately distributed in a uniform way around the signal optical frequency ν_S; then, beating of signal and an ASE component at frequency $\nu_{\text{ASE}} \in \left[\nu_S - \frac{1}{2}B_o, \nu_S + \frac{1}{2}B_o\right]$ would produce a rather flat spectrum over the electrical baseband $\left[-\frac{1}{2}B_o, \frac{1}{2}B_o\right]$ set by the range of $|\nu_{\text{ASE}} - \nu_S|$ values.

On the other hand, beating of ASE components at optical frequencies $\nu_{\text{ASE}}, \nu'_{\text{ASE}} \in \left[\nu_S - \frac{1}{2}B_o, \nu_S + \frac{1}{2}B_o\right]$ would produce a nonzero contribution over the electrical baseband $[-B_o, B_o]$ determined by the range of $|\nu_{\text{ASE}} - \nu'_{\text{ASE}}|$ values. Of course, all ASE components contribute at baseband zero frequency, but only one contribution is available at baseband border (corresponding to the unique $\nu_{\text{ASE}}, \nu'_{\text{ASE}}$ combination giving $|\nu_{\text{ASE}} - \nu'_{\text{ASE}}| = B_o$). At electrical frequencies $f \neq 0$, a different number of beat terms is available and this number is proportional to $(2B_o - f)/2B_o$, divided by a further factor of 2, because the baseband spectrum is two-sided.[23] Since ASE has a rather uniform profile, one can infer that the ASE–ASE beat spectrum is triangular in shape, having its maximum at zero baseband frequency and vanishing at baseband borders. Thence, considering that electrical frequencies f are passed through a lowpass filter of bandwidth B_e:

$$f_{\text{S-ASE}}(B_e, B_o) = 1 \quad \text{and} \quad f_{\text{ASE-ASE}}(B_e, B_o) = \frac{1}{2}\left(1 - \frac{B_e}{2B_o}\right) \tag{5.47}$$

At this point it is convenient to introduce the *responsivity* H of the photodetector:

$$H = \frac{e\eta}{h\nu_S} \quad [\mathrm{A\,W^{-1}}] \tag{5.48}$$

For mean values, the transition from optical to electrical domain is simply given by

$$I_{\text{coh}} \mapsto \mu_S = e\eta I_{\text{coh}} = HP_{\text{coh}}, \quad I_{\text{incoh}} \mapsto \mu_{\text{ASE}} = e\eta I_{\text{incoh}} = HP_{\text{ASE}} \tag{5.49}$$

[23] The reader will note we have a frequency repository $2B_o$ wide, which is baseband spectrum. At frequency f, the condition $|\nu_{\text{ASE}} - \nu'_{\text{ASE}}| = f$ selects a frequency subset $2B_o - f$ wide. This contributes to the positive *and* negative frequency parts of the ASE–ASE beat noise spectrum, except at its center.

Recalling all results from Equations 5.30, 5.31, 5.46 and 5.47, one finally has for the noise variances:

$$\begin{aligned}
\sigma^2_{\text{S-shot,opt}} &\mapsto \sigma^2_{\text{S-shot,el}} = 2B_\text{e}e^2\eta\sigma^2_{\text{S-shot,opt}} = 2B_\text{e}eHP_\text{coh} \\
\sigma^2_{\text{ASE-shot,opt}} &\mapsto \sigma^2_{\text{ASE-shot,el}} = 2B_\text{e}e^2\eta\sigma^2_{\text{ASE-shot,opt}} = 2B_\text{e}eHP_\text{ASE} \\
\sigma^2_{\text{S-ASE,opt}} &\mapsto \sigma^2_{\text{S-ASE,el}} = 2B_\text{e}(e\eta)^2\sigma^2_{\text{S-ASE,opt}} = \frac{4B_\text{e}H^2P_\text{coh}P_\text{ASE}}{m_\text{t}B_\text{o}} \\
\sigma^2_{\text{ASE-ASE,opt}} &\mapsto \sigma^2_{\text{ASE-ASE,el}} = 2B_\text{e}\frac{1}{2}\left(1-\frac{B_\text{e}}{2B_\text{o}}\right)(e\eta)^2\sigma^2_{\text{ASE-ASE,opt}} = \frac{2B_\text{e}H^2P^2_\text{ASE}}{m_\text{t}B_\text{o}}\left(1-\frac{B_\text{e}}{2B_\text{o}}\right) \\
\sigma^2_{\text{exc,opt}} &\mapsto \sigma^2_{\text{exc,el}} = (e\eta)^2\sigma^2_{\text{exc,opt}} = 2B_\text{e}(e\eta G_\text{a})^2\frac{\Sigma^2_\text{exc}}{\Delta\nu_\text{in}} \equiv 2B_\text{e}(e\eta G_\text{a})^2\sigma^2_{\text{in,opt}}
\end{aligned} \tag{5.50}$$

Further terms specific to the receiver are the circuit current i_d and the circuit and thermal noises:

$$\sigma^2_{\text{d,el}} = 2B_\text{e}ei_\text{d}, \quad \sigma^2_{\text{T,el}} = \frac{4B_\text{e}kT}{R_\text{l}} \tag{5.51}$$

All electrical noise contributions σ^2_el are measured in [A^2].

Applying Equations 5.50 and 5.51 to input quantities, the input SNR in the electrical domain is given by

$$L^{(\text{e})}_\text{in}\Big|_\text{QL} = \frac{\mu^2_\text{S}}{2B_\text{e}\left(e\mu_\text{S} + ei_\text{d} + \frac{2kT}{R_\text{l}}\right)} \tag{5.52}$$

which, for ideal photodetection (negligible circuit and thermal noise with respect to quantum noise; $\eta = 1$), reduces to $L^{(\text{e})}_\text{in}\big|_\text{QL} = \mu_S/2B_\text{e}e$, equivalent to Equation 5.17.

Similarly, applying Equations 5.50 and 5.51 to output quantities, one gets

$$L^{(\text{e})}_\text{out} = \frac{(HP_\text{coh})^2}{2B_\text{e}\left[eHP_{coh} + eHP_\text{ASE} + \frac{2H^2P_\text{coh}P_\text{ASE}}{m_\text{t}B_\text{o}} + \frac{H^2P^2_\text{ASE}}{m_\text{t}B_\text{o}}\left(1-\frac{B_\text{e}}{2B_\text{o}}\right) + (HG_\text{a}h\nu_\text{S})^2\sigma^2_{\text{in,opt}} + ei_\text{d} + \frac{2kT}{R_\text{l}}\right]} \tag{5.53}$$

From Equations 5.52 and 5.53, the SNR in the electrical domain can be calculated. Once again, it is easily seen that for ideal photodetection (negligible thermal and circuit noise, $\eta = 1$) and negligible ASE–ASE beat noise (because of the form factor in the electrical domain) the SNR evaluated in the electrical domain has the same value as that in the optical domain.

Equations 5.50–5.53 are the basis to discuss light path evaluation rules. From these expressions, the electrical SNR can be calculated according to Equation 5.9:

$$L^{(\text{e})} = \frac{(\mu_\text{S})^2}{\sum_{i=1}^{N}\sigma^2_{i,\text{el}}} \tag{5.54}$$

where the sum in the denominator is carried over all noise contributions in Equations 5.50 and 5.51.

5.2.3 Performance Parameters. Light Path Evaluation Rules

The discussion of Chapter 3 and of the previous section has given relevant definitions for OSNR and SNR:

$$\text{OSNR in bandwidth } B_o: \quad R(B_o) = \frac{P_S}{P_N(B_o)} = \frac{P_S}{\rho_N B_o} \tag{5.55}$$

$$\text{SNR in the optical domain}: \quad L^{(o)} = \frac{P_S^2}{(h\nu_S)^2 \sum_{i=1}^{N} \sigma_{i,\text{opt}}^2} = \frac{P_S^2}{(h\nu_S)^2 \sigma_{\text{tot,opt}}^2} \tag{5.56}$$

where the sum over all optical noise contributions in Equations 5.30 and 5.31 (times $2B$) appears in the denominator, and

$$\text{SNR in the electrical domain}: \quad L^{(e)} = \frac{(\mu_S)^2}{\sum_{i=1}^{N} \sigma_{i,\text{el}}^2} = \frac{(\mu_S)^2}{\sigma_{\text{tot,el}}^2} \tag{5.57}$$

where the sum over all electrical noise contributions in Equations 5.50 and 5.51 appears in the denominator (Figure 5.9).

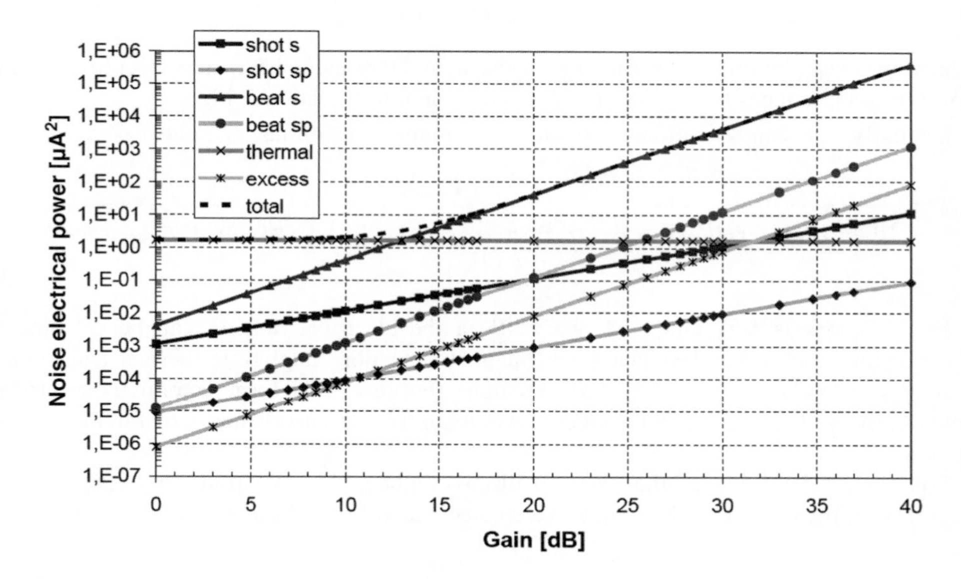

Figure 5.9 Noise electrical contributions and total noise power. Calculations done for $\eta = 0.9$, $n_{eq} = 2$, $\Gamma_{in} = \Gamma_{out} = 1$, $T = 300$ K, $\lambda_S = 1550$ nm, $P_{in} = -32$ dBm, $B_o = 2B_e = 10$ GHz, $R_l = 50\,\Omega$, $m_t = 2$, RIN $= -155$ dB/Hz

The definition of the *Q-factor* adapts the electrical SNR (Equation 5.57) to the case of a digital binary signal composed of '0' and '1' symbols (Figure 5.2). In digital optical telecom systems, the transmission signal is a sequence of marks ('1' symbols) and spaces ('0' symbols) carried by a modulated optical flux. After optical demodulation, if needed, the signal in the optical domain is detected, electrically preamplified, and sent to an electronic circuit to integrate the signal power over a bit period. Then the integrated signal is passed to a decision circuit that compares it with some assigned energy level (the *decision threshold*) to establish whether a mark or a space has been transmitted. As the optical signal power is affected by fluctuations, the energy of the integrated electrical signal may happen to fluctuate just around the decision threshold. Because of the receiver noise, there is always a finite probability for a bit to be erroneously identified by the receiver decision circuit, causing detection errors. The lower the SNR, the more probable these errors are.

A figure of merit for the whole detection process more direct than SNR is the error probability associated with it. This is the BER (Section 3.2) and is defined in terms of

i. the probability $p(\mathsf{m})$ that a symbol m is transmitted (m is '0' or '1'),
ii. the conditional probability $P(\mathsf{m}|\mathsf{s})$ to detect the symbol m by mistake, when the symbol s has been transmitted instead, together with the reversed probability $P(\mathsf{s}|\mathsf{m})$;
iii. the photoelectron statistics.

The BER probability is expressed as

$$P_{\mathrm{BER}} = P(1|0)p(0) + P(0|1)p(1) \tag{5.58}$$

The process of integrating the bit energy consists of counting the photoelectrons generated in the corresponding bit slot. If $P_n(\mathsf{s})$ is the probability of counting n photoelectrons when symbol s is received, and D_{th} is the threshold counting level (see Section 3.2 for details), then the conditional probabilities $P(\mathsf{m}|\mathsf{s})$ can be written as

$$P(1|0) = \sum_{n=D_{\mathrm{th}}}^{\infty} P_n(0), \qquad P(0|1) = \sum_{n=0}^{D_{\mathrm{th}}-1} P_n(1)$$

When n is very large the distribution $P_n(\mathsf{s})$ can be approximated with a function $P_x(\mathsf{s})$ of a continuous variable x and sums for the $P(\mathsf{s}|\mathsf{m})$ tend to integrals, as Figure 5.10 suggests. Actual standards for optical telecom systems are BER $\leq 10^{-12}$ – no more than one error on average during 100 s at 10 Gbit/s bit-rate; BER $\leq 10^{-15}$ is sometimes mentioned for specific cases: the performance of the transport layer usually achieves even this goal (less than one error on average every day of transmission at 10 Gbit/s!).

BER analysis for optically amplified digital signals can be done at various levels of detail, complexity, and generality (Section 3.2). In the semiclassical Gaussian model of photodetection developed by Personick [2], the continuous limiting case of many photoelectrons is described by Gaussian probability distribution functions $P_x(\mathsf{s})$. This is a natural choice because of the central limit theorem of statistics[24] and of the assumption that optical noise can be

[24] The theorem states that the sum of a large number of independent observations from the same distribution has, under rather general conditions, an approximate normal distribution.

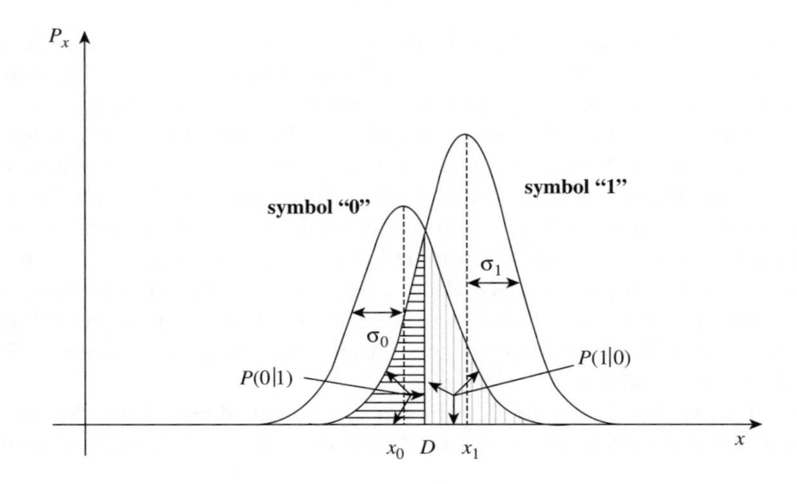

Figure 5.10 Continuous probability distributions for detection of symbols '0', '1'

modeled as Gaussian. So the reader can see that the Gaussian approximation is based on the assumptions that

i. noise statistics are reasonably described by the first and second moments of the error probability distribution function, and, consistently with this first hypothesis, that
ii. error probability distributions are Gaussian functions.

If μ_s and σ_s are the average value and standard deviation of the Gaussian distribution associated to the symbol s (Equations 5.49–5.51), then the conditional probabilities $P(m|s)$ are

$$\begin{aligned} P(1|0) &= \frac{1}{\sqrt{2\pi}} \int_{Q_0}^{\infty} \exp\left(-\frac{x^2}{2}\right) \mathrm{d}x \qquad Q_0 = \frac{D_{\mathrm{th}} - \mu_0}{\sigma_0} \\ P(0|1) &= \frac{1}{\sqrt{2\pi}} \int_{Q_1}^{\infty} \exp\left(-\frac{x^2}{2}\right) \mathrm{d}x \qquad Q_1 = \frac{\mu_1 - D_{\mathrm{th}}}{\sigma_1} \end{aligned} \tag{5.59}$$

The message is said to have a *balanced pattern* when it consists of a sequence of equally likely marks and spaces (on average, this happens for a sufficiently high word bit number, say, higher than 2^{15}, so one can assume $P(1|0) = P(0|1)$). In turn, this implies $Q_0 = Q_1$ in (5.59), determining the optimum decision level D as [19]

$$D_{\mathrm{th}} = \frac{-\mu_1\sigma_0^2 + \mu_0\sigma_1^2 + \sigma_0\sigma_1\sqrt{(\mu_1 - \mu_0)^2 + 2(\sigma_1^2 - \sigma_0^2)\ln(\sigma_1/\sigma_0)}}{\sigma_1^2 - \sigma_0^2}$$

For sufficiently high values of the SNR, the preceding expression simplifies to

$$D_{\mathrm{th}} = \frac{\mu_0\sigma_1 + \mu_1\sigma_0}{\sigma_0 + \sigma_1} \tag{5.60}$$

Personick's Q-factor is the common value $Q_0 = Q_1$ at the decision level (5.60):

$$Q = \frac{\mu_1 - \mu_0}{\sigma_1 + \sigma_0} \tag{5.61}$$

Because of the close relationship between Q-factor (Equation 5.61) and SNR (Equation 5.57), in the following, *logarithmic Q values* will be taken as $Q^* = 20 \log Q$.

Conveying the assumptions $p(0) = p(1) = 1/2$ and $P(0|1) = P(1|0)$ into Equation 5.58 for BER, one finally gets

$$P_{\mathrm{BER}} = \frac{1}{\sqrt{2\pi}} \int_Q^\infty \exp\left(-\frac{x^2}{2}\right) \mathrm{d}x = \frac{1}{2} \mathrm{erfc}\left(\frac{Q}{\sqrt{2}}\right) \underset{Q>3}{\cong} \frac{1}{\sqrt{2\pi}Q} \exp\left(-\frac{Q^2}{2}\right) \tag{5.62}$$

Since the Q-factor is intimately connected to BER, it is a basic quality parameter in the electrical domain. On the optical transport layer, the OSNR has a similar importance. It is interesting to work out an expression relating OSNR and Q-factor in an optical(ly amplified) network. For the sake of conciseness, let us define the *noise equivalent input factor* n_{eq}:

$$n_{\mathrm{eq}} = \frac{N}{G} = n_{\mathrm{sp}} \Gamma_{\mathrm{out}} \left(\frac{1}{\Gamma_{\mathrm{in}} \Gamma_{\mathrm{out}}} - \frac{1}{G}\right) \quad \text{so that} \quad P_{\mathrm{ASE}} = m_{\mathrm{t}} n_{\mathrm{eq}} G h \nu_{\mathrm{S}} B_{\mathrm{o}} \tag{5.63}$$

(from now on, suffix 'a' is dropped but G and N still refer to the amplifier; see Equations 5.18 and 5.21). Equation 5.63 is the equivalent number of photons as input to a noiseless amplifier with gain G in order to have at its output the same noise level as a real amplifier with the same gain. Though not very different from n_{sp}, n_{eq} behaves in a smoother way. Define two coefficients U and V as

$$U = 2m_{\mathrm{t}} \frac{B_{\mathrm{e}}}{B_{\mathrm{o}}} \left(\frac{1}{\eta n_{\mathrm{eq}} G} + 2\right), \qquad V = m_{\mathrm{t}} \frac{B_{\mathrm{e}}}{B_{\mathrm{o}}} \left[\frac{2}{\eta n_{\mathrm{eq}} G} + 2\left(1 - \frac{B_{\mathrm{e}}}{B_{\mathrm{o}}}\right) + \left(\frac{\sigma_{\mathrm{T,el}}}{\eta e n_{\mathrm{eq}} G}\right)^2 \frac{1}{m_{\mathrm{t}} B_{\mathrm{e}} B_{\mathrm{o}}}\right] \tag{5.65}$$

They depend on noise terms (here, excess noise is neglected) in a way that

$$U = \frac{1}{R} \frac{\sigma_1^2 - \sigma_0^2}{(\eta I_{\mathrm{N}})^2}, \qquad V = \left(\frac{\sigma_0}{\eta I_{\mathrm{N}}}\right)^2 \tag{5.66}$$

with the linear OSNR R and the average ASE photocurrent

$$I_{\mathrm{N}} = \frac{e P_{\mathrm{ASE}}}{m_{\mathrm{t}} h \nu_{\mathrm{S}}} \tag{5.67}$$

Then V are the fluctuations on the '0'-level measured in I_{N} units; similarly, U are the differential fluctuations on the '1'-level with respect to the '0'-level in the same units, and referred to the OSNR

$$R = \frac{G\bar{P}}{P_{\mathrm{ASE}}} = \frac{GP_1}{2P_{\mathrm{ASE}}}, \text{ since } \bar{P} = \frac{P_1 + P_0}{2} = \frac{P_1}{2} \tag{5.68}$$

In Equation 5.68, the mark optical power at the amplifier input is P_1 and the input optical power P_0 on space symbol is supposed to be zero (infinite extinction ratio): a power level $\bar{P}$ is fed into the OFA on average. Using Equation 5.65, the relation between OSNR and Q is

$$Q = \frac{m_t R}{\sqrt{V}} \frac{1}{1+\sqrt{1+\frac{U}{V}R}} = \frac{m_t\sqrt{V}}{U}\left(\sqrt{1+\frac{U}{V}R}-1\right), \qquad R = \frac{V}{U}\left[\left(\frac{QU}{m_t\sqrt{V}}+1\right)^2-1\right] \tag{5.69}$$

When signal–ASE beat noise is dominant (in telecommunications it is the usual condition with OFAs), the first part of Equation 5.69 becomes

$$Q = \frac{1}{4}\sqrt{m_t}\sqrt{\frac{2B_o}{B_e}-1}\left(\sqrt{1+\frac{B_o}{2B_o-B_e}8R}-1\right) \tag{5.70}$$

which, for $m_t = 2$, differs only by a small percentage from the usual OSNR–Q equation, deduced in quite a different way (see Section 4.4.6 in Ref. [19]):

$$Q = 2\sqrt{\frac{B_o}{B_e}}R\frac{1}{1+\sqrt{1+4R}} = \frac{1}{2}\sqrt{\frac{B_o}{B_e}}(\sqrt{1+4R}-1) \tag{5.71}$$

Equation 5.70 or Equation 5.71 can be used to relate OSNR in the optical band B_o to the Q-factor. For the conventional band B_{ref} of 12.5 GHz, the relation between intrinsic and conventional OSNR is simple for bit-rates not exceeding B_{ref}:

$$R_{ref} = \frac{B_o}{B_{ref}}R_{B_o} \quad \text{[lin.]}, \qquad \text{OSNR}_{ref} = 10\log\left(\frac{B_o}{B_{ref}}\right) + \text{OSNR}_{B_o} \quad \text{[dB]} \tag{5.72}$$

At higher bit-rates (40, 80, 160 Gbit/s) this relation is no longer valid because B_{ref} is narrower than the modulated signal linewidth: the reference OSNR value R_{ref} still has a comparative meaning, but its calculation requires knowing the signal spectral form factor.

Equations 5.70 and 5.71 may be applied in the limiting condition $B_o = 2B_e = B$ when the bit-rate B sets the minimum acceptable optical bandwidth B_o and the corresponding minimum acceptable electrical bandwidth $B/2$.

Just to be specific, Equation 5.71 becomes in the limiting condition $B_o = 2B_e = B$

$$Q = \frac{\sqrt{1+4R}-1}{\sqrt{2}} \tag{5.73}$$

This implies that the relation between Q and OSNR is independent of the bit-rate: the *intrinsic OSNR* (i.e., the OSNR in the bit-rate band B) needed to assure a given signal performance is the same at 2.5 Gbit/s, 10 Gbit/s, 40 Gbit/s, ... and higher is always the same. In view of Equation 5.72, it is equivalent to state that

$$R_{ref}(B') = \frac{B'}{B}R_{ref}(B) \quad \text{[lin.]}, \quad \text{OSNR}_{ref}(B') = 10\log\left(\frac{B'}{B}\right) + \text{OSNR}_{ref}(B) \quad \text{[dB]} \tag{5.74}$$

between *reference* OSNRs at two different bit-rates B' and B.

The OSNR can be used as a measure of light path baseline performance because it takes into account the unavoidable accumulation of optical noise along an optically amplified connection (Section 5.2.2.1). Every other transmission impairment degrades the quality of an ASE-limited optical circuit. The reader will note that the first part of Equations 5.69 and 5.70 and Equation 5.71 can all be cast under the common form

$$Q_R = \frac{\alpha R}{1+\sqrt{1+\beta R}} \tag{5.75}$$

where α and β are functions of the system electrical and optical bandwidths. It is straightforward to check that

$$\alpha = \frac{m_\mathrm{t}}{\sqrt{V}} \quad \text{and} \quad \beta = \frac{U}{V}$$

for the first expression of Equation 5.69; and

$$\alpha = 2B_\mathrm{o}\sqrt{\frac{m_\mathrm{t}}{B_\mathrm{e}(2B_\mathrm{o}-B_\mathrm{e})}} \quad \text{and} \quad \beta = \frac{8B_\mathrm{o}}{2B_\mathrm{o}-B_\mathrm{e}}$$

for Equation 5.70; and

$$\alpha = 2\sqrt{\frac{B_\mathrm{o}}{B_\mathrm{e}}} \quad \text{and} \quad \beta = 4$$

for Equation 5.71.

Equation 5.75 can be expanded in the form (see Equation 1.515.2 in Ref. [20]):

$$Q^* \equiv 10\log(Q_R^2) = 10\log\left(\frac{\alpha^2 B_{0,1\ \mathrm{nm}}}{\beta B_\mathrm{o}}\right) + \mathrm{OSNR}_{0,1\ \mathrm{nm}}$$

$$+\frac{20\log \mathrm{e}}{\sqrt{\dfrac{10^{\mathrm{OSNR}_{0,1\ \mathrm{nm}}/10}\beta B_{0,1\ \mathrm{nm}}}{B_\mathrm{o}}}} + O\left[\left(\frac{10^{\mathrm{OSNR}_{0,1\ \mathrm{nm}}/10}B_{0,1\ \mathrm{nm}}}{B_\mathrm{o}}\right)^{-3/2}\right] \tag{5.76}$$

Equation 5.76 describes a linear relationship between Q^* and OSNR, with both parameters expressed in decibels. The correction term $O(.)$ is limited to 0.5–1 dB as long as the OSNR values are at least in the range 12–18 dB (for 0.1 nm bandwidth).

5.2.4 Transmission Impairments and Enhancements: Simple Power Budgets

After having established basic relations for photodetection and quality parameters on the optical transport layer, the scope of the present section will be to discuss simple physical models of transmission penalties. According to the discussion of Section 5.2.1 about light path evaluation rules, these are useful both to get a global idea of the optical network design and to support real-time light path (re-)routing.

Generally speaking, *transmission impairments* are accounted for as penalties on one of the performance parameters listed above (power at receiver, Q-factor, eye closure, OSNR, SNR, etc.). To succeed in maintaining the performance of an ideal circuit, a higher value of the quality parameter is necessary in the real case, and this is the *penalty* associated with that circuit.

According to the specific degradation mechanism, it could be more natural to formulate penalties over one or the other of the quality parameters. In this book, transmission impairments and penalties will be almost always related to the quantity $Q^* = 20 \log Q$. In a more general context, penalties due to an impairment X affecting the quality parameter E will be denoted by the symbol $P_{\mathrm{X}}^{(E)}$.

The *penalty approach* consists of the following considerations.

1. QoS requirements (BER, throughput, packet loss) for services transported on a given light path determine a minimum target performance for that light path.;
2. The minimum target performance is translated into a minimum target value Q^*_{min} for the logarithmic Q-value at the receiver after an ideal light path.
3. The *baseline* performance (no penalties but for ASE accumulation) is translated into a corresponding logarithmic Q-value Q^*_{ASE}.
4. Check that $Q^*_{\mathrm{ASE}} \geq Q^*_{\mathrm{min}}$, otherwise the path is considered not feasible.
5. Subtract from the baseline Q^*_{ASE} every penalty due to transmission impairments other than ASE accumulation

$$Q^*_{\mathrm{ASE}} \rightarrow Q^*_{\mathrm{ASE}} - \sum_{i \in I} P_i^{(Q^*)} \quad [\mathrm{dB}]$$

   where the sum is over all penalties $\{P_i^{(Q^*)} | i \in I\}$ on Q^* along the light path;.
6. Aside from transmission impairments, also *enhancement mechanisms* may exist, and should be taken into account as a 'negative penalty' $E_j^{(Q^*)} = -P_j^{(Q^*)} < 0$. For instance, this is the case for the adoption of correction codes (FEC). We rewrite the previous relation and define the actual Q-value Q^*_{a} as

$$Q^*_{\mathrm{ASE}} \rightarrow Q^*_{\mathrm{ASE}} - \sum_{i \in I} P_i^{(Q^*)} - \sum_{j \in J} E_j^{(Q^*)} = Q^*_{\mathrm{a}} \quad [\mathrm{dB}] \tag{5.77}$$

7. Check that Q^*_{a} is sufficient to allocate the required QoS Q^*_{min} and any required system margins Q^*_{M} (usually for system *ageing*, possibly for higher layer impairments; see below): $Q^*_{\mathrm{a}} \geq Q^*_{\mathrm{min}} + Q^*_{\mathrm{M}}$.

From a physical point of view, the penalty approach is applicable when penalties themselves are weak enough that each of them can be analyzed separately from any other impairment and that they can be summed (logarithmically) along the light path. *The penalty approach is a linearization of the transmission problem around the baseline value.* The actual quality parameter is calculated by subtracting from the baseline the contribution of perturbations.

If the margin upper bound $Q^*_{\mathrm{ASE}} - Q^*_{\mathrm{min}}$ is very narrow, then the linear procedure (5.77) might be questionable: the linearization could affect an accurate estimate of optical path performance margins. However, the discussion of very critical light paths is not the major concern here, since very accurate simulations would be required in that case anyhow and performed, as discussed in Section 5.2.1, having in mind more accurate models, such as those described in Chapter 3.

The previous discussion suggests a 'layered approach' to light path evaluation:

1. simply check that $Q^*_{\mathrm{ASE}} \geq Q^*_{\mathrm{min}}$ at the baseline layer;
2. apply the linear procedure (5.77) and check whether $Q^*_{\mathrm{a}} \geq Q^*_{\mathrm{min}} + Q^*_{\mathrm{M}}$;

3. if any impairment cannot be properly assessed (then it might be desirable to reserve some role into Q_{M}^{*} for it), or when the linear procedure may appear questionable, a deeper investigation is required, introducing more elaborated models.

As an example of the implications of step 3, the question of the link PMD[25] is considered. Suppose the test $Q_{\mathrm{a}}^{*} \geq Q_{\min}^{*} + Q_{\mathrm{M}}^{*}$ is barely satisfied, having taken into account all transmission penalties, PMD included, through simplest criteria (namely, the usual $T_{\mathrm{bit}}/10$ criterion – see below). Because of the statistical nature of PMD itself, and of ageing,[26] the Q-factor performance of the particular light path should be evaluated for with systems operating at zero power margin; even a small additional performance degradation will render this system useless [22]. So, PMD fluctuations might lead the system out of margins.

This situation can be worth analyzing the PMD phenomenon to higher orders and/or to work out a baseline value more accurate from the operative point of view, around which the penalty approach could be used in a safer way (see Equation 5.103).

Another situation that may require a higher- layer approach happens when different optical multiplex sections (OMSs) are strongly coupled in the transparent network. Then models for nonlinear effect accumulation and transient phenomena that are more accurate than discussed in this section are required. They will lead to the determination of dynamical margins contributing to Q_{M}^{*}.

In this section, a set of simple models for transmission penalties and enhancements, all referred to $Q^{*} = 20 \log Q$, is described, to supply a framework for steps 1 and 2 listed above; that is, to a lowest layer description of light path evaluation rules. In our discussion, the goal of a system design is to obtain a Q-value at the receiver, say $Q_{\mathrm{a}}^{*} \geq Q_{\min}^{*} + Q_{\mathrm{M}}^{*}$, which is sufficiently high to guarantee a minimum performance $Q_{\min}^{*}$ net of suitable system margins Q_{M}^{*}. Any transmission impairment affects this target and must be taken into account to set the *operative margin* $Q_{\mathrm{a}}^{*} - (Q_{\min}^{*} + Q_{\mathrm{M}}^{*})$. As already said, degradation mechanisms are usually expressed in terms of different performance parameters; definitions are formulated according to the physical mechanism of the transmission impairment. Anyhow, definitions are mutually equivalent as long as they can be transformed one into another, say, from power penalties into Q^{*} penalties.

In the penalty approach, each degradation mechanism is investigated separately and referred to the worst-case situation, but assuming that the system is free of any other impairment. The effect $P_i^{(V)}$ of any penalty i on the performance parameter V (which can be Q, P, OSNR, SNR, EO, etc.) are summed together to estimate the performance of a real system. This amounts to making the transmission problem linear, seen as a mathematical relation $V = V(P_1^{(V)}, \ldots, P_m^{(V)})$ amongst the quality parameter V and the various degradation effects, around the ideal operation point $V_0 = V(0, \ldots, 0) \geq V_{\min}$ or around any suitably defined reference operation point $V_{\mathrm{ref}} = V(\eta, \ldots, \eta_m)$; see Section 5.2.4.3: each η_i represents a well-assessed, deterministic, physical impairment that affects the transport layer and the effect

[25] Or of any environment-dependent impairment; another example is the (seasonal, daily) dependence of fiber chromatic dispersion (CD) on temperature [21].

[26] Ageing margins are taken to cover allowance for future modifications to cable configuration (additional splices, increased cable lengths, etc.), fiber cable performance variations due to environmental factors (as PMD or temperature-driven CD variations) and degradation of any connectors, optical attenuators or other passive optical devices included in the optical path.

of which can be determined by means of a more rigorous treatment than the penalty approach itself (Section 3.2).

5.2.4.1 Basic Formulae

The basic formulae to investigate the relation between different performance parameters V and their related penalties are Equations 5.49 and 5.61:

$$Q = \frac{\mu_1 - \mu_0}{\sigma_1 + \sigma_0} = \frac{H(P_1 - P_0)}{\sigma_1 + \sigma_0} \tag{5.78}$$

where the fact that the photocurrent mean value is proportional to the signal power has been used, $\mu_i = HP_i$, $i = 0, 1$, through the responsivity H, defined with Equation 5.48.

The other basic formulae are BER relation (Equation 5.62)

$$P_{\mathrm{BER}} = \frac{1}{\sqrt{2\pi}} \int_Q^{\infty} \exp\left(-\frac{x^2}{2}\right) \mathrm{d}x = \frac{1}{2} \operatorname{erfc}\left(\frac{Q}{\sqrt{2}}\right) \underset{Q>3}{\cong} \frac{1}{\sqrt{2\pi}Q} \exp\left(-\frac{Q^2}{2}\right) \tag{5.79}$$

and the logarithmic definition of the Q-value

$$Q^* = 20 \log Q \quad [\mathrm{dB}] \tag{5.80}$$

In using Equation 5.78 we shall use primed quantities (P'_1, P'_0, σ'_1, σ'_0, Q') to describe a degraded situation and unprimed quantities (P_1, P_0, σ_1, σ_0, Q) to refer to the ideal situation. In relating different performance parameters, two important cases arise in practice: the *dominant thermal noise regime* (no OAs, use of PIN detectors) and the *dominant optical noise regime* (presence of OAs and of APD receivers).

For *dominant thermal noise*, independently of the optical signal level, one has $\sigma_1 = \sigma_0 = \sigma'_1 = \sigma'_0 = \sigma_{\mathrm{th}}$ and

$$\frac{Q'}{Q} = \frac{HP'_1 - HP'_0}{\sigma_{\mathrm{th}} + \sigma_{\mathrm{th}}} \frac{\sigma_{\mathrm{th}} + \sigma_{\mathrm{th}}}{HP_1 - HP_0} = \frac{P'_1 - P'_0}{P_1 - P_0} = \frac{P'_1}{P_1} \frac{1 - P'_0/P'_1}{1 - P_0/P_1} = \frac{P'_1}{P_1} \frac{1 - 1/r'}{1 - 1/r} \tag{5.81}$$

with $r = P_1/P_0$ and $r' = P'_1/P'_0$ (r and r' are extinction ratios – see below). Therefore:

$$P^{(Q^*)} \equiv 2P^{(Q)} = 2P^{(P)} + 20 \log\left(\frac{1 - r'^{-1}}{1 - r^{-1}}\right) \quad [\mathrm{dB}] \text{ (thermal noise)} \tag{5.82}$$

If $r, r' \gtrsim 13$ dB, then (5.81) and (5.82) give to a 5% accuracy $\frac{Q'}{Q} \cong \frac{P'_1}{P_1}$, which means $P^{(Q^*)} = 2P^{(Q)} = 2P^{(P)}$ [dB] (thermal noise).

When OAs and/or APD receivers are present, then *optical noise is dominant and is roughly proportional to the square root of the signal*, $\sigma_i = K\sqrt{P_i}$, $i = 1, 2$ (for OAs, recall Equation 5.26 for dominant signal–ASE beat noise – see also Figure 5.9. For APD receivers, note the analogies between the product $\mathbf{D}_{\mathrm{PIN}} \cdot \mathbf{M}$ of matrices (5.43) and (5.28) with the single matrix $\mathbf{D}_{\mathrm{APD}}$ (5.44)). In this condition the definition (5.78) gives

$$\frac{Q'}{Q} = \frac{HP'_1 - HP'_0}{K\sqrt{P'_1} + K\sqrt{P'_0}} \cdot \frac{K\sqrt{P_1} + K\sqrt{P_0}}{HP_1 - HP_0} = \frac{\sqrt{P'_1} - \sqrt{P'_0}}{\sqrt{P_1} - \sqrt{P_0}} = \sqrt{\frac{P'_1}{P_1}} \cdot \frac{1 - 1/\sqrt{r'}}{1 - 1/\sqrt{r}}. \tag{5.83}$$

This means that

$$P^{(Q^*)} = 2P^{(Q)} = P^{(P)} + 20\log\left(\frac{1-r'^{-1/2}}{1-r^{-1/2}}\right) \quad \text{[dB] (optical noise)} \tag{5.84}$$

If $r, r' \gtrsim 20$ dB, then (5.84) gives to a 10% accuracy $\frac{Q'}{Q} \cong \sqrt{\frac{P'_1}{P_1}}$, namely, $P^{(Q^*)} = 2P^{(Q)} = P^{(P)}$ [dB] (optical noise).

On the other hand, Equation 5.71 gives

$$\left(\frac{Q'}{Q}\right)^2 = \frac{1+2R'-\sqrt{1+4R'}}{1+2R-\sqrt{1+4R}} \cong \frac{2R'-\sqrt{4R'}}{2R-\sqrt{4R}} = \frac{R'}{R}\cdot\frac{1-1/\sqrt{R'}}{1-1/\sqrt{R}} \tag{5.85}$$

from which one obtains

$$P^{(Q^*)} = P^{(\mathrm{OSNR})} + 10\log\left(\frac{1-R'^{-1/2}}{1-R^{-1/2}}\right) \quad \text{[dB]} \tag{5.86}$$

If R and R' are not smaller than 20 dB, then the penalty on Q^2 is approximately equivalent to the penalty on OSNR, to a 10% accuracy:

$$P^{(Q^*)} \cong P^{(\mathrm{OSNR})} \quad \text{[dB]}$$

Some transmission impairments affect the eye diagram in a twofold way (Figure 5.11): the degradation both closes the eye and blurs its contour. In this situation, Equation 5.78 requires some manipulation. Moreover, the eye diagram degradation is not well represented by the relation between Q-factor and OSNR, as it stands with Equations 5.69–5.71. These expressions can be modified to introduce the eye-opening penalty $P^{(\mathrm{EO})}$ explicitly. It can be defined as the

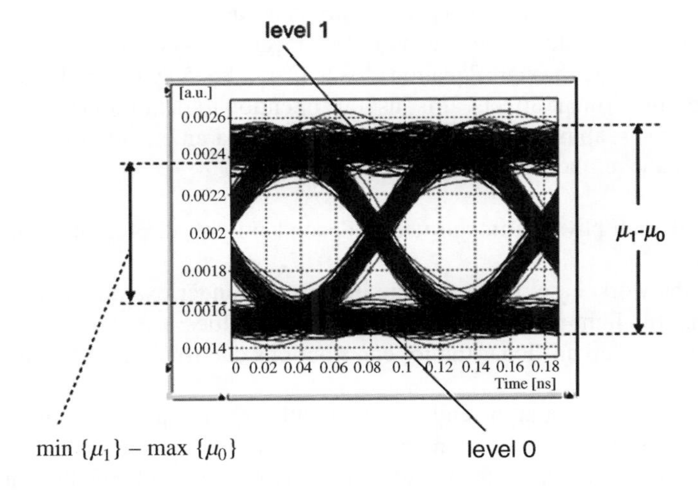

Figure 5.11 Eye diagram with waveform distortion. Levels '1' and '0' are split into many sublevels

ratio of the minimum eye opening, as set by the blurring of the contour lines, to the average eye opening described by the undistorted averaged '1' and '0' levels:

$$P^{(\mathrm{EO})} = \frac{\min\{\mu_1\} - \max\{\mu_0\}}{\mu_1 - \mu_0} \tag{5.87}$$

Besides $P^{(\mathrm{EO})}$, it is also convenient to introduce the noise figure F_{E} due to the possible presence of equalizers/compensators of signal distortions; F_{E} sets the relation between SNRs without and with that device. Starting from Equation 5.54:

$$F_{\mathrm{E}} = L_{\mathrm{noeq}} L_{\mathrm{eq}}{}^{-1} = \left(\frac{\sigma^2}{\mu^2}\right)_{\mathrm{eq}} \left[\left(\frac{\mu^2}{\sigma^2}\right)_{\mathrm{noeq}}\right] = \frac{(\sigma_{\mathrm{eq}})^2 (\mu_{\mathrm{noeq}})^2}{(\mu_{\mathrm{eq}})^2 (\sigma_{\mathrm{noeq}})^2} \approx \left(\frac{\sigma_{\mathrm{eq}}}{\sigma_{\mathrm{noeq}}}\right)^2 \tag{5.88}$$

The preceding expressions model an equalizer/compensator that improves the performance (e.g. reducing $P^{(\mathrm{EO})}$), but in the meantime, at the same signal level ($\mu_{\mathrm{eq}} = \mu_{\mathrm{noeq}}$), it can make the noise worse; to simplify the model further, it is assumed that $F_{\mathrm{E}} = (\sigma_{1,\mathrm{eq}}/\sigma_{1,\mathrm{noeq}})^2 = (\sigma_{0,\mathrm{eq}}/\sigma_{0,\mathrm{noeq}})^2$. According to these definitions for $P^{(\mathrm{EO})}$ and F_{E}, Equation 5.78 can be modified to

$$\begin{aligned} Q &= \frac{\mu_1' - \mu_0'}{\sigma_1' + \sigma_0'} \\ &\rightarrow \frac{\min\{\mu_1\} - \max\{\mu_0\}}{\mu_1 - \mu_0} \frac{\mu_1 - \mu_0}{\sigma_1' + \sigma_0'} = \frac{\min\{\mu_1\} - \max\{\mu_0\}}{\mu_1 - \mu_0} \frac{\mu_1 - \mu_0}{\sigma_1 + \sigma_0} \frac{\sigma_1 + \sigma_0}{\sigma_1' + \sigma_0'} \\ &= P^{(\mathrm{EO})} Q_{\mathrm{R}} \frac{1}{\sqrt{F_{\mathrm{E}}}} \end{aligned} \tag{5.89}$$

where Q_{R} is the value of Personick's factor (5.61) as set by OSNR (as the maximum system performance or system baseline) in terms of the μ_1 and μ_0 values defining the best eye-opening. Equations 5.87–5.89 improve the relation between Q-factor and OSNR when the eye diagram is blurred and when distortion compensation strategies are present.

Equations 5.62, 5.78 and 5.79 are used for a general system estimate; more precise calculations can be done through Equations 5.82, 5.84, 5.86, 5.89 or similar relations worked out for specific impairment effects. It is just our intention to detail transmission degradation due to various effects: all penalties will be referred to the parameter Q^2; that is, Equation 5.80. As already said above, the penalty approach consists of:

- an analysis of each physical impairment independently of all other possible transmission degradations;
- referring to the worst case of the specific impairment under examination;
- assuming that the light path is free from all other penalties;
- and then summing up all (logarithmic) contributions due to the different penalties.

From (5.62) one sees that at the end of a light path a QoS requirement of, say, BER better than 10^{-12} corresponds to the constraint $Q \geq 7$ in linear units, that is, $Q^* \geq 17\,\mathrm{dB}$. Every transmission penalty affects this reference performance and then the launch light path QoS must be correspondingly higher; compare with Equation 5.77.

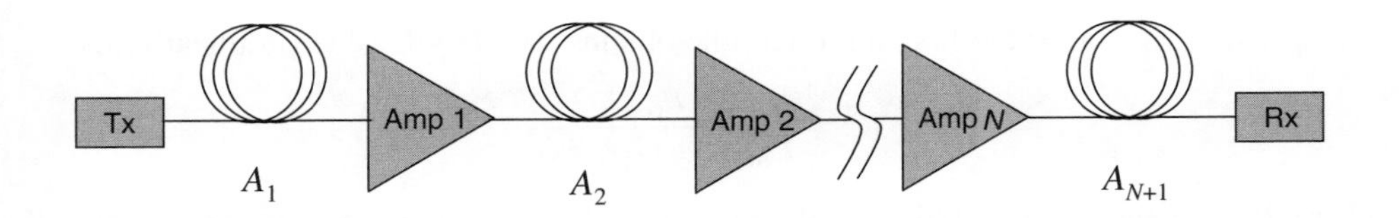

Figure 5.12 Chain of amplifiers and attenuating spans along a point-to-point link

5.2.4.2 Baseline Light Path Performance: Amplified Spontaneous Emission

As a starting point in the discussion of the system baseline, an equation for the OSNR along a chain of OAs will be worked out. Consider Equation 5.7 for the definition of the OSNR in a band B_o around the signal:

$$R(B_o) = \frac{P_S}{P_N}$$

where P_S is the signal power and P_N the noise power into the band B_o.

After amplifier i along the chain (Figure 5.12) the noise power will be given by

$$P_N = P_{ASE,i} + G_i P_{N,in} \tag{5.90}$$

where $P_{ASE,i}$ is the ASE noise generated by that amplifier, having gain G_i; $P_{N,in}$ is the noise power in band B_o, generated upstream of the i amplifier (by preceding amplifiers and by the source) and entering it. Equation 5.90 is a recurrence formula to calculate the OSNR evolution along the amplifier chain.

Note that the initial condition for the OSNR at the transmitter is given by the side-mode suppression ratio (SMSR) of the source, $R_0 \equiv \text{SMSR}|_{TX}$; Figure 5.13.

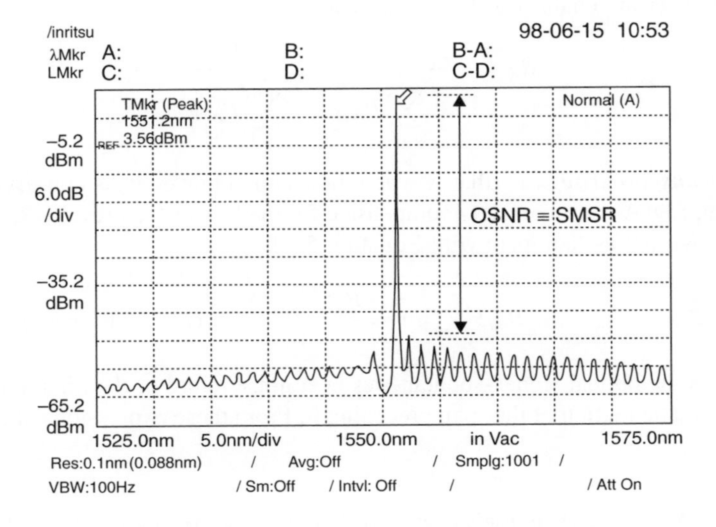

Figure 5.13 OSNR at transmitter

Applying (5.90) after the first amplifier, one obtains the OSNR value (in linear units):

$$R_1 = \frac{P_{\text{coh}}}{P_{\text{ASE},1} + G_1 A_1 P_{\text{SSE}}}$$

where P_{coh} is the power of the amplified signal, $P_{\text{ASE},1}$ is the ASE power in band B_{o} co-propagating with the signal, G_1 is the amplifier gain, A_1 is the attenuation of the (possible) link between TX and first amplifier, and P_{SSE} is the spontaneous emission of the source (the power emitted by all side modes) within the signal linewidth.

The preceding expression can be conveniently rewritten as

$$R_1 = \frac{R_0}{1 + \dfrac{P_{\text{ASE},1}}{G_1 A_1 P_{\text{SSE}}}} = \frac{R_0}{1 + \dfrac{P_{\text{ASE},1}}{G_1 A_1} \dfrac{R_0}{P_0}} \tag{5.91}$$

where $P_0 = R_0 P_{\text{SSE}}$ is the signal power emitted by the source. This equation can be easily iterated for application to the case shown in Figure 5.12. The OSNR (in linear units) after the Kth amplifier in the chain is given by

$$R_K = \frac{R_0}{1 + \frac{R_0}{P_0} \sum_{i=1}^{K} \left(\dfrac{P_{\text{ASE},i}}{\prod_{j=1}^{i} G_j A_j} \right)} = \frac{R_0}{1 + \frac{R_0}{I_0} \sum_{i=1}^{K} \left(\dfrac{m_{t,i} B_{\text{o}} F_i^{(\text{o})}}{2A_i \prod_{j=1}^{i-1} G_j A_j} \right)} \tag{5.92}$$

In the preceding expression, $I_0 = P_0/h\nu$ is the transmitter optical flux and an optical bandwidth B_{o} common to all amplifiers has been assumed. Suppose that the link is operated under transparent transport conditions; this means that $G_i A_i = 1$ over each span and, consequently, the launch signal power is reattained at the output of each (line) amplifier.

The OSNR after the chain of K OAs then reads

$$R_K = \frac{P_0}{1 + \sum_{i=1}^{K} P_{\text{ASE},i}} \cong \frac{P_0}{\sum_{i=1}^{K} P_{\text{ASE},i}} \tag{5.93}$$

In such a *transparent regime* (that is when the channel power P_0 is conserved span after span) – setting EDFA gain to exactly compensate for the loss of the previous span – the ASE power can be written as (compare with Equation 5.34)

$$P_{\text{ASE},i} = \frac{1}{2} G_i m_{t,i} B_{\text{o},i} F_{\text{s-sp},i}^{(\text{o})} h\nu = \frac{B_{\text{o}} F_{\text{s-sp},i}^{(\text{o})} h\nu}{A_i} \quad (G_i A_i = 1 \text{ for all } i)$$

where $F_{\text{s-sp},i}^{(\text{o})}$ is the optical signal–spontaneous beat noise figure of the ith amplifier,[27] G_i its gain, and A_i the attenuation of the span preceding it. From these expressions, the OSNR after

[27] It is the noise figure set by signal–spontaneous beat noise alone, supposed to be the dominant term. This result is valid for EDFAs; if SOAs are present, then some minor changes are required ($m_{t,i} = 1$).

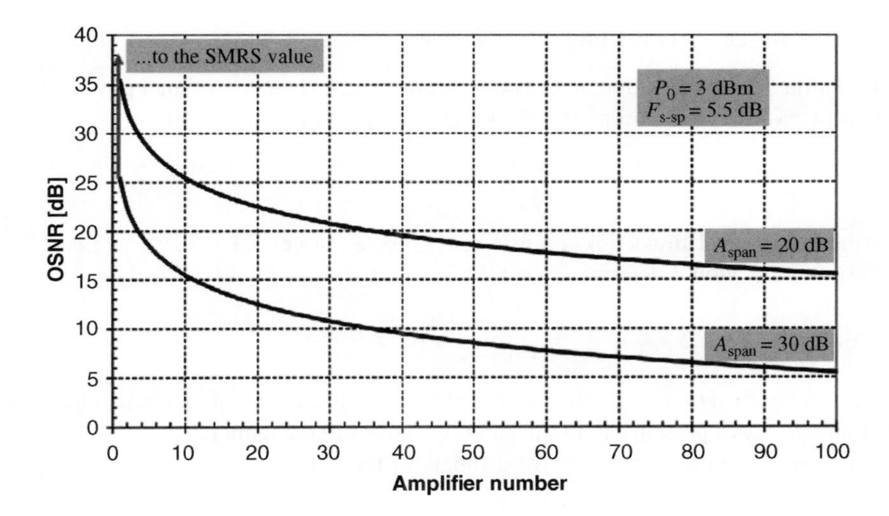

Figure 5.14 OSNR evolution along a chain of uniformly spaced, identical amplifiers, from Equation 5.94. The channel power is 3 dBm and the noise figure of each amplifier is 5.5 dB. Span attenuation 20/30 dB. Increasing P_0 (or decreasing either A or $F_{s-sp}^{(o)}$) by X dB, the curve shifts towards higher OSNR values by X dB

a chain of K equally operating amplifiers is

$$R_K \cong \frac{P_0 A_{span}}{K B_o h\nu F_{s-sp}^{(o)}} \quad \text{[linear]}, \quad R_K = P_0 - A_{span} - F_{s-sp}^{(o)} + 58 - 10\log K \quad \text{[dB]} \tag{5.94}$$

since $h\nu B_o \cong -58$ dBm in the reference bandwidth B_o corresponding to 0.1 nm.[28] Figure 5.14 shows the behavior of Equation 5.94.

For different operating conditions or different amplifier types, (5.93) transforms into

$$R_K = \frac{P_0}{B_o h\nu \sum_{i=1}^{K} \frac{F_{s-sp,i}^{(o)}}{A_i}} \quad \text{[linear]} \tag{5.95}$$

and

$$R_K = P_0 - 10\log\left(\sum_{i=1}^{K} \frac{F_{s-sp,i}^{(o)}}{A_i}\right) + 58 = P_0 - A_{tot} - F_{eq} + 58 \quad \text{[dB]} \tag{5.96}$$

where A_{tot} is the total link attenuation and F_{eq} is the *link equivalent noise figure*. In linear units:

$$A_{tot} = \prod_{i=1}^{K} A_i \quad \text{and} \quad F_{eq} = \sum_{i=1}^{K} \left(F_{s-sp,i}^{(o)} \frac{A_{tot}}{A_i}\right) \tag{5.97}$$

[28] At the 1550 nm wavelength this quantum noise plateau is −57.97 dB.

5.2.4.3 ASE: Baseline Sum Rule

From Equation 5.7 it is seen that, in the transparent regime (launched signal power recovered after amplification), degradations to OSNR sum up according to the rule

$$\frac{1}{R_{1+2}} = \frac{1}{R_1} + \frac{1}{R_2} \tag{5.98}$$

R_{1+2} is the OSNR level after a link comprising two noise sources of levels P_{N_1} and P_{N_2}. Each of them, taken alone, would determine the OSNR levels R_1 and R_2 respectively:

$$R_1 = \frac{P_S}{P_{N_1}} \quad \text{and} \quad R_2 = \frac{P_S}{P_{N_2}}$$

Equation 5.95 is useful to estimate the global OSNR when different, noise-like independent impairments $R_{eq,i}$ are present on a transparently transmitted channel.

In these conditions, a more operative estimate of the baseline is

$$\frac{1}{R_{out}} = \sum_i \frac{1}{R_i} + \sum_j \frac{1}{R_{eq,j}} \tag{5.99}$$

Sometimes, nonlinear effects can be modeled in such a way to determine an effective OSNR degradation, to be summed to the linear one through the above rule. Equation 5.96 can serve as a basis for the definition of a *light path evaluation metric that adds linearly* along an optical route (compare Figure 5.15) obtained from Equation 5.96, as a particular case of (5.99).

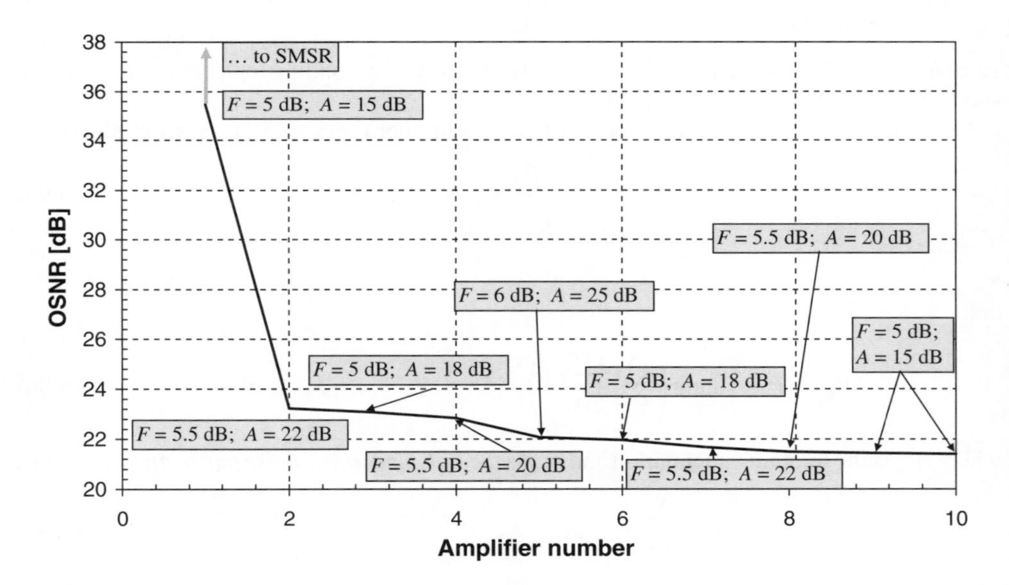

Figure 5.15 OSNR behavior along a nonuniform chain of amplifiers, from (5.96). The channel power is 3 dBm; noise figures and span values are shown in text boxes

5.2.4.4 Transmitter: Extinction Ratio

The *extinction ratio* (ER) is referred to the optical power of the levels '1' and '0', through the expression

$$r = \frac{P_1}{P_0} \tag{5.100}$$

the ideal situation corresponds to $r = \infty$ when no optical power is present on '0's. In a real case when the extinction ratio is finite, one assumes that the transmitter *average power* $\bar{P}$ would be the same as in the ideal case; thence:

$$\bar{P} = \frac{P_1}{2} = \frac{P'_1 + P'_0}{2} = \frac{r+1}{2r}P'_1 \Rightarrow \frac{P'_1}{P_1} = \frac{r}{r+1} \tag{5.101}$$

In the case of dominant thermal noise, from (5.81):

$$\frac{Q'}{Q} = \frac{P'_1 - P'_0}{P_1} = \frac{P'_1 - P'_1/r}{P_1} = \frac{P'_1}{P_1}\left(1 - \frac{1}{r}\right)$$

and then

$$\frac{Q'}{Q} = \frac{r-1}{r+1}$$

On the other hand, when APD or OA are present, Equations 5.83 and 5.101 give

$$\frac{Q'}{Q} = \sqrt{\frac{P'_1}{P_1}}\left(1 - \frac{1}{\sqrt{r}}\right) = \sqrt{\frac{r}{r+1}}\left(1 - \frac{1}{\sqrt{r}}\right) = \frac{\sqrt{r}-1}{\sqrt{r+1}}$$

One can conclude that

$$P_{r,\mathrm{T}}^{(Q^*)} = -20\log\left(\frac{r-1}{r+1}\right), \quad P_{r,\mathrm{O}}^{(Q^*)} = -20\log\left(\frac{\sqrt{r}-1}{\sqrt{r+1}}\right) \tag{5.102}$$

where $P_{r,\mathrm{T}}^{(Q^*)}$ is the penalty on Q^2 due to a finite extinction ratio r (in logarithmic units) in the case of dominant thermal noise (noise independent of signal level) and $P_{r,\mathrm{O}}^{(Q^*)}$ indicates that penalty when optical noise is dominant (noise dependent on signal level).

Equation 5.102 express the Q^2 penalty for an actual system with extinction ratio r, with respect to the ideal case $r = \infty$, assuming that the *average power* is the same for both systems.[29] The requirement $P^{(Q^*)} \leq 1$ dB due to a finite extinction ratio r implies $r \geq 12.4$ dB and $r \geq 19.6$ dB for signal-independent and signal-dependent noise respectively (Figure 5.16).

Another way to calculate the effect of finite r values consists of using Equations 5.49 and 5.50 for both mark and space levels. From Equation 5.49 we get a coherent contribution to μ_1 given by HP_{coh} and an incoherent contribution given by HP_{ASE}.

[29] Under the assumption that the *peak* optical power would be the same for both systems, one obtains penalty expressions that differ slightly from Equation 5.102; that is:

$$P_{r,\mathrm{T}}^{(Q^*)} = -20\log\left(\frac{r-1}{r}\right), \quad P_{r,\mathrm{O}}^{(Q^*)} = -20\log\left(1 - \frac{1}{\sqrt{r}}\right) \tag{5.102'}$$

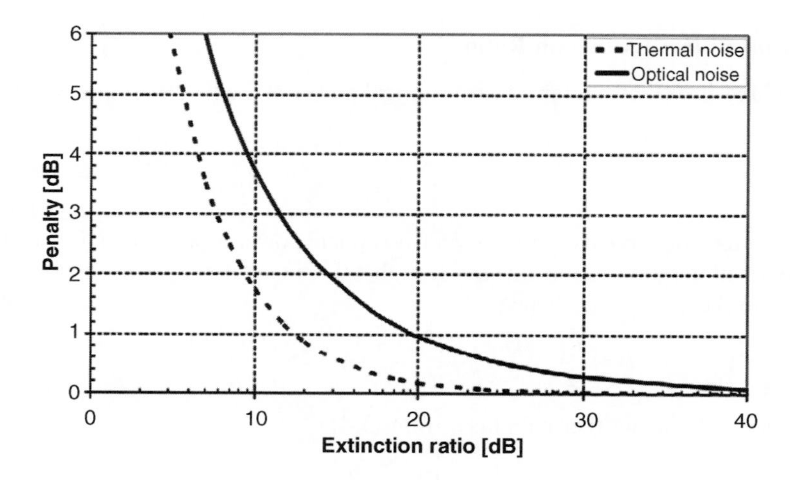

Figure 5.16 Penalty on Q^2 due to a finite extinction ratio. From Equation 5.102

As far as the space average photocurrent μ_0 is concerned, one has corresponding contributions $HP_{\rm coh}/r$ and $HP_{\rm ASE}$. Note that in taking the difference $\mu_1-\mu_0$ at the numerator of Q in the first part of Equation 5.78, ASE contributions cancel out. Identifying $P_{\rm coh}$ with P_1, fluctuations on mark are given by

$$\sigma_1^2 = 2B_{\rm e}eHP_1 + 2B_{\rm e}eHP_{\rm ASE} + \frac{4B_{\rm e}H^2P_1P_{\rm ASE}}{m_{\rm t}B_{\rm o}} + \frac{2B_{\rm e}H^2P_{\rm ASE}^2}{m_{\rm t}B_{\rm o}}\left(1-\frac{B_{\rm e}}{2B_{\rm o}}\right) + 2B_{\rm e}(e\eta G_{\rm a})^2\sigma_{\rm in,opt}^2 \tag{5.103}$$

and considering that for the '0' level $P_{\rm coh} = P_0 = P_1/r$, the variance on space is

$$\sigma_0^2 = 2B_{\rm e}eH\frac{P_1}{r} + 2B_{\rm e}eHP_{\rm ASE} + \frac{4B_{\rm e}H^2P_1P_{\rm ASE}}{m_{\rm t}rB_{\rm o}} + \frac{2B_{\rm e}H^2P_{\rm ASE}^2}{m_{\rm t}B_{\rm o}}\left(1-\frac{B_{\rm e}}{2B_{\rm o}}\right) + 2B_{\rm e}(e\eta G_{\rm a})^2\sigma_{\rm in,opt}^2 \tag{5.104}$$

After minor algebra one obtains

$$Q' \equiv Q(r,\xi)$$

$$= \frac{r-1}{\sqrt{r+1}}\,\frac{\xi R}{\sqrt{(1+r)\left(1-\dfrac{B_{\rm e}}{2B_{\rm o}}\right)+4\xi R}+\sqrt{(1+r)\left(1-\dfrac{B_{\rm e}}{2B_{\rm o}}\right)+4r\xi R}}\sqrt{2m_{\rm t}\frac{B_{\rm o}}{B_{\rm e}}} \tag{5.105}$$

from these expressions, where R is the OSNR referred to bandwidth $B_{\rm o}$ and to the average signal power (in linear units)

$$R = \frac{\bar{P}}{P_N(B_{\rm o})} = \frac{P_1+P_0}{2P_{\rm N}(B_{\rm o})} = \frac{P_1}{2P_{\rm N}(B_{\rm o})}\left(1+\frac{1}{r}\right)$$

according to Equation 5.7. Equation 5.105 is obtained directly from the first part of Equation 5.78 for Q and from Equations 5.103 and 5.104 where beat noise (both signal–ASE and ASE–ASE) is considered dominant. Indeed, from Equations 5.103 and 5.104 themselves, or from Figure 5.9, one sees that

$$\begin{aligned}(\sigma_{\mathrm{sh,s}})^2 &\approx R(\sigma_{\mathrm{sh,sp}})^2\\(\sigma_{\mathrm{b,s-sp}})^2 &\approx m_{\mathrm{t}} n_{\mathrm{eq}} G(\sigma_{\mathrm{sh,s}})^2\\(\sigma_{\mathrm{b,sp-sp}})^2 &\approx \frac{1}{2R}\left(1-\frac{B_{\mathrm{e}}}{2B_{\mathrm{o}}}\right)(\sigma_{\mathrm{b,s-sp}})^2 \approx \frac{G}{R}(\sigma_{\mathrm{sh,s}})^2 \approx G(\sigma_{\mathrm{sh,sp}})^2\end{aligned} \tag{5.106}$$

In obtaining Equation 5.105, OSNR has also been rescaled according to $R \mapsto \xi R$ to take into account in a heuristic way the *enhancement factor* $\xi = L_1/R$. This measures how much the OSNR is effective in determining a good SNR on the '1' level and depends on the modulation format. For instance, $\xi \cong 0.8$ for the CRZ format, 0.6 for RZ and 0.4 for NRZ [23]. Using the expressions worked out previously for μ_1 and σ_1, one sees at once that, when the signal–ASE beat noise is dominant:

$$L_1 = \left(\frac{\mu_1}{\sigma_1}\right)^2 = Q_1{}^2 \cong \frac{B_{\mathrm{o}}}{B_{\mathrm{e}}}\xi R$$

so that introducing the enhancement factor ξ means modifying further the system spectral form factor, depending upon the modulation format adopted [23]. Figure 5.17 shows that the RZ format has a robustness 2 dB stronger than NRZ; even stronger, from this point of view, would be the CRZ format. The ideal condition corresponds to a Q-factor $Q = Q(r = \infty, \xi = 1)$, in which case Equation 5.105 reduces to Equation 5.70.

On the other hand, using Equations 5.103 and 5.104 with no approximations on noise terms, a general expression for Personick's Q-factor is obtained in the form

$$Q(r,\xi) = \frac{N(r,\xi)}{D(r,\xi)} \tag{5.107}$$

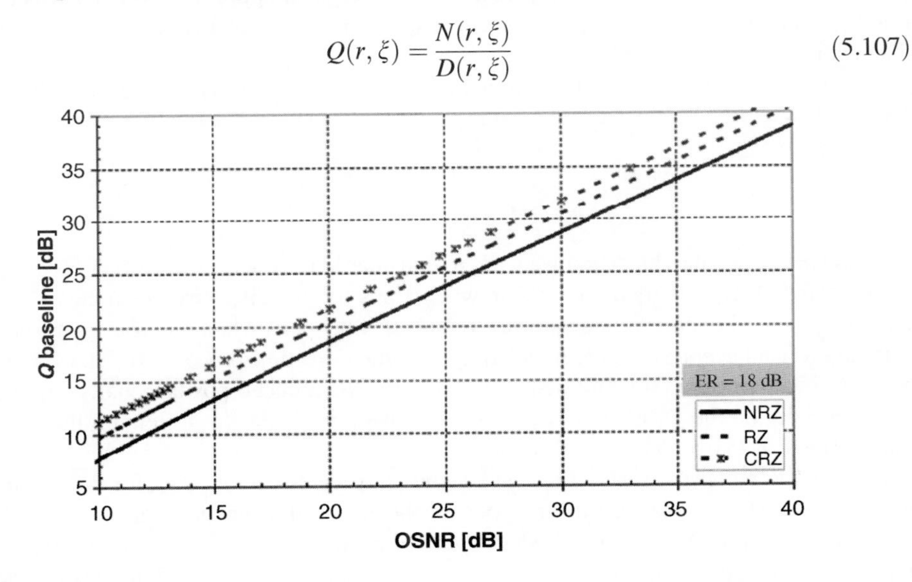

Figure 5.17 Baseline Q-factor versus OSNR for three different modulation formats: NRZ, RZ and CRZ. From Equation 5.105: the extinction ratio is assumed to be 18 dB

where

$$N(r,\xi) = \frac{r-1}{\sqrt{r+1}}\xi R\sqrt{\frac{2m_t B_o}{B_e}} \tag{5.108}$$

$$\begin{aligned} D(r,\xi) \quad = \quad & \sqrt{(1+r+2r\xi R)\frac{m_t B_o}{\eta P_{ASE}} + (1+r)\left(1-\frac{B_e}{2B_o}\right) + 4r\xi R + (1+r)\frac{2m_t B_o kT(h\nu_S)^2}{R_1(\eta e P_{ASE})^2}} \\ & + \sqrt{(1+r+2\xi R)\frac{m_t B_o}{\eta P_{ASE}} + (1+r)\left(1-\frac{B_e}{2B_o}\right) + 4\xi R + (1+r)\frac{2m_t B_o kT(h\nu_S)^2}{R_1(\eta e P_{ASE})^2}} \end{aligned} \tag{5.109}$$

In the ideal case $Q \equiv Q(r \to \infty, \xi = 1)$ (5.109) reduces to Equation 5.69.

Typical values for the extinction ratio with NRZ format at 10 Gbit/s range from 12 to 13 dB for simpler transmitters used with electrical equalizers at detection, to 20 dB for long-haul transmitters, up to at least 25 dB for devices with an external Mach–Zehnder modulator.

5.2.4.5 Receiver: Sensitivity, Intersymbol Interference, Overload

The elements of the study on electrical noise discussed in Section 5.2.2.2 can be applied to an analysis of the receiver performance in the Gaussian approximation. This section will briefly comment on three aspects of receiver performance: sensitivity, robustness to signal distortions, and overload.

When the extinction ratio of the transmitter is finite (see previous section), the receiver *sensitivity* should be referred to the average power (Equation 5.101):

$$\overline{P} = \frac{P_1 + P_0}{2} = \frac{1+r}{2r}P_1 \tag{5.110}$$

The aspect of signal distortion we want to consider here concerns the effect of *intersymbol interference* (ISI) that manifests itself when transmitting dispersive channels – channels that elongate and distort the transmitted signals. ISI happens when one transmission symbol spreads over adjacent symbols, interfering with the detection process. We shall treat ISI in Section 5.3.1, work out a simplified expression for ISI-induced power penalty, and translate it into penalty on the Q-factor. For the moment, we limit ourselves to remark that one of the main causes of ISI is CD.

Finally, the receiver *overload* has an obvious importance on its own for device integrity; however, it can also play a limiting role on the maximum reachable transparent distance ($\lesssim$30 *span*) for DWDM systems. If strong variations of spectral load were to happen, then the induced optical transients could adversely affect channels allocated in the red subband (1530–1549 nm). This aspect will be approached in Section 5.3.3, dedicated to optical transients.

5.2.4.6 Receiver: Excess Statistical Noise

The receiver penalty due to excess noise (optical source RIN, for instance) can be calculated by adding the corresponding terms in Equation 5.50 to the variances σ_1^2 and σ_0^2 under transparent transmission:[30]

$$\Delta_i = 2B_e(\eta e)^2\sigma_{\mathrm{exc},i}^2 = B_e(HP_i)^2\mathrm{RIN}, \qquad i = 0, 1 \tag{5.111}$$

From (5.78), assuming $P_1' = P_1$ and $P_0' = P_0$, one obtains

$$\frac{Q'}{Q} = \frac{H(P_1' - P_0')}{\sigma_1' + \sigma_0'}\frac{\sigma_1 + \sigma_0}{H(P_1 - P_0)} = \frac{\sigma_1 + \sigma_0}{\sqrt{\sigma_1^2 + \Delta_1} + \sqrt{\sigma_0^2 + \Delta_0}} \tag{5.112}$$

In the case of dominant thermal noise:

$$\frac{Q'}{Q} = \frac{2}{\sqrt{1 + \dfrac{\Delta_1}{\sigma_T^2}} + \sqrt{1 + \dfrac{\Delta_0}{\sigma_T^2}}} \tag{5.113}$$

where σ_T^2 is given by the second part of Equation 5.51.

When optical noise is dominant, Equation 5.112 gives

$$\frac{Q'}{Q} = \frac{K(\sqrt{P_1} + \sqrt{P_0})}{\sqrt{K^2P_1 + \Delta_1} + \sqrt{K^2P_0 + \Delta_0}} = \frac{K\sqrt{P_1}(1 + 1/\sqrt{r})}{\sqrt{K^2P_1 + \Delta_1} + \sqrt{K^2P_1/r + \Delta_0}}$$

From Equation 5.111 one obtains $\Delta_1/\Delta_0 = r^2$, so that

$$\frac{Q'}{Q} = \frac{K\sqrt{P_1}(1 + 1/\sqrt{r})}{\sqrt{K^2P_1 + \Delta_1} + \sqrt{K^2P_1/r + \Delta_1/r^2}} = \left(1 + \frac{1}{\sqrt{r}}\right)\frac{1}{\sqrt{1 + \frac{\Delta_1}{\sigma_1^2}} + \frac{1}{\sqrt{r}}\sqrt{1 + \frac{\Delta_1}{r\sigma_1^2}}} \tag{5.114}$$

where

$$\frac{\Delta_1}{\sigma_1^2} \cong \frac{P_1B_o\mathrm{RIN}}{2P_{\mathrm{ASE}}} \cong \frac{r}{r+1}R_{\mathrm{ref}}B_{\mathrm{ref}}\mathrm{RIN}$$

In both cases:

$$P_{\mathrm{SNR}} = -20\log\left(\frac{Q'}{Q}\right) \tag{5.115}$$

From Figure 5.18 it is apparent that to reduce the Q^2 penalty due to source RIN to less than 0.5 dB (1 dB), RIN itself must be lower than −140 dB/Hz (−137 dB/Hz), when transmission conditions are good (ER = 20 dB, OSNR = 30 dB).

The worse the transmission conditions are, the weaker the RIN constraint is; for ER = 10 dB, OSNR = 25 dB, the Q^2 penalty is less than 0.5 dB (1 dB), when RIN is lower than −135 dB/Hz (−130 dB/Hz).

---

[30] According to the Gaussian model for photodetection and the white-noise Gaussian additive model for penalties.

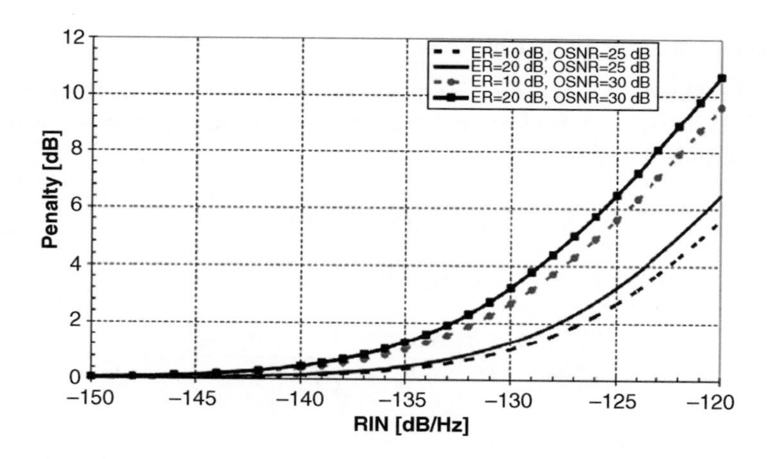

Figure 5.18 Q^2 penalty due to source RIN. From Equations 5.114 and 5.115

5.2.4.7 Jitter

Time fluctuations Δt in the arrival of optical pulses at the receiver can degrade its performance. Consider a PIN receiver (dominant thermal noise σ_T) in the ideal conditions of infinite ER: μ_1 denotes the average mark photocurrent. Jitter induces photocurrent fluctuations described by an average value μ_J and a standard deviation σ_J. Then the Q-factor, without and with jitter, is given by

$$Q = \frac{\mu_1}{2\sigma_T}, \quad Q' = \frac{\mu_1 - \mu_J}{\sqrt{\sigma_T{}^2 + \sigma_J{}^2} + \sigma_T} \tag{5.116}$$

where μ_J is subtracted from μ_1 because jitter prevents the pulse from being sampled at the optimum point $t = 0$. Denoting the current pulse with $H_{out}(t)$, one can write [24]

$$I_J = \mu_1 [H_{out}(0) - H_{out}(\Delta t)] \tag{5.117}$$

In Equation 5.116 $\mu_J = \langle I_J \rangle$; on the other hand, the standard deviation σ_J depends upon the signal pulse shape around the decision threshold. Using the pulse shape for a raised-cosine filter

$$H_{out}(t) = \frac{\sin(2\pi Bt)}{2\pi Bt} \frac{1}{1-(2Bt)^2} \tag{5.118}$$

at bit rate B, for not too high penalties ($B\Delta t \ll 1$) one can work out the following expression for the current I_J:

$$I_J = \mu_1 [H_{out}(0) - H_{out}(\Delta t)] = \mu_1 \left[1 - \frac{\sin(2\pi B\Delta t)}{2\pi B\Delta t} \frac{1}{1-(2B\Delta t)^2} \right]$$

$$\cong \mu_1 \left\{ 1 - \left[1 - \frac{(2\pi B\Delta t)^2}{6} \right] \left[1 + (2B\Delta t)^2 \right] \right\} \cong \mu_1 \left[1 - 1 - (2B\Delta t)^2 + \frac{(2\pi B\Delta t)^2}{6} \right]$$

$$\cong \left(\frac{2\pi^2}{3} - 4\right)(B\Delta t)^2 \mu_1 \equiv k(\Delta t)^2 \mu_1 \tag{5.118}$$

The calculation for σ_J is based on the assumption [24] that time fluctuations have a Gaussian distribution $F(\Delta t)$ with standard deviation τ_J:

$$F(\Delta t) = \frac{\exp\left[-\frac{(\Delta t)^2}{2(\tau_J)^2}\right]}{\sqrt{2\pi}\tau_J} \tag{5.119}$$

Using Equation 5.118, the distribution function $F(\Delta t)$ (5.119) for time jitter Δt can be transformed into the distribution function $F(I_J)$ for the current I_J:

$$F(I_J) = F(\Delta t)\left(\frac{dI_J}{d\Delta t}\right)^{-1}, \quad \text{where } \Delta t = \Delta t(I_J)$$

Therefore, one can write

$$F(I_J) = 2\frac{\exp\left[-\frac{I_J}{2k\mu_1(\tau_J)^2}\right]}{\sqrt{2\pi}\tau_J} \frac{1}{2k\mu_1}\sqrt{\frac{k\mu_1}{I_J}} = \frac{\exp\left[-\frac{I_J}{2k\mu_1(\tau_J)^2}\right]}{\sqrt{2\pi k\mu_1 I_J}\tau_J} = \frac{\exp\left(-\frac{I_J}{b\mu_1}\right)}{\sqrt{b\pi\mu_1 I_J}}$$

where

$$b = \left(\frac{4\pi^2}{3} - 8\right)(B\tau_J)^2 \tag{5.120}$$

In the preceding calculation, the factor of two in the expression for $F(I_J)$ takes into account the fact that, for a given jitter value Δt, the corresponding current value satisfies $I_J(\Delta t) = I_J(-\Delta t)$.

The mean value $\mu_J = \langle I_J \rangle$ reads

$$\mu_J = \int_0^{+\infty} \frac{I_J e^{-I_J/b\mu_1}}{\sqrt{b\pi\mu_1 I_J}} dI_J = \frac{b\mu_1}{\sqrt{\pi}} \int_0^{+\infty} \sqrt{u}\, e^{-u}\, du = \frac{b\mu_1}{2} \tag{5.121}$$

and the standard deviation σ_J is

$$(\sigma_J)^2 = \int_0^{+\infty} \frac{(I_J)^2 e^{-I_J/b\mu_1}}{\sqrt{b\pi\mu_1 I_J}} dI_J - (\mu_J)^2 = \frac{(b\mu_1)^2}{\sqrt{\pi}} \int_0^{+\infty} u\sqrt{u} e^{-u} du = \frac{3}{4}(b\mu_1)^2 - \frac{1}{4}(b\mu_1)^2 = \left(\frac{b\mu_1}{\sqrt{2}}\right)^2 \tag{5.122}$$

Equations 5.121 and 5.122 allow us to write

$$\frac{Q'}{Q} = \frac{\mu_1 - \mu_J}{\sqrt{\sigma_T^2 + \sigma_J^2} + \sigma_T} \frac{2\sigma_T}{\mu_1} = \frac{2-b}{1 + \sqrt{1 + \frac{b^2}{2}\frac{\mu_1^2}{\sigma_T^2}}} = \frac{2-b}{1 + \sqrt{1 + (\sqrt{2}bQ)^2}}$$

and the penalty over $Q^* = 20\log Q$ due to jitter is

$$P_{\mathrm{J,T}}^{(Q^*)} = -20\log\frac{2-b}{1+\sqrt{1+(\sqrt{2}bQ)^2}} \tag{5.123}$$

Some considerations on (5.123) are needed. The jitter model presented here is very simple: in practice the penalty is influenced by the pulse shape and the effective jitter distribution. Using (5.123) one obtains very steep penalty curves, because they make reference to an ideal value $Q^* = 20\log Q$ that would be extremely large (say, 180 dB!). Though weak it might be, every degradation effect has a drastic impact on such an enormous value.

Leaving aside more complete models for time-jitter, one can proceed as sketched in Figure 5.19, calculating the degradation $P_{\mathrm{J,T}}^{(Q^*)}$ with reference to more reasonable operating Q^* values at the receiver (as set by other transmission impairments, ASE accumulation, for instance). In such a way, the usual $T_{\mathrm{bit}}/10$ criterion is obtained [24]: as long as the jitter parameter $B\tau_{\mathrm{J}} < 0.1$ (i.e., as long as $\tau_{\mathrm{J}} < T_{\mathrm{bit}}/10$), penalties due to jitter are lower than 1 dB over a vast range of operative Q^* values (say, up to $Q^* = 20$). For higher values of Q^*, $P_{\mathrm{J,T}}^{(Q^*)}$ curves become steeper and steeper, as even very weak impairments consistently affect an ideal system.

Consider now the case when signal–ASE bit noise is dominant; then, using the same symbols as in the previous case:

$$Q = \frac{\mu_1}{\sigma_1}, \quad Q' = \frac{\mu_1 - \mu_{\mathrm{J}}}{\sqrt{\sigma_1{}^2 + \sigma_{\mathrm{J}}{}^2}} \Rightarrow \frac{Q'}{Q} = \frac{\mu_1'}{\sigma_1'}\frac{\sigma_1}{\mu_1} = \frac{1-\dfrac{b}{2}}{\sqrt{1+\left(\dfrac{b}{\sqrt{2}}\dfrac{\mu_1}{\sigma_1}\right)^2}} = \frac{1-\dfrac{b}{2}}{\sqrt{1+\left(\dfrac{bQ}{\sqrt{2}}\right)^2}}$$

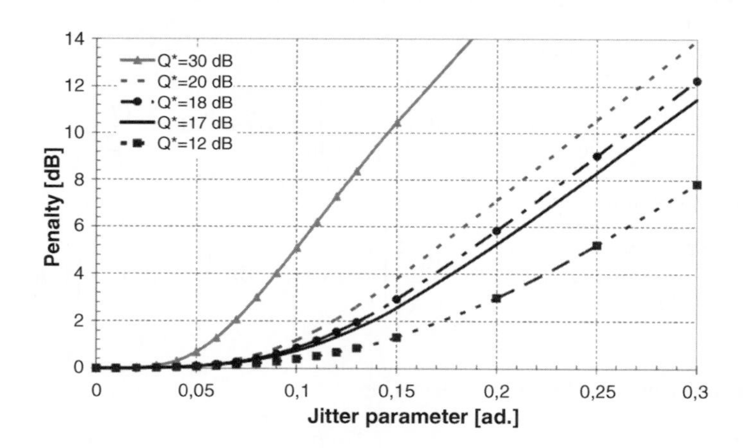

Figure 5.19 Penalty on Q^2 due to source time-jitter for a PIN receiver. From Equation 5.123, calculated for different Q^2 values at receiver, from large margins down to zero margin (say, 17 dB without FEC; 12 dB with FEC)

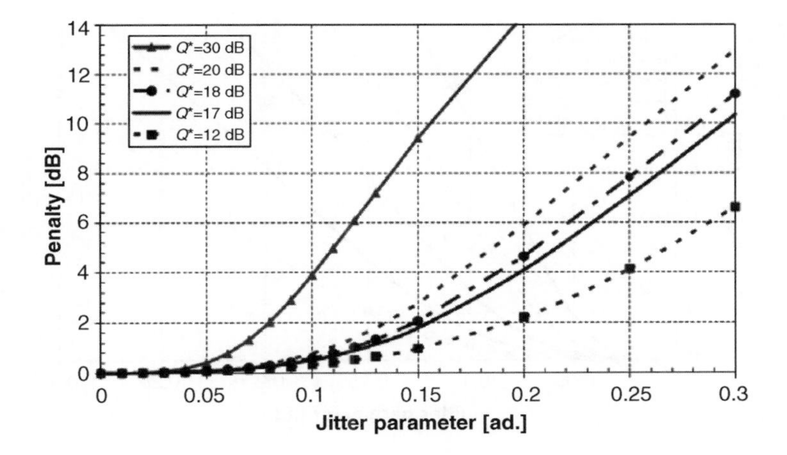

Figure 5.20 As Figure 5.19, but for dominant optical noise

Then:

$$P_{\mathrm{J,O}} = -20\log\frac{1-\frac{b}{2}}{\sqrt{1+\left(\frac{bQ}{\sqrt{2}}\right)^2}} \tag{5.124}$$

In this case also the $T_{\mathrm{bit}}/10$ criterion is substantially valid (Figure 5.20).

For dominant optical noise, a different description of the jitter penalty can be developed. Assuming limiting detection conditions, say $B_{\mathrm{o}} = 2B_{\mathrm{e}} = B$, with B the bit rate, Equation 5.124 can be rewritten thus:

$$P_{\mathrm{J,O}} = -20\log\frac{1-\frac{b}{2}}{\sqrt{1+\frac{b}{\sqrt{2}}\frac{P_1 m_{\mathrm{t}} B_{\mathrm{o}}}{4B_{\mathrm{e}}P_{\mathrm{ASE}}}}} = -20\log\frac{1-\frac{b}{2}}{\sqrt{1+b^2R(B)}} \tag{5.125}$$

where $R(B)$ is the intrinsic OSNR; therefore, the jitter penalty can be plotted with the $R(B)$ as a parameter, as done in Figure 5.21. This confirms that the jitter penalty does not exceed 1 dB as long as τ_{J} is shorter than 10% of the bit slot, for (intrinsic) OSNR values up to 20 dB. When $R(B)$ is larger, the 1 dB threshold decreases and at 'infinite' OSNR it stabilizes around 4% of the bit slot (OSNR ~35 dB corresponds to $Q^* \approx 38$ dB).

5.2.4.8 Amplification Noise

The characteristics of amplification noise have already been discussed in Section 5.2.2.1. They have been inserted in the equations for receiver performance; see Equations 5.49 and 5.50, for instance. On a light path operating at transparency, it will be considered that (line) amplifier gain exactly compensates for span attenuation. Taking the nonuniform gain distribution of EDFAs into account, the transparency condition will only hold for some 'average' channels within the DWDM comb. Over the whole transmission spectrum, one can consider a gain variation $G \pm \Delta G$ (G is the transparency gain value for average channels). For actual

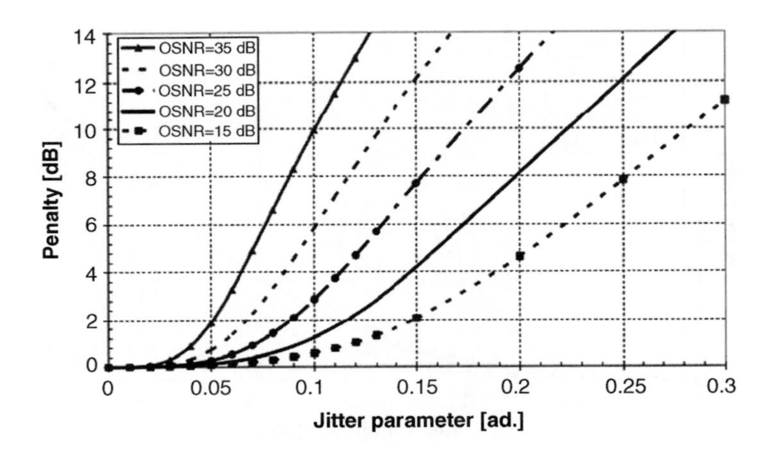

Figure 5.21 Penalty on Q^2 due to time jitter. From Equation 5.125

amplifiers, the gain variation ΔG is 0.5–1 dB per device. This gain variation induces a power variation ΔP from the transparency value P_{ch}: some channels will experience lower power and OSNR levels and other channels will have higher values than average channels.

In a stiff chain, gain variations tend to sum linearly. A global link optimization permits one to moderate this behavior: the baseline degradation is much weaker, as shown in Figure 5.22.

5.2.4.9 Crosstalk

When one or more disturbing signals (interferers) superimpose on the transmission channel, one speaks of crosstalk effects. Relating to the spectral separation between

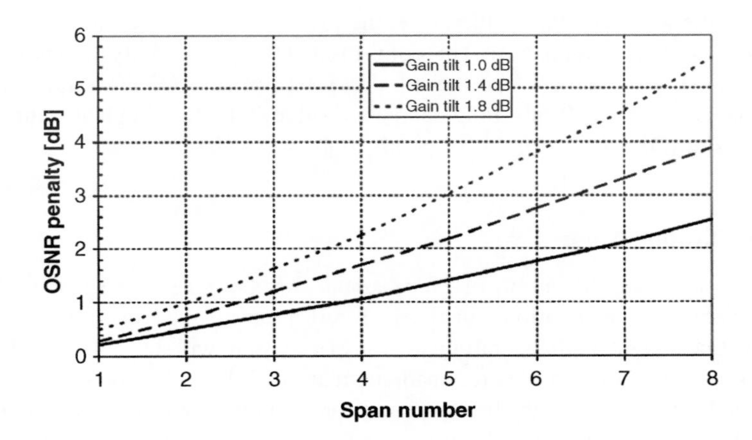

Figure 5.22 Penalty due to gain nonuniformity of amplifiers

transmission channel and interferers, a distinction is made between interchannel and intrachannel crosstalk.

Intrachannel crosstalk is the situation in which one or more interferers affect the signal channel at the same wavelength (*coherent crosstalk*) or when they are so close to the transmission frequency that frequency differences fall within the receiver electrical bandwidth (*in-band crosstalk*). These events happen at multiplexers located downstream of a demultiplexer, where there are optical switches or when nonlinear optical effects are present. The transmission impairment is highest if the interferers have the same optical polarization as the transmission channel and if the crosstalk signal is out of phase with respect to the channel.

Interchannel crosstalk happens when interferers have a wavelength different from the transmission wavelength in such a way that all beat tones fall outside the electrical bandwidth of the receiver (*out-of-band* or *incoherent crosstalk*). Interchannel crosstalk may be present in demultiplexers and spectral filters. This incoherent effect is weaker than the coherent one considered previously.

Since terms used to describe optical crosstalk and its effects are not completely consistent in the usage of various authors, we find it convenient to make reference to ITU-T Study Group 15 for definitions, as done in Table 5.1 [25–28].

The term *crosstalk* is maintained for the description of system effects, while the properties of the discrete components are given the term *isolation*; see Figure 5.23.

Table 5.1 ITU-T definitions for terms relating to crosstalk

System parameters
From Rec. G.692:
Interchannel crosstalk, ratio of total power in the disturbing channels to that in the wanted channel (wanted and disturbing channels at different wavelengths)
Interferometric (or intrachannel) crosstalk, ratio of the disturbing power (not including ASE) to the wanted power within a single channel (wavelength)
Interchannel crosstalk penalty, penalty assigned in the system budget to account for interchannel crosstalk
Interferometric crosstalk penalty, penalty assigned in the system budget to account for interferometric crosstalk
From Rec. G. 959.1:
Channel power difference, the maximum allowable power difference between channels entering a device
From Rec. G. 691:
Extinction ratio, ratio of power at the centre of a logical ‘1’ to the power at the center of a logical ‘0’
Eye-closure penalty, receiver sensitivity penalty due to all eye-closure effects
Component parameters
From Rec. G. 671:
Insertion loss, the reduction in power from input to output port at the wanted channel wavelength
Unidirectional isolation, the difference between the device loss at a disturbing channel wavelength and the loss at the wanted channel wavelength
Adjacent channel isolation, the isolation of the device at the wavelengths one channel above and below the wanted channel
Nonadjacent channel isolation, the isolation of the device at the wavelengths of all disturbing channels except for the adjacent channels

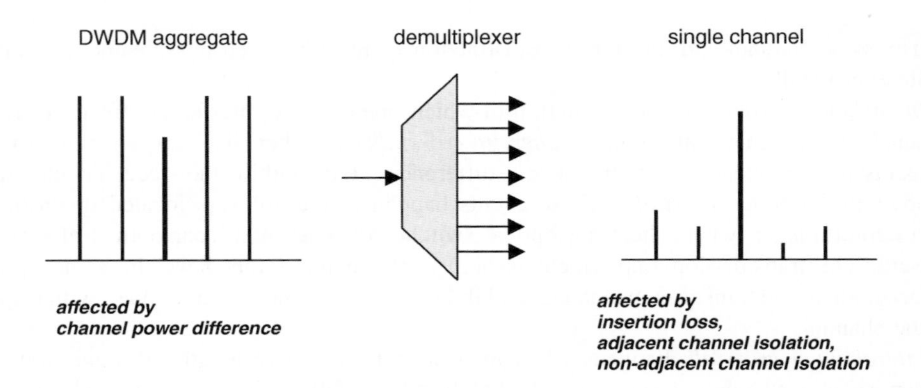

Figure 5.23 Schematics of a demultiplexer. The definition of parameters involved is given in Table 5.1

Coherent (Intrachannel) Crosstalk

Coherent crosstalk is stronger if transmission signal and interferers have the same polarization and when they are out of phase; in a worst-case calculation, both these conditions are assumed to be valid. Let P denote the mean signal power and εP the average power of an interferer: ε is called the *crosstalk level*. Supposing that both signals are at the same wavelength, the electrical field $E(t)$ at receiver will be [19]

$$E(t) = \sqrt{2P}d_{\mathrm{S}}(t)\cos[2\pi f_{\mathrm{C}}t + \phi_{\mathrm{S}}(t)] + \sqrt{2\varepsilon P}d_{\mathrm{X}}(t)\cos[2\pi f_{\mathrm{C}}t + \phi_{\mathrm{X}}(t)]$$

In this expression one has $d_S(t) = 0, 1$ according that the transmission channel being on space or on mark; similarly, $d_{\mathrm{X}}(t) = 0, 1$ whenever the interferer is on space or mark (the interferer itself is originated by modulated signals).

The frequency of the optical carrier is f_{C}; $\phi_{\mathrm{S}}(t)$ and $\phi_{\mathrm{X}}(t)$ are the optical phases for the channel and the interferer respectively. For the sake of simplicity, both channel and interferer are assumed to have an infinite ER; the (further) penalty for finite r values can be calculated independently, according to the very penalty approach at the beginning of Section 5.2.4. The photocurrent is proportional to the average optical power on the receiver:

$$P_{\mathrm{RX}} = Pd_{\mathrm{S}}(t) + \varepsilon Pd_{\mathrm{X}}(t) + 2\sqrt{\varepsilon}Pd_{\mathrm{S}}(t)d_{\mathrm{X}}(t)\cos[\phi_{\mathrm{S}}(t) - \phi_{\mathrm{X}}(t)]$$

The crosstalk level ε is assumed to be small, $\varepsilon \ll 1$, so that the term proportional to ε can be neglected with respect to that of order $\sqrt{\varepsilon}$. For a worst-case estimate, the harmonic term is put to -1: now the mark power is $P'_1 = P(1-2\sqrt{\varepsilon})$ while that on space is still vanishing, $P'_0 = 0 = P_0$.

For dominant thermal noise, (5.78) gives

$$\frac{Q'}{Q} = \frac{HP'_1}{\sigma_{\mathrm{T}} + \sigma_{\mathrm{T}}}\frac{\sigma_{\mathrm{T}} + \sigma_{\mathrm{T}}}{HP_1} = \frac{P(1-2\sqrt{\varepsilon})}{P} = 1-2\sqrt{\varepsilon}$$

while for dominant optical noise

$$\frac{Q'}{Q} = \frac{HP'_1}{K\sqrt{P'_1}}\frac{K\sqrt{P_1}}{HP_1} = \sqrt{\frac{P'_1}{P_1}} = \sqrt{1-2\sqrt{\varepsilon}}$$

Calling $P^{(Q^*)}_{\mathrm{X,T}}$ the crosstalk penalty when noise is signal independent and $P^{(Q^*)}_{\mathrm{X,O}}$ the same penalty when noise depends on signal level, the preceding expressions give

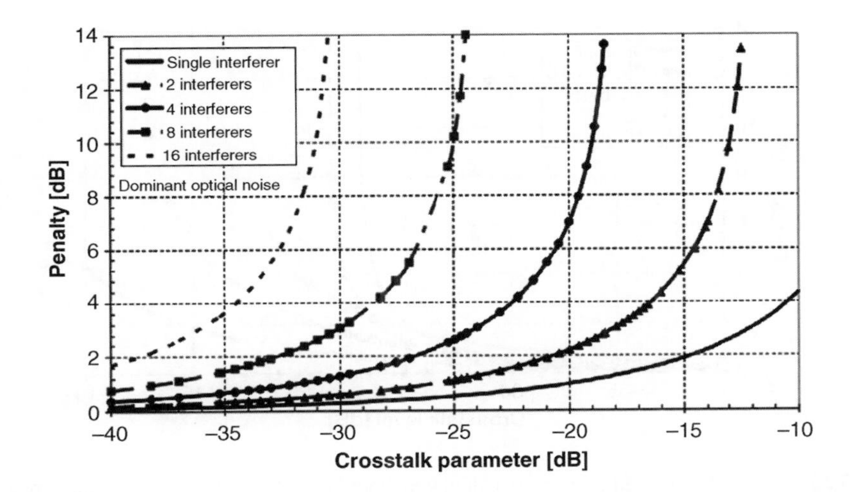

Figure 5.24 Penalty on Q^2 due to intrachannel crosstalk when optical noise is dominant. The extinction ratio is assumed to be infinite

$$P_{\mathrm{X,T}}^{(Q^*)} = -20\log(1-2\sqrt{\varepsilon}), \qquad P_{\mathrm{X,O}}^{(Q^*)} = -10\log(1-2\sqrt{\varepsilon}) \tag{5.126}$$

For the crosstalk penalty on $Q^* = 20 \log Q$ does not exceed 0.5 dB (1 dB), the crosstalk level must be lower than 31 dB (25 dB) (dominant thermal noise) or lower than 25 dB (20 dB) (dominant optical noise) (Figure 5.24).

The technical literature contains plenty of models for the optical crosstalk; this is no surprise given the importance of such a quantity. We will be content with Equation 5.126; but anyhow, we will take a concise tour around other, more sophisticated, models proposed so far. For dominant optical noise, and for a decision threshold D_{th} set at the mean power level[31] $\frac{1}{2}(P_1 + P_0)$, the model proposed in [29] gives results similar to Equation 5.126:

$$P_{\mathrm{X,O}}^{(Q^*)} = -10\log(1-m\sqrt{\varepsilon}), \quad m = \begin{cases} 4, & \text{dc-coupled RX} \\ 2, & \text{ac-coupled RX, best case} \\ 6, & \text{ac-coupled RX, worst case} \end{cases} \tag{5.127}$$

At low crosstalk levels, the photocurrent statistics are considered Gaussian [29] and one obtains

$$P_{\mathrm{X,O}}^{(Q^*)} = -5\log(1-4\varepsilon\bar{Q}^2) \tag{5.128}$$

$\bar{Q}$ is approximately the Q-value that corresponds to a fixed minimum acceptable level for BER (say, $\bar{Q} = 5.9$ at BER $= 10^{-9}$, $\bar{Q} = 6.94$ at BER $= 10^{-12}$;[32] Figure 5.25).

One can see experimentally that, in the case of a decision threshold set at average received power, crosstalk effects are reproduced by the Gaussian approximation 5.128 rather well for $\varepsilon \lesssim -25$ dB (Figure 5.25). For larger ε values, the penalty curve should approximate the

[31] Results of [29] are expressed as system power penalties. Using Equation 5.84 they can be translated directly into penalties over Q^*, as done in Equations 5.127 and 5.128.

[32] Since differences are so low, Q has been used instead of $\bar{Q}$ in Figure 5.25.

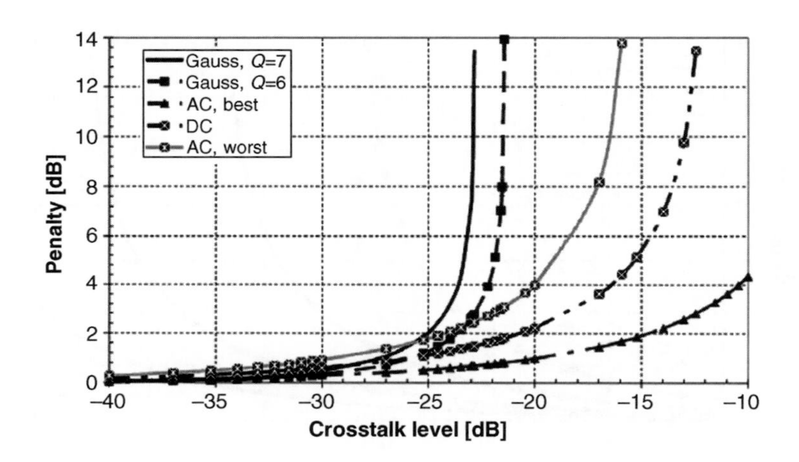

Figure 5.25 Crosstalk-induced Q^2 penalty calculated from (5.127) and (5.128). Decision threshold set at average received power

condition of a dc-coupled receiver, since the laser linewidth is much wider than the lowpass cutoff frequency of the receiver. However, this behavior is strongly affected by receiver saturation, which may shift the tolerance region (the region on the left of the vertical asymptote for penalty curves) towards lower ε values [29].

If there are N_X interferers against the transmission channel, each having average power $\varepsilon_i P$, $i = 1, \ldots, N_X$, Equation 5.126 can be straightforwardly modified to the form

$$P_{X,T}^{(Q^*)} = -20 \log\left(1-2\sum_{i=1}^{N_X}\sqrt{\varepsilon_i}\right), \quad P_{X,O}^{(Q^*)} = -10 \log\left(1-2\sum_{i=1}^{N_X}\sqrt{\varepsilon_i}\right) \tag{5.129}$$

for dominant thermal or optical noise respectively. The generalization $\sqrt{\varepsilon} \mapsto \sum \sqrt{\varepsilon}$ can also be applied to the case of Equation 5.127 and 5.128.

Up to now we have assumed an ideal extinction ratio (ER). For finite ER values, the penalty calculated with previous formulae can be summed to the penalty due to the finite r value, as calculated with Equation 5.102[33] (see Figure 5.26).

For finite ERs, complete models have also been proposed, as described in Ref. [30]. In the case of dominant optical noise they give

$$P_{(r,X),O}^{(Q^*)} = -10 \log\left[1 + \frac{r+1}{r-1}\left(\varepsilon - 4\sqrt{\frac{r\varepsilon}{r+1}}\right)\right] \tag{5.130}$$

for a decision threshold D_{th} set at average received power. For an optimized threshold, they give

$$P_{(r,X),O}^{(Q^*)} = -10 \log\left\{1 - 2\left[\frac{(1+\sqrt{r})\sqrt{\varepsilon(r+1)}}{r-1}\right]\right\} \tag{5.131}$$

[33] The penalty given by the sum of Equations 5.102 and 5.126 agrees with the results of (5.131) within 0.5 dB, for crosstalk levels in the range −40 to −8 dB and ER = 25 dB; Figure 5.26. The agreement is worse for lower ER values.

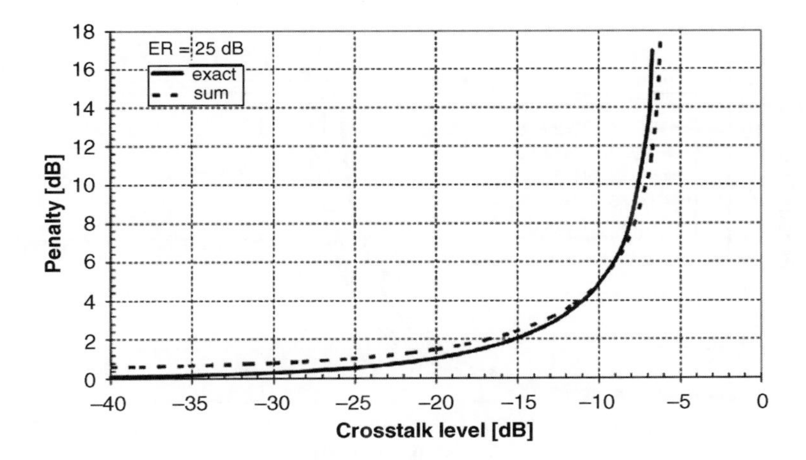

Figure 5.26 A comparison between the 'exact' model (5.131) and the 'sum' model of Equations 5.102 and 5.126

The reader will recognize that the second expression of Equation 5.126 is the limit of Equation 5.131 for $r \rightarrow \infty$. The results from Equations 5.130 and 5.131 are shown in Figure 5.27. A comparison between this figure and Figure 5.24 confirms that Equations 5.126 and 5.129 give a good estimate to the coherent crosstalk in the case of an optimized decision threshold.

In the case of multiple interferers, a model alternative to Equation 5.129 is formulated in terms of a Gaussian probability density function for the penalty.

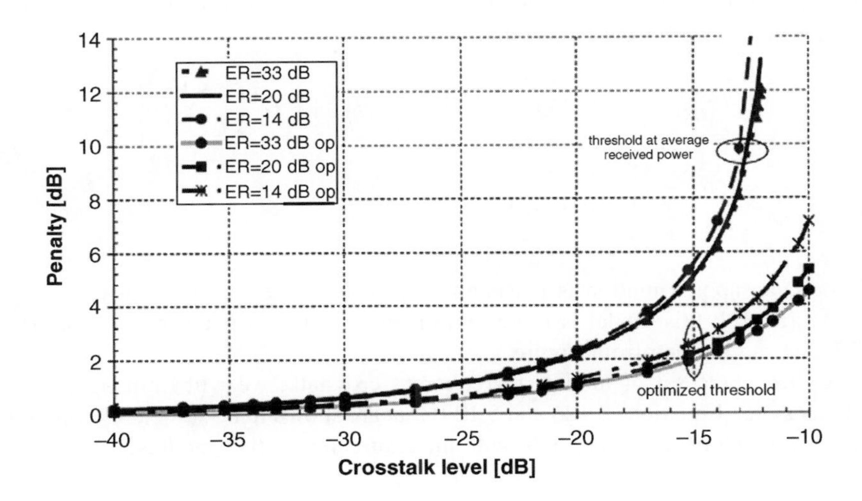

Figure 5.27 Crosstalk penalty from Equations 5.130 and 5.131

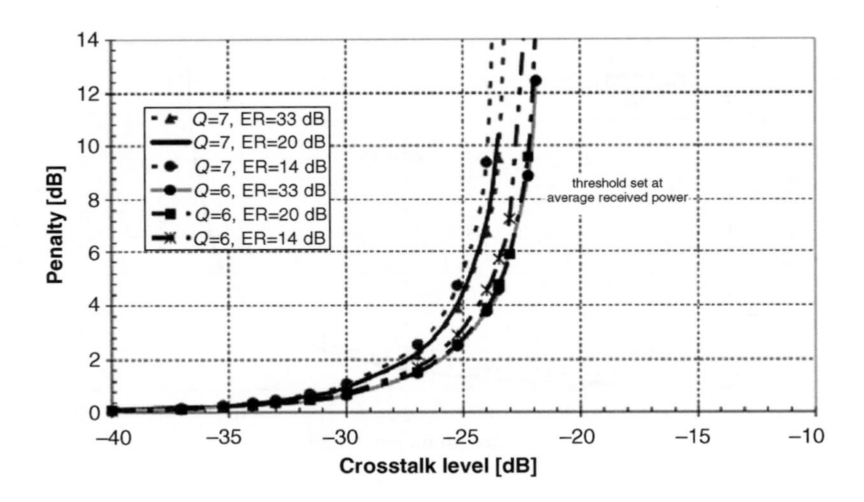

Figure 5.28 Crosstalk penalty expressed by (5.132)

Independently of the noise regime, this Gaussian model [31] gives[34]

$$P^{(Q^*)}_{(r,\mathrm{X})} \cong -10\log\left[1-4\varepsilon\bar{Q}^2\frac{r(r+1)}{(r-1)^2}\right] \tag{5.132}$$

for decision threshold at average received power. In (5.132), $\bar{Q}=\sqrt{2}\,\mathrm{erfc}^{-1}(4P_{\mathrm{BER}})$ instead of the usual relation $Q=\sqrt{2}\,\mathrm{erfc}^{-1}(2P_{\mathrm{BER}})$; for $\mathrm{BER}=10^{-12}$, this amounts to a shift from $Q=7.03$ to $\bar{Q}=6.94$[32].

Results from Equation 5.132 are shown in Figure 5.28 for different Q values and extinction ratios. For an optimized decision threshold:[34]

$$\begin{aligned} P^{(Q^*)}_{(r,\mathrm{X}),\mathrm{T}} &\cong -10\log\left[1-2\varepsilon Q^2\left(\frac{r+1}{r-1}\right)^2+\varepsilon^2Q^4\left(\frac{r+1}{r-1}\right)^2\right] \\ P^{(Q^*)}_{(r,\mathrm{X}),\mathrm{O}} &\cong -10\log\left[1-2\varepsilon Q^2\frac{r+1}{(\sqrt{r}-1)^2}\right] \end{aligned} \tag{5.133}$$

The Gaussian approximation is reasonable if, in the presence of N_{X} disturbing signals, $N_{\mathrm{X}} \gtrsim 12$–16 [31,32]; this model is more optimistic than the estimate from Equations 5.126 and 5.129, as can be seen from Figure 5.29.

Among this whole set of models for coherent crosstalk, we will limit ourselves to the simplest ones (i.e., Equations 5.126 and 5.129) that are shown to be sufficiently accurate for a system analysis (compare Figure 5.24 with the results from other models).

[34] Results from [31] are power penalties; here, they have been translated into Q^* penalties; compare footnote 31.

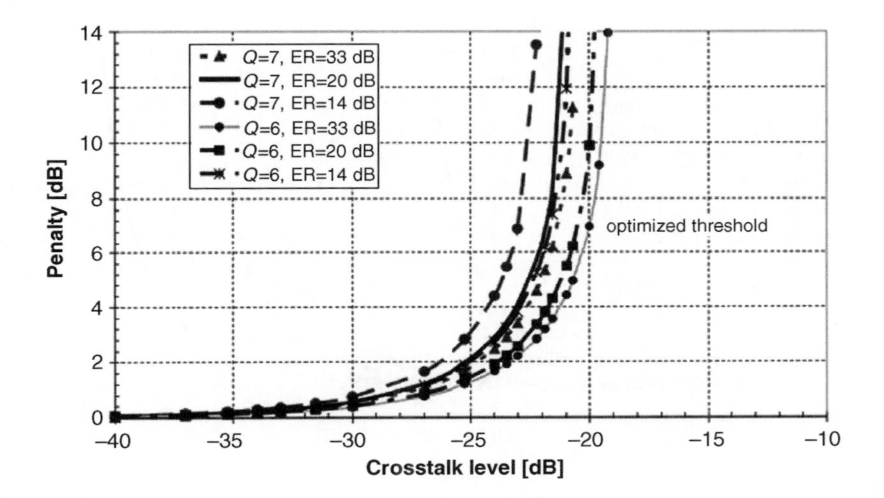

Figure 5.29 From the second expression of Equation 5.133, dominant optical noise

Incoherent (Interchannel) Crosstalk

Also in the case of interchannel crosstalk, let P be the average signal power and εP be the average interferer power at the receiver; the parameter ε is once again the crosstalk level. Now the beat frequencies between these components fall outside the receiver electrical bandwidth and the received power is simply the sum of the two powers: [19]

$$P_{\mathrm{RX}} = Pd_{\mathrm{S}}(t) + \varepsilon P d_{\mathrm{X}}(t)$$

For an infinite extinction ratio in the (ideal) reference situation, upon receiving a '1' bit the power $P_1' = P$ is detected, while on receiving a '0' bit a power $P_0' = \varepsilon P$ is detected. Therefore, when thermal noise is dominant:

$$\frac{Q'}{Q} = \frac{H(P_1' - P_0')}{\sigma_{\mathrm{T}} + \sigma_{\mathrm{T}}} \frac{\sigma_T + \sigma_{\mathrm{T}}}{HP_1} = \frac{P(1-\varepsilon)}{P} = 1 - \varepsilon$$

It must be stressed that, even in the case of dominant optical noise, the interferer power $P_0' = \varepsilon P$ cannot be attributed to OAs and, for infinite ER

$$\frac{Q'}{Q} = \frac{HP_1'}{\sigma_1} \frac{\sigma_1}{HP_1} = \frac{P(1-\varepsilon)}{P} = 1 - \varepsilon$$

also in this case, so that, as a general result, the penalty $P_{\tilde{\mathrm{X}}}(Q^*)$ on Q^* due to incoherent crosstalk $\tilde{\mathrm{X}}$ reads

$$P_{\tilde{\mathrm{X}}}(Q^*) = -20 \log(1 - \varepsilon) \tag{5.134}$$

We see that $P_{\tilde{\mathrm{X}}}^{(Q^*)} < 0.5$ dB (1 dB) if $\varepsilon \leq 12.5$ dB (10 dB).

If N_{X} signals disturb the transmission channel out-of-band, and each interferer has an average power $\varepsilon_i P$, $i = 1, \ldots, N_{\mathrm{X}}$, at detection, then Equation 5.134 becomes

$$P_{\mathrm{X}}^{(Q^*)} = -20 \log\left(1 - \sum_{i=1}^{N_{\mathrm{X}}} \varepsilon_i\right) \tag{5.135}$$

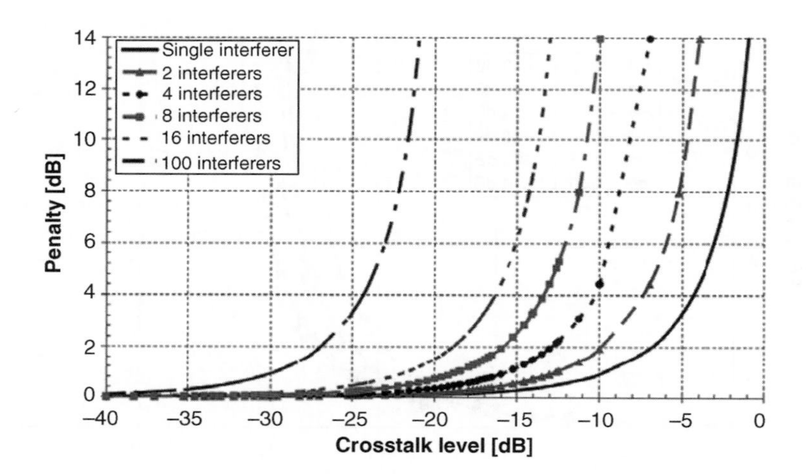

Figure 5.30 Penalty on Q^2 due to interchannel crosstalk. From Equations 5.134 and 5.135; infinite extinction ratio

Figure 5.30 shows Q^2- penalties due to interchannel crosstalk by single and multiple interferers when ER is infinite.

Other models show that, for a finite ER r, the intrachannel crosstalk penalty is given by [19,31,32]

$$P^{(Q^*)}_{(r,\tilde{X}),O} = -20\log\left(1-\varepsilon\frac{r+1}{r-1}\right) \tag{5.136}$$

It should be noted that Equation 5.134 is the limit of Equation 5.136, when $r \to \infty$. In the case of N_X equivalent interferers, assuming that the penalty is distributed according to a Gaussian function, (5.136) is generalized into the following form:

$$P^{(Q^*)}_{(r,\tilde{X}),O} = -10\log\left[1-\varepsilon^2 N_X Q^2\left(\frac{r+1}{r-1}\right)^2\right] \tag{5.137}$$

Figure 5.31 compares the results of Equation 5.136 with the sum of penalties from Equations 5.135 and 5.102 (we use the first one of Equation 5.102, since we saw that, in the case of incoherent crosstalk according to the simpler model, the penalty always corresponds to the situation when noise is independent from signal). The difference between the two curves is lower than 1 dB for ER$\gtrsim$15 dB (lower than 0.5 dB for ER$\gtrsim$18 dB]).

Crosstalk: General Remarks

In the investigation of crosstalk, the large number of available models hits the attention. This should not come as a surprise, since crosstalk is present everywhere in a transmission system.

From a system point of view, the simplest models such as Equations 5.129 and 5.135 seem to be sufficiently accurate. They give estimates for the ideal case of infinite ER, but we have just seen that, for reasonably small penalties, these can be summed together (see Figures 5.26 and 5.31). More refined models have mainly been discussed here because they are mentioned or used in the ITU-T sector.

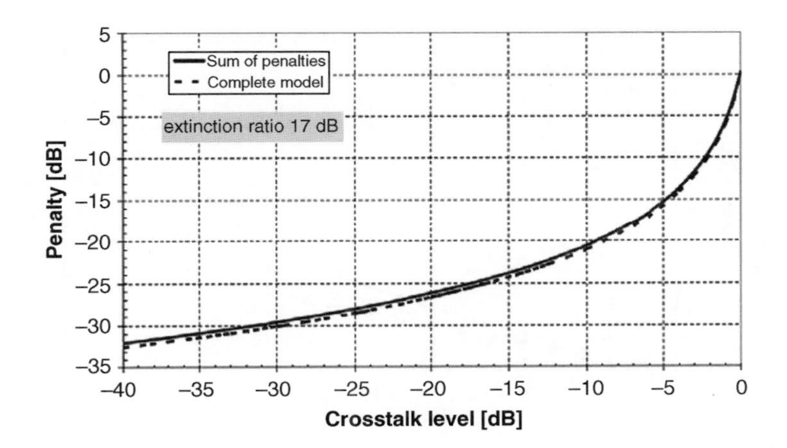

Figure 5.31 Comparison of penalties for incoherent crosstalk, as calculated from (5.136) (complete model) and the sum of Equation 5.135 with Equation 5.102 (sum of penalties); ER = 17 dB

Crosstalk suppression is important in complex optical networks, where the signal propagates through many nodes and accumulates interference effects from different components at each node. If the signal accumulates crosstalk from N_X interferers, each one having a crosstalk level ε, then Equations 5.129 and 5.135 give us an estimate of tolerances on a single perturbing effect. In an optically amplified network where a light path crosses 10 nodes, the total in-band crosstalk penalty on Q^* can be lower than 1 dB only if the crosstalk level for every coherent disturbance is $\lesssim$40 dB. On the other hand, if we consider the impact of a hundred signals interfering out-of-band with the transmission channel, the crosstalk level must be lower than 30 dB. *Crosstalk substantially depends on the number of network elements, line terminals, add/drop multiplexers and cross-connects (that contain filters, muxes, demuxes) crossed by the optical circuit.* Also, nonlinear optical effects may give rise to crosstalk; in that case, *nonlinear crosstalk depends on the transmission fiber, on the number of loaded DWDM channels and on their spacing*; see Section 5.3.2. A simple model for linear crosstalk due to an optical node can be formulated by considering that:

- a level ε_{if} of coherent crosstalk [due to the optical switch and to the mux in the $n-$th node following a demux in the $(n-1)-$node];
- a level ε_{adj} of incoherent crosstalk [due to the demux, from the two channels adjacent to the one under consideration];
- a level $\varepsilon_{\text{nadj}}$ of incoherent crosstalk [due to the demux, from the $(M-3)$ channels non adjacent to the one under consideration, M being the system load].

The corresponding penalty can be calculated from the relations in Equations 5.129 and 5.135:

$$P_X^{(Q^*)} = -10\log(1-2N\sqrt{\varepsilon_{\text{if}}})-20\log[1-2N\varepsilon_{\text{adj}}-N(M-3)\varepsilon_{\text{nadj}}] \\ \cong -10\log\{1-2N[\sqrt{\varepsilon_{\text{if}}}-2\varepsilon_{\text{adj}}-(M-3)\varepsilon_{\text{nadj}}]\} \tag{5.138}$$

It accounts for N coherent interferers and $N(M-1)$ incoherent effects; (5.138) foresees a penalty lower than 2 dB with $N = 20$ and $M = 40$, if $\varepsilon_{\text{if}} \lesssim -45$ dB and $\varepsilon_{\text{adj}} \approx \varepsilon_{\text{nadj}} \lesssim -38$ dB; compare Figure 5.32.

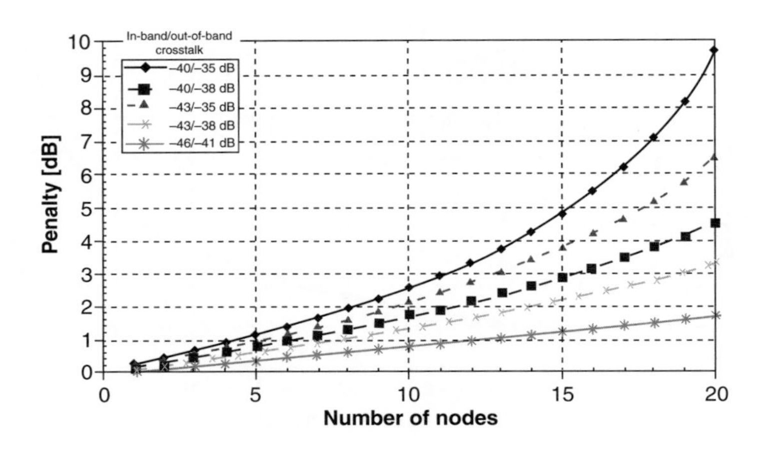

Figure 5.32 Q^2 penalty as a function of the number of optical nodes

5.2.4.10 Cumulative Filtering

In a DWDM system, a transmission channel crosses a number of spectrally selective elements. A cascade of many filters has degrading effects not only due to the attenuation of components, but also due to the fact they can have different characteristics, they can be misaligned in their peak wavelength, and they can cause problems due to insufficient isolation between adjacent channels.

The general problem is rather complex to investigate; however, it is possible to given suitable requirements on network components, like a tolerance mask for filters, in order to maintain the impact of cumulative filtering within acceptable limits; see Figure 5.33. Indicative values for mask parameters (at 10 Gbit/s bit rate) are the following, according to qualification requirements of TELCORDIA GR-1209:

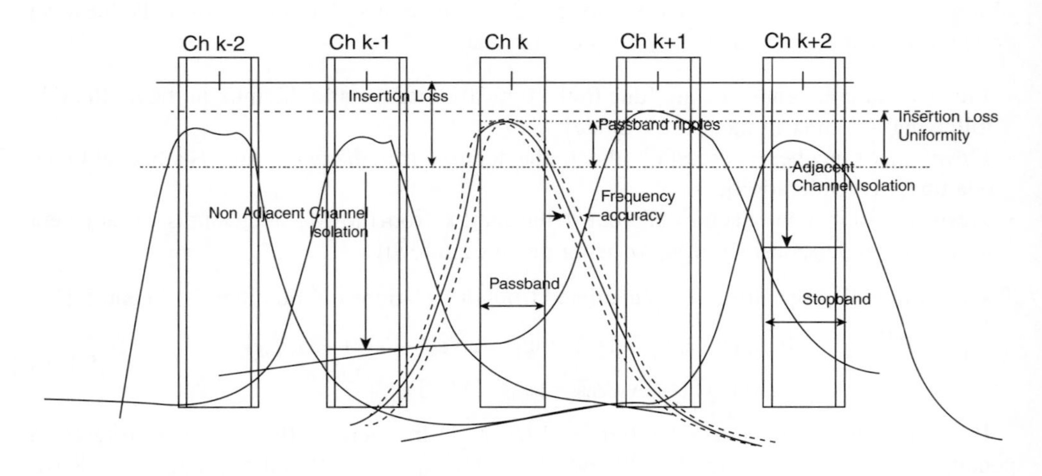

Figure 5.33 Filter tolerance mask

insertion loss	0.5–1.5 dB
insertion loss uniformity	0.5 dB
passband	>35 GHz at 1 dB >50 GHz at 3 dB <150 GHz at 20 dB <220 GHz at 30 dB
passband ripple	0.5 dB
peak frequency accuracy	±2.5 GHz (±0.02 nm)
tolerance with respect to ITU-T grid	±2.5 GHz (±0.02 nm)
wavelength thermal stability	<±2.5 GHz (<±0.02 nm)
polarization dependence	±0.6 GHz (±0.005 nm)
adjacent channel isolation	>25 dB (but typically, 33 dB)
nonadjacent channel isolation	>25 dB (but typically, 40 dB)
backreflection	>40 dB
stopband (backreflections)	±25 GHz (±0.2 nm)
CD contribution	±10–50 ps/nm
polarization-dependent loss	≤0.2 dB
PMD	≤0.2 ps

These parameters should be stable on a life cycle of 20–25 years. On the base of mask parameters it is possible to quantify suitable margins without the need of taking into account the presence of each filter explicitly.

The following three figures show the results of a simulation campaign on a metropolitan network performed within the European Project NOBEL: the nodes crossed by light paths are (R)OADMs, with a typical insertion loss of 8–12 dB. All refer to input conditions into an FEC stage. Figure 5.34 shows the evolution of the Q^*-factor as a function of the number of crossed OADMs: the modulation format is NRZ with external modulator (EAM); the CD is completely compensated over each span.

Figure 5.35 is similar, but for a duobinary (DB) signal and totally compensated dispersion. Since the DB format is robust to dispersion, it should be employed in the absence of compensation strategies for CD. However, in this second case the penalties have the same portrait displayed in Figure 5.35, but they increase by ~3.5 dB after 20 nodes. For partially compensated dispersion, the situation improves slightly, but anyhow it is impossible to cross more than 12 nodes with penalties lower than 2 dB. Finally, Figure 5.36 refers to a directly modulated (DML) NRZ format.

These figures can be commented on as follows:

a. for a ±10 GHz tolerance on the transmitter central frequency (a commercially available feature), penalties for NRZ-EAM format are lower than 1 dB to cross up to 12 nodes;
b. NRZ-EAM modulation takes advantage of a moderate filtering – for a laser frequency deviation smaller than 5 GHz, penalties are negative or vanishing to cross up to 20 nodes;[35]

[35] This robustness to filtering is confirmed by the possibility of doing vestigial sideband modulation (VSB) with NRZ formats: nearly half the spectrum is eliminated with modest quality degradation.

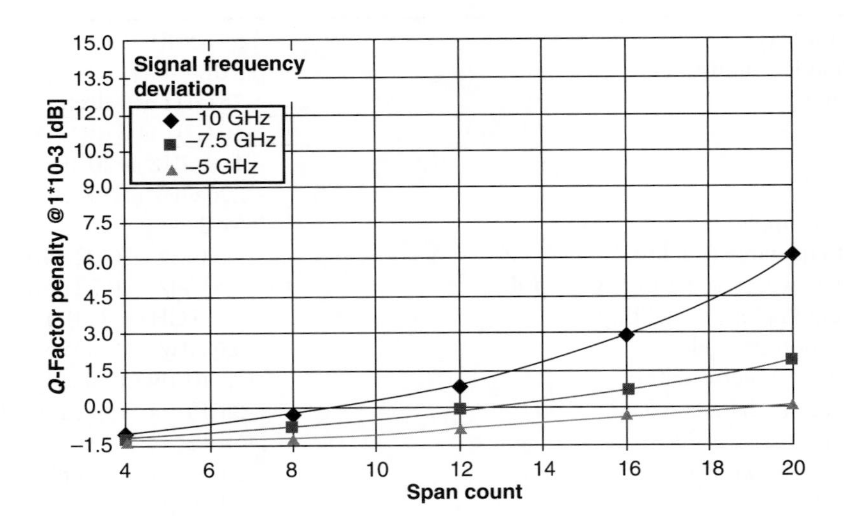

Figure 5.34 Q^*-factor penalty (insertion loss IL = 12 dB) as a function of the node number for an NRZ-EAM signal. Different values of frequency difference amongst the lasers are considered. Successive filters are shifted by ±12.5 GHz

c. the DB format also takes advantage of a moderate filtering, though less than the NRZ-EAM;
d. the DML signal is already filtered at the transmitter – further filtering at nodes induces transmission degradation;
e. The situation can be improved by adopting electronic equalizers at the receiver; see Section 5.4.

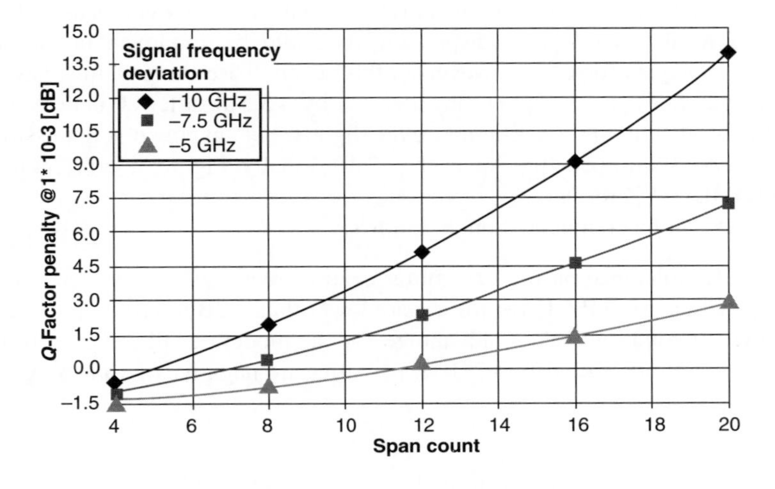

Figure 5.35 As Figure 5.34, but for a duobinary signal. IL = 12.7 dB

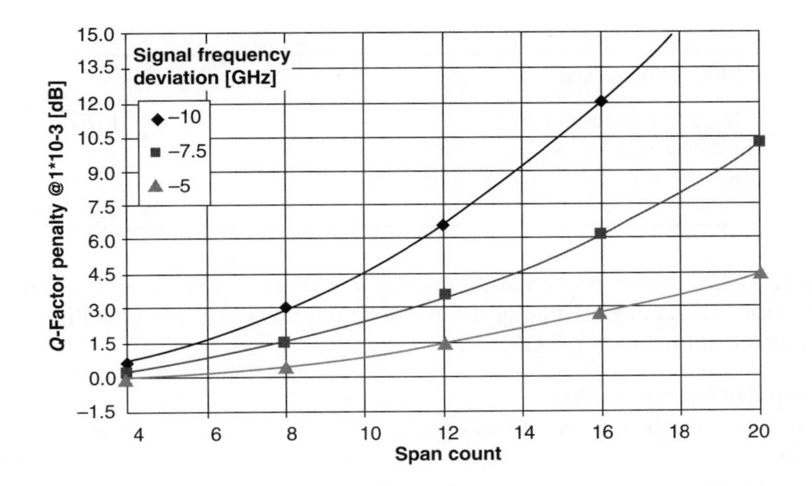

Figure 5.36 As for Figure 5.34, but for an NRZ-DML signal. IL = 12 dB

5.2.4.11 Impairments Due to Optical Polarization Effects

The performance of optical telecommunication systems can be affected by the optical polarization of the signal. A first impairment is caused by PMD. PMD is related to birefringence induced by internal and external mechanical stress and by elliptical deformation of the transmission fiber core. PMD is described by a differential group delay (DGD) between the two polarizations of optical pulses, after they have travelled a certain fiber length.

For newly installed fibers, the PMD coefficient is of order $0.1\,\text{ps km}^{-1/2}$ and the fiber contribution is negligible. However, lumped effects may arise from other optical components, such as isolators, WDM couplers, filters, and amplifiers [8]. PMD, the impact of which is negligible at low bit rates, can set severe limitations to transmission at bit rates equal or higher than 10 Gbit/s. Optical components available on the market have DGD of order 0.1 ps and are used to fabricate EDFAs with similar PMD properties [33].

A second, polarization-dependent effect, is dichroism or polarization-dependent loss (PDL). Experimental results have shown that a 0.12–0.13 dB/span PDL may produce noise floors in BER curves, with strong penalties that accumulate with path length and can limit it to ~6000 km [8].

Moreover, PDL can have a dynamical aspect, the time evolution of which causes fading related to SNR and BER fluctuations. In EDFAs, the active fiber itself can show a behavior dependent upon the polarization of the input signal (polarization hole-burning), because transition cross-sections for erbium ions are not isotropic in the polarization plane. Polarization-dependent hole-burning in the gain is due to the fact that, when signal saturates the active medium, erbium ions aligned to signal state-of-polarization (SOP) become more heavily saturated than others. In this situation, two probe beams polarized in phase and at square angles with the signal SOP exploit a different gain (PDG) [4]. The differential gain between the two polarizations substantially depends on signal SOP, while its variance is also affected by pump SOP. PDG is negligible for a single amplifier (typically, it is 0.1 dB/amp), but it can give rise to strong effects in long optically amplified chains.

Table 5.2 Form factors for the parameter A in Equation 5.139

Modulation format	A
NRZ rectangular – PIN	21–26
RZ Gaussian – PIN	25
RZ raised-cosine ($\alpha = 0.5$) – PIN	21
APD	44

PDL and PMD induce performance fluctuations in ultra long-haul systems; PDG is responsible for a direct degradation of the eye diagram. Polarization scrambling can be used to combat PDG and increase the OSNR [5].

Polarization-Mode Dispersion

This section is devoted to a short PMD-penalty analysis. In a digital NRZ system, pulse energy is separated into two components having orthogonal polarizations, which are mutually retarded by a DGD $\Delta\tau$. For *small* penalties, the following expression for the logarithmic penalty on the eye diagram is obtained [34]:

$$P_{\mathrm{EO}} = A\gamma(1-\gamma)\left(\frac{\Delta\tau}{T}\right)^2 \quad [\mathrm{dB}] \tag{5.139}$$

where the quantity γ $(0 \leq \gamma \leq 1)$ is the power division factor between the two polarizations, T is the half maximum pulse duration (the reciprocal of the bit rate B for the NRZ format), and A is a numerical parameter that depends on the optical pulse shape and the receiver characteristics; typical values for nonamplified receivers are $A \approx 12$–25 [22]; see Table 5.2.

In real transmission systems the penalty (5.139) changes at random with parameters γ and $\Delta\tau$, which assume a *stochastic nature*. The maximum tolerance to PMD is conventionally fixed to a 1 dB penalty; with analogy with fading effects in radiofrequency systems, such a 1 dB penalty is considered equivalent to *system outage*. Still conventionally, the maximum tolerable outage probability (OP) is fixed in $1/18\,000 \cong 5.6 \times 10^{-5}$ (30 min/year): this probability would correspond to a 4σ deviation for a Gaussian distribution function.[36]

The probability for having an eye-closure penalty ε larger than $\varepsilon_0 = 1$ dB is obtained by integrating the probability density function for ε itself. Under the hypotheses that:

1. the stochastic parameter $\Delta\tau$ is ruled by a *Maxwellian distribution function*

$$F(\Delta\tau) = \sqrt{\frac{2}{\pi}}\frac{1}{\alpha^3}(\Delta\tau)^2 \exp\left[-\frac{(\Delta\tau)^2}{2\alpha^2}\right]$$

   with average μ and variance σ^2, expressed by

$$\mu = 2\alpha\sqrt{\frac{2}{\pi}}, \qquad \sigma^2 = \alpha^2\left(3 - \frac{8}{\pi}\right)$$

2. the stochastic parameter γ is uniformly distributed over the interval $[0, 1]$
3. parameters $\Delta\tau$ and γ are statistically independent

[36] A tighter conventional requirement fixes such a maximum outage probability into 21 min/year.

Equation 5.139 leads to a simple exponential expression for the EOP distribution function F_{EOP}:

$$F_{\mathrm{EOP}}(\varepsilon) = \eta \mathrm{e}^{-\eta\varepsilon} \tag{5.140}$$

where

$$\eta = \frac{16T^2}{A\pi\langle\Delta\tau\rangle^2} \tag{5.141}$$

and $\langle\Delta\tau\rangle$ is the DGD mean value. Then the OP is

$$P_{\mathrm{oa}} \equiv \mathrm{Prob}\{\varepsilon \geq \varepsilon_0\} = \int_{\varepsilon_0}^{\infty} \eta \mathrm{e}^{-\eta\varepsilon}\, \mathrm{d}\varepsilon = \mathrm{e}^{-\eta\varepsilon_0} \tag{5.142}$$

where $\varepsilon_0 = 1$ dB, as considered above. The maximum tolerance on PMD can be obtained from (5.142), by equating it to the probability 1/18 000 established above for a maximum of 30 min/year outage. This fixes the value $\eta = 9.8$,[37] which, upon substitution into Equation 5.141, gives in turn

$$\frac{\langle\Delta\tau\rangle}{T} = \sqrt{\frac{16}{A\pi\eta}}$$

For a typical value $A = 25$:

$$\left.\frac{\langle\Delta\tau\rangle}{T}\right|_{\max} = k_{\max} = 0.14 \tag{5.143}$$

In the case of amplified receivers, $A \approx 44$ values have been given in the literature. Then the maximum tolerable PMD value (intended as the average DGD value caused by PMD) (5.143) decreases to the fraction $k_{\max} = 0.10$ that is generally assumed as a *typical acceptance criterion for PMD impairments*.

For a Maxwellian probability distribution function, one also has

$$\mathrm{Prob}\{\Delta\tau > 3\langle\Delta\tau\rangle\} \cong 4 \times 10^{-5}$$

which stays at the base of the tighter maximum outage of 21 min/year, discussed in footnote 36. The value $\Delta\tau = 3\langle\Delta\tau\rangle$ just corresponds to a 'deterministic' 1 dB penalty with Equation (5.139), in the worst case of power division $\gamma = 0.5$ given as a reference by ITU-T. Other values for the ratio $\Delta\tau_{\max}/\langle\Delta\tau\rangle$, corresponding to different outage probabilities, are shown in Table 5.3.

From the threshold relation Equation 5.143, a maximum transmission length, as limited by PMD, can be established. Introducing the fiber PMD coefficient

$$C_{\mathrm{PMD}} = \frac{\langle\Delta\tau\rangle}{\sqrt{L}}$$

where L is the total connection length, since the bit slot T is the reciprocal of the bit rate B,

[37] With the tighter convention of footnote 36, one would obtain the value $\eta = 10.0$.

Table 5.3 Outage probability thresholds from NOBEL analysis

Probability threshold P_{th}	Gaussian: standard deviations away from the mean (σ)	Maxwell: Ratio of maximum to mean (S)
10^{-3}	3.1	2.5
10^{-5}	4.3	3.2
10^{-7}	5.2	3.7
10^{-9}	6.0	4.2

Table 5.4 Transmission limit from PMD

C_{PMD} [ps km$^{-1/2}$]	L_{max} [km] at 10 Gbit/s	L_{max} [km] at 40 Gbit/s
0.50	400	25
0.25	1 600	100
0.10	10 000	625
0.05	40 000	2 500

Equations 5.141 and 5.143 lead to

$$B^2 L \leq B^2 L_{PMD} = \left(\frac{k_{max}}{C_{PMD}}\right)^2 \tag{5.144}$$

See Table 5.4 and Figure 5.37.

More generally, given the PMD contributions from the elements crossed by the light path under examination, say $\{PMD_{APP,1}, PMD_{APP,2}, \ldots, PMD_{APP,N_{MAX}}\}$, from ITU-T

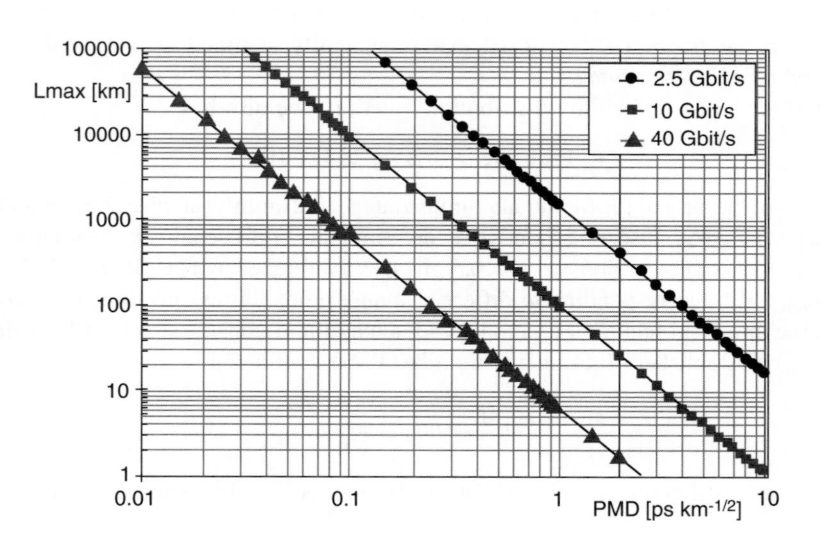

Figure 5.37 PMD-induced limits at various bit rates

Recommendations G.691 [27] and G.692 [28], the basic transmission limitation is fixed by the fact that the total accumulated PMD on the optical circuit, $\mathrm{PMD_{tot}}$, does not exceed the value $k_{\max} = 0.10$ (or 10 ps, at the bit rate of 10 Gbit/s). The PMD value on the circuit can be calculated on the basis of

$$\mathrm{PMD_{tot}} = \sqrt{\sum_j (\mathrm{PMD}_{\mathrm{APP},j})^2 + \sum_j (\mathrm{PMD}_{\mathrm{FIBRE},j})^2} \tag{5.145}$$

where

$$\mathrm{PMD}_{\mathrm{FIBRE},j} = C_j \sqrt{L_j} \tag{5.146}$$

in terms of the PMD coefficient C_j[ps km$^{-1/2}$] and of the length L_j of the jth fiber span.

DGD contributions from transmission equipments are typically around 1 ps for nodes and 0.5 ps for amplifiers;[38] as observed previously, in (5.145) each PMD quantity equals one-third of the corresponding maximum DGD value.[39]

The presence of a DGD between principal states of polarization distorts the optical pulses and then produces ISI, with an eye diagram closure. The corresponding EOP (5.139) can be translated into a penalty on the Q-factor; the reader should recall that (5.139) itself is valid for dominant first-order PMD; that is, when DGDs are shorter than the bit period.

System outage was defined as the condition when DGD exceeds a fixed threshold, which corresponds to a minimum acceptable Personick's factor Q_{lim}. It has been shown that the Maxwellian distribution allows us to obtain an OP P_{oa} due to the wings of the statistical distribution that exceeds such a threshold: the weaker the limit on DGD, the larger the OP will be. This relation can be investigated to set an expression for OSNR and Q^*-factor penalties, based on (5.139). The starting point is (5.140) for the probability distribution function of EOP. It is to be transposed from the original logarithmic formulation to the corresponding linear one, obtaining

$$F(Y) = F(P_{\mathrm{EOP}}) \left(\frac{\mathrm{d}Y}{\mathrm{d}P_{\mathrm{EOP}}} \right)^{-1}, \quad P_{\mathrm{EOP}} = f^{-1}(Y),\ Y \geq 1$$

where

$$Y = f(P_{\mathrm{EOP}}) = 10^{P_{\mathrm{EOP}}/10}, \quad P_{\mathrm{EOP}} = f^{-1}(Y) = 10 \log Y$$

Now:

$$\begin{aligned} F[f^{-1}(Y)] &= \eta \exp[-\eta f^{-1}(Y)] = \eta \exp(-10\eta \log Y) \\ &= \eta \exp\left[\ln 10 \frac{\log(Y^{-10\eta})}{\ln 10} \right] = \eta Y^{10\eta / \ln 10} \end{aligned}$$

and

$$\frac{\mathrm{d}Y}{\mathrm{d}P_{\mathrm{EOP}}} = \frac{\ln 10}{10} Y$$

---

[38] A more accurate version of Equation 5.145 is obtained weighting each deterministic contribution $(\mathrm{PMD}_{\mathrm{APP},j})^2$ by a factor $8/(3\pi)$.

[39] For other outage probabilities, see Table 5.3.

so that

$$F(Y) = \begin{cases} \dfrac{10}{\ln 10}\dfrac{\eta}{Y} Y^{-10\eta/\ln 10} & Y \geq 1 \\ 0 & \text{otherwise} \end{cases}$$

Taking into account that the penalty Y is interpreted as a quantity larger than unity, in the absence of excess noise F_{E} Equation 5.89 gives the relation $Q = Q_{\mathrm{R}}/Y$. This allows us to calculate the probability distribution function for the Q-factor as

$$F(Q) = F(Y)\left(\frac{\mathrm{d}Q}{\mathrm{d}Y}\right)^{-1}, \quad Y = Y(Q) \quad 0 \leq Q \leq Q_{\mathrm{R}}$$

In conclusion:

$$F(Q) = \begin{cases} \dfrac{10\eta}{Q \ln 10}\left(\dfrac{Q_{\mathrm{R}}}{Q}\right)^{-10\eta/\ln 10} & 0 \leq Q \leq Q_{\mathrm{R}} \\ 0 & \text{otherwise} \end{cases}$$

The outage probability P_{oa} is related to the fact that the DGD does exceed a maximum tolerable instantaneous $\Delta\tau_{\max}$; thence:

$$P_{oa} = \int_{\Delta\tau_{\max}}^{+\infty} F(\Delta\tau)\,\mathrm{d}(\Delta\tau) = \int_{0}^{Q_{\lim}} F(Q)\,\mathrm{d}Q = \left(\frac{Q_{\lim}}{Q_{\mathrm{R}}}\right)^{10\eta/\ln 10} \equiv (P_Q)^{10\eta/\ln 10} \tag{5.147}$$

where $Q_{\lim} = Q(\Delta\tau_{\max})$. The inversion of the preceding equation gives the penalty P_Q in terms of P_{oa}:

$$P_Q = (P_{\mathrm{oa}})^{\ln 10/10\eta}, \quad P_{\mathrm{SNR}} \equiv P_{Q^*} = -\frac{2\ln 10}{\eta}\log P_{\mathrm{oa}} \quad [\mathrm{dB}]$$

Finally, taking Equation 5.142 into account:

$$P_{\mathrm{PMD}}^{(Q^*)} = 2\varepsilon = 2A\gamma(1-\gamma)\left(\frac{\Delta\tau}{T}\right)^2 \quad [\mathrm{dB}] \tag{5.148}$$

In this approximation of first-order PMD, and coherently with the OSNR-Q Equation 5.71 and with Equations 5.86 and 5.89, one finds that the penalty $P_{\mathrm{PMD}}^{(Q^*)}$ has a negligible impact on the OSNR.

This analytic approach has a good accuracy in the calculation of the PMD-induced penalty (Figure 5.38) in correspondence to different P_{oa} values and for reasonable values of the baseline Q_{R} that, to this approximation, is not influenced by PMD. Figure 5.38 shows the behavior of the Q^2 penalty versus the average DGD, as calculated by Equation 5.148. Different values for the parameter A have been used and these are discussed in the literature; compare Table 5.2.

What has been discussed so far pertains substantially to the NRZ format. Indeed, one could expect PMD also depends on the duty cycle of the optical pulses: for RZ modulation, short optical pulses are more robust to PMD since their energy is more confined to the center of each bit slot and the pulses themselves have more room to spread during transmission, at least in the approximation of first-order PMD. The issue is further complicated by the fact that shorter

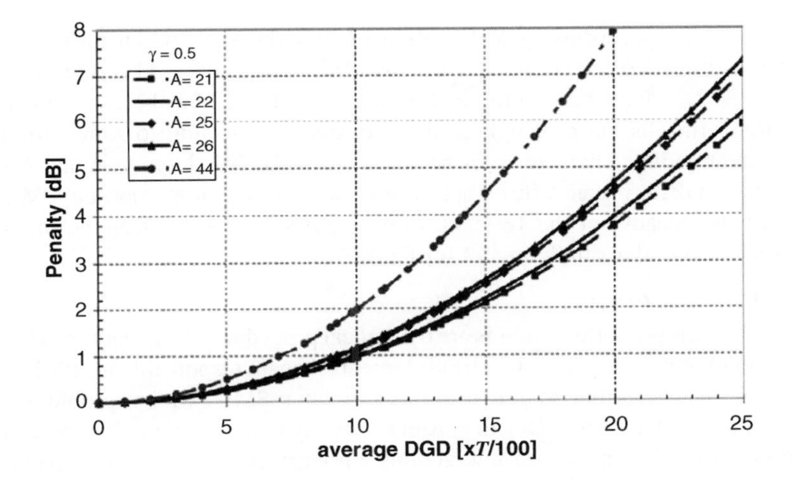

Figure 5.38 Q^* penalty calculated from Equation 5.148

pulses have broader spectra and are more exposed to *higher order* PMD. These considerations may indicate the existence of an optimum RZ pulse width with respect to PMD tolerance. It has been shown [35] that this is really the case; but since this optimum occurs so close to zero for realistic PMD values and that it is very shallow, one can expect a better performance as the RZ pulses become shorter and shorter, for reasonable operating conditions.

In line with this discussion, Figure 5.39 shows the results obtained for the PMD-induced penalty when Equation 5.148 is slightly modified as follows:

$$P_{\text{PMD}}^{(Q^*)} = 2A\gamma(1-\gamma)\sqrt{f_{\text{dc}}}\left(\frac{\Delta\tau}{T}\right)^2 \quad [\text{dB}] \tag{5.149}$$

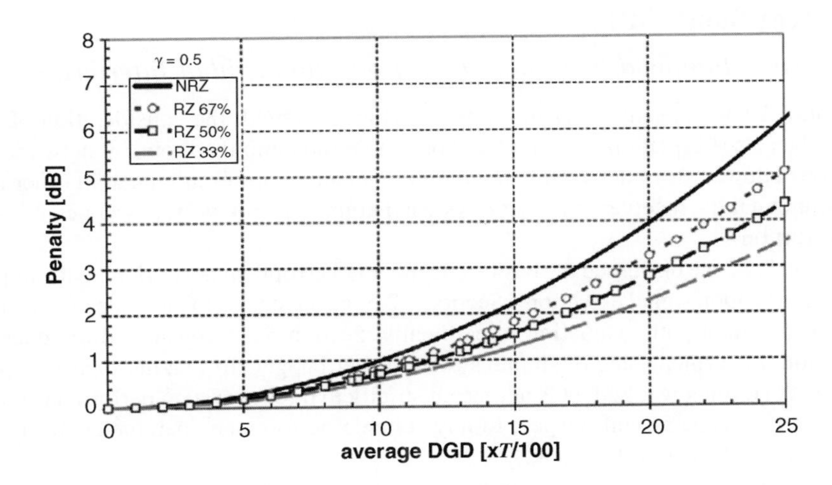

Figure 5.39 Tolerance to PMD for NRZ and RZ modulation formats

where f_{dc} is the optical pulse duty cycle. Equation 5.149 is a heuristic way to take care of the fact that shorter pulses are less influenced by ISI. NRZ ($f_{dc} = 1$) and RZ formats ($f_{dc} = 1/3,\ 1/2,\ 2/3$) have been compared; in all cases $A = 22$ has been assumed.

The CS-RZ format is more complicated to consider. According to some simulations, it should have intermediate characteristics between RZ with 50% duty cycle and NRZ. Other results from the NOBEL project affirm that CS-RZ should be the more robust to PMD amongst all OOK formats, because of the 180° bit-to-bit phase shift that reduces the impact of ISI. It is not possible to find a simple model for CS-RZ.

Dichroism (PDL and PDG)

The set-up and routing of optical circuits are dynamical procedures; light paths can be as long as thousands of kilometers and cross 100–1000 elements that are responsible for PMD- and PDL-induced impairments, further to the transmission fiber PMD. Major problems are SNR fluctuations, PMD, PDL, and PDG distortions, and dynamical crosstalk due to nonlinear propagation of optical signals with interactions amongst different DWDM channels, influenced by the gain dynamics of amplifiers and by wideband CD compensators. Since the PDL cumulated along an optical path increases with the number of components and that PDL contributions, expressed in logarithmic units, sum each other according to a quadratic law [36], it should be checked that PDL and PDG are lower than, say

$$\mathrm{PDL_{tot}} = \sqrt{\sum_{i=1}^{N_{\mathrm{tot}}} \mathrm{PDL}_i^2} \lesssim 1.5\ \mathrm{dB} \tag{5.150}$$

which is an estimate based on the results from Ref. [8], quoted above. The constraint due to PDL/PDG does not depend on bit rate. If PDL/PDG contributions are of order 0.10–0.15 dB/component, then (5.150) corresponds to a constraint $N = 100$–225 for the number of dichroic components along the light path.

5.3 System Bandwidth

5.3.1 System Bandwidth, Signal Distortion, Intersymbol Interference

In Section 5.2.1 we began analyzing system performance with the consideration of system power budgets; see Equations 5.1 and 5.2. For simple (nonamplified) optical networks, those expressions allowed us to determine the minimum channel power at launch in order to cope with link attenuation, receiver sensitivity, system margins and, possibly, power penalties due to signal distortion.

In Section 5.2.2 we discussed the basic concept of noise (optical noise along with amplifiers, electrical noise upon signal detection). Section 5.2.3 introduced us to the issue of performance parameters for light path evaluation rules; finally, Section 5.2.4 considered the question of transmission impairments and of simple system power budgets from a more general point of view. Now the term power budget is not meant strictly to refer to system performance in terms of power, but in general with respect to any suitable performance parameter; for example, the Q-factor, as chosen in this book.

It is now time to recognize that another important aspect of system performance is the 'band budget'; that is, whether the verification that the system time response, as determined by all system components, is sufficient to carry out the transmission signal without significant distortion. Signal distortion can derive from a limited time response of transmitters and receivers, as well as from a differential delay introduced by the transmission fiber amongst different frequency components of the optical pulse (CD) or between its two polarizations (PMD; see preceding section).

In very general terms, the system time response T_{sys} can be expressed as

$$T_{sys} = \sqrt{T_T^2 + T_F^2 + T_R^2} \tag{5.151}$$

where each component T_i represents, for instance, the fraction from 10% to 90% of the rise time in the impulse response of the ith subsystem to a step variation in its input conditions. In Equation 5.151, T_T refers to the transmitter, T_F to the transmission fiber (more generally, to the transit between system transmitter and receiver) and T_R to the receiver. The time T_F depends on light path characteristics at the physical layer (type of fiber, length, bit rate, dispersion management, etc.) while T_T and T_R are generally known to the system designer as input data in network engineering. The transmitter time response is determined by the electronics driving the optical source and by its electrical parameters. For laser transmitters to be used in optical telecommunications, T_T can be as low as ~0.1 ns; using external modulators, even shorter time responses can be reached, so that the 40 Gbit/s bit rate is technically mature and research and development are now considering bit rates up to 200 Gbit/s (these high speeds would imply T_T = 5–25 ps). The receiver time response T_R is fixed by the full electrical band at half maximum (FWHM) referred to the front-end of the receiver itself.

The rise time T_{sys} of a linear system is reciprocally related to the system band B_{sys}. This relationship can be described with the example of an RC circuit, used as an electrical filter (the receiver electrical filter, for instance). When a constant voltage V is applied at time $t = t_0$, the voltage across the capacitance varies in time according to

$$V_{out}(t) = V\left[1 - \exp\left(-\frac{t}{RC}\right)\right], \quad t \geq t_0$$

and the corresponding rise time T_{sys} is

$$T_{sys} \cong 2.2RC \tag{5.152}$$

Of course, the circuit transfer function is

$$H(f) = \frac{1}{1 + 2\pi i f RC} \tag{5.153}$$

where f denotes the frequency in the electrical bandwidth. According to (5.153), one obtains the familiar result for the system FWHM bandwidth B_{sys}:

$$B_{sys} = \frac{1}{2\pi RC} \tag{5.154}$$

Combining Equations 5.152 and 5.154 one finally obtains

$$T_{sys} = \frac{2.2}{2\pi B_{sys}} \cong \frac{0.350}{B_{sys}} \tag{5.155}$$

Equation 5.155 can be used to estimate the system time response; different filters will give slightly different results, but for fiber-optic systems the constant 0.350 gives a prudent estimate, which is frequently used in practice [24]. For the pulse shape to be returned at the filter (or system) output satisfactorily, the band B_{sys} should be large enough to accommodate all spectral components of the pulse; that is, to satisfy the condition $B_{sys}\tau \sim 1$, where τ is the pulse duration. Taking Equation 5.155 into account, one can draw the following estimate:

$$T_{sys} \cong 0.350\tau \tag{5.156}$$

This means that the system bandwidth $B_{sys} = T_{sys}^{-1}$ should be sufficient to accommodate for up to the third harmonic of the fundamental frequency τ^{-1}, which is related to pulse modulation (the source linewidth is assumed to be negligible with respect to the modulation bandwidth). The maximum bit rate B tolerated in the optical circuit is given by these considerations.

Generally speaking, the most critical component of the light path, from the bandwidth point of view, is the *receiver*, but this also depends upon the pulse duty cycle. In the RZ format each optical pulse for the mark is shorter than the bit slot $T = B^{-1}$ and its amplitude returns to zero *before* the end of the slot. On the contrary, with the NRZ format each mark stays activated over the whole bit period, its amplitude does not return to zero during sequences of marks, and the effective pulse width depends upon data pattern, while it is unchanged for RZ pulses. An advantage with the NRZ format is that the bandwidth associated with the bit flux is nearly one-half of the bandwidth needed by the RZ format, since on–off transitions are correspondingly less numerous. In other words, in the received RZ pulses there is a spectral component at $f = B$, whereas for NRZ pulses a frequency $f \cong B/2$ is present. So the NRZ format is generally preferred for bandwidth economics, while RZ (and its variants) are more robust with respect to receiver sensitivity, PMD, and optical nonlinearity.

For an RZ modulation format with 50% duty cycle, the bit rate B is given by the available electrical bandwidth B_e. Then the system time response must satisfy

$$T_{sys} \leq \frac{0.350}{B} \quad \text{(RZ)} \tag{5.157}$$

and Equation 5.155 implies that

$$B_{sys} \geq 3B_e \equiv 3B \quad \text{(RZ)}$$

At 10 Gbit/s, for instance, the system bandwidth must be at least 30 GHz.

On the other hand, for an NRZ code $B \cong 2B_e$, and then

$$T_{sys} \leq \frac{0.700}{B} \quad \text{(NRZ)} \tag{5.158}$$

Therefore, the condition

$$B_{sys} \geq 3B_e \equiv 1.5B \quad \text{(NRZ)}$$

states that NRZ transmission at 10 Gbit/s requires a system bandwidth in excess of 15 GHz.

This discussion gives an idea of the bandwidth B_{sys} needed to transmit in a satisfactory manner a narrow-linewidth signal (by which we mean a signal the line of which is much narrower than B) modulated at the bit rate B. At such a bit rate, a sinusoidal modulation substantially broadens by $2B$ the source linewidth.

Moreover, in the case of a square-wave modulation, an infinite number of harmonics of the fundamental frequency appear in the pulse spectrum. In practice, there can even be an appreciable level of the fifth harmonic, but it is reasonable to preserve the signal spectrum up to the third harmonic included, $B_{\text{sys}} \geq 3B_{\text{e}}$, with

$$B_{\text{e}} = \begin{cases} B & \text{RZ format} \\ B/2 & \text{NRZ format} \end{cases} \tag{5.159}$$

Equations 5.157 and 5.158 are empirical rules to check the system bandwidth with respect to signal bit rate and format. Coming back to Equation 5.151 for T_{sys}, one can consider the contribution $T_{\text{s}} = \sqrt{T_{\text{T}}^2 + T_{\text{F}}^2}$ associated with the optical signal at the receiver and the contribution T_{R} of the receiver itself. If $T_{\text{s}} < T_{\text{R}}$, then the system is limited by the receiver. On the contrary, if T_{s} is the dominant term, then system is limited elsewhere;[40] for instance, by all transmission effects that impact on the propagating signal. When the optical signal transition time is long, the bit symbol is at risk of spreading outside the bit slot, producing ISI and degrading BER.

The rise time T_{F} for the fiber is determined by dispersion effects, such as CD and PMD. In the preceding section we saw how the PMD-induced DGD between the two polarizations affects BER; that is, the Q-factor. Here the impact of CD will be considered.

When CD acts, the rise time T_{F} is determined by[41]

$$T_{\text{F}} = D\sigma_{\lambda}L \tag{5.160}$$

where D is the absolute value of the (average) CD coefficient on a given fiber span, σ_{λ} is the source spectral width and L is the fiber length. Expressing D in [ps/(nm km)], σ_{λ} in nanometers and L in kilometers, the fiber transition time – or group velocity dispersion (GVD) – is obtained in picoseconds. This originates from the fact that different frequency components of the optical signal travel the fiber in different times, because they are guided in a different fashion (waveguide dispersion) along a different optical path – i.e., the product of refractive index times the geometrical path – (material dispersion) (Section 4.4.1). The corresponding group delay τ_{GVD} will be tolerable if it does not exceed a fraction ε of the *time per symbol*[42]

$$T_{\text{F}} \equiv \tau_{\text{GVD}} \leq \frac{\varepsilon}{B} \tag{5.161}$$

at bit rate B. When symbols spread out of their slot, interference amongst different symbols will appear (ISI). For NRZ pulses and single longitudinal mode lasers, the ISI induces a power

[40] The rise time T_{sys} fixes the tolerance to distortions cumulated along the system. From Equations 5.157 and 5.158, dividing T_{sys} by the linewidth $\sigma_{\lambda} = (\lambda^2/c)\sigma_{\nu} \approx (\lambda^2/c)B$, a tolerance can be estimated, which is given by 7000–14 000 ps/nm at 2.5 Gbit/s, 450–900 ps/nm at 10 Gbit/s, and 25–50 ps/nm at 40 Gbit/s, depending on the modulation format.

[41] In single-mode fibers, only intramodal CD exists; it substantially depends both on waveguiding and on material properties.

[42] *Time per symbol* and *bit interval* are different quantities. They coincide only in the case of *binary* transmission of pulses having a single quantization level (for amplitude, phase, frequency, or polarization). However, for an M-ary transmission (where a pulse with M symbols is employed) the time per symbol equals a fraction $1/\log_2 M$ of the bit interval [24]. In this section, we shall be concerned with binary transmission only.

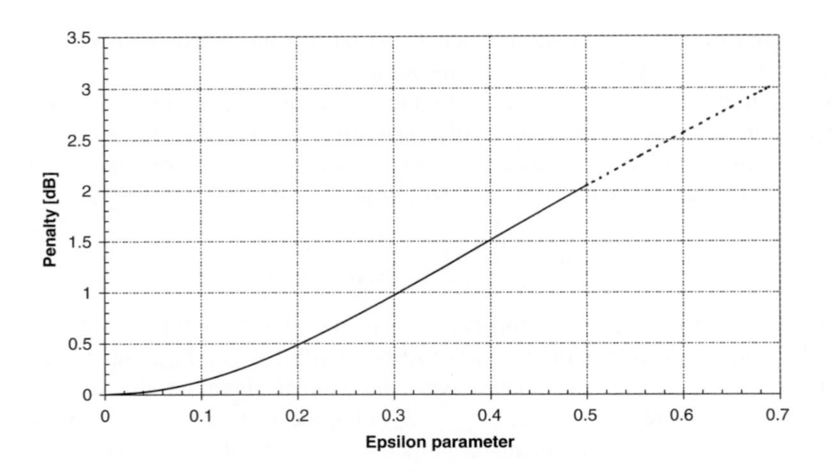

Figure 5.40 Power penalty with the 'epsilon'-model for ISI

penalty $P_{\mathrm{ISI}}^{(\mathrm{P})}$ given by the so called 'epsilon'-model [24] of ITU-T Rec. G.957 [37]:

$$P_{\mathrm{ISI}}^{(\mathrm{P})} = 5\log(1 + 2\pi\varepsilon^2) \quad [\mathrm{dB}] \tag{5.162}$$

According to the epsilon-model, the penalty (5.162) introduced by CD does not exceed 0.5 dB when $\varepsilon \leq 0.203$; for $\varepsilon \leq 0.306$, $P_{\mathrm{ISI}}^{(\mathrm{P})} \leq 1$ dB; and for $\varepsilon \leq 0.491$, $P_{\mathrm{ISI}}^{(\mathrm{P})} \leq 2$ dB [19,24] (Figure 5.40).[43] The quantity $P_{\mathrm{ISI}}^{(P)}$ was defined as the logarithmic value of the increase in power at the receiver needed to compensate for the decrease in peak power (and the decrease of the power integrated over the sampling slot B^{-1}) due to pulse spreading. This power decrease equals the ratio of the pulse duration σ_t with ISI, to the pulse duration $\sigma_{t,0}$ without ISI. Calling this ratio the pulse time-spread factor f_{b}:

$$f_{\mathrm{b}} = \frac{\sigma_t}{\sigma_{t,0}} \tag{5.163}$$

one can immediately write [24]

$$P_{\mathrm{ISI}}^{(\mathrm{P})} = 10\log f_{\mathrm{b}} \tag{5.164}$$

Now the relation linking (5.164) and the epsilon-model (5.162) will be discussed. This relation fixes the *link dispersion limit*. The group delay (5.160) depends on fiber dispersion D, fiber length L, and source spectral width $\sigma_{\lambda,0}$ in wavelength units (on the other hand, $\sigma_{v,0}$ will indicate the source spectral width in frequency units):

$$\tau_{\mathrm{GVD}} = DL\sigma_{\lambda,0} = D(\lambda)L\frac{\lambda^2}{c}\sigma_{v,0} \tag{5.165}$$

[43] To tell the truth, $P_{\mathrm{ISI}}^{(\mathrm{P})}$ depends slightly on the BER level; for a level 10^{-12}, the values for ε mentioned in the text are approximated to 0.2, 0.3, and 0.48 respectively.

for a signal at (peak) wavelength λ; c is the light speed in vacuum. The source spectral width $\sigma_{v,0}$ is related to its time coherence (i.e., its characteristic width $\sigma_{t,0}$ in the time domain) by the *reciprocity inequality* (it could also be called the *Fourier–Heisenberg relation*) [38]:

$$\sigma_{v,0}\sigma_{t,0} \geq \frac{1}{4\pi} \tag{5.166}$$

Condition (5.166) can be satisfied as an equality only by a Gaussian pulse.[44] The Gaussian pulse with peak intensity I_0 and (time) standard deviation $\sigma_{t,0}$ is written as

$$I(t) = I_0 \exp\left(-\frac{t^2}{2\sigma_{t,0}^2}\right) \tag{5.167}$$

with total power given by

$$P = \int I(t)\,\mathrm{d}t = I_0 \int_{-\infty}^{+\infty} \exp\left(-\frac{t^2}{2\sigma_{t,0}^2}\right) \mathrm{d}t = I_0\sqrt{2\pi}\sigma_{t,0} \tag{5.168}$$

Next, consider rectangular pulses of duration f_{dc}/B^{-1}, where f_{dc} is the duty cycle ($f_{dc} = 1$ for NRZ pulses and $0 < f_{dc} < 1$ for RZ pulses). Taking into account up to the fourth harmonic of the fundamental frequency in the spectrum of a rectangular pulse,[45] the reciprocity inequality (5.166) can be written as

$$\sigma_{v,0}\sigma_{t,0} = \frac{1}{\pi} \quad \text{(rectangular pulse at 10 \% band peak)} \tag{5.169}$$

Moreover, equating the power P of the Gaussian pulse (Equation 5.168) to the power of the rectangular pulse having the same amplitude and duration f_{dc}/B^{-1}, one has

$$P = I_0\sqrt{2\pi}\sigma_{t,0} = I_0 f_{dc}B^{-1} \Rightarrow \sigma_{t,0} = \frac{f_{dc}}{\sqrt{2\pi}B} \tag{5.170}$$

Then, from Equations 5.169 and 5.170 it can be drawn that

$$\sigma_{v,0} = \frac{1}{\pi\sigma_{t,0}} = \sqrt{\frac{2}{\pi}\frac{B}{f_{dc}}} \tag{5.171}$$

Recalling Equations 5.160 and 5.161, one finally obtains

$$\tau_{GVD} = D(\lambda)L\frac{\lambda^2}{c}\sqrt{\frac{2}{\pi}\frac{B}{f_{dc}}} \leq \frac{\varepsilon}{B} \tag{5.172}$$

Therefore:

$$D(\lambda)L \leq \sqrt{\frac{\pi}{2}\frac{c\varepsilon f_{dc}}{\lambda^2 B^2}} \quad \text{(narrow spectrum/high bit rate)} \tag{5.173}$$

[44] With respect to time, and then also to frequency.

[45] Because of the 'ringing' effect in the Fourier transform of a rectangular pulse.

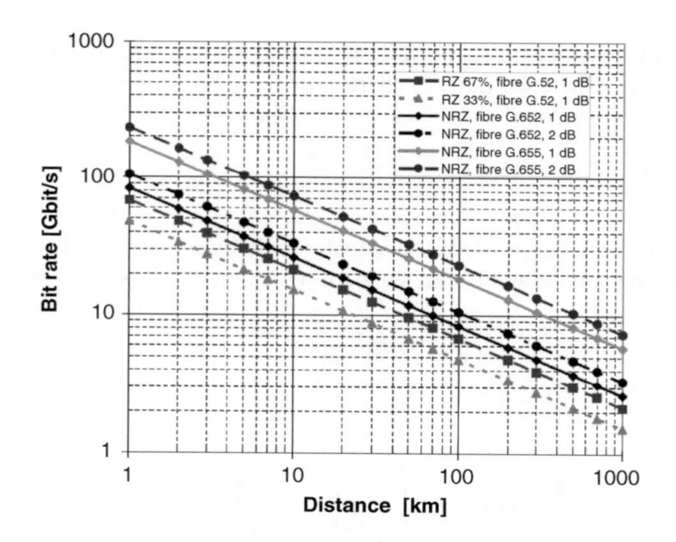

Figure 5.41 Dispersion limits for various kinds of links and narrow linewidth sources

This is the link *dispersion limit* for rectangular pulses limited by the reciprocity relation. At 10 Gbit/s and at a wavelength $\lambda = 1550$ nm, Equation 5.173 gives a dispersion limits of 70 km on G.652 fiber (assuming $D(\lambda) = 17\,\text{ps}/(\text{nm km})$), 600 km on G.653 fiber ($D(\lambda) \sim 2\,\text{ps}/(\text{nm km})$), and 340 km on G.655 fiber ($D(\lambda) \sim 3.5\,\text{ps}/(\text{nm km})$). All these values correspond to a 1 dB penalty with respect to the optimal transmission condition. If a 2 dB penalty can be tolerated at the receiver, then these previous values become 115 km, 960 km, and 550 km respectively. Figure 5.41 shows the results from Equation 5.173 for various bit rates, transmission formats, tolerated penalty, and span length.

Further, it should be observed that

$$\sigma_t = \sqrt{(\sigma_{t,0})^2 + \tau_{\text{GVD}}^2} \tag{5.174}$$

so that the pulse time-spread factor (Equation 5.163) can be written as

$$f_{\text{b}} = \frac{\sigma_t}{\sigma_{t,0}} = \sqrt{1 + \left(\frac{\tau_{\text{GVD}}}{\sigma_{t,0}}\right)^2} = \sqrt{1 + \left(\frac{D(\lambda)L\sigma_{\lambda,0}}{\sigma_{t,0}}\right)^2}$$

$$= \sqrt{1 + 2\pi\left[\frac{BD(\lambda)L\sigma_{\lambda,0}}{f_{\text{dc}}}\right]^2} = \sqrt{1 + 2\pi\left[\frac{BD(\lambda)L\lambda^2\sigma_{v,0}}{cf_{\text{dc}}}\right]^2} = \sqrt{1 + \left[\frac{2B^2D(\lambda)L\lambda^2}{cf_{\text{dc}}^2}\right]^2}$$

Thence:

$$P_{\text{P}}^{(\text{ISI})} = 5\log\left[1 + \left(\frac{\varepsilon}{B\sigma_{t,0}}\right)^2\right] = 5\log\left[1 + 2\pi\left(\frac{\varepsilon}{f_{\text{dc}}}\right)^2\right], \quad \varepsilon = BD(\lambda)L\sigma_{\lambda,0}$$

This latter expression is just Equation 5.162 for the epsilon-model, also incorporating the case of RZ formats.

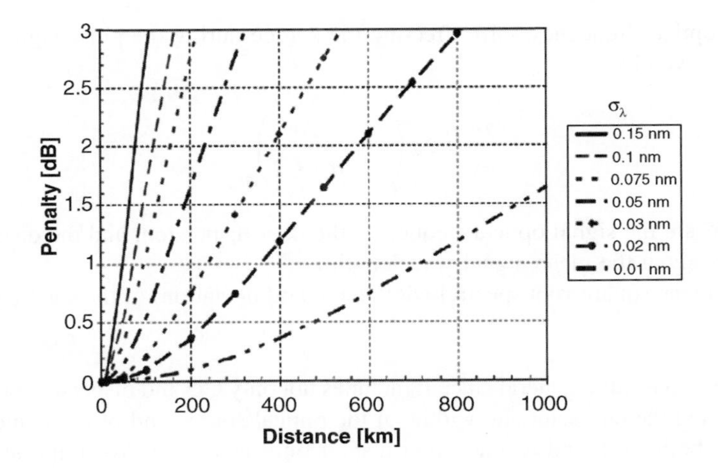

Figure 5.42 Dispersion limits for broadband sources

Combining Equations 5.160 and 5.161 one can also obtain the following relation:

$$BLD(\lambda)\sigma_{\lambda} \leq \varepsilon, \quad \text{(broad spectrum/low bit-rate)} \tag{5.175}$$

which is valid in the opposite limit of a broadband source, namely a source characterized by a product $\sigma_{\nu}\sigma_{t}$ much larger than the basic limit indicated by the reciprocity inequality.[46] Figure 5.42 shows some results calculated with Equation 5.175.

It is possible to give a sounder theoretical basis to the preceding estimates, for a deeper comprehension of the parameters describing the pulse width in the time and in the frequency domain, concurring with Equations 5.173 and 5.175. Following [24,40], the mean square-root duration σ_t of a generic pulse at the transmission fiber output is given by

$$\sigma_t^2(L) = \left(\sigma_{t,0}^2 - \frac{\lambda^2 D(\lambda) L \Delta\nu}{c\sqrt{2}}\right)^2 + \left(\frac{\lambda^2 D(\lambda) L}{c}\right)^2 \left[\frac{1}{(4\pi\sigma_{t,0})^2} + \sigma_{\nu,0}^2\right]$$
$$+ \frac{L^2}{8}\left\{\left(\frac{\lambda^2}{c}\right)^2 \left[S(\lambda) + \frac{2D(\lambda)}{\lambda}\right]\left[\frac{1}{(4\pi\sigma_{t,0})^2} + \sigma_{\nu,0}^2 + \frac{(\Delta\nu)^2}{2}\right]\right\}^2 \tag{5.176}$$

In this expression:

$\sigma_{t,0}$ is the mean square-root duration (standard deviation) of a Gaussian pulse at launch ($L = 0$);

[46] One can operatively define wide band and narrow band sources in connection with the adopted bit rate. Let Γ_{ν}, B, and f_{dc} be the source spectral width at 20 dB, the bit rate, and the duty cycle respectively. The source is called *wide band* if $\Gamma_{\nu} > 14B/f_{\mathrm{dc}}$; it is called *narrow band* if a $\Gamma_{\nu} < B/4f_{\mathrm{dc}}$. Compare ITU-T Recommendations G.663 and G.957 [39,37].

$\Delta\nu$ is the optical frequency shift affecting the source during the pulse signal, so that the chirp is given by

$$2\pi\left(\nu+\frac{t\Delta\nu}{\sigma_{t,0}\sqrt{2}}\right)$$

ν, D, and S are the signal optical frequency, the fiber dispersion, and the dispersion slope respectively at the mean signal wavelength λ;

$\sigma_{\nu,0}$ is the mean square-root spectral width (standard deviation) of the source in frequency units.

Equation 5.176 is quite general since it includes not only CD and dispersion slope, but also chirp effects and the characteristic widths of the optical source and of the launched pulse.

Neglecting both chirp and second-order dispersion, Equation 5.176 simplifies to

$$\sigma_t^2(L)=\sigma_{t,0}^2+\sigma_{\mathrm{D}}^2(L) \tag{5.177}$$

where the dispersion-induced broadening, which takes both pulse modulation and source spectrum into account, is

$$\sigma_{\mathrm{D}}(L)=\frac{\lambda^2 D(\lambda)L}{c}\sqrt{\frac{1}{(4\pi\sigma_{t,0})^2}+\sigma_{\nu,0}^2}=D(\lambda)L\sqrt{\left(\frac{\lambda^2}{4\pi c\sigma_{t,0}}\right)^2+\sigma_{\lambda,0}^2} \tag{5.178}$$

while, as in Equation 5.165:

$$\sigma_{\lambda,0}=\frac{\lambda^2}{c}\sigma_{\nu,0}$$

is the root-mean-square width (standard deviation) of the source in wavelength units. When the source width prevails, Equation 5.178 gives

$$\sigma_{\mathrm{D}}(L)=D(\lambda)L\sigma_{\lambda,0} \tag{5.179}$$

while for a highly coherent source one has

$$\sigma_{\mathrm{D}}(L)=\frac{\lambda^2 D(\lambda)L}{4\pi c\sigma_{t,0}} \tag{5.180}$$

Consider now an unchirped pulse train at bit rate B (bit slot B^{-1}). For an RZ format, the launched waveform has time duration equal to a fraction f_{dc} ($0<f_{\mathrm{dc}}<1$) of the pulse duration; for the NRZ synchronous format $f_{\mathrm{dc}}=1$. The relation

$$N\sigma_{t,0}=\frac{f_{\mathrm{dc}}}{\sqrt{2\pi}B} \tag{5.181}$$

(compare Equation 5.170) establishes that the root-mean-square duration of the rectangular pulse at launch is contained N times in the RZ bit window, as it is reduced by the duty cycle. The value of the numerical form factor N depends on the kind of pulse launched. Upon substituting Equation 5.181 back into Equation 5.178 one has

$$\sigma_{\mathrm{D}}(L) = \frac{\lambda^2 D(\lambda)L}{c}\sqrt{\left(\frac{\sqrt{2\pi}NB}{4\pi f_{\mathrm{dc}}}\right)^2 + \sigma_{\nu,0}^2} = D(\lambda)L\sqrt{\left(\frac{NB\lambda^2}{\sqrt{8\pi}cf_{\mathrm{dc}}}\right)^2 + \sigma_{\lambda,0}^2} \tag{5.182}$$

With the epsilon-model, already discussed above, one can tolerate a maximum dispersion value given by (compare Equation 5.161)

$$(\sigma_{\mathrm{D}})_{\mathrm{max}} \leq \frac{\varepsilon}{B} \tag{5.183}$$

As we saw in the previous, simplified, deduction, it is common usage [24] to take $N = 4$ ($N = 3.46$ contains in practice all the power of an NRZ rectangular pulse, whereas $N = 4$ contains 95.4% of the power of a Gaussian pulse). Using Equations 5.183 and 5.181 in Equation 5.182, one finds

$$D(\lambda)L \leq \frac{\varepsilon c}{\lambda^2 B\sqrt{\left(\sqrt{\frac{2}{\pi}}\frac{B}{f_{\mathrm{dc}}}\right)^2 + \sigma_{\nu,0}^2}} \tag{5.184}$$

For a highly monochromatic source:

$$D(\lambda)LB^2\lambda^2 \leq \sqrt{\frac{\pi}{2}}c\varepsilon f_{\mathrm{dc}} \tag{5.185}$$

Recall Equation 5.173. On the other hand, for a broadband source (see Equation 5.175):

$$D(\lambda)LB\lambda^2\sigma_{\nu,0} \leq c\varepsilon \quad \text{or} \quad D(\lambda)LB\sigma_{\lambda,0} \leq \varepsilon \tag{5.186}$$

This conclusion ascertains the equivalence between the two approaches to dispersion penalty that we have proposed.

For practical estimates, the root-mean-square width σ of the pulse (a quantity more prone to theoretical calculations) can be replaced with the full waveform width Γ at -20 dB, considering that, in the Gaussian approximation [19]

$$\Gamma \approx 6.0697\sigma \tag{5.187}$$

Typical values for the CD coefficient D are[47]

$$\bar{D}(\lambda = 1550\,\mathrm{nm}) = \begin{cases} 17\ \mathrm{ps/(nm\ km)} & \text{fibre G.652} \\ 2.5\ \mathrm{ps/(nm\ km)} & \text{fibre G.653} \\ 4\text{–}8\ \mathrm{ps/(nm\ km)} & \text{fibre G.655} \end{cases} \tag{5.188}$$

[47] For G.655 fiber, negative values for D are also possible.

For a given DWDM link, the dispersion penalty is not uniform for all associated light paths; that is, the different transmission channels. Indeed, the dispersion coefficient at wavelength λ can be written

$$\bar{D}(\lambda) = \bar{D}(\lambda_0) + S(\lambda - \lambda_0) \tag{5.189}$$

where $\bar{D}(\lambda_0)$ is a reference value – the one given by (5.188), for instance – and S is the dispersion slope; on the C- and L-bands, one typically has

$$S = \begin{cases} 0.060\ \text{ps}/(\text{nm}^2\ \text{km}) & \text{fibre G.652} \\ 0.070\ \text{ps}/(\text{nm}^2\ \text{km}) & \text{fibre G.653} \\ 0.058\text{–}0.085\ \text{ps}/(\text{nm}^2\ \text{km}) & \text{fibre G.655} \end{cases} \tag{5.190}$$

In summary, CD induces a power penalty by ISI that is simply described by Equations 5.162 and 5.164 of the epsilon-model:

$$P_{\text{ISI}}^{(\text{P})} = 5\log(1 + 2\pi\varepsilon^2) = 10\log\frac{\sigma'}{\sigma}$$

where σ' is the pulse duration with ISI and σ the pulse duration without ISI. For different types of source one has

$$DLB^2\lambda_{\text{S}}^2 = \sqrt{\frac{\pi}{2}}c\varepsilon f_{\text{dc}} \quad \text{(narrow linewidth)} \tag{5.191}$$

$$DLB\lambda_{\text{S}}^2\sigma_\nu = c\varepsilon \quad \text{or} \quad DLB\sigma_\lambda = \varepsilon \quad \text{(broadband)} \tag{5.192}$$

Since for an infinite extinction ratio Equation 5.78 implies

$$\frac{Q'}{Q} = \frac{HP'}{\sigma'}\frac{\sigma}{HP} = \frac{P'\sigma}{P\sigma'} = \frac{K^2\sigma'^2\sigma}{K^2\sigma^2\sigma'} = \frac{\sigma'}{\sigma} \tag{5.193}$$

then the ISI-induced penalty on $Q^* \equiv 10\log Q^2$ is given by

$$P_{\text{ISI}}^{(Q^*)} = 2P_{\text{ISI}}^{(\text{P})} = 10\log(1 + 2\pi\varepsilon^2) \quad [\text{dB}] \tag{5.194}$$

Thence, from Equation 5.191 one obtains

$$\begin{aligned} DL &= \sqrt{\frac{\pi}{2}}\frac{cf_{\text{dc}}}{B^2\lambda_{\text{S}}^2}\varepsilon = 156.5\frac{f_{\text{dc}}}{\boldsymbol{B}^2(\Lambda/1.550)^2}\varepsilon \quad [\text{s/m}] \\ &= 1.565\times10^5\frac{f_{\text{dc}}}{\boldsymbol{B}^2(\Lambda/1.550)^2}\varepsilon \quad [\text{ps/nm}] \end{aligned}$$

where practical units has been introduced for bit rate $\boldsymbol{B}$ [Gbit/s] and wavelength Λ [μm].

This equation gives penalties that are too pessimistic for unchirped modulated signals. To obtain a more reasonable tolerance to CD-induced ISI, it is convenient to adopt a condition which is intermediate between $B_{\text{sys}} \geq 1.5B$, discussed with Equation 5.158 for a generic NRZ

Table 5.5 Tolerance to CD, from Equation 5.194. Signal wavelength 1550 nm

$P_{\mathrm{ISI}}^{(Q^*)}$ [dB]	10 Gbit/s [ps/nm]		40 Gbit/s [ps/nm]	
	NRZ	RZ 50%	NRZ	RZ 50%
0.5	±355	±178	±22	±11
1.0	±519	±259	±32	±16
1.5	±654	±327	±40	±20
2.0	±779	±390	±48	±24
2.1	**±800**	**±400**	**±50**	**±25**
2.5	±899	±450	±56	±28
3.0	±1017	±508	±64	±32
3.5	±1134	±567	±70	±35
4.0	±1255	±627	±78	±39

pulse, and $N = 4$ used for Gaussian pulses [24]. Since N enters the relevant formulae as a factor, we will take the geometric mean of these two values, namely $N = 2.45$, and obtain

$$DL = \frac{\sqrt{8\pi}}{2.45}\frac{cf_{\mathrm{dc}}}{B^2\lambda_S^2}\varepsilon = 2.555 \times 10^5 \frac{f_{\mathrm{dc}}}{\boldsymbol{B}^2(\Lambda/1.55)^2}\varepsilon \quad [\mathrm{ps/nm}] \qquad (5.195)$$
$$\boldsymbol{B}\ \text{band}\ [\mathrm{Gbit/s}],\ \Lambda\ \text{wavelength}[\mu\mathrm{m}]$$

The tolerances given by Equations 5.194 and 5.195 are listed in Table 5.5.[48]

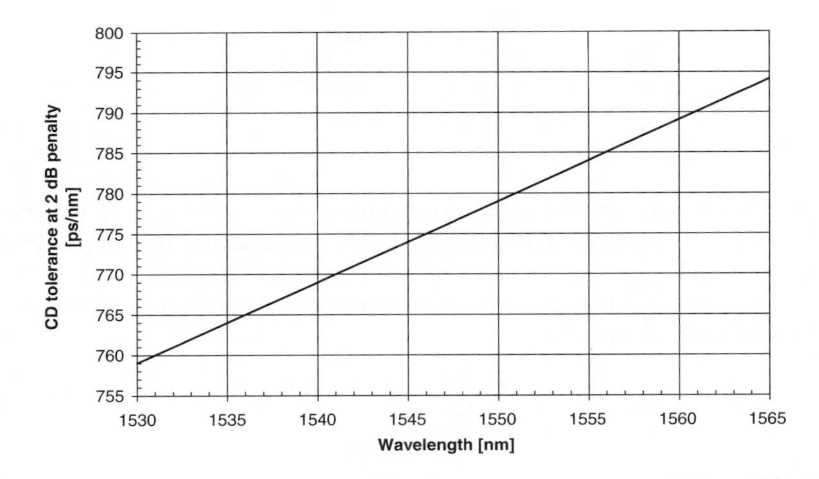

Figure 5.43 Tolerance to CD (at 2 dB Q^2 penalty) versus signal wavelength. The behavior is linear and corresponds to a variation of 1 ps/nm

[48] The bold row in Table 3.5 refers to the CD tolerance at 2 dB penalty, as commonly quoted in the literature.

Finally, this simple model implies that the dependence of the CD tolerance DL upon signal wavelength is linear in the C-band, as shown in Figure 5.43. We stress that the epsilon-model described here is simple, analytical, and capable of giving accurate values for the CD-induced ISI penalty on the quality parameter $Q^* \equiv 10 \log Q^2$.

Dispersion strategies are discussed in Appendix 5.7.

5.3.2 Fiber-Optical Nonlinear Effects

Since in DWDM long-haul transport systems we are interested in transmitting a bunch of optical channels at large distances without regeneration, we must use high levels of optical power, both as aggregate power (say, 20 dBm) and as single tributary power (say, 3 dBm/channel). Such power levels available at launch and at the amplifier's output can give rise to nonlinear interactions in the transmission fiber and amongst the transmitted channels.

In approaching optical nonlinear effects, it is convenient first to introduce two parameters related to the physical size over which such effects manifest themselves. The first parameter is the *effective length* L_{eff}:

$$L_{\text{eff}} = \frac{1-e^{-\alpha L}}{\alpha} \tag{5.196}$$

referred to a fiber the length of which is L and the attenuation of which is α.[49] The second parameter is the *effective area* A_{eff} that substantially coincides with the area of the fiber mode field radius (it approximately coincides with the fiber core geometric area):

$$A_{\text{eff}} = \frac{\left[\iint\limits_F I_S(r,\theta) r \, dr \, d\theta\right]^2}{\iint\limits_F I_S^2(r,\theta) r \, dr \, d\theta} \cong \pi w_c^2 \tag{5.197}$$

where $I_S(r,\theta)$ is the transverse intensity distribution of the signal guided in the fiber, on its cross-section F (r and θ are polar coordinates on the fiber cross-sectional plane); w_c is the *mode field radius* of the guided wave.

In the second place, the fiber nonlinear behavior is expressed frequently by means of its *nonlinear coefficient* γ, defined as

$$\gamma = \frac{2\pi}{\lambda} \frac{n_2}{A_{\text{eff}}} \quad [\text{W}^{-1}\,\text{km}^{-1}] \tag{5.198}$$

[49] A fiber having attenuation A [dB] has a linear attenuation α given by

$$-A = \log(e^{-\alpha L}) \Rightarrow A = 10\alpha L \log e = \alpha L = \frac{A}{10 \log e}.$$

Therefore:

$$\alpha = \frac{A}{10 L \log e} \cong 0.230 \frac{A\,[\text{dB}]}{L\,[\text{m}]}\,[\text{m}^{-1}].$$

in terms of the signal wavelength, the effective area, and the nonlinear refractive index n_2 of the fiber glass; its measurement is not simple (depending on the specific condition of silica glass from fiber type to fiber type) and covers a somewhat large range of $n_2 = 2.2 \times 10^{-8}$ to $n_2 = 3.4 \times 10^{-8}$ $\mu m^2/W$. The nonlinear refractive index is the macroscopic quantity responsible of optical nonlinear effects.

As far as the specific nonlinear effects are concerned, we shall mention:

- stimulated Brillouin scattering (SBS), on each channel;[50]
- stimulated Raman scattering (SRS), amongst different channels;
- the Kerr effect, and specifically:
  - self-phase modulation (SPM), on each channel
  - cross-phase modulation (XPM), amongst different channels
  - XPM with ASE, on each channel;
- FWM, amongst different channels.

In this section these various effects will be discussed briefly, as long as they produce penalties on the system performance.

5.3.2.1 Brillouin Effect

The optical power backreflected into the Stokes line depends on signal modulation format and bit rate. The Brillouin gain coefficient is given by [41]

$$\tilde{g} = \tilde{g}_{B}\left\{\frac{1}{\varepsilon} - \frac{B}{\varepsilon^{2}\Delta\nu_{B}}\left[1 - \exp\left(\frac{\Delta\nu_{B}}{B}\right)\right]\right\}, \quad \varepsilon = \begin{cases} 2 & \text{for ASK(NRZ) and FSK} \\ 1 & \text{for PSK} \end{cases} \tag{5.199}$$

In the preceding expression, $\tilde{g}_B$ and $\Delta\nu_B$ are respectively the peak gain nonlinear coefficient and the SBS band, which, in silica fibers, have the values $\sim 2.2 \times 10^{-11}$ m/W and 15–20 MHz respectively. The threshold condition for SBS is given by

$$P_{th} \approx \frac{21 b A_{eff}}{\tilde{g} L_{eff}}\left(1 + \frac{\Delta\nu_{S}}{\Delta\nu_{B}}\right), \qquad L_{eff} = \frac{1 - \exp(-\alpha^{(p)} L)}{\alpha^{(p)}}$$

where P_{th} is the launched signal power that pumps SBS, $1 \leq b \leq 2$ is a factor depending upon the relative polarizations of pump and Stokes waves, and L_{eff} is the effective length (5.196) at the pump wavelength, the attenuation of which is $\alpha^{(p)}$.

The factor $1 + (\Delta\nu_S/\Delta\nu_B)$ shows that SBS can be contrasted by broadening the source linewidth (dithering) and, possibly, concatenating transmission fibers having a different frequency shift ν_B (having a different GeO_2 content). If dithering were absent, the SBS threshold would be around 1 mW power level; with a $\Delta\nu_S = 200$ MHz dithering, the threshold increases to $\sim$14 mW. Since all DWDM systems employ dithering, we shall not be concerned with SBS anymore.

[50] Single-channel effects are obviously present also in a WDM system, affecting every optical carrier.

5.3.2.2 Raman Effect

The Raman effect alters the spectral profile of the DWDM comb. In the presence of a massive channel load, the ratio

$$R(\lambda) = \frac{P'(\lambda)}{P(\lambda)}$$

(which can be larger or smaller than unity) between channel power at wavelength λ, with and without SRS, can be approximated by [42]

$$R(\lambda) = \frac{\beta P_{\mathrm{agg}} L_{\mathrm{eff}} (\lambda_{\mathrm{max}} - \lambda_{\mathrm{min}}) \exp[\beta P_{\mathrm{agg}} L_{\mathrm{eff}} (\lambda - \lambda_{\mathrm{min}})]}{\exp[\beta P_{\mathrm{agg}} L_{\mathrm{eff}} (\lambda_{\mathrm{max}} - \lambda_{\mathrm{min}})] - 1} \tag{5.200}$$

where

$$\beta = \frac{g_{\mathrm{R}}}{2 A_{\mathrm{eff}} \Delta\lambda_{\mathrm{R}}}$$

for the Raman gain coefficient $g_{\mathrm{R}} = (6\text{–}7) \times 10^{-14}$ m/W, the Raman band $\Delta\lambda_{\mathrm{R}}$ (corresponding to about 15 THz) [41] of the aggregate DWDM power P_{agg} and the minimum and maximum load wavelengths λ_{min} and λ_{max}. Assuming that ER is infinite, $R(\lambda)$ directly transforms into a power penalty:

$$P_{\mathrm{SRS}}^{(\mathrm{P})}(\lambda) = 10 \log R(\lambda) \tag{5.201}$$

This simple model supplies a linear estimate on a logarithmic scale. It underestimates the penalty charging 'blue' channels (in the short wavelength wing of the C-band) and overestimates the impact in 'red' channels. Its merit is to give an expression for the global power imbalance over the DWDM comb.

As long as it is formulated, this argument applies to a single (amplified) span. However, each line amplifier restores the optical power at levels capable of exciting SRS and therefore the effect of $R(\lambda)$ cumulates along an amplifier chain. In this model approximation, this means that in Equation 5.201 the effects in dB over each amplified span sum up each other. This fact can be grasped using Equation 5.200 with a re-definition for L_{eff} that is convenient for a fiber length L with amplifiers spaced by l:

$$L_{eff} = \frac{1 - e^{-\alpha L}}{\alpha} \cdot \frac{L}{l} \tag{5.202}$$

With respect to the original definition (Equation 5.196), Equation 5.202 substantially states that the presence of a number $N_{\mathrm{s}} = L/l$ of cascaded spans multiplies the value of the effective length by the same number: after each amplifier, the optical power raises to the launch value again and so nonlinear effects are present again. Figure 5.44 shows the entity of Raman power imbalance due to SRS in the case of a single span.

The worst penalty (the one affecting the shortest wavelength channel; say, channel 1) can be calculated in an alternative way. In a WDM system with N channels spaced $\Delta\lambda$ apart, the power fraction $p_1(k)$ that is transferred from the shortest wavelength channel, having power P_{ch}, to the kth channel is given by [19]

$$p_1(k) = g_{\mathrm{R}} k \frac{\Delta\lambda}{\Delta\lambda_{\mathrm{R}}} \frac{P_{\mathrm{ch}} L_{\mathrm{eff}}}{2 A_{\mathrm{eff}}}$$

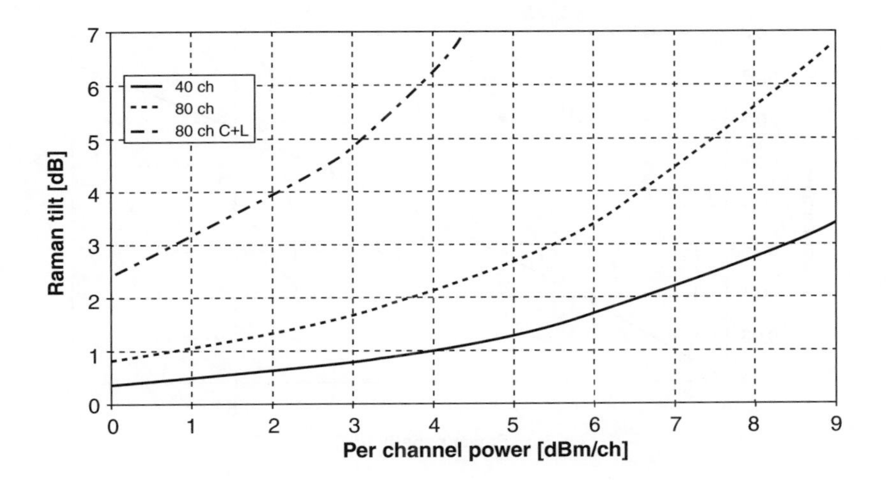

Figure 5.44 Raman power imbalance for 40- and 80-channel C-band systems and for 80-channel systems over the C + L bands. Single span

Thence the total power extracted from channel 1 is[51]

$$p_1 = \sum_{k=1}^{N-1} p_1(k) = g_R \frac{\Delta\lambda}{\Delta\lambda_R} \frac{P_{ch} L_{eff}}{2A_{eff}} \frac{N(N-1)}{2}$$

The power penalty affecting channel 1, therefore, is

$$P_{SRS}^{(P)} = -10\log(1-p_1)$$

To reduce this power penalty due to SRS below ρ dB, one must ascertain that

$$p_1 \le 1-10^{-\rho/10}$$

This is plotted in Figure 5.45.

As an example, for $\rho \tilde{<} 0.5$ dB, $p_1 \tilde{<} 0.1$; that is:

$$bP_{ch}N(N-1)\Delta\lambda L_{eff} = bP_{agg}\Delta\lambda_{agg}L_{eff} \le \frac{4p_1\Delta\lambda_R A_{eff}}{g_R} \equiv 4\times 10^4 \text{ mW nm km} \tag{5.203}$$

$P_{agg} = NP_{ch}$ is the total launched power and $\Delta\lambda_{agg} = (N-1)\Delta\lambda$ is the transmission band; b is a numerical factor which takes care of whether CD exists ($b \cong 1/2$) or not ($b \cong 1$): the presence of a residual CD reduces the effect of SRS because different channels rapidly get a phase mismatch that detune each other.[52] Equation 5.202 states that to maintain the power penalty due to SRS within 0.5 dB in a DWDM system with 40 channels 100 GHz apart, the maximum

[51] The fraction of optical power coupled from channel 1 to another channel is of order $2/N$ the total power subtracted from channel 1 itself. Generally, this crosstalk effect can be neglected.

[52] Expressions (5.203) and (5.202) once more show that SRS penalties cumulate along amplifier chains.

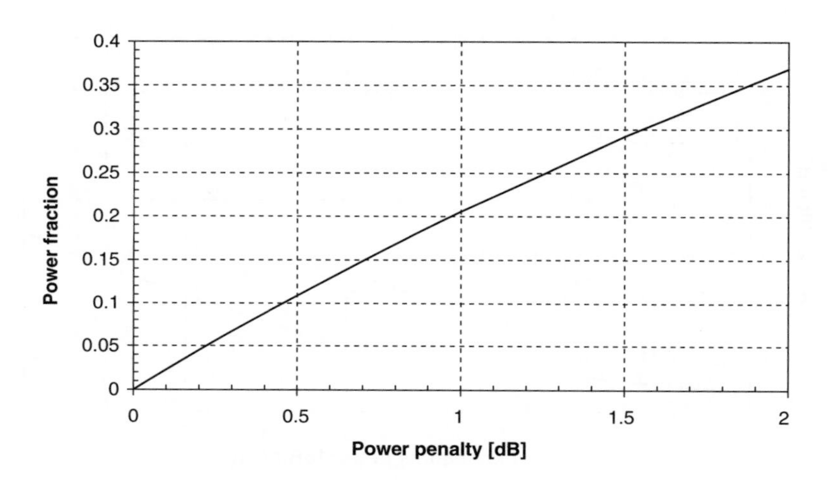

Figure 5.45 Relation between power penalty and fraction of power extracted from channel 1 due to SRS

channel power is ~3 mW/channel (namely, ~4.7 dBm/channel). For an 80-channel system with 50 GHz spacing the maximum channel power is one-half of the previous one: ~1.5 mW/channel (1.8 dBm/channel). These estimates hold in the presence of CD (G.652 fibers, $b \cong 1/2$ in Equation 5.202). Without CD ($b \cong 1$ in Equation 5.202, as for G.653 fibers and, in part, for G.655 fibers) the Raman threshold is 3 dBm lower (Figure 5.46).

The estimates given by these simple models are in reasonable agreement with dedicated numerical simulations.

In this discussion of SRS, we spoke of *power penalties*; in this particular case, $P_{\mathrm{SRS}}^{(P)}$ is relevant because it can be used for a power imbalance estimate.

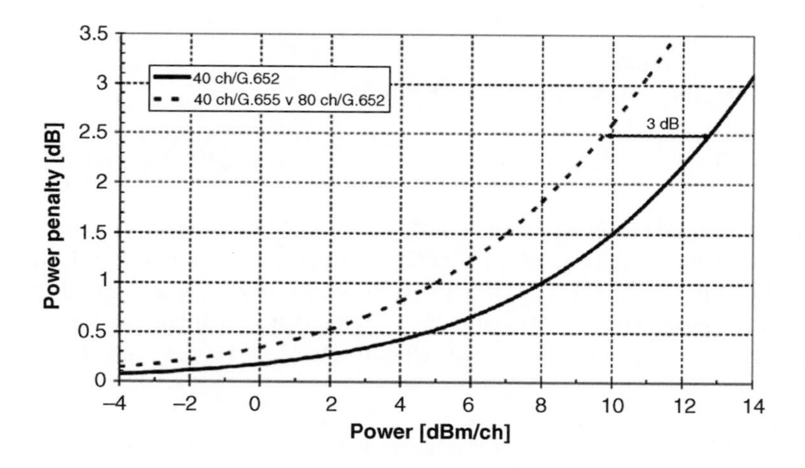

Figure 5.46 Maximum SRS-induced power penalty

5.3.2.3 Four-Wave Mixing

The FWM effect is due to a nonlinear coupling that may happen amongst different channels in the transmission fiber. When the angular frequencies of channels i, j, and k satisfy the condition

$$\omega_i + \omega_j - \omega_k = \omega_x$$

they produce nonlinear crosstalk on channel x with a crosstalk power P_{ijk} given by

$$P_{ijk} = \left(\frac{\omega_x n_2}{3cA_{\text{eff}}} L_{\text{eff}}\right)^2 \eta_{ijk} d_{ijk} P_i P_j P_k \mathrm{e}^{-\alpha L} \tag{5.204}$$

In the preceding expression, d_{ijk} is the *degeneracy factor*

$$d_{ijk} = \begin{cases} 3 & i = j \\ 6 & i \neq j \end{cases}$$

while η_{ijk} is the coupling efficiency [43]

$$\eta_{ijk} = \frac{\alpha^2}{\alpha^2 + \Delta\beta^2}\left\{1 + 4\frac{\exp(-\alpha L)\sin^2\left(\frac{\Delta\beta L}{2}\right)}{[1-\exp(-\alpha L)]^2}\right\} \tag{5.205}$$

for a fiber of length L and attenuation α, at a phase mismatch $\Delta\beta$ amongst the coupled optical carriers

$$\Delta\beta = \beta_i + \beta_j - \beta_k - \beta_x$$

$\Delta\beta$ is influenced by fiber dispersion: FWM is phase-matched only near the zero-dispersion wavelength λ_0 for a given span. The phase mismatch between the optical carriers, due to their spectral allocation with respect to λ_0, may be calculated by means of the CD coefficient D with the equation [43]

$$\Delta\beta(\omega_i, \omega_j, \omega_k) = \frac{\lambda_S^2}{2\pi c}\left|(\omega_i - \omega_k)(\omega_j - \omega_k)\right|\left[D(\lambda_S) + (|\omega_i - \omega_k| + |\omega_j - \omega_k|)\frac{\lambda_S^2}{4\pi c}\left.\frac{\partial D(\lambda)}{\partial \lambda}\right|_{\lambda=\lambda_S}\right] \tag{5.206}$$

Previous expressions are simplified since they do not take polarization effects and pump depletion into account. Moreover, they refer to a single (amplified) span.

In the case of an optical route with q OAs in chain, the FWM efficiency η_{ijk} (Equation 5.205) is generalized to [43,44]

$$\eta_{ijk}^{(q)} = \frac{1}{q^2}\frac{\alpha^2}{\alpha^2 + \Delta\beta^2}\frac{\sin^2\left(\frac{q\Delta\beta l}{2}\right)}{\sin^2\left(\frac{\Delta\beta l}{2}\right)}\left\{1 + 4\frac{\exp(-\alpha l)\sin^2\left(\frac{\Delta\beta l}{2}\right)}{[1-\exp(-\alpha l)]^2}\right\} \tag{5.207}$$

in which l is the amplifier spacing; thence the power P_{ijk} of the intermodulation product is given by Equation 5.203 times q^2 [8].

In extremely succinct terms, the features of FWM in DWDM systems are the following ([8] (Chapter 7), [19]):

- on G.653 fiber spans with uniform dispersion characteristics, FWM penalties reach intolerable levels when one WDM channel is exactly allocated at the zero-dispersion wavelength λ_0; then one should
  - use the L-band
  - use a nonuniform channel spacing;
- on G.652 fiber, FWM is substantially negligible (unless very high bit rates are employed, beyond 10 Gbit/s);
- on G.655 fiber, FWM is substantially negligible if the channel power is moderated.

Hence, counteractions to FWM are based on unequal channel spacing (UCS) and on the optimization of amplifier spacing and channel power (see chapter 7 in Ref. [8]), or, more drastically, on the use of the L-band. The penalty due to FWM can be evaluated supposing that all amplifiers work at transparency and all channels have the same power and are uniformly separated over the transmission spectrum; for a worst-case evaluation, a maximum efficiency is assumed:

$$\eta_{ijk,\,\max} = \frac{\alpha^2}{\alpha^2 + \Delta\beta^2}\left\{1 + \frac{4\exp(-\alpha l)}{[1-\exp(-\alpha l)]^2}\right\}$$

Hence, the crosstalk power on channel x amounts to

$$P_{\mathrm{FWM}} = \sum_{\substack{\omega_i+\omega_j+\\ -\omega_k=\omega_x}} P_{ijk} = \left(\frac{\omega_x n_2}{3cA_{\mathrm{eff}}} L_{\mathrm{eff}}\right)^2 (P_x)^3 \sum_{\substack{\omega_i+\omega_j+\\ -\omega_k=\omega_x}} \eta_{ijk,\max} d_{ijk}$$

where sums are performed over all combinations of optical frequencies such that condition $\omega_i + \omega_j - \omega_k = \omega_x$ is fulfilled.

The penalty estimate follows the general scheme for crosstalk, as discussed in Section 5.2.4.9, where the crosstalk level is now

$$\varepsilon = \frac{P_{\mathrm{FWM}}}{P_x} \approx (P_{\mathrm{ch}} N_{\mathrm{s}})^2$$

in terms of channel power P_{ch} and number of amplified spans N_{s}; see Equation 5.202. This behavior implies a constraint of the type $P_{\mathrm{ch}} N_{\mathrm{s}} \approx$ const for a given level of penalty. The more accurate result that can be obtained from simulations can be cast in the form

$$P_{\mathrm{ch}}^2 N_{\mathrm{s}} = P_{\mathrm{FWM}}^2 \quad [\mathrm{lin.}], \qquad P_{\mathrm{ch}} = P_{\mathrm{FWM}} - 5\log N_{\mathrm{s}} \quad [\mathrm{dB}] \tag{5.208}$$

and the parameter P_{FWM} is obtainable from the simulation campaign and a statistical data analysis [45]. The FWM penalty on Q^* will be lower than, say, 1 dB as long as Equation 5.208 is satisfied. This *scaling law* (and similar others we will see for SPM and XPM) are very useful in system design, since the knowledge of the parameter P_{FWM} and the choice of a channel power level P_{ch} immediately allows the designer to get the maximum number of transmission spans N_s to maintain the FWM penalty lower than 1 dB on that optical circuit.

For systems designed accurately, FWM penalties are negligible; but for optical circuits travelling on C-band systems on G.653 fibers, those routes will require an ad hoc analysis. Moreover, FWM may also act as an intrachannel FWM (IFWM) effect in very high bit-rate systems (40 Gbit/s and beyond); see Appendix 5.7.

5.3.2.4 Self-Phase Modulation

SPM causes positive chirp to optical pulses and contributes to pulse broadening in the normal dispersion regime ($\beta'' > 0$, or $\lambda < \lambda_0$).[53] The behavior in the anomalous dispersion ($\beta'' < 0$, or $\lambda > \lambda_0$) regime may be more complex [45]. To investigate SPM versus GVD, it is useful to define the *dispersion length*

$$L_D = \frac{T_0^2}{|\beta''(\omega_S)|} = \frac{\pi c T_B^2}{2\lambda_S^2 |D(\lambda_S)|}$$

where $T_0 \approx T_B/2$ is the launch pulse duration (T_B is the bit slot), and the *nonlinear length*

$$L_{NL} = \frac{1}{\gamma P_1} = \frac{\lambda_S A_{eff}}{2\pi n_2 P_1}$$

where P_1 is the peak pulse power and γ is given by Equation 5.198.

In terms of the ratio between the two:

$$N^2 = \frac{L_D}{L_{NL}} = \frac{\gamma}{|\beta_2|} T_0^2 P_1$$

one can conclude (see Figure 5.47):

- for $N \gg 1$, GVD effects overwhelm SPM and pulses may envisage compression;
- for $N \approx 1$, SPMN and GVD are comparable and (soliton-like) pulses tend to be stable;
- for $N \ll 1$, SPM dominates and produces pulse amplitude modulation.

For bit rates of 10 Gbit/s and beyond, SPM limits the transmitted power to few dBm/channel. With the above notations, after a propagation distance L the pulse broadening due to SPM/GVD can be written as

$$\frac{T(L)}{T_0} = \sqrt{1 + \sqrt{2}\frac{L_{eff} L}{L_{NL} L_D} + \left(1 + \frac{4}{3\sqrt{3}} \frac{L_{eff}^2}{L_{NL}^2}\right) \frac{L^2}{L_D^2}} \tag{5.209}$$

[53] $\beta''(\omega)$ is the second derivative of the propagation constant β as a function of the optical pulsation ω.

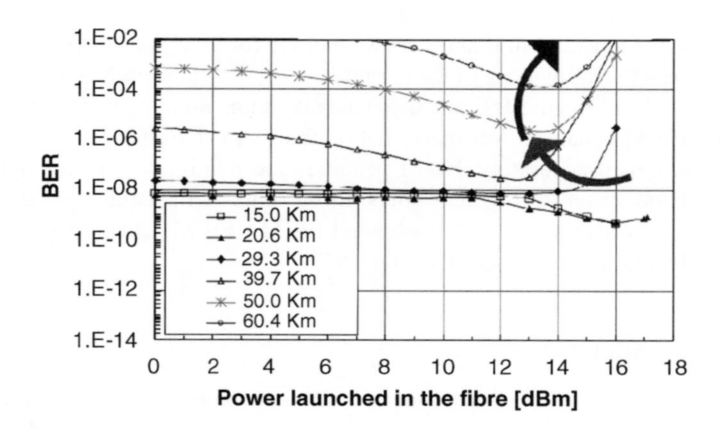

Figure 5.47 The SPM-dominated regime gradually passes through into the GVD-dominated regime, at different launched powers

Recalling what was said about ISI in Section 5.3.1, a power penalty can be calculated as in Equation 5.164:

$$P^{(P)}_{(SMP)} = 10\log\left[\frac{T(L)}{T_0}\right] = 5\log\left[1+\sqrt{2}\frac{L_{\text{eff}}L}{L_{\text{NL}}L_{\text{D}}}+\left(1+\frac{4}{3\sqrt{3}}\frac{L_{\text{eff}}^2}{L_{\text{NL}}^2}\right)\frac{L^2}{L_{\text{D}}^2}\right]$$

Using Equations 5.196 and 5.202, the SPM-induced power penalty can be calculated both for single-span and multispan cases, but the calculus would require a deep knowledge of pulse and fiber parameters that enter the characteristic lengths L_{NL} and L_{D}.

An alternative way to proceed is that of setting another scaling law as we did for FWM. Assuming a simple accumulation of penalties span after span, we arrive at a constraint of the type $P_{\text{ch}}N_{\text{s}} \approx$ const for a given level of penalty. This result is confirmed by numerical simulations (Figure 5.48). Therefore, one has

$$P_{\text{ch}}N_{\text{s}} = P_{\text{SPM}} \quad [\text{lin.}], \qquad P_{\text{ch}} = P_{\text{SPM}} - 10\log N_{\text{s}} \quad [\text{dB}] \tag{5.210}$$

P_{SPM} being a design parameter that can be determined for various penalty values; when Equation 5.210 is satisfied, the penalty on Q^* is lower than, say, 1 dB.

Various authors have experimentally shown that in systems adopting a *final optimization full-compensation scheme* (FOCS):

$$P_{\text{SPM}} \approx -10\log\left(\frac{B\,[\text{Gbit/s}]}{1\ \text{Gbit/s}}\right) + 31.5\ \text{dBm}, \quad \text{per}\, B \le B_{\text{lim}} \tag{5.211}$$

The model is accurate as far as the system is substantially limited by SPM; that is, when $L_{\text{NL}} \ll L_{\text{D}}$. In the situation $L_{\text{NL}} \sim L_{\text{D}}$ when SPM and GVD effects become comparable, the

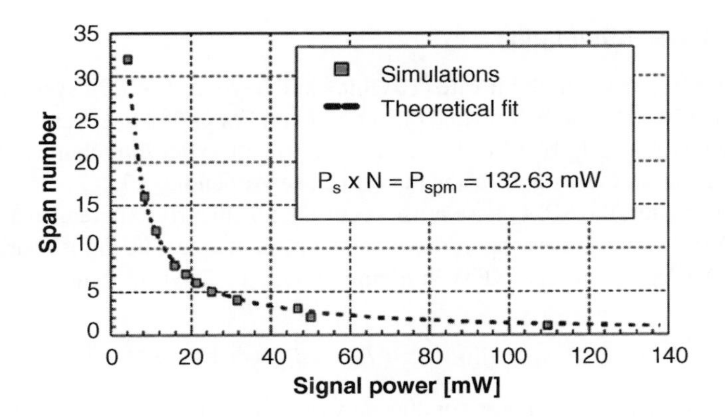

Figure 5.48 Maximum channel power versus span number [45]

parameter P_{SPM} rapidly decreases. Starting from $L_{\mathrm{NL}} \sim L_{\mathrm{D}}$ one obtains the following upper limit for the bit rate:

$$B_{\mathrm{lim}} \sim D^{-1/2}$$

For a dispersion coefficient of 17 ps/(nm km), B_{lim} amounts to 22 Gbit/s; for a dispersion parameter of 4.5 ps/(nm km), $B_{\mathrm{lim}} \sim 0.3\ \mathrm{Tbit/s}$.

The scaling law (Equation 5.210) can also be extended to other dispersion compensation schemes. For example, in the *distributed undercompensation scheme* (DUCS), this scaling law still holds with the transformation:

$$P_{\mathrm{max}}^{\mathrm{DUCS}} = KP_{\mathrm{max}}^{\mathrm{FOCS}}$$

At 10 Gbit/s the scale factor $K = 2.2$, showing a good improvement in terms of reachable distance or usable channel power. To exploit these theoretical advantages, the DUCS must be configured suitably. An optimum residual dispersion value of $D_u = 760 \pm 20$ ps/nm is determined, in correspondence to which the dispersion per transmission span is simply

$$u = \frac{D_{\mathrm{u}}}{N_{\mathrm{s}}}$$

The eye-diagram closure may become rather strong along the optical circuit. Hence, it is convenient to adopt a conservative approach to be extended to the whole light path:

$$P_{\mathrm{SPM}}^{(\mathrm{EOP})} = \begin{cases} 1\ \mathrm{dB} & \text{if } N_{\mathrm{s}} \leq N_{\mathrm{s}}(P_{\mathrm{SPM}}) \\ +\infty & \text{otherwise} \end{cases}$$

Using (5.89) one can equivalently state

$$P_{\mathrm{SPM}}^{(Q^*)} = \begin{cases} 2\ \mathrm{dB} & \text{if } N_{\mathrm{s}} \leq N_{\mathrm{s}}(P_{\mathrm{SPM}}) \\ +\infty & \text{otherwise} \end{cases} \qquad (5.212)$$

ISPM is amongst the most important optical nonlinear effects in high bit-rate systems; see Appendix 5.7.

5.3.2.5 Cross-Phase Modulation

XPM can put limitations to high bit-rate (10 Gbit/s and beyond) DWDM systems with narrow channel spacing (50 GHz and below) and on G.653 fiber. For G.652 fibers and channel spacing of 100 GHz, XPM is negligible; but for manifesting itself as an intrachannel XPM (IXPM) effect in systems at 40 Gbit/s and faster than this, see Appendix 5.7.

On multispan routes, the SPM phase noise accumulates linearly, whereas in a WDM system the SPM and XPM phase noises superpose in a quadratic manner (i.e., noise variances sum up linearly) [8]. One can approach XPM in a similar way to SPM, obtaining a corresponding scaling law:

$$P_{\mathrm{ch}}N_{\mathrm{s}} = P_{\mathrm{XPM}} \quad [\mathrm{lin.}], \qquad P_{\mathrm{ch}} = P_{\mathrm{XPM}} - 10\log N_{\mathrm{s}} \quad [\mathrm{dB}] \tag{5.213}$$

where P_{XPM} is a design parameter for the network that can be obtained from numerical simulations and statistical data analysis of the infrastructure.

The problem of the interplay between dispersion and nonlinear effects for the high bit rate (>40 Gbit/s) is elaborated in Appendix 5.7.

5.3.3 Optical Transients

In transparent networks the elimination of express traffic transponders may couple different OMSs together. Sudden spectral load variations on an OMS can trigger optical instabilities in adjacent line systems; the power excursions at the input of line amplifiers propagate the instability along the whole network. A typical case is as follows. Consider two OMSs (with a similar channel consistency) converging towards a network node N: the OMS1 comes from west and crosses N towards east; OMS2 is issued north; and it is also due east through N along the same DWDM system as OMS1. Suppose OMS2 is broken (by a fiber cut, for instance): the sudden change in system load also impacts the performance of OMS1.

Indeed, faults in a DWDM network, capable of producing sudden load changes, can also affect residual channels. Because of the Raman effect on the DWDM comb and spectral hole-burning acting in EDFAs, load changes may cause important optical power excursions, even with EDFAs having fast gain control mechanisms (Figure 5.49).

In contrast to intrisic amplifier dynamics, having characteristic times of order *tens of microseconds*, these power excursions can persist for longer times, until channel equalization is reached in OXCs or in EDFAs themselves. These equalization times may last as long as a *few seconds*.

EDFA impulse response consists of three distinct regions: (a) initial perturbation, (b) relaxation oscillations, and (c) final steady state. The pattern is quite general:

- the deeper the saturation, the shorter the relaxation constant;
- the heavier the spectral load, the shorter the relaxation constant;
- the stronger the load variation, the longer the relaxation constant.

The amplifier dynamics are approximately described by the law [46]

$$P(t) = P_{\mathrm{ss}}\left[\frac{P(0)}{P_{\mathrm{ss}}}\right]^{\exp(-t/\tau_{\mathrm{c}})} \tag{5.214}$$

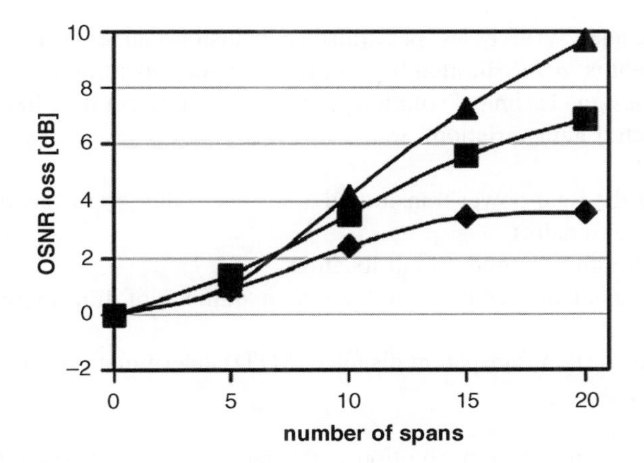

Figure 5.49 OSNR degradation for sudden spectral load changes (3 dB, diamonds; 6 dB, squares; 9 dB, triangles). Worst-case estimate

where $P(0)$ and P_{ss} are the optical power levels at the start and end of the transient and τ_c is the effective lifetime of the upper laser level in the erbium-ion energy levels, averaged along the amplifier length; for a single-channel EDFA, τ_c can be shorter than 10 μs. EDFAs must have suitable control mechanisms (feedback on pump power, in the first place) to moderate the transitory effects. Pump-power-controlled EDFAs can manage up to 6 dB input power excursions to an output power excursion as low as 0.5 dB within ~10 μs[54] [46]. Smarter EDFAs are commercially available that are capable of stabilizing input power variations as large as 15 dB into an output excursion of 1 dB within ~200 μs.

In amplifier chains, this behavior is further exacerbated. The chain effect gives rise to faster and stronger oscillations; also in this case, a common qualitative pattern can be recognized:

- in region (a), the total gain linearly increases with time – gain and power at the chain output increase with the number N of devices;
- in region (b), the first power overshoot is reached in a time proportional to N^{-1}, with a slope proportional to $N-1$.

In a 10-amplifier chain, the time constant may reduce to 1–2 μs. These considerations give a rule of thumb to estimate transient margins: an input load variation of 6 dB (15 dB with smarter EDFAs) can produce, at worst, an output transient of $\sim N/2$ dB (N dB) downstream of a cascade of N amplifiers; compare Figure 5.49.

[54] The variation in spectral load L_c, expressed in decibels, refers to the corresponding variation in total power. As an example, if in a set of $N_{tot} = 40$ channels (with same power per channel) only $N_{sop} = 10$ will survive, the final aggregate power is a quarter of the initial power,; that is, 6 dB lower. Therefore:

$$L_c = 10 \log\left(\frac{N_{sop}}{N_{tot}}\right)$$

A more accurate analysis gives less pessimistic conclusions, under the assumption that every EDFA gain transient is *locally* limited by gain feedback circuits; this is the state of the art in advanced EDFA design. Technical solutions to control/counteract the effects due to the fiber and to *global* spectral gain variations are:

1. stabilization of the total power transmitted along the route, by means of:
   a. compensation signal(s)
   b. aggregate rerouting around a fault location
   c. fast power imbalance control, at least at a number of line amplifiers (equalizing amplifiers);
2. flexible agile maximum transparent distance (MTD) determination (MTD as a function of signal wavelength).

In the following of this section, solution 2 will be considered in some detail. The typical DWDM system is assumed as an 80-channel, C-band system with line EDFAs, but no Raman amplifiers. A typical Raman imbalance of 1 dB/span is assumed; see Figure 5.44. Specific constraints are obtained depending on channel power, OSNR margin, EDFA design, and RX design. However, it is possible to draw some general conclusions on the behavior of the worst DWDM channel.

The reference scenario envisages a fully loaded 80-channel system. Figure 5.50 shows typical power excursions for a range of load-drop events in a 20-span system.

Two characteristic regions are found.

1. Region I – from 1530 to 1549 nm (blue subband, channels 1–50), where:
   a. power variations are largely independent on wavelength;
   b. power variations substantially depend on span number N_s;
   c. the region boundary is defined by two coefficients $a_{\pm}(N_s)$ that give the maximum (positive and negative) power variation.
2. Region II – from 1549 to 1562 nm (red subband, channels 51–80), where:
   a. power variations are negative and strongly dependent on wavelength;

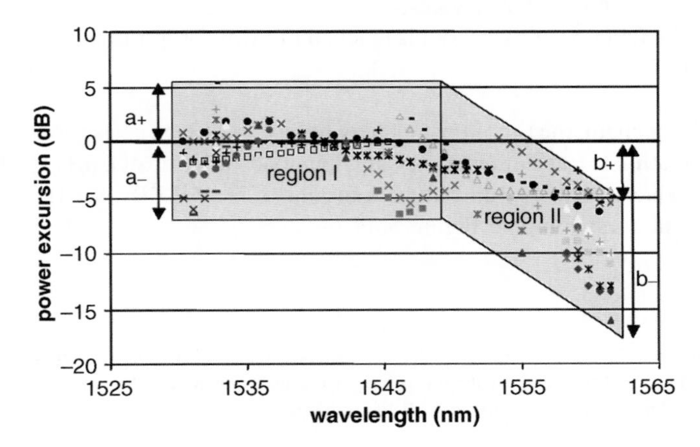

Figure 5.50 Power excursion regions; 20-span light path on G.652 fiber, 80 channels at 1 dBm/ch. Fast control on EDFA gain is assumed

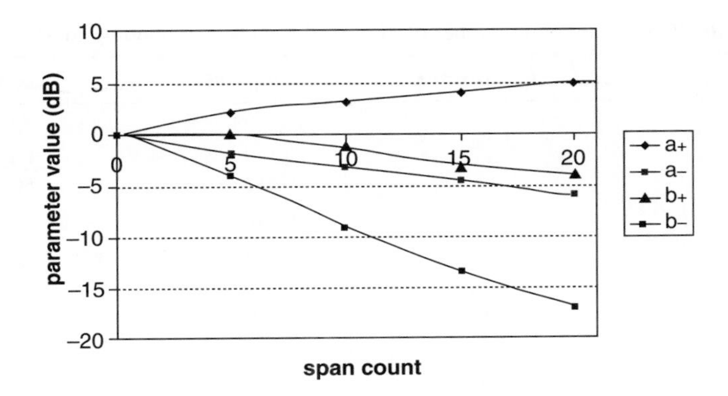

Figure 5.51 The dependence of mask coefficients on span number

b. the region boundary is defined by two coefficients $b_{\pm}(N_s)$ that determine extreme values for the power variation of the worst channel (that is always the longest wavelength channel).

Figure 5.51 plots the dependence of the mask coefficients of Figure 5.50 on the span number N_s along a transparent route. Starting from such data it is possible to translate the wavelength-dependent power excursion during transients into an agile MTD.

5.3.3.1 Nonlinear and Overload Limit

This limitation stems from the scaling laws (5.208), (5.210), and (5.213) describing Kerr-type nonlinear mechanisms:

$$P_{ch} + 10 \log N_s \leq P_{max}$$

where P_{ch} is the average channel power launched over each span for a total of N_s spans and P_{max} is defined in correspondence with a penalty of X dB (typically $X = 1$). A real system is designed to work with margins M; transient power excursion ΔP_{ch} will not cause error at detection as long as

$$\Delta P_{ch} \leq P_{max} + M - (P_{ch} + 10 \log N_s) \tag{5.215}$$

Consider the following set of values: $P_{max} = 18$ dBm, $M = 2$–3 dB, $P_{ch} = 1$–3 dBm and $\Delta P_{ch} = 0.5$–1 dB; the corresponding MTDs are reported in Table 5.6.

With margins higher than 5 dB it is difficult to exceed 30 spans, while it is easily accomplished for $M \leq 3$ dB.

Figure 5.50 shows *total* power variations (as measured after the last span); average power variations are weaker, as shown in Figure 5.52, lowest curve. In the envisaged example, the fiber nonlinearity would limit the MTD to ~30 spans (extrapolation of the upper curve). Positive transients can also lead to system outage because of receiver overload (assumed at 9 dB in Figure 5.52). For small margins, such those shown in the example, the overload limit could also be critical.

Table 5.6 Maximum number of amplified spans corresponding to MTD

M [dB]	P_{ch} [dBm]	ΔP_{ch} [dB]	Max N_s
2	1	0.5	70
		1	63
		2	50
	3	0.5	44
		1	39
		2	31
3	1	0.5	89
		1	79
	3	0.5	56
		1	50
		2	39

Since Region II features by negative power excursions, it does not present limits from the point of view of nonlinear or overload effects.

5.3.3.2 Noise and Sensitivity Limit

Experimental results and simple considerations show that in Region I OSNR degradation is half the power excursion in decibels: an inspection of Figures 5.50 and 5.51 shows that the power variation is centered at the average value in Region I, while OSNR is degraded for negative channel power variations only. Figure 5.53 illustrates such a degradation effect; a minimum performance of OSNR $>$ 13 dB (over 0.1 nm) is considered for equalized receivers and/or advanced modulation formats, and limitations inherent to Region I are seen typically to go beyond 30 spans (upper curve). Power excursions may also affect receiver sensitivity, as indicated by the lower curve; this second effect clearly depends on the receiver dynamics.

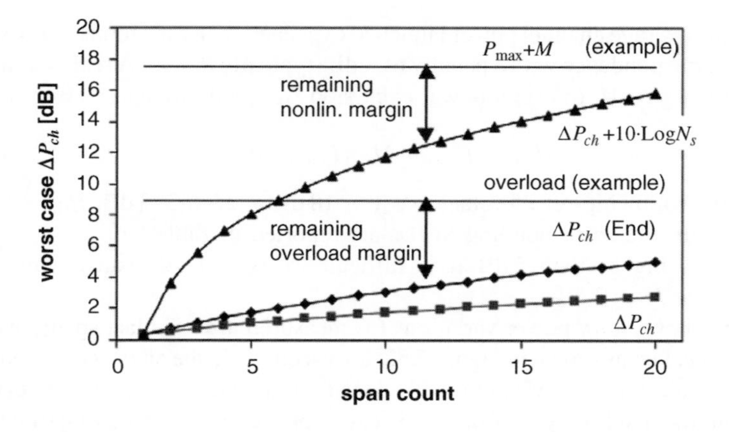

Figure 5.52 MTD limitations due to *positive transients* of power in Region I. Upper curve: fiber nonlinearity; middle curve: receiver overload; lower curve: average transient

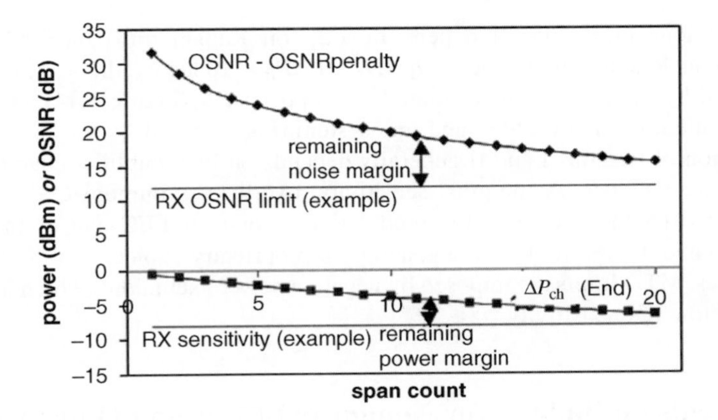

Figure 5.53 MTD limitations due to *negative transients* of power in Region I. The upper curve refers to the limit imposed by optical noise; the lower curve refers to receiver sensitivity

The reader should note that positive power excursions do not imply any noise or sensitivity limitation. Because of this, Figure 5.53 only shows negative variations that are nearly equal, in absolute value, to positive variations considered in the preceding figure.

The relation $\Delta\text{OSNR} \cong \frac{1}{2}\Delta P_{\text{ch}}$ stems from these considerations. Similar issues hold for the plot of the total power variation (lower curve), which in Figure 5.53 has an opposite behavior with respect to Figure 5.52.

On the other hand, Region II can set severe limitations, especially as far as the RX sensitivity is concerned (Figure 5.54). Since this spectral region is defined by negative power excursions only, here $\Delta\text{OSNR} \cong \Delta P_{\text{ch}}$.

As a commentary to preceding figures, some general conclusions can be drawn.

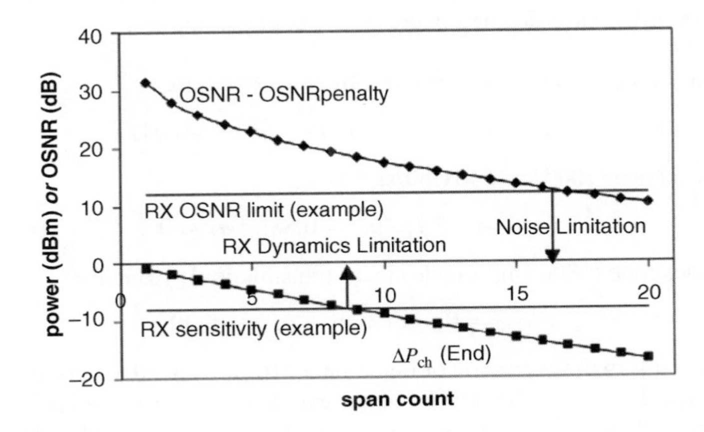

Figure 5.54 MTD limitations due to *negative transients* of power in Region II. The upper curve refers to the limit imposed by optical noise; the lower curve refers to receiver sensitivity

- Power variations in Region II depend directly on Raman imbalance ΔP_R over each amplified span. Results simply scale as $b_-(N_s) \cong \Delta P_R + 10 \log N_s$; this scale transformation is influenced by the level of the average launch power P_{ch} (Figure 5.44) and by different choices for the transmission fiber and line system (Figure 5.46).
- The definition of Regions I and II generally depends on line amplifier characteristics.
- Other parameters, such as margins, sensitivity, overload, minimum OSNR, and so on, are influenced by the choice of the modulation format, of FEC and of the RX design (PIN, APD) and, therefore, they are generally a proprietary choice.
- In most cases, MTD limits are imposed by negative power excursions, which lead Region II channels below RX sensitivity.

5.4 Comments on Budgets for Nonlinear Effects and Optical Transients

The matter presented in Section 5.3.2 is sufficient for a system analysis of the network. In particular, Figure 5.44 permits one to check the power channel at launch in such a way to avoid excessive Raman imbalance and penalty.

This basic approach to launch power must be verified with respect to the occurrence of the other nonlinear effects (SPM, XPM, FWM); if the design parameters P_{SPM}, P_{XPM}, and P_{FWM} are known, then the scaling laws (5.208), (5.210), and (5.213) immediately give this second-order check. Finally, Section 5.3.3 allows us to establish if the network is sufficiently guaranteed against possible optical transients.

With a correct choice for the channel power, the relation between Q-factor and OSNR (Equation 5.75) can be reasonably approximated with a polynomial expression:

$$Q^* \equiv 20 \log Q = a_0 + a_1 \mathrm{OSNR} + a_2 N_s + f(N_f) \qquad [\mathrm{dB}] \qquad (5.216)$$

where OSNR is expressed in decibels; N_s and N_f are the number of amplified spans and the number of filters crossed by the signal respectively; a_0, a_1, and a_2 are numerical coefficients that depend on the network; finally, f is a function of filter number that will be discussed in a subsequent section. Simulations performed within the framework of European projects MUFINS and NOBEL have detailed interesting sets of parameters:

mid-haul backbone (1000 km, linear transmission), 10 Gbit/s:

$$a_0 = 0.40; \quad a_1 = 0.96; \quad a_2 = -0.041$$

mid-haul backbone (1000 km), 40 Gbit/s:

$$a_0 = -5.1; \quad a_1 = 0.88; \quad a_2 \approx 0$$

long-haul backbone (1800 km, pseudo-linear transmission), 10 Gbit/s:

$$a_0 = 0; \quad a_1 = 1; \quad a_2 = -0.30$$

The first two terms in the expansion of Equation 5.216 correspond to the expression of light path baseline Equation 5.75 in the absence of any eye-diagram closure penalty or excess noise factor (compare with Equation 5.79) and the coefficients a_0 and a_1 can be identified starting from the OSNR-Q formulae of type of Equation 5.75 (see Equations 5.69–5.71): a_0 depends solely on the optical and electrical receiver bandwidths, while a_1 is unity.

If any eye-diagram closure penalty is present, then it sums to the baseline value of a_0, as can be seen in the mid-haul case at 40 Gbit/s cited above. In that case, $P^{(\mathrm{EO})}$ is largely determined by PMD. On the other hand, the parameter a_2 is a feature of further penalties, especially those due to nonlinear effects (the reader should remember the scaling laws). It is the case of the long-haul backbone quoted above, where the pseudo-linear transmission regime is well designed and PMD is negligible. It is evident that the sensitivity of a_2 to network structure is rather strong: passing from linear to pseudo-linear transmission, its absolute value increases by an order of magnitude. Even if long-haul backbone data for 40 Gbit/s are not available, it is likely that a_2 also increases (in absolute value) with bit rate.

The last contribution $f(N_\mathrm{f})$ in (5.216) will be discussed in detail in following sections.

5.4.1 Compensators/Equalizers

The performance increase[55] due to the presence of compensators/equalizers can be expressed in the following form:

$$Q_{\mathrm{nocomp}} = \frac{\mu_1'-\mu_0'}{\sigma_1'+\sigma_0'} \rightarrow Q_{\mathrm{comp}} = \frac{\mu_1-\mu_0}{\sigma_1+\sigma_0}$$

$$\frac{Q_{\mathrm{comp}}}{Q_{\mathrm{nocomp}}} = \frac{\dfrac{\mu_1-\mu_0}{\sigma_1+\sigma_0}}{\dfrac{\mu_1'-\mu_0'}{\sigma_1'+\sigma_0'}} = \frac{\dfrac{\mu_1-\mu_0}{\sigma_1+\sigma_0}}{\dfrac{\mu_1'-\mu_0'}{\sigma_1+\sigma_0}} \cdot \frac{\sigma_1'+\sigma_0'}{\sigma_1+\sigma_0} = \left(\frac{Q_{\mathrm{comp}}}{Q_{\mathrm{nocomp}}}\right)_{\text{constant noise}} \frac{1}{\sqrt{F_\mathrm{E}}} \qquad \text{[lin. un.]}$$

$$Q^*_{\mathrm{comp}} - Q^*_{\mathrm{nocomp}} = 20 \log\left(\frac{Q_{\mathrm{comp}}}{Q_{\mathrm{nocomp}}}\right) = \Delta Q^*_\mathrm{E} - F_\mathrm{E} \qquad \text{[dB]}$$

For example, the distortion compensation can alleviate the eye-diagram closure with respect to the uncompensated case. Therefore, an increment of performance E_E can be defined, which is related to the effect of the compensator or equalizer, by means of

$$E_\mathrm{E} = Q^*_{\mathrm{comp}} - Q^*_{\mathrm{nocomp}} = \Delta Q^*_\mathrm{E} - F_\mathrm{E} \quad \text{[dB]} \tag{5.217}$$

This expression takes into account two facts: (i) on one side, the compensator increases the value of Q^* as a whole, but (ii) its presence may degrade the SNR. To quantify the global effect of compensation strategies, the term ΔQ^*_E must be known – F_E, the excess noise factor, being generally limited to less than 2 dB. Another important feature of the transport system is the modulation format, which strongly influences the effects of equalized receivers.

5.4.2 CD Equalization

Table 5.7 and Figure 5.55 describe the case of NRZ, CSRZ, and DB signals with three different equalization schemes: ATC, adaptive threshold control; VE, Viterbi equalizer; FFE + DFE, *feed-forward* plus *decision feedback* schemes. If cumulative filtering is present besides transmission distortions, then the pattern changes appreciably, as will be discussed in Section 5.4.4.

[55] Namely a negative penalty, $P = -E_\mathrm{E}$.

Table 5.7 Equalization performance in the case of three specific modulation formats

	Max. Q penalty [dB]	NRZ 10.7 Gbit/s	DB 10.7 Gbit/s	CSRZ 10.7 Gbit/s		
		FFE + DFE	VE		VE	VE
		Model 2L	Model 2	Model 1	Model 1	Model 1
Max. CD	2	1400	1900	2100	3400	830
[ps/nm]	3	1660	3060	2800	3700	1050
Max.	2	18	20	17	19	21
PMD [ps]	3	22	25	21	21	24

5.4.3 PMD Equalization

Table 5.8 lists Q^2 penalties when PMD equalization schemes are adopted obtained by simulations that were performed by the NOBEL European Project. The penalty clearly depends, further than on the average DGD, also on the baseline factor Q, as described for an NRZ modulation format at 10.7 Gbit/s, in correspondence to a 10^{-5} outage probability.

5.4.4 Simultaneous Presence of Distortions, Electronic Equalization, and Cumulative Filtering

A situation that deserves specific discussion corresponds to the simultaneous equalization of CD and PMD effects and of cumulative filtering by means of electronic equalization (VE, for instance). In this section, a linear transmission regime is assumed, as it is the case of

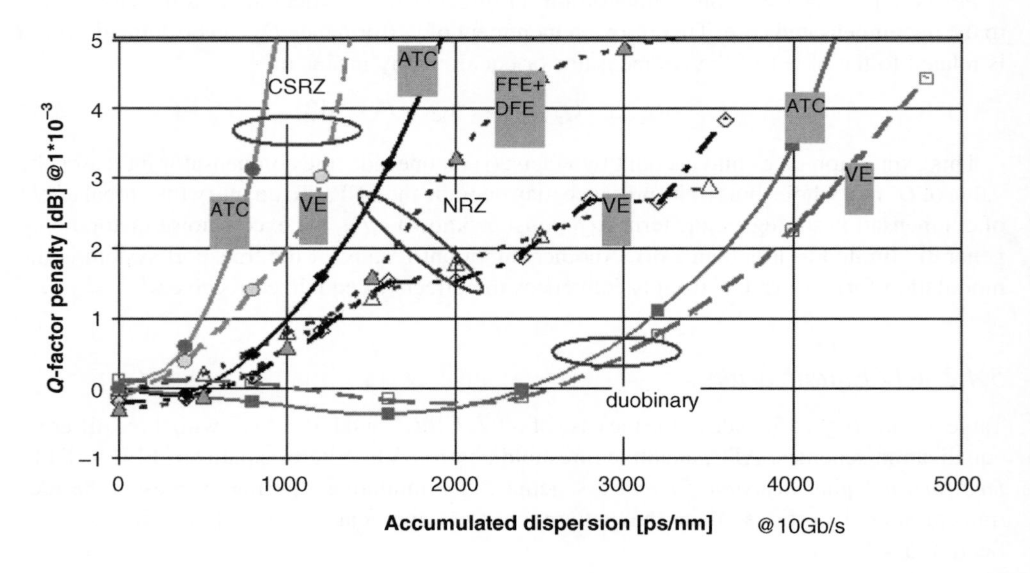

Figure 5.55 CD-induced penalty on Q^2 at the input of an FEC stage. Equalization schemes: ATC, VE, FFE + DFE. Modulation formats: NRZ, CSRZ, DB

Table 5.8 Q^2 penalties due to PMD [dB] in correspondence to an outage 10^{-5} probability for FFE + DFE equalization scheme (10.7 Gbit/s NRZ ASK)

	PMD [ps] = mean DGD				
Q [dB] w/o PMD	5	10	15	20	25
13.2	0.41	1.03	1.91	3.86	5.58
14.3	0.40	1.01	2.05	4.04	6.16
15.3	0.20	0.86	2.07	3.98	6.27
16.3	0.28	0.90	2.11	4.19	6.50
17.4	0.20	1.03	2.25	4.52	6.79
18.4	0.30	1.04	2.32	4.58	7.45
19.5	0.50	1.14	2.45	4.67	7.50
20.5	0.46	1.15	2.49	5.07	8.14

metropolitan networks. The situation is described by the results of simulations done within the NOBEL European Project, with the following system parameters:

bit rate	10.7 GHz
residual CD	3000 ps/nm
DGD	34 ps
filter shift	±12.5 GHz
filter detuning to TX	±10, ±7.5, ±5 GHz
VE: three ADC bits, 2-bit memory (four states), completely redundant sampling	
linear transmission in the fiber (metro applications)	
Fourth-order Bessel filter, 3 dB bandwidth	7.5 GHz

A line OADM is considered to be equivalent to two filters (mux/demux) according to a model that considers a given SNR at the input of the network element, with no optical losses for switching matrices. The behavior of the insertion loss and of GVD for mux/demux is shown in Figure 5.56. Black curves show such quantities for one OADM (a couple of mux/demux filters). The 3 dB bandwidth of each OADM is 60 GHz. The worst-case simulation of cumulative filtering is done assuming a shift of ±12.5 GHz between the peak frequencies of mux and demux filters (this is indicated in the figures with the 'shifted' curves).

We remind the reader that Section 5.2.4.10 discussed the problem of cumulative filtering alone, that Section 5.4.2 presented CD equalization, and that Section 5.4.3 related to PMD equalization. This section assumes that all these phenomena are simultaneously present in the system: the filtering effect is due to the presence of OADMs. Q^2 penalties are calculated downstream of a VE, using frequency detuning between TX and filter as a parameter: ±10, ±7.5 and ±5 GHz. All simulations assume the presence of residual PMD, 10.55 ps, and of residual CD, 3000 ps/nm; these are the worst values considered in the reference networks for the NOBEL Project.

Simulation results supply conservative estimates that can be used by wavelength assignment and routing criteria, constrained by transmission effects. The simultaneous equalization of CD,

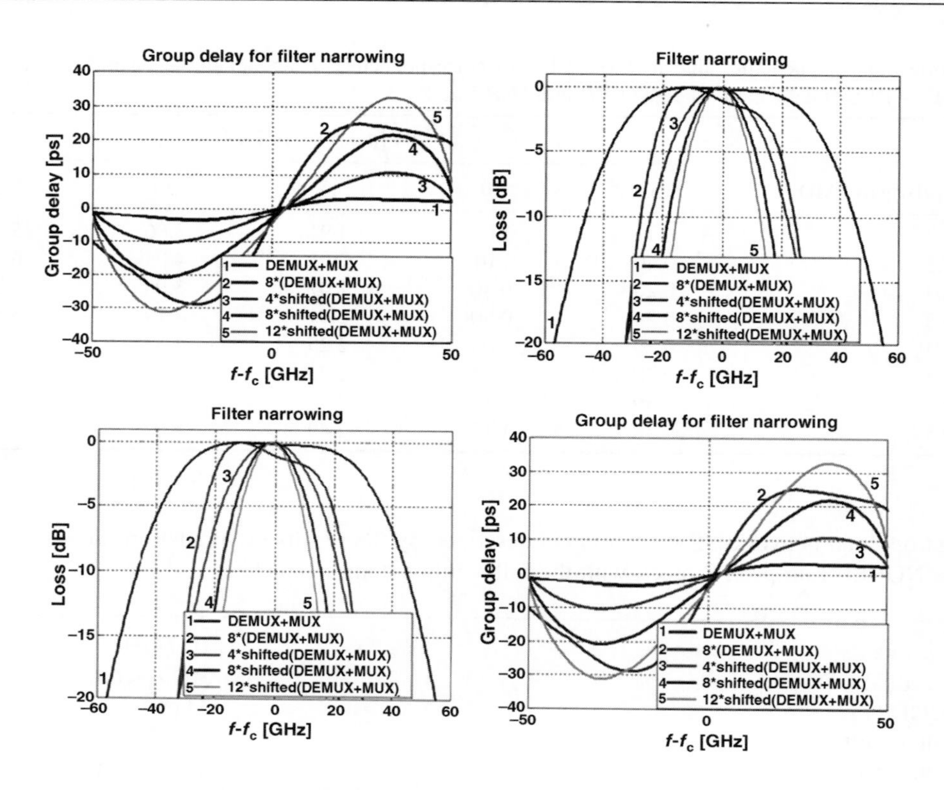

Figure 5.56 Insertion loss and GVD for one OADM

PMD, and cumulative filtering evidences VE capabilities: three ADC bits, 2-bit memory (four states) and completely redundant oversampling is assumed, which is state-of-the-art technology for optical fluxes at 10.7 Gbit/s. Figures 5.57–5.59 show Q^2 penalties versus number of OADMs, when a PMD-induced DGD of 34 ps and a CD of 3000 ps/nm are present on the line. Results are distinguished by the frequency detuning between laser and filters, as a percentage of the 3 dB bandwidth of the OADM. NRZ, optical duobinary (ODB) and directly modulated laser (DML) [56] formats are considered.

With a residual CD of 3000 ps/nm, VE is capable of giving a net improvement in Q^2 even with respect to the back-to-back configuration.

In fact, the penalty reference is the corresponding back-to-back condition in the presence of ATC. For low values of frequency offset (5 GHz) and the NRZ formats, Q^2 penalties decrease when the number of OADMs increases; this seemingly paradoxical result is due to the fact that with more and more muxes the effective filtering bandwidth narrows and reduces the noise power on the receiver. In this condition, distortion-induced penalties are overcompensated by in-band noise reduction.

Table 5.9 reports the number of in-line OADMs tolerated for a 1 dB and 2 dB Q^2 penalty with NRZ, ODB, and DML pulses. Penalty curves are well approximated by parabolas; Table 5.10 shows the corresponding coefficients for the three modulation formats and the three frequency

[56] It is a direct modulation of the laser transmitter, but the chirp is managed through a very narrow filter.

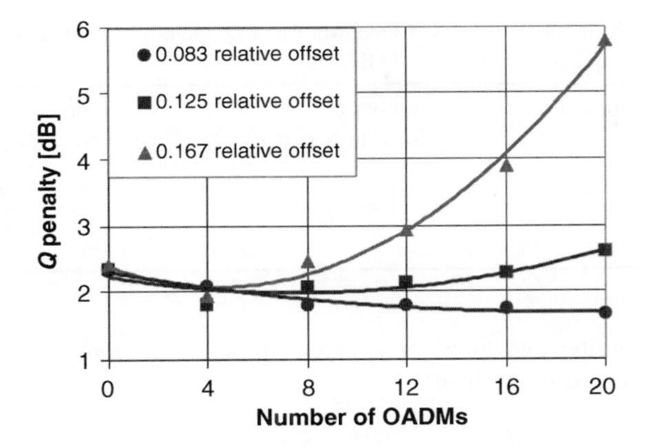

Figure 5.57 Q^2 penalties as a function of OADM number for NRZ signals

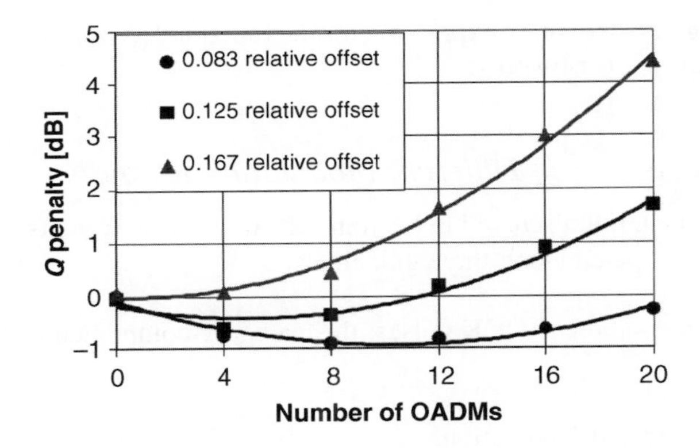

Figure 5.58 Q^2 penalties as a function of OADM number for DML formats

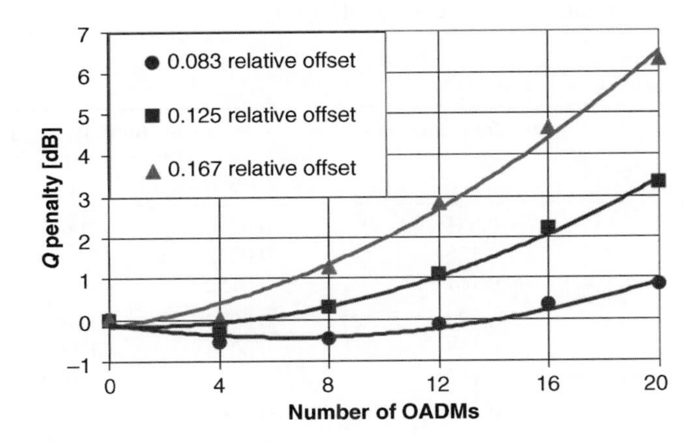

Figure 5.59 Q^2 penalties as a function of OADM number for ODB signals

Table 5.9 Number of cascaded OADMs for a Q^2 penalty lower than 1 dB, at a transmitter frequency shift of ±7.5 GHz

Modulation format	Number of OADMs for a 1 dB penalty	Number of OADMs for a 2 dB penalty
NRZ	—	9
DML	16	22
ODB	12	15

shifts between transmitter and filters considered in the simulations. In such a way, the Q^2 penalty can be calculated in the form

$$P_{\mathrm{OADM}}^{(Q^*)} \equiv f(N_{\mathrm{OADM}}) = b_2(N_{\mathrm{OADM}})^2 + b_1 N_{\mathrm{OADM}} + b_0 \tag{5.218}$$

where b_0, b_1, and b_2 are the coefficients for quadratic trinomials and N_{OADM} is the number of cascaded OADMs.

Equation 5.218 gives an explicit expression for the function $f(N_{\mathrm{f}})$ introduced previously, at least for the case of metro networks.

5.4.5 General Features of Different Modulation Formats

On the basis of the material discussed in Section 5.2, modulation formats alternative to NRZ and RZ may be investigated along three guidelines.

- Suitable phase modulation of OOK pulses; this category comprehends the
  - CRZ format
  - CSRZ format
  - single side-band (SSB) modulation.
- Coherent-type modulation, with simplified receiver (without local oscillator), as for differential phase-shift keying (DPSK) format.
- Electrical signal correlation code, such as ODB format.

Table 5.10 Polynomial coefficients for Equation 5.218

Modulation format	Relative frequency shift	Coefficients for polynomial approximation		
		b_2	b_1	b_0
NRZ	0.083	0.0018	−0.0667	2.304
	0.125	0.0042	−0.0644	2.233
	0.167	0.0152	−0.1368	2.383
DML	0.083	0.0078	−0.163	−0.08844
	0.125	0.0095	−0.0937	−0.1792
	0.167	0.0122	−0.0165	−0.0117
ODB	0.083	0.0076	−0.1009	−0.0914
	0.125	0.01	−0.0219	−0.1398
	0.167	0.012	0.0946	−0.1658

Table 5.11 Qualitative pattern of different modulation format features (CD, chromatic dispersion; NL, nonlinearity; PMD, polarization-mode dispersion)

Modulation format	CD	NL	PMD	RX sensitivity
NRZ	+	−	−	0
RZ 33/50	− −	+	+	+
CSRZ 67	0	+	+	+
DB	++	− −	−	−
RZ-DPSK	+	+	+	++
RZ-DQPSK	++	+	+	+
APRZ 67	−	++	+	+

Following this classification,[57] it is possible to give some rules of thumb (Table 5.11). Performance is always referred to NRZ. In the light of investigations quoted in the literature, CSRZ offers a modest performance in terms of CD, though it is better than RZ. PMD robustness, which depends substantially on pulse shape, for a 67% duty cycle (as in CSRZ) would be intermediate between RZ (better) and NRZ (worse), but CSRZ may exhibit the bit-to-bit parity inversion that makes it better than RZ. CSRZ could be a candidate for medium-haul networks and very high bit rates.

Concerning CRZ, chirp dependence strongly constrains the performance: to get the best from each format, adaptive solutions seem necessary, and they require a new design for the optical system as a whole.

SSB modulations now seem slightly more than a curiosity. Indeed, the concept of halving the bandwidth (taken from coherent modulations) to increase CD robustness is valid, but the proposed solutions for intensity modulation with direct detection (IMDD) systems are such that they simply produce a spectrum asymmetry, shifting the central frequency, but with a wider bandwidth. As a consequence, the performance is often worse than with NRZ and RZ.

DPSK stems from the more efficient format. In a linear regime it guarantees an ~3 dB gain together with a dispersion tolerance comparable to IMDD formats. In a nonlinear regime, since at constant signal power the energy of a single pulse is half the power of corresponding OOK formats, an analogous gain should be expected.

The fundamental DPSK limit stays in receiver complexity, particularly with the interferometer. Whether a cheap solution for the RX or whether the cost for such a complex RX were considered worth it, then it would be natural considering also the differential quadrature phase-shift keying (DQPSK) variant, which doubles the spectral efficiency.

ODB has the great advantage of introducing a minimal complexity added to TX (a precodification stage, substantially), in front of the advantages related to halving the bandwidth. However, experimental verifications of ODB performance in the nonlinear, multichannel, regime do not confirm expectations.

As a concluding comment, one can say that:

- for metro networks at 10 Gbit/s, DML-NRZ is interesting together with electronic equalization; at 40 Gbit/s, ODB may become convenient;

[57] This classification does not consider explicitly vestigial modulation, which can be thought of as an issue in the question of reducing DWDM channel spacing.

- in medium-haul (~1000 km) backbones NRZ is surely the more convenient one; at 40 Gbit/s, CSRZ may become a candidate with CD compensation on a channel-by-channel basis and enhanced FEC;
- in very long-haul core networks, NRZ can be used at 10 Gbit/s with FEC; for 40 Gbit/s the best candidate is DPSK.

5.5 Semianalytical Models for Penalties

The subject of Section 5.2 can be formulated in an alternative way. We saw that the Gaussian model of photodetection leads to defining a factor of merit Q as the ratio

$$Q = \frac{\mu_1 - \mu_0}{\sigma_1 + \sigma_0}$$

between the average values μ_1 and μ_0 of the electrical signal on '1' and '0' symbols and the corresponding standard deviations σ_1 and σ_0. This Gaussian model has a good accuracy in the definition of a transmission channel with *additive white Gaussian noise* (AWGN) for transmission penalties. The actual transmission signal S_{r} is the sum of an ideal signal S and a white (i.e., with a uniform spectral density) noise N, with Gaussian distribution around a zero mean value:

$$S_{\mathrm{r}} = S + N \tag{5.219}$$

At each actual value $\bar{S}$ of the ideal signal, the effective quantity $\bar{S} + N$ is distributed in a Gaussian way around the value $\bar{S}$ (having itself its own statistics). If S is also a Gaussian stochastic process, then the sum of two independent Gaussian statistical variables is still a Gaussian statistical variable with a mean value that is the sum of the mean values of the components and a variance that is the sum of the variances of the component [47, Chapter 3]:

$$\overline{S_{\mathrm{r}}} = \bar{S} + \bar{N} = \bar{S}, \quad (\sigma_r)^2 = (\sigma_S)^2 + (\sigma_N)^2 \tag{5.220}$$

Therefore, a penalty η_{p} associated with an AWGN process only affects the optical signal variance and not its mean value. Therefore, the basic equation, equivalent to (5.81), (5.83), and (5.85), in the new approach is

$$\frac{Q'}{Q} = \frac{\sigma_1 + \sigma_0}{\sqrt{(\sigma_1)^2 + (\sigma_{\mathrm{p},1})^2} + \sqrt{(\sigma_0)^2 + (\sigma_{\mathrm{p},0})^2}} \tag{5.221}$$

With this equation, Q is Personick's factor value corresponding to the ideal case, Q' is the one degraded by penalty η_{p}; σ_1 and σ_0 are the standard deviations of the electrical signals associated with '1s' and '0s' in ideal conditions. In the AWGN model, the penalty η_{p} affecting both '1s' and '0s' is described by the standard deviations $\sigma_{\mathrm{p},1}$ and $\sigma_{\mathrm{p},0}$.

5.6 Translucent or Hybrid Networks

Up to this point we have considered transparent transport networks: they are characterized by a certain MTD value. To overcome this distance, the concept of hybrid network, or translucent network, must be introduced. In a translucent network the signal is transported onto the optical layer as long as possible; when it has been degraded to a level that cannot be further weakened, the regeneration functionality must be introduced. At present, two technological solutions are available: (i) optical–electrical–optical (OEO) 3R regenerators, or transponder units, which is a well-assessed technology, and (ii) 2R regenerators, possibly in all-optical technology.

5.6.1 Design Rules for Hybrid Networks

The design of hybrid networks can exploit the matter discussed for transparent networks, in application to transparent subconnections that constitute the hybrid connection under exam.

An OEO 3R regenerator uses the degraded optical signal frame for a complete regeneration of the signal itself. Therefore, it can be said that when the degraded signal can still be detected with an acceptable Q factor (say, not lower than 12 dB with FEC) at the 3R device, the regenerator is capable of restoring a new signal having the same quality as the original one. Then the unique limitation (of a technical, and not economical, nature) to using OEO 3R devices is given by the requirement that the cumulative BER, (i.e., the total BER summed over all subconnections between 3R devices – along the multiply regenerated route; Figure 5.60) is lower than a given QoS requirement. For example:

$$\mathrm{BER_{tot}} = \sum_{i=1}^{N_{3R}+1} \mathrm{BER}_{(i)} \leq \mathrm{BER}_{(\max)}, \quad \text{where, say,} \ \mathrm{BER}_{(\max)} = 10^{-12} \ \text{without FEC} \tag{5.222}$$

The calculation of each term $\mathrm{BER}_{(i)}$ can be done according to the techniques described so far in the book.

A 2R regenerator cannot work this way. It cannot restore a degraded signal; it can only act on the noise background and increase the eye-diagram opening through the use of a nonlinear transfer function to redistribute power between levels '1' and '0'. Moreover, this nonlinear power redistribution is affected by an ASE noise increase (an optical 2R regenerator always incorporates an OA). Hence, the action of a 2R (optical) regenerator can be described as follows (Figure 5.61).

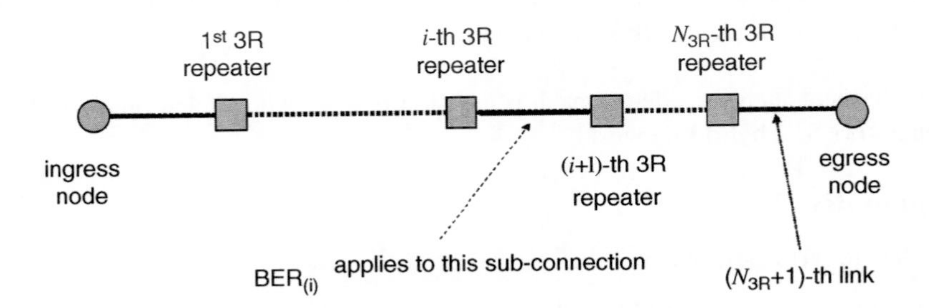

Figure 5.60 Schematics of an optical connection, multiply regenerated at the 3R level

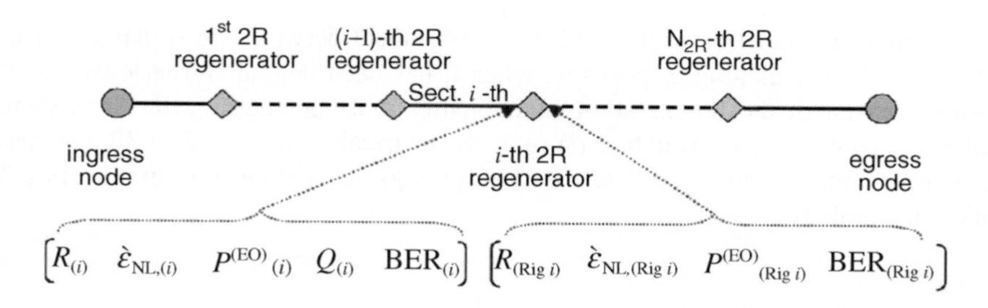

Figure 5.61 Schematics of an optical connection, multiply regenerated at the 2R level. Here, $\varepsilon_{NL} = 1 - P^{(EO)}$ (in linear units)

1. One establishes the relation between Q-factor and OSNR, for instance with the Equation 5.216:
$$Q_{[dB]} = a_0 + a_1 \mathrm{OSNR}_{[dB]} + a_2 N_s \tag{5.223}$$
2. Similarly, the baseline is established:
$$Q_{R[dB]} = a_0 + a_1 \mathrm{OSNR}_{[dB]} \tag{5.224}$$
3. ASE accumulation is calculated with usual expressions; for example, Equation 5.96:
$$\mathrm{OSNR}\ [\mathrm{dB}] = P_0\ [\mathrm{dBm}] + 58\ \mathrm{dBm} - A_{tot}\ [\mathrm{dB}] - F_{eq}\ [\mathrm{dB}]$$
4. The linear behavior of the optical circuit is assumed to be well compensated and that its degradation is substantially due to optical nonlinearity; with these assumptions and Equation 5.89:
$$Q = P^{(EO)} Q_R \quad \in [\text{linear units}]$$
the relation between Q-factor and baseline may be expressed as
$$\Delta Q_{NL,[dB]} = a_2 N_{span} \quad [\mathrm{dB}] \quad \text{thence} \quad a_2 = 20 \log(P^{(EO)}_{span})$$
5. The relation input–output intensity for the 2R regenerator has the form
$$I_{out} \sim \tanh \gamma I_{in}$$
where, in practice, the nonlinearity parameter $\gamma \approx 0.65$ is for an interference structure of the Mach–Zehnder type.
6. Under reasonable assumptions, see for instance [48], a 2R device should
   a. reduce the eye-diagram closure penalty by a factor γ,
   b. reduce ASE noise by a factor γ^2 thence improving the OSNR by the same factor
   c. to leave the cumulated BER unchanged.

On the basis of such considerations, it is easy to implement an allocation algorithm for 2R regenerators in a hybrid network.

5.7 Appendix

5.7.1 Dispersion Managed Links

The vast majority of installed fiber links consist of several amplified fiber spans, the lengths of which are determined by the fiber loss factor and are generally a few tens of kilometers.

As discussed in Section 5.4, CD becomes an issue at 10 Gbit/s, for which the dispersion length is typically below 50 km. This means that, for multispan links, dispersion has to be compensated for. This is most commonly done by the use of *dispersion-compensating modules* (DCM*s*) whose dispersion parameter D has opposite sign compared with the standard single-mode fiber (SSMF). Although new technologies are starting to appear and be tested [49,50], DCMs in installed systems today are implemented as a section of fiber specially designed to have negative dispersion. This special fiber is called *dispersion-compensating fiber* (DCF) [51], and is generally characterized by

$$\begin{aligned} D_{\rm DCF} &\cong -5D_{\rm SSMF} \\ a_{\rm DCF} &\cong 3a_{\rm SSMF} \\ \gamma_{\rm DCF} &\cong 2\gamma_{\rm SSMF} \end{aligned} \tag{5.225}$$

Normally, a DCF module of length $L_{\rm DCF} \cong L_{\rm SSMF}/5$ is placed at each amplifier node (*in-line* compensation), so that the accumulated dispersion is brought back to zero (or close to it). In order to minimize the impact of noise, an extra amplifier can be placed between the SSMF and the DCF.

An SSMF span followed by a node with EDFAs and DCF makes a *dispersion-managed* (DM) cell. From a fiber-propagation point of view, a DM link is therefore modeled as a waveguide of length $L = L_1 + L_2 + \ldots$, where the parameters α, D, and γ periodically switch between the SSMF and DCF values. As will be seen later, it may be convenient to place part of the DCF, of length $XL_{\rm DCF}$, at the transmitter (precompensation). In this case, the last cell should have a DCF of length $(1-X)L_{\rm DCF}$. The parameter X is called the *launch position.*

Figure 5.62 shows a typical DM link (if WDM is used to transmit several channels, then a wavelength multiplexer (MUX) will be placed right after the transmitter and a wavelength demultiplexer (DEMUX)). Transmission on such DM links takes place in a transmission regime in which pulses periodically broaden over several bit slots and then compress back, in a sort of *breathing* process. As will be seen in Section 5.7.2, this has important consequences for the way in which nonlinear effects impair transmission, especially at bit rates of the order of 40 Gbit/s, for which tens of pulses overlap with each other throughout most of the transmission distance.

It should be said that DM is not the only transmission regime thinkable, and dispersion compensation could be concentrated at one point (transmitter, receiver, or midpoint), which would give rise to a highly dispersed regime [52] with high accumulated dispersion throughout most of the link, and consequent very large overlap and no breathing. While either approach is

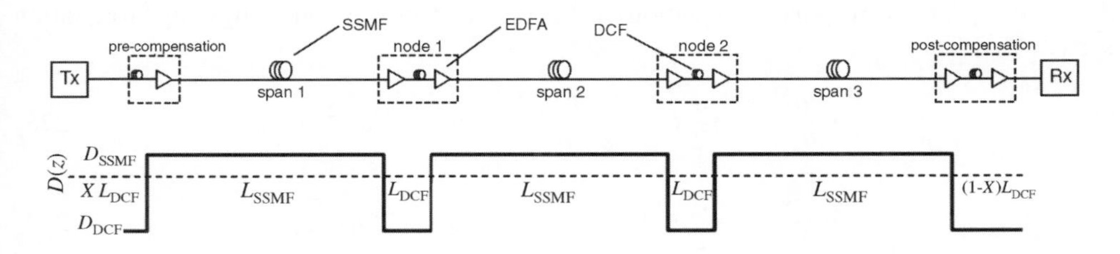

Figure 5.62 A dispersion-managed (DM) link with three spans. Tx indicates the transmitter, and Rx the receiver. The dispersion profile along the link is shown below

fine in a perfectly linear situation, this is not the case for real fibers, for which the Kerr effect is small but not negligible. Which of the two transmission regimes is better is questionable [53]. The main reason why in-line dispersion is normally used is the fact that it provides a more robust configuration, especially in a network scenario, in which the path followed by different channels is different, so having an accumulated dispersion close to zero for any path is a desirable feature.

5.7.2 Intrachannel Nonlinear Effects

In typical DM links, the local dispersion is relatively high, so interchannel impairments such as FWM and XPM are low. On the other hand, high local dispersion causes pulses within each channel to overlap during a significant part of the transmission distance and to interact nonlinearly. These intrachannel nonlinear effects are the main nonlinear impairments in DM transmission at bit rates in the order of 40 Gbit/s [54].

The signal $A(t,z)$ at any specific location will in general be the sum of different data pulses:

$$A(t,z) = \sum_q a_q(t,z) \tag{5.226}$$

where q runs through the *one* pulses. Inserting the Fourier transform of Equation 5.226 into the Nonlinear Schrödinger equation (NLSE), and solving for bit slot q, one obtains

$$\mathrm{j}\frac{\partial a_q}{\partial z} = -\mathrm{j}\frac{a}{2}a_q + \frac{\beta_2}{2}\frac{\partial^2 a_q}{\partial t^2} - \gamma\left(|a_q|^2 + 2\sum_r |a_r|^2 + \sum_{r;s} a_r a_s^*\right)a_q \tag{5.227}$$

The first nonlinear term, $|a_q|^2$, is responsible for SPM, while the second term, $\sum |a_r|^2$, generates a contribution to the nonlinear phase shift that depends on the power of other pulses, and is known as IXPM. IXPM causes different data pulses to experience different time shifts, and it results in serious impairments in 40 Gbit/s transmission, as will be seen in Section 5.7.2.1 The third term, $\sum a_r a_s^*$, is responsible for IFWM, a phase-sensitive phenomenon through which power is transferred from each pulse group, $a_q a_r a_s^*$, to a specific temporal location n. IFWM is a significant source of impairments in 40 Gbit/s transmission systems and will be discussed in Section 5.7.2.2.

5.7.2.1 Intrachannel Cross-Phase Modulation

The propagation of two pulses in an optical fiber is governed by the NSLE, in which now $A = a_1 + a_2$. The corresponding equation can be separated into the following coupled equation system [55]:

$$\mathrm{j}\frac{\partial a_1}{\partial z} = -\mathrm{j}\frac{a}{2}a_1 + \frac{\beta_2}{2}\frac{\partial^2 a_1}{\partial t^2} - \gamma(|a_1|^2 + 2|a_2|^2)a_1 \tag{5.228}$$

$$\mathrm{j}\frac{\partial a_2}{\partial z} = -\mathrm{j}\frac{a}{2}a_2 + \frac{\beta_2}{2}\frac{\partial^2 a_2}{\partial t^2} - \gamma(|a_2|^2 + 2|a_1|^2)a_2 \tag{5.229}$$

It can be seen that the presence of pulse 2 affects the propagation of pulse 1 through the $2|a_2|^2$ term, and vice versa. Analysis of IXPM impact on DM systems can be found, for example, in [56], but a more intuitive explanation can be obtained by drawing a parallel with quantum mechanics [57]. Having reversed the roles of time and space, the term in parentheses can be interpreted as a potential to which the pulse reacts. Thus, for example, pulse 1 will experience a potential

$$V_1(t) = -\gamma(|a_1|^2 + 2|a_2|^2) \tag{5.230}$$

The SPM term creates a shallow potential well for the pulse. (For high enough power levels the potential well will eventually be able to localize the pulse energy, which is an alternative way to view SPM-induced soliton formation [55] and pulse compression [58]. In the quasi-linear regime, this is far from the case and the pulse will broaden throughout transmission.) The IXPM term introduces an asymmetry in the potential. Because the derivative of the potential is a force, the pulse will be accelerated towards or away from the other pulse, depending on the sign of β_2. The acceleration is zero when the pulses do not overlap, since the term $-2\gamma|a_2|^2$ will then be zero across pulse 1. When the pulses begin to overlap, the slope and depth of $-2\gamma|a_2|^2$ will be significant and a_1 will be accelerated, gaining a certain velocity. However, the acceleration decreases as the pulses broaden further, and the peak power $|a_2|^2$ will decrease correspondingly. From this point on, the pulse will continue to drift towards (or away from) the other pulse at constant velocity, until the signal reaches the next fiber section, for which the dispersion sign is reversed, and the pulse starts drifting in the opposite direction, at the same time as it starts narrowing.

The pulse will keep drifting at constant velocity until the slope and depth of $-2\gamma|a_2|^2$ again become strong enough to accelerate the pulse to a new velocity. This is repeated several times throughout transmission, and the final location of the pulse will depend on where the accelerations take place. If the final location of the pulse is not at the center of the bit slot (*timing jitter*), then the pulse will be sampled in a nonoptimal position (where the intensity is not maximum), possibly leading to a bit error. However, since the locations where the accelerations take place are a function of the accumulated dispersion, IXPM can be successfully suppressed by launch-position optimization [59,60].

5.7.2.2 Intrachannel Four-Wave Mixing

FWM arises when three waves at different wavelengths combine nonlinearly to generate a fourth wave at a specified wavelength, as seen in Section 5.3.2.3. FWM is a consequence of the presence of the nonlinear term $\gamma|A|^2A$, in the NLSE, when A is a sum of at least three waves. In IFWM [61,62] the same principle holds, although now A is a sum of three pulses in the time domain that overlap due to dispersion broadening. A fourth pulse, rather than a fourth wave, is then generated.

In order to analyze IFWM it is convenient to decompose the signal envelope in the NLSE as $A = A_l + A_p$, where A_l is the solution to the linear Schrödinger equation (i.e., with $\gamma = 0$, and A_p is a non-inear perturbation. For the power ranges encountered in practical systems, a quasi-linear transmission regime can be assumed, for which $A_p \ll A_l$. Thus, the NLSE becomes

$$\mathrm{j}\frac{\partial A_p}{\partial z} = -\mathrm{j}\frac{a}{2}A_p + \frac{\beta_2}{2}\frac{\partial^2 A_p}{\partial t^2} - \gamma|A_l|^2 A_l \tag{5.231}$$

if higher order perturbation terms, $A_l|A_p|^2$, $A_l^* A_p^2$, and $A_p|A_p|^2$, are neglected for a first-order analysis. In a binary transmission system the envelope signal is the sum of *zero* and *one* pulses:

$$A_l(z,t) = \sum_n a_n(z,t) = \sum_n a_0(z,t-nt) \tag{5.232}$$

where $a_0(z,t)$ is a single pulse propagating in the linear regime ($\gamma = 0$, T is the time slot, which is 25 ps at 40 Gbit/s, and n runs through the *one* bit slots.

Using Equation 5.232 one obtains $\gamma|A_l|^2 A_l = \gamma \sum a_q a_r a_s^*$; and since Equation 5.231 is linear in A_p, the perturbation envelope can be defined as the sum of perturbation contributions; that is:

$$A_p = \sum a_{qrs} \tag{5.233}$$

where each contribution a_{qrs} is determined by

$$j\frac{\partial a_{qrs}}{\partial z} = -j\frac{a}{2}a_{qrs} + \frac{\beta_2}{2}\frac{\partial^2 a_{qrs}}{\partial t^2} - \gamma a_q a_r a_s^* \tag{5.234}$$

It is relatively easy, in the frequency domain, to solve Equation 5.234 for a_{qrs} at distance z in a fiber (with $a_{qrs}(t,0) = 0$, since the perturbation must be zero at the transmitter):

$$a_{qrs}(z,t) = j\gamma \int_0^z F^{-1}[F(a_q a_r a_s^*) e^{k(z-z')/2}]\, dz', \tag{5.235}$$

where F and F^{-1} denote the Fourier and inverse Fourier transforms respectively, and the propagation parameter k is defined as

$$k = j\omega^2 \beta_2 - \alpha \tag{5.236}$$

If several fiber sections are cascaded after each other, as is the case in a DM link, then the perturbation contribution at the end of the link will be

$$a_{qrs}(L,t) = \sum_i j\gamma_i \int_0^{L_i} F^{-1}[F[a_q a_r a_s^*] e^{k_i(L_i - z')/2}]\, dz' \tag{5.237}$$

where L_i is the length of the ith fiber section, and $L = \sum L_i$. Accordingly, k_i and γ_i refer to the propagation and nonlinear parameters for the ith fiber section.

Equation 5.237 can only be solved numerically. Further insight can, however, be gained by an approximate analysis [57,63], through which it can be shown that the contribution a_{qrs} after one DM cell[58] will be a pulse centered around

$$t_{qrs} = (q + r - s)T \tag{5.238}$$

[58] Some simplifying assumptions are made in [57] and [63]: the data pulses are assumed to be Gaussian, the link is lossless ($\alpha = 0$), and the DM cell is symmetric ($L_1 = L_2$, $\beta_{2,1} = -\beta_{2,2}$, and $\gamma_1 = \gamma_2$).

The intensity of the perturbation in the middle of a specific bit slot n will, therefore, be the sum of all contributions a_{qrs} for which $q + r - s = n$:

$$A_{\mathrm{p}}(L, t-nT) = \sum_{q,r,(s=q+r-n)} a_{qrs} \tag{5.239}$$

The perturbation in bit slot n, $A_{\mathrm{p}}(L, t-nT)$, will cause a *ghost pulse* if bit n is a *zero* and *amplitude jitter* if bit n is a *one*. It should be noted from Equation 5.237 that each contribution a_{qrs} is determined by three generating pulses. Specifically, a_{qrs} must be proportional to

$$a_q a_r a_s^* = |a_q||a_r||a_s|\mathrm{e}^{\mathrm{j}(\phi_q + \phi_r - \phi_s)} \tag{5.240}$$

given the linearity of Equation 5.237. It follows that changing the relative phase of the driving pulses will change the phase of a perturbation contribution and, therefore, the amplitude of the ghost pulse (or amplitude jitter intensity) in each bit slot. Several IFWM-suppressing techniques proposed in the literature stem from this observation.

5.7.2.3 IFWM Suppression Methods

During the past few years IFWM has received considerable attention in the research community and a number of suppression techniques have been proposed. Many of them use the extra degree of freedom represented by the signal phase to increase nonlinear tolerance in OOK transmission. These techniques are reviewed in Section 6.2.

Transmitting neighboring bits with *alternate polarization* [64] is another way to reduce the impact of IFWM. The idea here is that pulses on orthogonal polarizations will not interact nonlinearly with each other (in fact, this is not completely true [55], but the interaction efficiency is indeed reduced). Alternate polarization has been numerically and experimentally shown to improve nonlinear tolerance in combination with different pulse types [65,66]. This transmission improvement comes at the cost of increased complexity at the transmitter.

A different approach is taken by *subchannel multiplexing* [67], in which the full-bit-rate signal is obtained by time-division multiplexing two signals at half the bit rate, each at a wavelength slightly different from each other. The result is that neighboring pulses in the resulting signal have slightly different wavelengths. This breaks the FWM frequency-matching condition for the build-up of the IFWM contributions. The main problem with this approach is that the signal's spectrum is broadened considerably. Implementation complexity is also an issue. One can also think of reducing the efficiency of IFWM by making the *spacing between pulses unequal*, as proposed in [68], so that the IFWM contribution does not accumulate all in the middle of the bit slot. Unfortunately, this is accompanied by some spectral broadening, and especially by a rather complex implementation.

Some researchers have followed a different strategy, and propose line coding techniques [69,70]. Here, the worst bit combinations for IFWM (such as a space followed by many marks) are simply avoided by coding the bit sequence before being transmitted. Although these techniques are successful and achieve good improvement in IFWM tolerance, a considerable overhead – from 10% to more than 40% [71] – is the price to pay.

During the past couple of years, an *electric-domain approach* has gained popularity for dispersion compensation [72], and very recently researchers have started investigating the possibility of applying this electronic equalization to IFWM suppression [73,74]. A limiting factor here, besides today's limited power of digital signal processing, is the fact that the phase information is lost at the receiver, so full compensation cannot be performed in the electrical domain. A proposed solution to this problem is *predistortion*, in which the signal is equalized in the optical domain at the transmitter, where both intensity and phase can be modulated. This technique was shown to be successful in dispersion compensation [75,76] and it has recently been numerically tested for IFWM suppression [77].

References

[1] Tomkos, I., Vogiatzis, D., Mas, C. *et al.* (2004) Performance engineering of metropolitan area optical networks through impairment constraint routing. *IEEE Communications Magazine*, **42**, 540–547.

[2] Personick, S.D. (1973) Receiver design for digital fiber optic communication systems, I. *Bell Systems Technical Journal*, **52** (6), 843–874.

[3] Personick, S.D. (1973) Receiver design for digital fiber optic communication systems, II. *Bell Systems Technical Journal*, **52** (6), 875–886.

[4] Leners, R., Georges, T., François, P.L. and Stéphan, G. (1994) Analytic model of polarization dependent gain in erbium doped fibre amplifiers. Conference on Optical Amplifiers and Their Applications, pp. 24–26, Yokohama (Japan); Tech. Dig. Post-deadline papers.

[5] Latellier, V., Bassier, G., Marmier, P. *et al.* (1994) Polarisation scrambling in 5 Gbit/s 8100 km EDFA based system. *Electronics Letters*, **30** (7), 589–590.

[6] Marcuse, D. (1970) *Engineering Quantum Electrodynamics*, Harcourt, Brace & World, New York, NY.

[7] Shimoda, K., Takahashi, H. and Townes, C.H. (1957) Fluctuation in amplification of quanta with application to maser amplifiers. *Journal of the Physical Society of Japan*, **12** (6), 686–700.

[8] Desurvire, E. (1994) *Erbium-Doped Fiber Amplifiers – Principles and Applications*, John Wiley & Sons, Inc., New York, NY.

[9] Okoshi, T. and Kikuchi, K. (1988) *Coherent Optical Fiber Communications*, KTK-Kluwer.

[10] Risken, H. (1989) *The Fokker-Planck Equation*, 2nd edn, Springer-Verlag, Berlin.

[11] Shimada, S. and Ishio, H.(eds) (1994) *Optical Amplifiers and their Applications*, John Wiley & Sons, Ltd, Chichester.

[12] Haken, H. (1983) *Laser Theory*, Springer, Berlin.

[13] Heffner, H. (July 1962) The fundamental noise limit of linear amplifiers. Proceedings of the IRE, pp. 1604–1608.

[14] Maitland, A. and Dunn, M. H. (1969) *Laser Physics*, North-Holland, Amsterdam.

[15] Olsson, N.A. (1989) Lightwave systems with optical amplifiers. *Journal of Lightwave Technology*, **7** (7), 1071–1082.

[16] Perina, J. (1984) *Quantum Statistics of Linear and Nonlinear Optical Phenomena*, Reidel, Dordrecht.

[17] Marcuse, D. (1991) Calculation of bit-error probability for a lightwave system with optical amplifiers and post-detection gaussian noise. *Journal of Lightwave Technology*, **9** (4), 505–513.

[18] Chan, B. and Conradi, J. (1997) On the non-Gaussian noise in erbium-doped fiber amplifiers. *Journal of Lightwave Technology*, **15** (4), 680–687.

[19] Ramaswami, R. and Sivarajan, K.N. (2002) *Optical Networks – A Practical Perspective*, 2nd edn, Morgan Kaufmann, San Francisco, CA.

[20] Gradshteyn, I. S. and Ryzhik, I.M. (1965) *Table of Integrals, Series, and Products.*

[21] Kato, T., Koyano, Y. and Nishimura, M. Temperature dependence of chromatic dispersion in various types of optical fibers, OFC'2000, paper TuG7.

[22] Kissing, J., Gravemann, T. and Voges, E. (2003) Analytical probability density function for the Q factor due to PMD and noise. *IEEE Photonics Technology Letters*, **15** (4), 611–613.

[23] Lima, I.T. Jr, Lima, A.O., Zweck, J. and Menyuk, C.R. (2003) Performance characterization of chirped return-to-zero modulation format using an accurate receiver model. *IEEE Photonics Technology Letters*, **15** (4), 608–610.

[24] Agrawal, G.P. (1997) *Fiber-Optic Communication Systems*, John Wiley & Sons, Inc., New York, NY.

[25] ITU-T (2003) Optical transport network physical layer interfaces, Rec. G.959.1.

[26] ITU-T (2003) Transmission characteristics of optical components and subsystems, Rec. G.671.

[27] ITU-T (2003) Optical interfaces for single-channel STM-64, STM-256 and other SDH systems with optical amplifiers, Rec. G.691.

[28] ITU-T (2004) Optical interfaces for multichannel systems with optical amplifiers, Rec. G.692.

[29] Goldstein, E.L., Eskildsen, L. and Elrefaie, A.F. (1994) Performance implications of component crosstalk in transparent lightwave networks. *Photonics Technology Letters*, **6** (5), 657–660.

[30] Legg, P.J., Tur, M. and Andonovic, I. (1996) Solution paths to limit interferometric noise induced performance degradation in ASK/direct detection lightwave networks. *Journal of Lightwave Technology*, **14** (9), 1943–1954.

[31] Liu, F., Rasmussen, C.J. and Pedersen, R.J.S. (1999) Experimental verification of a new model dscribing the influence of incomplete signal extinction ratio on the sensitivity degredation due to multiple interferometric crosstalk. *Photonics Technology Letters*, **11** (1), 137–139.

[32] Takahashi, H., Oda, K. and Toba, H. (1996) Impact of crosstalk in an arrayed-waveguide multiplexer on $N \times N$ optical interconnection. *Journal of Lightwave Technology*, **14** (6), 1097–1105.

[33] Caponi, R., Potenza, M., Schiano, M. *et al.* (1998) Deterministic nature of polarisation mode dispersion in fibre amplifiers. ECOC '98, Madrid, Spain.

[34] Poole, C.D., Tkach, R.W., Chraplyvy, A.R. and Fishman, D.A. (1991) Fading in lightwave systems due to polarization-mode dispersion. *IEEE Photonics Technology Letters*, **1**, 68–70.

[35] Sunnerud, H., Karlsson, M. and Andrekson, P.A. (2001) A comparison between NRZ and RZ data formats with respect to PMD-induced system degradation. *Photonics Technology Letters*, **13** (5), 448–450.

[36] Mecozzi, A. and Shtaif, M. (2002) The statistics of polarization-dependent loss in optical communication systems. *IEEE Photonics Technology Letters*, **14**, 313–315.

[37] ITU-T, (2006) Optical interfaces for equipments and systems relating to the synchronous digital hierarchy, Rec. G.957.

[38] Born, M. and Wolf, E. (1964) *Principles of Optics*, 2nd rev. edn, Pergamon Press, Oxford.

[39] ITU-T (2003) *Application Related Aspects of Optical Amplifier Devices and Subsystems*, Rec. G.663.

[40] Marcuse, D. (1981) Pulse distortion in single-mode fibers. 3: Chirped pulses. *Applied Optics*, **20** (20), 3573–3579.

[41] Agrawal, G.P. (1989) *Nonlinear Fiber Optics*, Academic Press, San Diego, CA.

[42] Zirngibl, M. (1998) Analytical model of Raman gain effects in massive wavelength division multiplexed transmission systems. *Electronics Letters*, **34** (8), 789–790.

[43] Shibata, N., Braun, R.-P. and Waarts, R.G. (1987) Phase-mismatch dependence of efficiency of wave generation through four-wave mixing in a single-mode optical fiber. *IEEE Journal of Quantum Electronics*, **23**, 1205.

[44] Inoue, K. (1992) Phase mismatching characteristic of four-wave mixing in fiber lines with multistage amplifiers. *Optics Letters*, **17** (11), 810.

[45] Percelsi, A., Riccardi, E. and Sordo, B. (2001) Metodologie per la progettazione trasmissiva di collegamenti con sistemi DWDM, Fotonica 2001, Ischia, 23–25 maggio.

[46] Sun, Y., Srivastava, A.K., Zhou, J. and Sulhoff, J.W. (1999) Optical fiber amplifiers for WDM optical networks. *Bell Labs Technical Journal*, **4**, 187–206.

[47] Levine, B. (1973) *Fondements Théoriques de la Radiotechnique Statistique*, Vol. **I**, Mir, Moscow.

[48] Mørk, J., Öhman, F. and Bischoff, S. (2003) Analytical expression for the bit error rate of cascaded all-optical regenerators. *IEEE Photonics Technology Letters*, **15** (10), 1479–1481.

[49] Mårtensson, J., Djupsjobacka, A., Li, J. *et al.* (2006) Field transmission using fiber Bragg grating-based dispersion compensating modules with continuous response over the full C-band. European Conference on Optical Communications (ECOC), Cannes, France, Th3.3.6.

[50] Fews, H.S., Stephens, M.F.C., Straw, A. *et al.* (2006) Experimental comparison of fibre and grating-based dispersion compensation schemes for 40 channel 10Gb/s DWDM systems. European Conference on Optical Communications (ECOC), Cannes, France, Th3.2.5.

[51] Kaminow, I.P. and Koch, T.L. (1997) *Optical Fiber Telecommunications IIIB*, Academic Press.

[52] Mecozzi, A., Clausen, C. and Shtaif, M. (2000) Analysis of intrachannel nonlinear effects in highly dispersed optical pulse transmission. *IEEE Photonics Technology Letters*, **12** (4), 392–394.

[53] Pizzinat, A., Schiffini, A., Alberti, F. *et al.* (2002) 40-Gb/s systems on G.652 fibers: comparison between periodic and all-at-the-end dispersion compensation. *IEEE Journal of Lightwave Technology*, **20** (9), 1673–1679.

[54] Winzer, P.J. and Essiambre, R.J. (2006) Advanced modulation formats for high-capacity optical transport networks. *IEEE Journal of Lightwave Technology*, **24** (12), 4711–4728.

[55] Agrawal, G.P. (1989) *Nonlinear Fiber Optics,* Quantum Electronics – Principles and Applications, Academic Press, San Diego, CA.
[56] Mårtensson, J., Berntson, A., Westlund, M. *et al.* (2001) Timing jitter owing to intrachannel pulse interactions in dispersion-managed transmission systems. *Optics Letters*, **26** (2), 55–57.
[57] Johannisson, P. (2006) Nonlinear intrachannel distortion in high-speed optical transmission systems, PhD thesis, Chalmers University of Technology.
[58] Siegman, A.E. (1986) *Lasers*, University Science Books, Mill Valley, CA.
[59] Mårtensson, J., Westlund, M. and Berntson, A. (2000) Intra-channel pulse interactions in 40Gbit/s dispersion-managed RZ transmission system. *Electronics Letters*, **36** (3), 244–246.
[60] Mecozzi, A., Clausen, C., Shtaif, M. *et al.* (2001) Cancellation of timing and amplitude jitter in symmetric links using highly dispersed pulses. *IEEE Photonics Technology Letters*, **13** (5), 445–447.
[61] Essiambre, R.-J., Mikkelsen, B. and Raybon, G. (1999) Intra-channel cross-phase modulation and four-wave mixing in high-speed TDM systems. *Electronics Letters*, **35**, 1576–1578.
[62] Mamyshev, P.V. and Mamysheva, N.A. (1999) Pulse-overlapped dispersion-managed data transmission and intrachannel four-wave mixing. *Optics Letters*, **24** (21), 1454–1456.
[63] Ablowitz, M.J. and Hirooka, T. (2000) Resonant nonlinear intrachannel interactions in strongly dispersion-managed transmission systems. *Optics Letters*, **25** (24), 1750–1752.
[64] Matera, F., Romagnoli, M. and Daino, B. (1995) Alternate polarisation soliton transmission in standard dispersion fibre links with no in-line controls. *Electronics Letters*, **31**, 1172.
[65] Appathurai, S., Mikhailov, V., Killey, R. and Bayvel, P. (2004) Effective suppression of intra-channel nonlinear distortion in 40 Gbit/s transmission over standard singlemode fibre using alternate-phase RZ and alternate polarisation. *Electronics Letters*, **40** (14), 897–898.
[66] Xie, C., Kang, I., Gnauck, A. *et al.* (2004) Suppression of intrachannel nonlinear effects with alternate-polarization formats. *IEEE Journal of Lightwave Technology*, **22** (3), 806–812.
[67] Zweck, J. and Menyuk, C.R. (2002) Analysis of four-wave mixing between pulses in high-data-rate quasi-linear subchannel-multiplexed systems. *Optics Letters*, **27** (14), 1235–1237.
[68] Kumar, S. (2001) Intrachannel four-wave mixing in dispersion managed RZ systems. *Photonics Technology Letters, IEEE*, **13** (8), 800–802.
[69] Alic, N. and Fainman, Y. (2004) Data-dependent phase coding for suppression of ghost pulses in optical fibers. *IEEE Photonics Technology Letters*, **16** (4), 1212–1214.
[70] Vasic, B., Rao, V., Djordjevic, I. *et al.* (2004) Ghost-pulse reduction in 40-Gb/s systems using line coding. *IEEE Photonics Technology Letters*, **16** (7), 1784–1786.
[71] Djordjevic, I.B. and Vasic, B. (2006) Constrained coding techniques for the suppression of intrachannel nonlinear effects in high-speed optical transmission. *IEEE Journal of Lightwave Technology*, **24** (1), 411–419.
[72] Bülow, H., Franz, B., Buchali, F. and Klekamp, A. (2006) Electronic mitigation of impairments by signal processing. European Conference on Optical Communications (ECOC), Cannes, France, We2.5.1.
[73] Djordjevic, I.B. and Vasic, B. (2006) Adaptive BCJR equalizer in suppression of intrachannel nonlinearities. European Conference on Optical Communications (ECOC), Cannes, France, We1.5.5.
[74] Xia, C. and Rosenkranz, W. (2006) Mitigation of optical intrachannel nonlinearity using nonlinear electrical equalization. European Conference on Optical Communications (ECOC), Cannes, France, We1.5.3.
[75] McGhan, D., Laperle, C., Savchenko, A. *et al.* (2005) 5120 km RZ-DPSK transmission over G652 fiber at 10 Gb/s with no optical dispersion compensation. Optical Fiber Communication Conference (OFC), Aneheim, CA, USA, PDP27.
[76] Killey, R., Watts, P., Mikhailov, V. *et al.* (2005) Electronic dispersion compensation by signal predistortion using digital processing and a dual-drive Mach–Zehnder modulator. *IEEE Photonics Technology Letters*, **17** (3), 714–716.
[77] Weber, J.K., Fischer, C., Bunge, C.-A. and Petermann, K. (2006) Electronic precompensation of intra-channel nonlinearities at 40 Gbit/s. European Conference on Optical Communications (ECOC), Cannes, France, We1.5.4.

6

Combating Physical Layer Degradations

Herbert Haunstein, Harald Rohde, Marco Forzati, Erwan Pincemin, Jonas Martensson, Anders Djupsjöbacka and Tanya Politi

6.1 Introduction

Today's core networks rely on optical transmission for high capacity and long reach. Since the late 1970s, when the first optical transmission systems were introduced in real systems, doubling/quadrupling of the modulation rate was used to achieve higher capacities while engineering of the optical channel was deployed for reach expansion with minimum capital/operational expenditure. The quest for a higher capacity times length product (BL), that the wide bandwidth/low-loss fiber medium seemed so promising in fulfilling, has been only incrementally achieved. In an effort to use the existing infrastructure for higher capacity systems, fiber impairments and component performance limitations were limiting factors that would only unveil themselves when wider bandwidths, higher power or even more channels were about to be introduced through wavelength-division multiplexing (WDM). Although the advent of erbium-doped fibre amplifiers (EDFAs) dramatically assisted towards BL enhancement 20 years ago, they still remain what we call 1R regenerators. Today, limitations arise mainly from dispersive and nonlinear effects; therefore, amplifiers do not seem to combat the impairments.

Presently, transmission systems will deploy 40 Gbit/s technology, a development that has been made possible by advances in the technology of transponders and, as with previously attempted BL enhancements, hinges on the promise of capital expenditure reductions for wavelength-division multiplexed long-haul and metropolitan transmission systems. Such systems are expected to satisfy the practical requirements of the existing carriers' fiber infrastructure, and the constraints imposed by typical functionalities of next-generation all-optical networks. Critical issues remain system compatibility with deployed standard

Core and Metro Networks Edited by Alexandros Stavdas

single-mode fiber (SSMF) lines (which are frequently hampered by relatively large polarization-mode dispersion (PMD)) and optical add/drop multiplexer (OADM) capabilities. To this end, it is critical to compensate the physical degradations that were explained in Chapter 5. In this context, it is crucial to combine the efforts of the research community with those of the operators in order to guide the practical implementation of new high-speed optical communication technologies, not only in the long-awaited 40 Gbit/s, but also for future higher ones.

In previous chapters the different physical degradations were explained together with ways to compensate for example losses and amplifiers. Evidently, as the bit rate increases, special enabling technologies may be adopted, such as compensating components for chromatic dispersion (CD) and PMD, or modulation formats resilient to transmission impairments and OADM filtering, distributed and/or dynamic channel equalization. Practical implementations of 40 Gbit/s and beyond systems may completely relay on these subsystems. In this chapter, practical ways for combating physical layer impairments are discussed with emphasis on the ones that limit the introduction of 40 Gbit/s systems.

6.2 Dispersion-Compensating Components and Methods for CD and PMD

6.2.1 Introduction on Optical CD and PMD Compensator Technology

In Section 5.3 the phenomenon of CD has been explained with emphasis on impairments caused by CD on an optical signal. In Section 5.6 we outlined the ways the influence of the CD can be eliminated (in the linear case) or at least reduced (when nonlinear fiber effects are accounted for) by dispersion compensators. In this chapter, three different classes of dispersion-compensating devices will be presented:

- static dispersion-compensating devices, which are able to compensate a fixed (but usually wavelength-dependent) value of dispersion;
- dynamic dispersion-compensation devices, in which the dispersion value can be precisely set by a control signal within a relatively small range;
- dynamic dispersion compensators, in which the dispersion value can be set over a wide range.

Static dispersion compensators are used in point-to-point links for bit rates of up to 40 Gbit/s. The dispersion values of the link or of the single spans are known and each span or link is compensated according to the dispersion map (Section 5.6).

As the sensitivity of an optical transmission link towards CD increases proportionally to the square of the data rate, transmission systems with high data rates of, for example, 100 Gbit/s or 160 Gbit/s are extremely vulnerable to insufficiently adjusted dispersion compensators. Figure 6.1 shows a comparison of the dispersion tolerances of data streams modulated by simple on–off keying (OOK; see Section 2.6.1) at different transmission data rates between 10 and 160 Gbit/s. The figure plots the optical signal-to-noise ratio (OSNR) penalty[1] against the accumulated dispersion value at the position of the receiver.

[1] OSNR penalty: the amount of additional OSNR needed to achieve the same bit error rate (BER) of 10^{-9} as without dispersion.

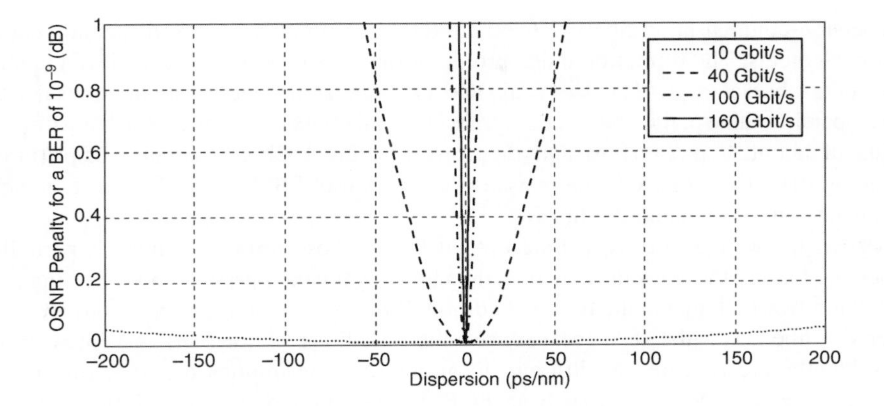

Figure 6.1 OSNR penalty for different data rates and dispersion values

For a bit rate of 10 Gbit/s, the OSNR penalty is less then a tenth of a decibel within a range of about ±300 ps/nm, while for 160 Gbit/s an OSNR penalty of 1 dB is already reached when the dispersion mismatch has a value of about ±3 ps/nm. Evidently, a dispersion mismatch of the equivalent of some 10 m of SSMF can significantly reduce the system performance at such bit rates.

In order to avoid large stock requirements for different dispersion compensation modules (DCMs), tunable DCMs are used to compensate the remaining dispersion which is left after inserting the closest matching fixed DCM.

The typical variation of the optical transmission time of a pulse through a fiber due to temperature change is 40 ps/(km K). Assuming a link length of 200 km and a temperature change of about 6 K, this small value converts into a change of effective fiber length of 10 m, justifying the need for dynamically fine-tunable dispersion compensators for high-speed transmission links due to temperature variations along these links alone. As the transmission power of single channels within a dense WDM (DWDM) system might vary, the self-phase modulation (SPM)-induced optimal dispersion setting might also vary and might have to be adjusted by a tunable compensator with a small but precisely settable compensation range.

Transparent optical networks allow in some configurations arbitrary switching of wavelengths through a fiber mesh. A wavelength arriving at an arbitrary receiver can have different accumulated dispersion values, depending on the path it has been switched through. Therefore, at the receiver, a device which is able to compensate a wide range of dispersion is needed.

6.2.2 Optical Compensation Schemes

This section presents briefly the most commonly used optical dispersion compensation schemes. Some of the envisaged optical processing schemes are capable of mitigating different distortions simultaneously. Owing to the processing of the optical field containing amplitude, phase, and polarization information, all linear distortions can theoretically be completely compensated by applying the inverse optical transfer function of the transmission channel.

- Dispersion-compensating fibers (DCFs) are the most commonly used dispersion compensation elements. The refraction index profile of the fiber is designed to achieve the desired dispersion characteristics [1]. A typical DCF has dispersion values of about minus five times the dispersion value of the fiber used for transmission; for example, an SSMF has a dispersion value of about 17 ps/(nm km) and the corresponding DCF has about −85 ps/(nm km), meaning that 20 km of DCF compensates for 100 km of SSMF. A DCF is a static device; its dispersion value cannot be tuned.
- Fiber Bragg gratings (FBGs) [2] are optical fibers whose refraction index is periodically modulated along the transmission axis, thus forming a Bragg grating and reflecting certain wavelengths which match the Bragg condition. If the refraction index variations are chirped over the fiber (i.e., the distance between the variations is not constant), then different wavelengths are reflected at different locations. As this implies different run-times for different optical frequencies, such an FGB can be tailored to have arbitrary dispersion properties.
- Etalons (Fabry–Perot resonators) [3] consist of a pair of facing (dielectric) mirrors that create a cavity. If this cavity length is an integer multiple of the wavelength, then a standing-wave optical field builds up in between the two mirrors, resulting in a wavelength-dependent transmission. The amplitude and phase response of such an etalon shows (cyclic) spectral parts so it is equivalent to a dispersive element; thus, it can be used as a dispersion compensating element.
- Tunable chirped FBGs (CFBGs) [2–5] are already commercially available as multichannel devices at 10 Gbit/s and 40 Gbit/s. These devices are thermally tuned. A temperature change leads to a change in the length of the FBG, changing the length grating constant and, therefore, the dispersion properties (Figure 6.2). Double grating devices allow for a wider and symmetric tuning range (tuning range referenced below).
- Tunable GT etalons (Gires–Tournois etalons) [6,7] consist of two cascaded multilayer dielectric cavities in the reflection mode which is thermally tuned to generate a varying dispersion (Figure 6.2). They are multichannel devices and operating prototypes have already been demonstrated.
- Ring resonators are infinite impulse response (IIR) filters which have been realized as an integrated optical circuit structure (planar lightwave circuit (PLC)). With only a few rings (e.g., four) a channelized all-pass filter can be generated which exhibits a high CD (Figure 6.3). Tuning is accomplished with the help of planar heaters. Only laboratory demonstrations have been reported so far [8].

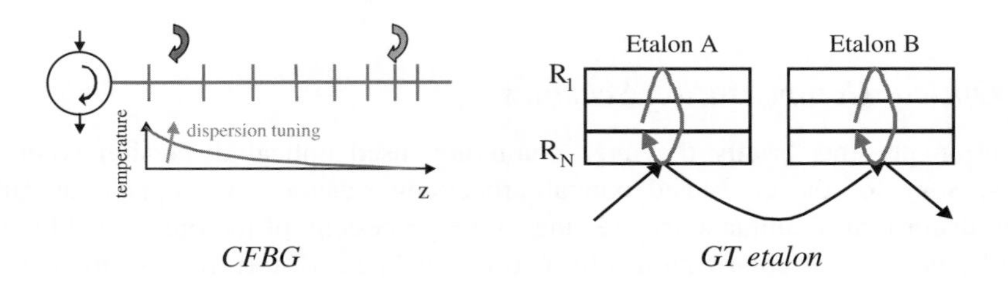

Figure 6.2 Tunable CD compensators

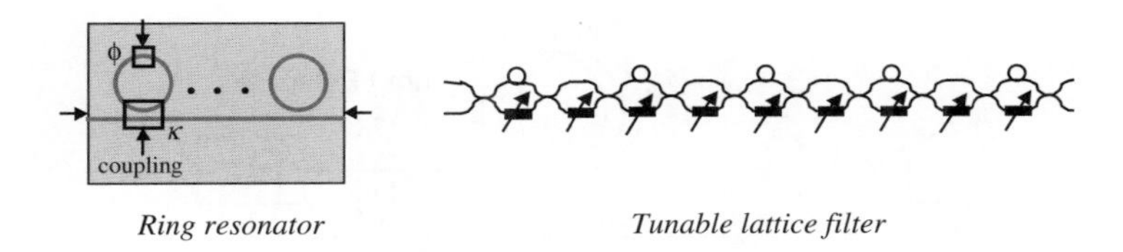

Ring resonator *Tunable lattice filter*

Figure 6.3 Integrated optic (PLC) CD compensators

- Tunable lattice filters (finite impulse response filters) have also been realized as PLCs with thermo-optic tuning elements (Figure 6.3). Laboratory prototypes with up to 13 stages have been shown.

6.2.2.1 Polarization-Mode Dispersion Compensation

As optical communications systems are driven towards increasing bit rates and longer transmission distances, PMD starts affecting the transmission quality. PMD is caused by the different signal velocities of the orthogonal states of polarization in birefringent media (so-called differential group delay (DGD)). Its effects can be compensated in the optical domain with one of the following principles:

- The "standard" fiber-based PMD compensator consists of a cascade of polarization controllers and birefringent polarization-maintaining fibers (PMFs) (see Figure 6.4).
- If the signal light travels in one of the two principle states of polarization (PSPs) of the transmission link, then no DGD and, therefore, no signal distortions occur. Figure 6.5 shows the schematic setup of a PMD mitigation scheme using this effect. Directly after the transmitter Tx a polarization controller (PC) adjusts the polarization state of the transmitted signal such that it equals one of the two PSPs of the transmission link. After the link, an (optional) polarizer filters the incoming light before the receiver. The receiver generates an error signal which is fed back to the transmitter-side PC to be able to adapt the PC settings.

The scheme, although simple and easy to realize, has a number of disadvantages. It needs a feedback (FB) channel, it can only compensate for first-order PMD (i.e., the DGD has to be sufficiently constant over the optical spectrum of the signal), and it cannot cope with circularly polarized parts of the signal.

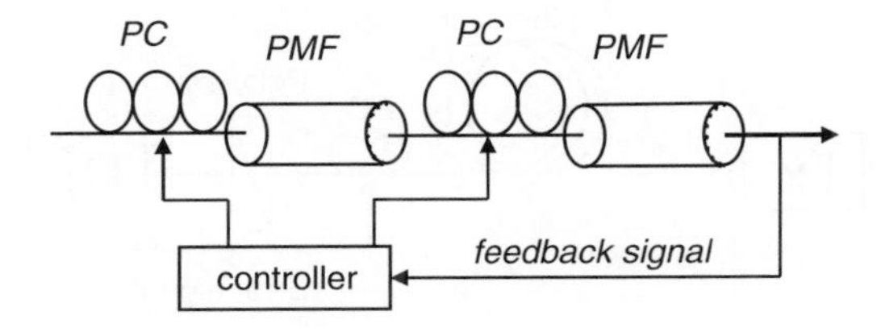

Figure 6.4 Schematic setup of the standard PMD compensator setup

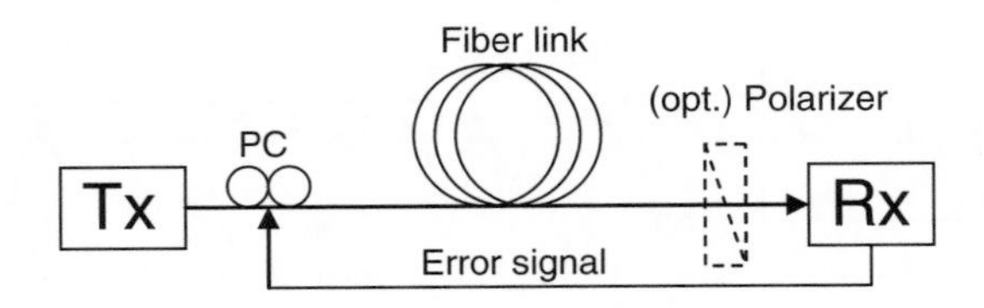

Figure 6.5 Schematic setup of the PSP PMD mitigation scheme

- PMD mitigation by fast distributed polarization scrambling. A further approach for PMD mitigation has recently been proposed based on fast polarization scrambling (20 MHz) of the signal along the link and forward error correction (FEC) in the receiver. Experiments at 10.7 Gbit/s indicated a PMD limit increase beyond ~30 ps can be tolerated for an allocated 3 dB OSNR margin. All WDM channels are polarization scrambled simultaneously by fast scramblers, as shown in Figure 6.6. Hence, the PMD statistics of the link, which might cover the Maxwellian DGD distribution in the time scale of years for terrestrial fiber, is strongly accelerated to the time scale shorter than an FEC frame (3 μs). Hence, outage events are mapped to error events which can completely be corrected by the FEC, provided the total amount of errors per frame does not exceed the correction capability ($\sim 10^{-5}$ for the standard FEC and 10^{-3} for the ultra FEC) and the maximum burst correction length. This scheme has been experimentally verified for bit rates of up to 43 Gbit/s [9–13].
- A cheap and simple scheme, although with limited performance, is shown in Figure 6.7. After the transmission the signal passes a polarization controller and a polarizer. If the PMD-induced distortion is not too big, then filtering out just the component of the light which gives the best received signal is a possibility to reduce the PMD distortions.

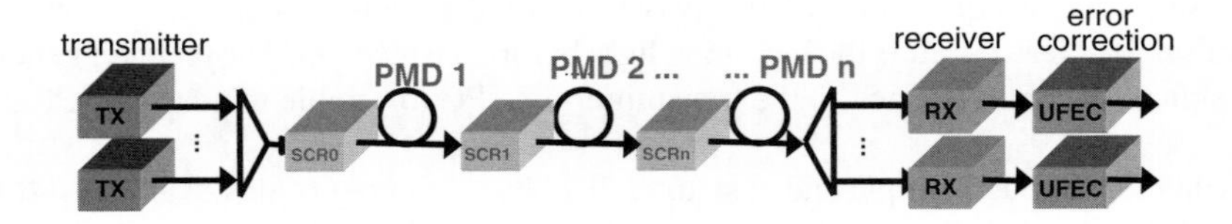

Figure 6.6 Schematic of multichannel PMD mitigation with distributed fast polarization scramblers (SCR) and enhanced error correction (EFEC)

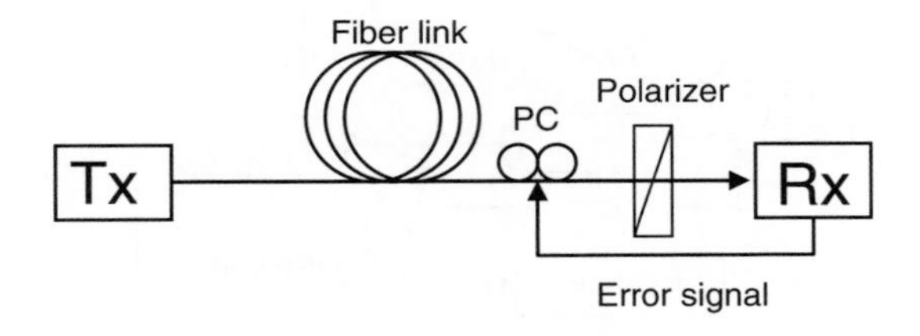

Figure 6.7 Schematic of setup of the polarizer PMD mitigation scheme

The inline PMD compensator is a first-order compensator, although it has some slightly beneficial effects also on higher order PMD. As a consequence, it is designed to be used as a distributed compensator. The major advantage of this PMD compensator is indeed its simplicity, which makes it easy to implement and integrate in EDFA modules.

6.2.3 Key Parameters of Optical Compensators

The performance improvement which can be achieved through the use of optical compensators depends on the mitigation scheme. For example, a single-channel compensator might achieve perfect compensation in the point of optimum setting, whereas in multichannel compensators a common tuning of all wavelength channels often means a nonoptimum setting of at least some of the channels. In general, the more degrees of freedom (DOFs) there are reserved per channel, the lower the residual penalties that can be achieved.

For dynamically adapting structures (in general) the adaptation speed and the residual penalty are intimately linked: relaxing the speed requirements for adaptation will allow for sufficient time:

i. to observe even slight deviations of the FB signal from optimum
ii. to manage more tuning parameters (if existent) to adjust precisely the signal processing part to the actual distortion.

The adaptation speed of an optical compensator/equalizer is the time the device needs to adjust to a distorted signal (dispersion different from the initial setting) ending up with a low remaining penalty (Q penalty; e.g., 1 or 2 dB).

The adaptation time will be determined by three mechanisms.

1. **Tuning technology:** Thermal tuning with time constants of *several seconds* is applied in CFBG and GT-etalon dispersion compensators. Thermal heating electrodes of PLC phase controllers and liquid-crystal polarization (LQ) controllers operate in the time range of 1 ms to a few milliseconds. With piezo fiber squeezer polarization controllers and electrooptic phase controllers in lithium niobate ($LiNbO_3$), speeds of 100 μs down to submicroseconds respectively can be achieved.
2. **Acceptable signal quality:** Several measures of signal quality can be used with different performance metrics. The electrical spectrum of the detected signal (spectral line) and the degree of polarization (for PMD only) can be used for fast FB (10 μs and faster), but with reduced sensitivity, whereas FEC error count allows very precise and low penalty adaptation but requires a longer time for the acquisition of a sufficient number of errors. Acquisition time spans in the range of 1 ms to more than 0.1 s seem to be appropriate. An eye monitor enables a trade-off between sensitivity and acquisition time.
3. **Adaptation scheme:** In general, we can distinguish between the commonly used FB and the feed-forward (FF) adaptation scheme.

- The FB scheme is based on consecutive variation of the compensator's tuning parameters (dithering) towards an optimum quality signal. Dithering of many tuning parameters can add up to many dither steps and thus long adaptation times. For PMD compensators with

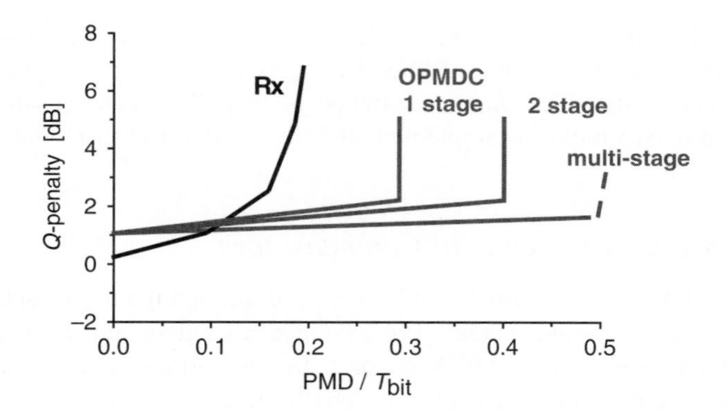

Figure 6.8 *Q*-penalty versus PMD (mean DGD) of optical PMD compensators

optimized dither procedures, proper adaptation is reached after *tens or even hundreds* of dither steps.

- FF adaptation allows for *one-step setting* of the compensator to the actual signal distortion. However, it needs a quality signal which provides a measure of the actual PMD or CD of the signal and it needs phase-stable optical processing to enable a calibrated setting of the tuning elements. It has been discussed for PMD compensation by using a wavelength scanning polarimeter for measuring the actual PMD distortion. With phase-stable PLC and LQ optics, a setting speed (adaptation speed) of less than 10 ms might be possible.

The following penalty assessment rules are applied and included in Figures 6.8 and 6.9: As additional adaptation penalty we assume:

- A 1 dB penalty for the (single-channel) PMD compensators (up to 2 dB at PMD limit) due to several adaptation parameters and time-consuming dithering.

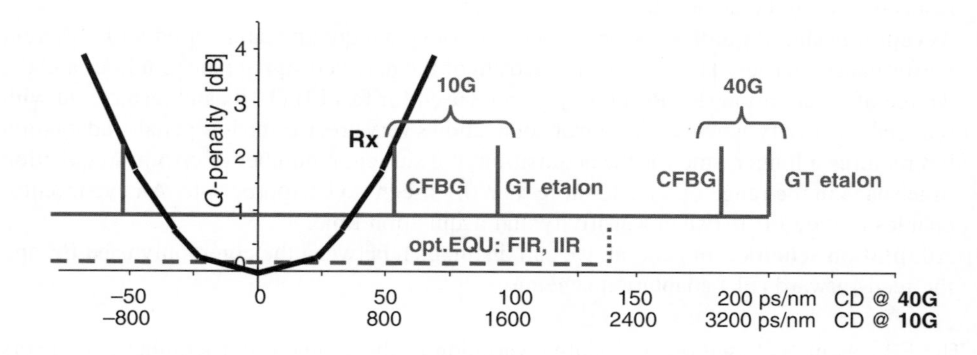

Figure 6.9 *Q*-penalty versus CD of dispersion compensators

- Also 1 dB for the FF-controlled PMD compensator which has a reduced adaptation penalty but must live with some mismatch of the commanded signal processing setting due to phase drift and incomplete FF distortion measurement.
- Negligible additional adaptation penalty (0 dB) for CFBG and etalon CD compensators due to relaxed time requirements on only one or two tuning parameters.
- A 0.2 dB penalty for slow optical equalizer dithering.

The CD and PMD operation ranges are summarized in Figures 6.8 and 6.9 showing the Q-penalty versus CD or PMD (nonreturn to zero (NRZ) signal) for the different options. Dither penalty/adaptation mismatch, as mentioned above, were taken into account. Specifically, the tuning ranges are also listed in more detail in Table 6.1.

The key parameters of the different compensator options can be found in Table 6.1. Rough guesses of the resulting adaptation speeds are listed in the right column; if not explicitly indicated by FF, an FB adaptation is assumed.

6.2.4 Compensators Suitable for Translucent Networks

Considering all the aforementioned parameters of the different CD and PMD compensator options, as well as the requirements from the network and the distortion dynamics, we come to the following selection for transparent networks.

6.2.4.1 CD Compensator

- **CFBG:** The most mature concept with a large tuning range is the thermally tuned CFBG. Unfortunately, it hardly meets the speed requirements; even the few seconds time scale for "loss of service" might be too fast for the actual tuning technology. Optimization of the thermal tuning or application of fast piezo actuator tuning might be inevitable if these devices were to be applied.
- **PLC-type (integrated-optic):** Lattice-filter structures designed for RZ bandwidth requirements and with not significantly more than about 10 stages (and phase tuning elements) tend to have a strongly reduced tuning range. Therefore, cascaded ring resonators remain as the only option. Adaptation down to the 50 ms limit might be possible.

6.2.4.2 PMD Compensator Combined with CD Compensator (Single Channel)

The speed requirements are mainly determined from the likely worst-case link PMD dynamics. Currently, adaptation speeds of a few milliseconds seem to be regarded as sufficient.

- **Fiber-based with $LiNbO_3$ polarizer or piezo controllers (1–2 stages):** With the appropriate FB signal (e.g., spectral lines), millisecond speed demand can be fulfilled. Sufficiently fast adaptation with piezo tuning is not obvious for multistage structures.
- **PLC-type, LQ technology (multistage):** Only with FF adaptation might these technologies have a chance to meet the speed requirements. The polarimeter quality signal does not allow for CD adaptation.

Table 6.1 Key parameters of optical CD and PMD compensators

Mitigation capability	Technology	Bit rate	Tuning range	Channel spacing	Tuning (speed)	Adaptation speed (FB)
Tunable CD compensation	CFBG	10G	±700 ps/nm	100 GHz	Thermal tuning (seconds)	10 s
		40G	±700 ps/nm	100 GHz		
	GT-etalon	10G	±1500 ps/nm	50 GHz	Thermal tuning (seconds)	10 s
		40G	±200 ps/nm	200 GHz		
Multidistortion equalizer, mainly CD	Ring resonator, PLC	10G	±2000 ps/nm	23 GHz	Thermal heater (ms)	100 ms
		40G	±150 ps/nm	74 GHz		
	n-stage lattice filter, PLC	40G	NRZ bandwith: ±120 ps/nm (13 stages)	Single channel, multichannel?	Thermal heater (ms)	100 ms
		10G?				
PMD compensation	1–2 stages: polarization control + PMF		25%T to 40%T (2nd stage)	Single channel	Piezo (10 μs)/LC (ms)	FB: ms–10 ms FF: <10 ms
					$LiNbO_3$ (μs)	1 ms
	1–2 stages: PLC with PBS				Thermal heater	FB: >10 ms FF: <10 ms
PMD + CD compensation	3 − *n* stages: polarization control + PMF		>40%T	Single channel	$LiNbO_3$ (μs)	1 ms
					Piezo (10 μs)/LC (ms)	FB: >10 ms FF: <10 ms
	PLC with PBS				Thermal heater (ms)	FB: >10 ms FF: <10 ms

6.2.5 Impact of Group-Delay Ripple in Fiber Gratings

6.2.5.1 Ripple Sensitivity

An ideal dispersion compensator exhibits dispersion and dispersion slope characteristics that exactly match the fiber it is supposed to compensate for. However, most dispersion compensators do not have this characteristic. For example, a dispersion-compensating fiber may show a correct dispersion value but will often show a dispersion slope that does not match the fiber that it is meant to compensate. A dispersion compensating FBG can easily be matched to both dispersion and dispersion slope but shows group-delay ripple (GDR) instead. This ripple originates from small, undesirable reflections inside the grating. The ripple that originates from the near and the far ends of the grating can be suppressed to some extent with apodization. However, residual reflections and "stitching errors" inside the grating are difficult to suppress. These reflections will cause a ripple in the phase response and in the amplitude response of the grating.

The above-mentioned reflections, which are seen as a ripple in the transfer curve of the device, give rise to the following question: To what extent can this ripple be tolerated in a system environment? A number of authors have been interested in this question [14–27].

Below we will analyze the ripple situation with a sinusoidal model. We will anticipate that the measured, residual, time delay is of sinusoidal type with chosen amplitude, phase, and period:

$$\Delta\tau = \frac{T_r}{2}\sin\left(\varphi_r + 2\pi\frac{\lambda}{\lambda_r}\right) \tag{6.1}$$

where T_r is the peak–peak amplitude, φ_r is the phase, and λ_r is the period. This time-delay function is also known as the residual group-delay (RGD) and it is a common performance metric for a fiber grating. However, the RGD function itself is of little use in a computer model since filters in computer models often rely on phase- and amplitude-response curves. The RGD function has to be converted into a phase curve. We now use a well-known formula from signal theory:

$$\Delta\tau = -\frac{\partial\varphi}{\partial\omega} = \frac{\lambda_c^2}{2\pi c}\frac{\partial\varphi}{\partial\lambda} \tag{6.2}$$

Equation 6.1 is now substituted into Equation 6.2 and by integrating with respect to λ we get

$$\begin{aligned}\varphi &= \frac{2\pi c}{\lambda_c^2}\int_{\lambda_c}^{\lambda}\frac{T_r}{2}\sin\left(\varphi_r + 2\pi\frac{\lambda}{\lambda_r}\right)\mathrm{d}\lambda \\ \varphi &= \frac{T_r\lambda_r c}{2\lambda_c^2}\left[\cos\left(\varphi_r + 2\pi\frac{\lambda_c}{\lambda_r}\right) - \cos\left(\varphi_r + 2\pi\frac{\lambda}{\lambda_r}\right)\right]\end{aligned} \tag{6.3}$$

where λ_c is the center wavelength. By changing the variable

$$\phi_r = \varphi_r + 2\pi\frac{\lambda_c}{\lambda_r} \tag{6.4}$$

where φ_r is the phase relative to λ_c. We can now write the final expression for the phase deviation as

$$\varphi = \frac{T_r \lambda_r c}{2\lambda_c^2}\left[\cos(\phi_r) - \cos\left(\phi_r + 2\pi\frac{\lambda-\lambda_c}{\lambda_r}\right)\right] \tag{6.5}$$

and distinguish between two special cases: $\varphi_r = 0$ and $\varphi_r = \pi/2$.

For $\varphi_r = 0$ the phase distortion function represents an even function round λ_c:

$$\varphi = \frac{T_r \lambda_r c}{2\lambda_c^2}\left[1 - \cos\left(2\pi\frac{\lambda-\lambda_c}{\lambda_r}\right)\right] \approx \frac{\pi^2 T_r \lambda_r c}{\lambda_c^2}\left(\frac{\lambda-\lambda_c}{\lambda_r}\right)^2 \tag{6.6}$$

which can be interpreted as a second-order phase-distortion function around λ_c or as a deviation of the first-order dispersion term *D*.

For $\varphi_r = \pi/2$ the phase distortion function represents an odd function around λ_c which can be written as

$$\varphi = \frac{T_r \lambda_r c}{2\lambda_c^2}\sin\left(2\pi\frac{\lambda-\lambda_c}{\lambda_r}\right) \approx -\frac{2\pi^3 T_r \lambda_r c}{3\lambda_c^2}\left(\frac{\lambda-\lambda_c}{\lambda_r}\right)^3 \tag{6.7}$$

This can be interpreted as a first- and third-order phase-distortion function around λ_c. In this case the first-order distortion is a pure time delay and can be neglected. The relevant distortion here is the third-order distortion, which works as a deviation of the second-order dispersion term *S*. For a fairly well dispersion-compensated signal, the case in Equation 6.7 is the most severe. This case is sometimes used to calculate a direct measure for the impact of RGD on a system.

In Figure 6.10 we have calculated optical eye-closure (EC) for a 10 Gbit/s chirp-free NRZ data signal with a raised-cosine pulse form. The gray-scale coding represents the EC and each gray-scale step represent a closure step of 0.1 dB. The EC is here defined as how much the FBG itself closes the eye when impacts from receiver-filters and so on are compensated for. As can be seen in Figure 6.10, the maximum sensitivity occurs at different ripple periods for the

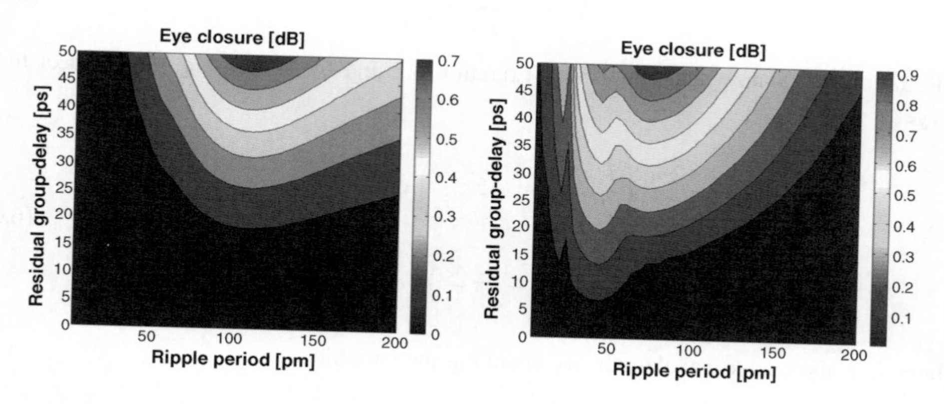

Figure 6.10 EC, in optical dB, as a function of sinusoidal group-delay (*x*-axis) and amplitude (*y*-axis) for an NRZ-modulated 10 Gbit/s signal. Left: $\varphi_r = 0$. Right: $\varphi_r = \pi/2$

two values of φ_r. The exact value of the ripple period depends on the modulation format, modulation speed, pulse shape, and so on.

6.2.5.2 Figure of Merit

A more ambitious approach to calculating the system impact of RGD is to establish a figure of merit (FOM) for the FBG based on its ripple function. The traditional way to estimate EC has been to measure the RGD over a wavelength interval that is comparable to the optical signal bandwidth. Lately [26,27] it has been suggested that an estimate based on the phase ripple for the FBG will give better values. In the literature there are six major routes to calculating the FOM:

- measuring the peak–peak value of the RGD over a specific bandwidth;
- measuring the variance of the RGD over a specific bandwidth;
- measuring the standard deviation of the RGD over a specific bandwidth;
- measuring the peak–peak value of the phase ripple over a specific bandwidth;
- measuring the variance of the phase ripple over a specific bandwidth;
- measuring the standard deviation of the phase ripple over a specific bandwidth.

The EC is then calculated by the formula

$$EC = C_1 \times FOM + C_2 \tag{6.8}$$

where C_1 and C_2 are constants that have to be fitted to the modulation format, modulation speed, pulse shape, and so on.

As indicated by its label "figure of merit," the measures cited above are not exact, but a reasonably good correlation can be achieved when EC is estimated with the aid of an FOM. In Figure 6.11, an example of an FOM/EC calculation is given. As x-coordinate, an FOM value based on the standard deviation of the phase ripple with 60 pm bandwidth and 10 pm resolution has been calculated for the entire C-band; that is, 1530–1565 nm. As y-coordinate, the EC was calculated with traditional eye-diagram analysis with the same resolution. The modulation format was set to 10 Gbit/s NRZ. The dotted line is the linear-fit function; in this case the correlation coefficient is 0.9.

6.2.5.3 Fiber Bragg Gratings in Field Experiment

The impact of RGD in FBGs has been investigated in a field experiment [28]. In this case an 820 km long looped-back transmission link between Stockholm and Hudiksvall in Sweden was used. The link in the experiment was equipped with a commercial DWDM system designed for 10 Gbit/s with respect to amplifier configuration and dispersion compensation. In the original configuration, dispersion was compensated by standard DCFs, but for this test 500 km of dispersion compensation was replaced by commercial DCMs based on fixed CFBGs. These modules were designed to cover the entire C-band; that is, 1530–1565 nm. Measurement of RGD for such a fiber grating and the depicted system setup are given in Figures 6.12 and 6.13.

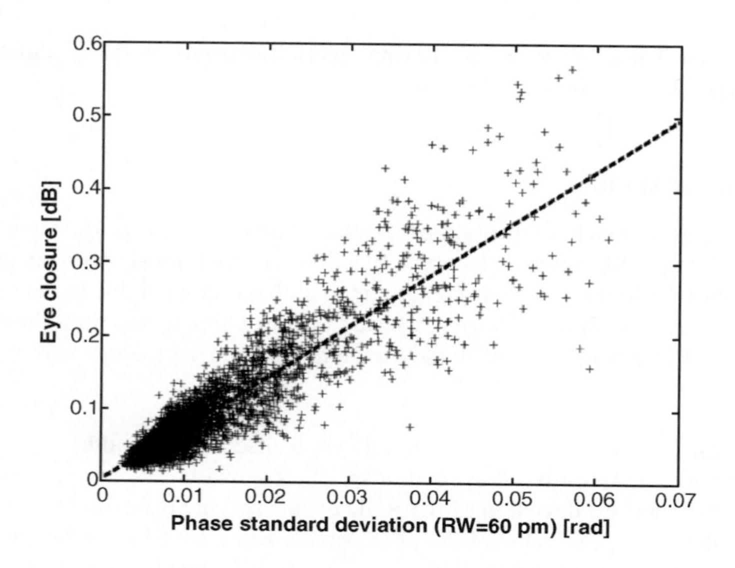

Figure 6.11 FOM/EC calculation based on standard deviation of the phase response for a fibre grating

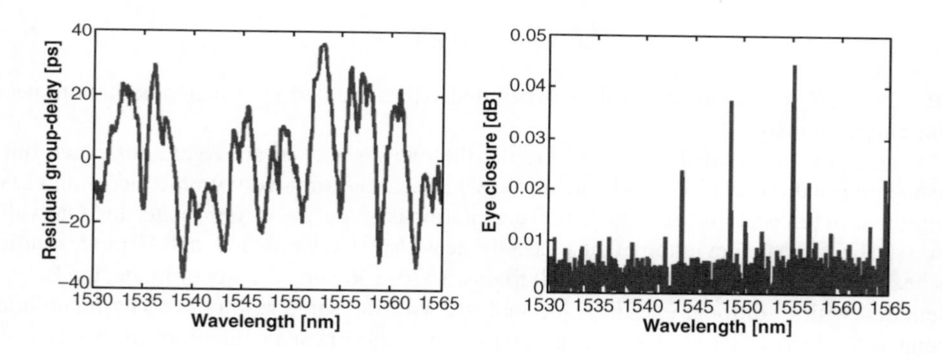

Figure 6.12 RGD measurement on a CFBG compensating for 120 km of SSMF (left) and its calculated EC at 10 Gbit/s NRZ (right)

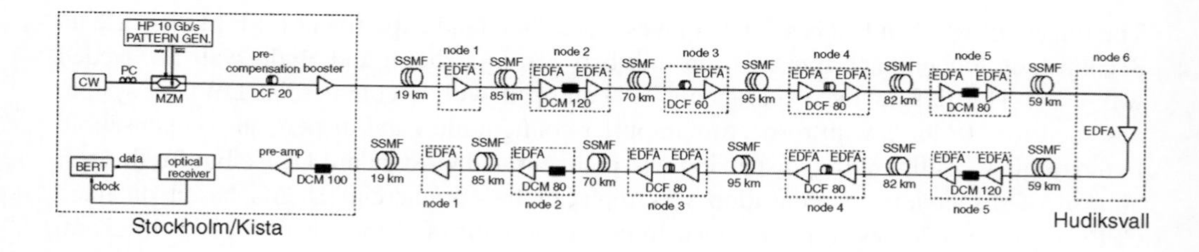

Figure 6.13 System setup for the 10 Gbit/s field experiment

In the experiment, a 10 Gbit/s NRZ-modulation format was used. The 10 Gbit/s signal was generated by modulating a double-arm Mach–Zehnder modulator (MZM) with $2^{31}-1$ pseudo-random bit sequence (PRBS) data using push–pull driving directly by the data and data-bar (2 V peak-to-peak each) output from a pulse pattern generator, thereby creating an almost chirp-free signal. The bias of the modulator was adjusted so that the minimum of the transfer function of the modulator was reached by data modulation, this to ensure a high extinction ratio for the transmitted signal. At the receiver side, an EDFA followed by an optical attenuator was used to add different levels of amplified spontaneous emission (ASE) noise to the transmitted signal through a 3 dB coupler (not shown in figure). Thereby, the BER of the transmitted signal could be measured for different OSNRs. Figure 6.14 summarizes the measurement results for the 10 Gbit/s experiment.

The 16 DWDM channels were distributed on a 100 GHz grid starting from 1546.92 nm for channel 1 and to 1558.98 nm for channel 16. The general trend of negative OSNR penalty seen in the measurement is explained by the original link not being fully dispersion compensated. For the original link, a residual dispersion of 240 ps/nm was measured with a dispersion slope of 6 ps/nm^2; the latter figure depends on the DCFs used in the setup and does not fully compensate for the dispersion slope in the transmission fiber. The input power into the transmission fibers was 3 dBm/channel and the input power to the DCMs and the DCFs was −3 dBm/channel.

In conclusion, commercial FBG-based DCMs which cover the entire C-band (i.e., 1530 to 1565 nm) were tested in a field experiment over 820 km of G.652 fiber with very low penalty figures for data transmission rates of 10 Gbit/s NRZ. In the test, 500 km of fiber was compensated for by DCMs and 320 km by DCFs.

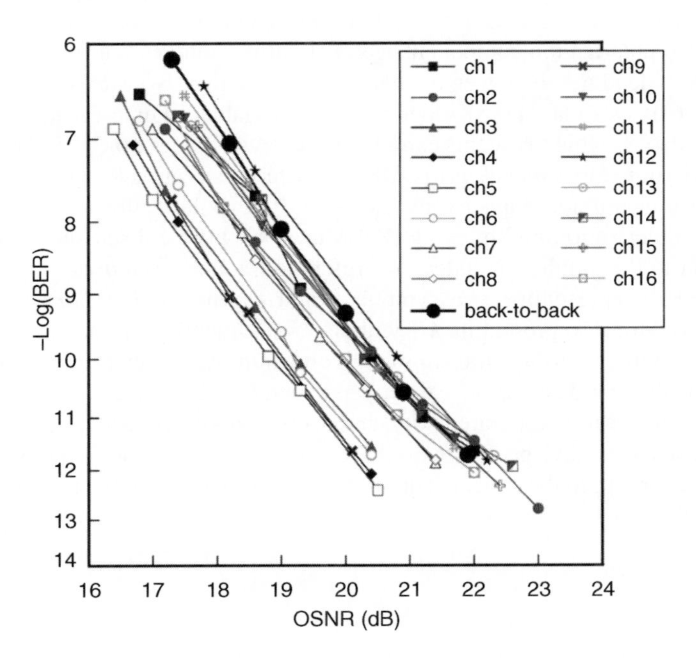

Figure 6.14 BER versus OSNR for all the 16 10 Gbit/s channels (NRZ) with DCMs along the link

6.3 Modulation Formats

As seen in Chapter 4, in digital communications the information is transmitted as digital data – that is, as a set of discrete *symbols* chosen from a finite set (an *alphabet*) [29,30]. In order to be transmitted over a physical medium (e.g., the ether in radio communications or the silica waveguide in fiber-optic communications) each symbol is represented by a specific value of a complex signal A [31], which describes a physical property – specifically, light for fiber-optic communications. This operation is referred to as data modulation. There are different ways to achieve data modulation, depending on which parameter of the complex signal A changes with time and the number of states that are allowed. The choice of signal parameters and number of states characterizes the *modulation format*.

As far as the signal parameters are concerned, one can decide to let the amplitude of the complex signal A carry the information, as in amplitude-shift keying (ASK) modulation formats that are discussed in more detail in Sections 6.3.1 and 6.3.2. Alternatively, one can let the phase, or its time derivative, carry the information, as in phase-shift keying (PSK; Section 6.3.4) and frequency-shift keying (FSK) respectively. A combination of more than one signal characteristic can be modulated; for example, the amplitude *and* the phase of the signal to represent different information symbols, as in amplitude and phase-shift keying (APSK) widely used in radio communications. In fiber-optic communications, APSK modulation formats are not very popular due to higher implementation complexity; however, demonstrations of such transmission systems have been implemented [32,33].

One could also use the polarization property of light to encode data, as in polarization shift keying (PolSK) [34,35]. Here, a bit value 0 could be represented, for example, by a horizontally polarized signal and a bit value 1 by a vertically polarized signal. However, the state of polarization of an optical signal is not trivial to control and, moreover, due to imperfections in the fiber, the two polarization components generally travel at different speeds (birefringence; see Section 5.7) leading to power exchange between the two states (cross-polarization interference [36]); so, today, PolSK does not seem a viable scheme for fiber-optic communications. It is also possible to use this extra DOF, however, to make the signal more tolerant to specific transmission impairments, as in OOK alternate-polarization RZ (APol-RZ) [37,38], or to double the transmission capacity by using the two polarizations as two transmission channels as in polarization multiplexing (Pol-Mux),[2] as will be discussed in Section 6.3.5.

With respect to the number of states, one refers to *binary transmission* if only two states are allowed for A, representing two symbols 0 and 1, whereas in *M-ary transmission*, M states are allowed for A, representing M symbols. Consequently, $\log_2 M$ times information is carried by symbols in M-ary transmission compared with binary transmission, which translates into higher transmission capacity per hertz of bandwidth. The price to pay is higher transmitter and receiver complexity, and decreased tolerance to noise. A large number of modulation schemes have been proposed recently. Figure 6.15 shows a modulation format classification according to the concepts introduced above, some of which are discussed in this section.

[2] Cross-polarization interference is an issue here too [1], although it is easier to take care of this on a channel base rather than on a bit-to-bit base. Pol-Mux is currently heavily used in front-edge research experiments.

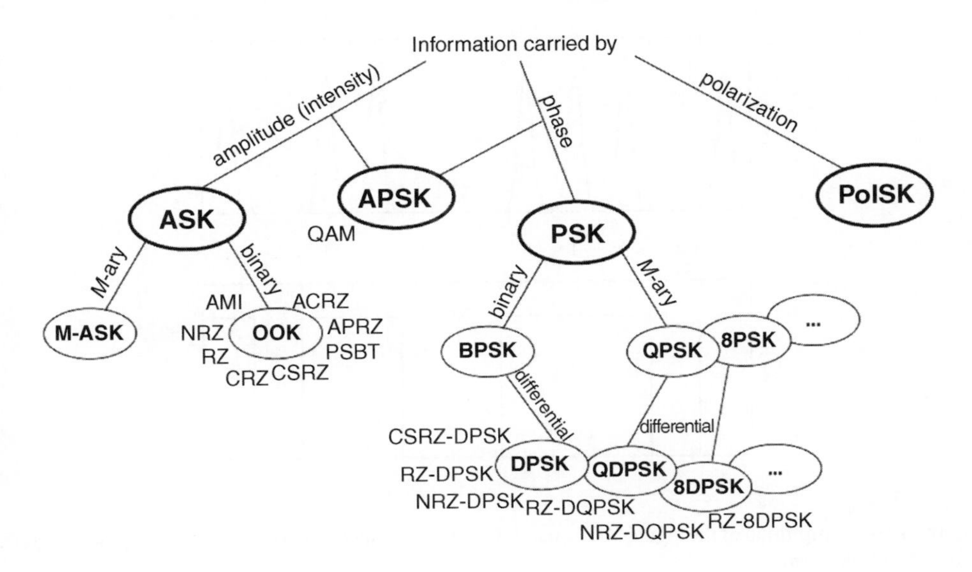

Figure 6.15 Classification of modulation formats

6.3.1 On–Off Keying Modulation Formats

Most of the currently installed fiber-optic systems use basic modulation schemes, in which the information is carried by the amplitude of the light signal A: ASK. These are also referred to as intensity modulation schemes in the literature. M-ary ASK for fiber-optic transmission has been studied numerically and experimentally [39,40], but its bandwidth efficiency gain does not counterbalance the increased complexity and noise tolerance issues [41]. Consequently, most optical ASK modulation formats proposed to date are binary.

In binary ASK, only two amplitude levels are used for information coding. The two levels are usually "*light on*", representing a 1, and "*light off*", representing a 0; therefore, binary ASK is often referred to as OOK. OOK takes different forms depending on whether the signal stays on between two marks or whether it is switched off in between. The former scheme is termed NRZ and the latter return-to-zero (RZ). Mathematically, the information-carrying signal will be

$$A(t-nT) = b_n p(t), \qquad -T/2 \leq t-nT < T/2 \tag{6.9}$$

where b_n is the nth transmitted symbol, which can be 0 or 1, and T is the time it takes to transmit the bit (the bit duration). The function $p(t)$ is simply a constant for NRZ and a pulse for RZ. The amplitude $|A(t)|$ of an RZ-OOK and an NRZ-OOK signal is shown in Figure 6.16.[3] In transmission analysis the pulse shape is often assumed to be Gaussian, which is mathematically very convenient, and usually a good approximation:

$$p(t) = \sqrt{P_0}\mathrm{e}^{-(t/\tau_\mathrm{P})^2} \tag{6.10}$$

where P_0 is the peak power and τ_P is the pulse half-width (the time it takes for the pulse power to drop from its peak to P_0/e). Another measure of the pulse width is the full width at half

[3] In the literature it is customary to refer to RZ-OOK simply as RZ and to NRZ-OOK as NRZ.

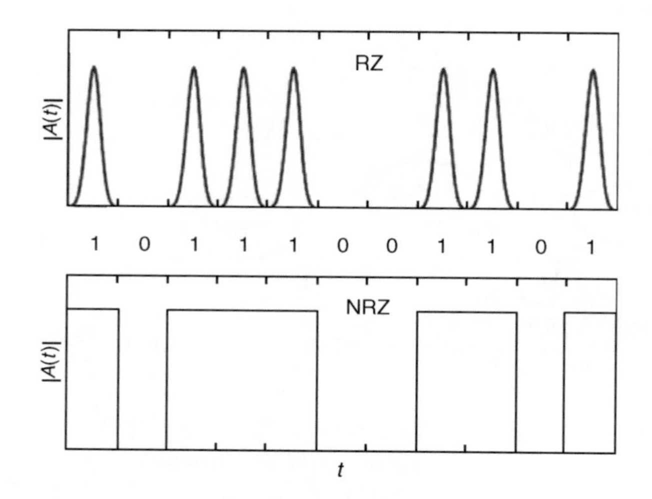

Figure 6.16 Amplitude of the signal $|A(t)|$ encoding the data sequence [1 0 1 1 1 0 0 1 1 0 1] for RZ-OOK and NRZ-OOK formats

maximum τ_F [1]. For Gaussian pulses, $\tau_F = 1.665\tau_P$. The ratio between the full width at half maximum and the bit duration, τ_F/T, is referred to as the duty cycle.

NRZ is cheaper to implement, since it only needs a continuous-wave (CW) source (a CW laser) and one intensity modulator to modulate the electrical data stream onto the CW. To generate RZ, one needs either an extra modulator, properly driven by a periodic signal [42], or to substitute the CW laser with a pulsed laser. Moreover, complexity is further increased because the pulse train and the data stream must be synchronized and aligned. For this reason, commercial fiber-optic communication systems up to 10 Gbit/s per channel mostly employ NRZ. However, pulsed signals are more tolerant to system impairments [43–51], and today most research and development activity on high-speed OOK transmission (40 Gbit/s and above) focuses on RZ. NRZ remains, nonetheless, a good option for applications where cost is the driving force, like in short-reach transmission.

Another pulsed OOK modulation format which is rather common is carrier-suppressed RZ (CSRZ) [52] (or CSRZ-OOK to be specific), in which consecutive bit slots are given a π phase shift. Note that here the phase of the signal is not used to carry information; rather, it is an extra DOF used to make the signal more robust to transmission impairments. Indeed, this π phase shift is responsible for a somewhat better tolerance to intrachannel four-wave mixing (IFWM) than RZ [52]. This is partly related to the narrower signal spectrum and the resulting higher dispersion tolerance, and to a better tolerance to SPM-induced pulse broadening, since neighboring pulses interfere destructively when overlapping.[4] CSRZ is characterized by a transmitter and receiver complexity that is comparable to standard RZ, and has become

[4] An additional explanation is that the CSRZ format is more tolerant to the existence of IFWM-generated ghost pulses (despite the generation of IFWM being equally efficient for RZ and CSRZ). Indeed, signal pulses will become broadened at the receiver due to the interplay between CD and SPM during transmission. The power level in a zero bit slot will then be the result of the linear superposition between the ghost pulse and the tails of neighboring pulses, which is minimized by a π relative phase shift [53].

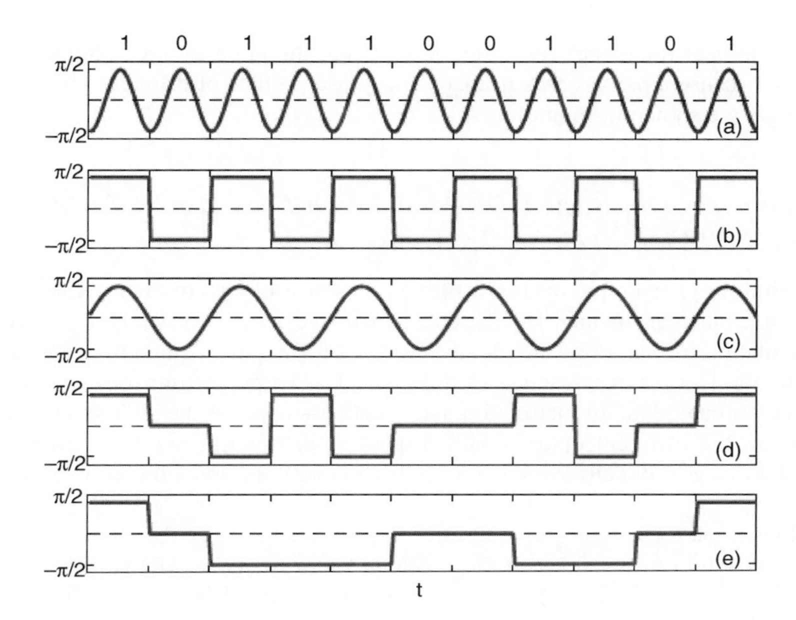

Figure 6.17 Phase of the signal *A* for the data sequence [1 0 1 1 1 0 0 1 1 0 1] for five different OOK formats: (a) CRZ, (b) CSRZ, (c) ACRZ, (d) AMI, (e) optical duobinary, or PSBT

a popular modulation format. Several other phase-modulated OOK modulation formats have been introduced in which the phase of the signal is modulated to increase robustness to specific transmission impairments.

The chirped RZ (CRZ)-OOK modulation format, for instance, is generated by bit-synchronous sinusoidal phase modulation of the RZ signal, giving the pulses a frequency chirp [54,55]. Figure 6.17a shows the phase profile of a CRZ signal. CRZ possesses increased nonlinear tolerance compared with RZ, but is also accompanied by a broader spectrum, which can easily become a problem in DWDM systems, where the signal undergoes significant filtering.

In the alternate-chirp RZ (ACRZ)-OOK modulation format [56], neighboring pulses are given a chirp[5] with opposite sign, by filtering a phase-modulated CW. ACRZ was also shown to bring about an improvement in nonlinear tolerance with respect to CSRZ and RZ. Incidentally, the pulses have a fixed alternating phase, which is potentially what gives the format the improved IFWM tolerance. (This effect is indeed exploited in alternate-phase RZ (APRZ)-OOK, which will be discussed in more detail in Section 6.3.3.)

In alternate-mark inversion (AMI) [57], the phase of the signal is π shifted every consecutive mark, while in *optical duobinary* [58,59], sometimes referred to as phase-shift binary transmission (PSBT) [60], the phase of a mark is given a π shift with respect to the previous mark if there is an odd number of zeros in between. Duobinary is analyzed in more detail later in this Section 6.2. In AMI, the phase is shifted by π between neighboring pulses, rather than neighboring bits, and the pulse width is generally narrower than in CSRZ. This results in

[5] A pulse is said to be chirped if it has a time-dependent frequency and, hence, a nonlinear phase profile. A positive (negative) chirp corresponds to the sign of the second time derivative of the pulse phase being positive (negative).

improved nonlinear tolerance, better than CSRZ, but the price to pay is higher transmitter complexity, because a pre-coder is needed. The phase profiles of different phase-modulated OOK formats are shown in Figure 6.17.

6.3.2 Comparison of Basic OOK Modulation Formats: NRZ, RZ, and CSRZ for 40 Gbit/s Transmission

In this section we give a rapid overview of the robustness of the commonly used modulation formats at 40 Gbit/s per channel. The targeted network segment is very important to the choice of the modulation format. For example, for the NRZ format, the competitive cost and simple implementation is a major advantage, but drawbacks like poorer transmission performance make it a better candidate for metropolitan area network applications. In this section we want to benchmark the different options with respect to their implementation complexity and expected tolerance to impairments, based on both laboratory and simulation experiments at 40 Gbit/s.

Regarding implementation, the simplest of these formats is NRZ. It requires only one modulator at the transmitter side and a low-voltage radio-frequency (RF) driver.[6] Neither pre-coding nor electrical filtering are necessary on the electrical binary sequence. Regarding performance, the NRZ is also one of the less resilient modulation formats with respect to PMD, which is often very detrimental at 40 Gbit/s per channel, or intrachannel nonlinear impairments. On the other hand, because of its low spectral occupancy, the NRZ format is more robust to the accumulation of CD.

The conventional RZ format is more complicated and more expensive to implement because it requires an additional pulse carver combined with the corresponding RF driver. But, as the RZ pulses are short with respect to the bit duration, its resistance to first-order PMD is better than that of NRZ. On the other hand, because of its large spectral occupancy, accumulation of CD degrades its performance. When considering intrachannel nonlinear effects, the RZ formats are globally more tolerant than the NRZ [43–51].

CSRZ is particularly interesting because it represents a good trade-off between the resistance to intrachannel nonlinear effects, to residual CD and to first-order PMD, as already explained above. Note, however, that its practical implementation requires an additional pulse carver, driven at half the bit rate and double the voltage, hence requiring a high-output power driver.[7]

In order to evaluate the transmission performance of these systems and compare the different modulation formats it is important to develop an experimental configuration combined with simulation results. For the methodology to be valid one has to take care of the following points:

(a) In developing an experimental configuration for a 40 Gbit/s transmission system it is important to investigate the role of key practical constraints that may lead to serious obstacles to the successful introduction of 40 Gbit/s WDM EDFA-based transmission systems.

[6] The RF driver for the NRZ modulator has to supply only 5–6 V_{pp} to the single-electrode MZM (biased at the transmission null point).

[7] The RF driver for the 20 GHz carver of the CSRZ modulator has to supply 10–12 V_{pp}.

The experimental configurations used for 40 Gbit/s deployment experiments are chosen to reflect closely the practical conditions of existing terrestrial fiber link infrastructures. For instance, SSMF is used, rather than nonzero dispersion-shifted fiber, since the former is the most largely deployed fiber on terrestrial transmission networks of European carriers. In addition, EDFAs are used without any Raman amplification (also see Appendix 6.6).[8]

(b) An interplay between simulation and experiments is used in a complementary way. Simulations are used to find optimum areas and to optimize tunable parameters; experiments are used for evaluation of systems.

Note that the transmission quality of the optical signals is subject to the following main impairments: OSNR degradation, intrachannel nonlinearities, residual CD, first-order PMD, and optical filtering at the add/drop sites. In order to investigate those, an experimental setup and steps for experimental evaluation and numerical evaluation are shown in Appendix 6.6.

Figure 6.18 shows a temporal and spectral characterization of the transmitter for the three modulation formats under test. The measured extinction ratios of the NRZ, CSRZ, and 33% RZ formats are 12.6 dB, 14.1 dB, and 15.3 dB respectively. Their relative temporal and spectral characteristics can be directly observed.

Figure 6.19 shows the simulation results that compare the transmission performance of the NRZ, 33% RZ, and CSRZ modulation formats. The contour level plots of Q-factor values (in decibels) are shown for the central channel out of a comb of five simulated channels. Figure 6.19 illustrates the dependence of the Q-factor at the output of the 4×100 km spans of SSMF as a function of both the pre-chirp and the launch signal average power P_S. Note that, for each pre-chirp value, the post-chirp is optimized in order to obtain the highest possible Q value.

As can be seen in Figure 6.19, whenever the launch signal power has a relatively low value, for all modulation formats the selection of a particular pre-chirp is not critical for system performance. On the other hand, as the input power of the signal P_S increases above 1 dBm, the best performance is obtained for negative values of the pre-chirp. The optimal pre-chirp is equal to -500 ps/nm, -600 ps/nm, and -700 ps/nm for 33% RZ, CSRZ, and NRZ respectively. The results in Figure 6.19 predict as well that, for well-spaced (200 GHz) channels, the CSRZ format leads to about 2 dB performance improvement (in terms of Q-factor) over NRZ format and about 1 dB improvement over the 33% RZ format.

In order to evaluate the robustness *to intrachannel nonlinearities* [61–65], the experimental configuration of Figure 6.53 (see Appendix 6.6 for details) with only the eight even channels was used, with a spectral granularity of 200 GHz as explained in detail in Appendix 6.6. Figure 6.20 (top) compares the BER of the central channel after transmission as a function of the channel power injected into SSMF spans. The input OSNR is equal to 25 dB. The optimal span input power (around 4 dBm) is virtually the same for all the modulation formats

[8] Clearly, from the research viewpoint it is important to carry out comprehensive investigations of the performance improvements introduced by distributed Raman amplifiers (DRA). However, the potential advantages offered by DRAs should be weighed along with practical issues of importance for operators, which may ultimately hamper deployment of DRA systems; namely: (a) the need for carriers to measure fiber losses at Raman pump wavelengths (before the installation of the transmission system); (b) the uncertain reliability of optical connectors located between Raman pumps and fiber spans (involving system outage in case of failure); (c) the ocular safety of people in charge of maintenance (because of the presence of high optical powers in the system). Finally, reflecting the practical conditions of fiber links, no optimization of the dispersion map is performed. As a result, the residual dispersion per span is not accurately tuned to a predefined optimal value.

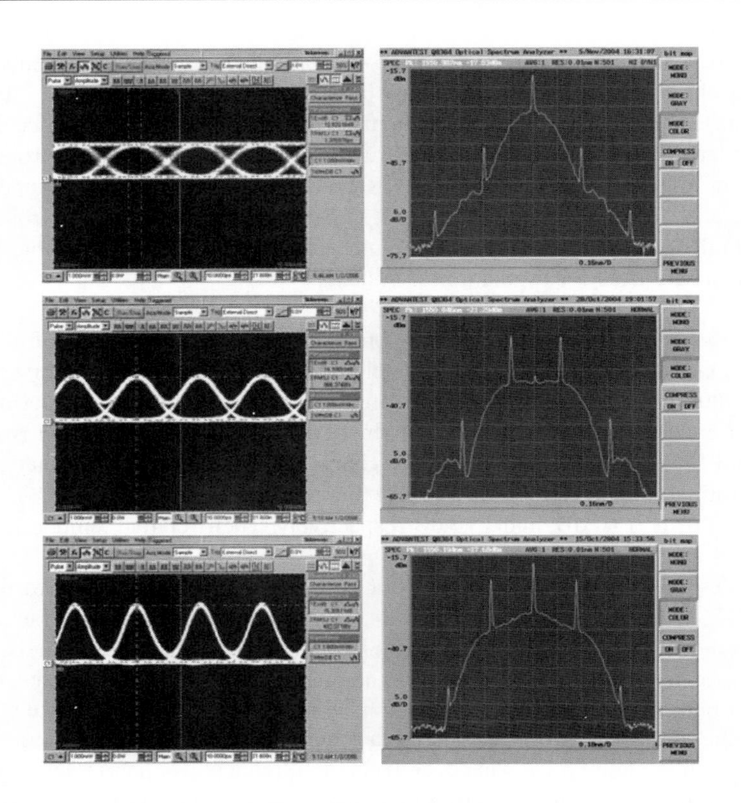

Figure 6.18 Temporal and spectral characterization of our transmitters: NRZ (top), CSRZ (middle), 33% RZ (bottom), from the experimental setup of Appendix 6.6

considered here. In contrast, the various modulation formats show different BER values at the optimum span input power: the 33% RZ slightly outperforms the CSRZ, and it is definitely better than the NRZ format. Nonetheless, in order to evaluate the nonlinear penalty corresponding to each modulation format accurately, it is also important to measure the BER versus OSNR at the receiver, as it is obtained both in back-to-back and after 400 km transmission. These measurements are made at the optimum input power previously determined. The superior resilience of CSRZ to intrachannel nonlinearities is clear in Figure 6.20 (bottom). Indeed, for a BER of 10^{-9} the OSNR penalty is only 0.75 dB for CSRZ, whereas it is 1.5 dB for 33% RZ, and it is higher than 2 dB for NRZ. The 1 dB margin of the 33% RZ in back-to-back OSNR sensitivity (when compared with CSRZ) is erased after transmission owing to the larger sensitivity of this format to IFWM. The higher resilience of CSRZ to this impairment is due to its relatively large duty cycle, as well as to its stronger pulse confinement (owing to the periodic π-phase shifts) which reduces pulse overlapping when CD accumulates. Figure 6.20 (bottom) also shows that 33% RZ is the most resistant format to OSNR degradation: in back-to-back and for a BER of 10^{-9}, 33% RZ has an OSNR margin of 0.75 dB (2.75 dB) when compared with CSRZ (NRZ).

The measurements of back-to-back sensitivities and resilience to intrachannel nonlinearities presented here are globally in line with previous available results [62–65].

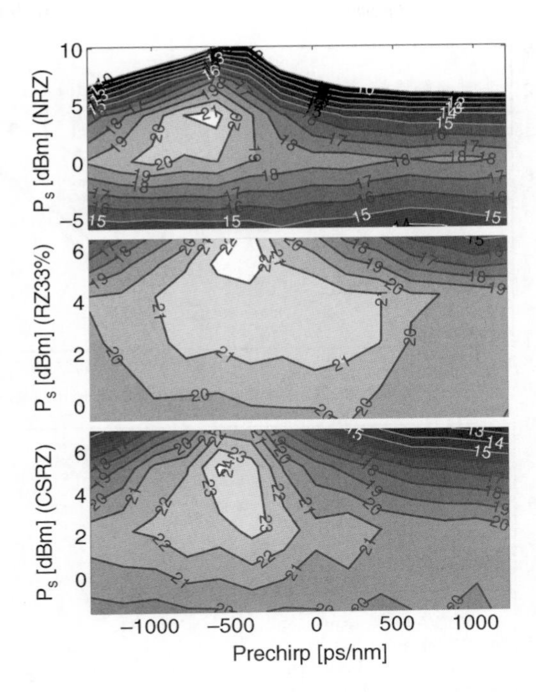

Figure 6.19 Numerical simulation of Q-factor levels (in dB) versus pre-chirp and signal average power (PS), for NRZ (top), 33% RZ (middle) and CSRZ (bottom) formats with 200 GHz channel spacing. Post-chirp is adjusted for each pre-chirp value to obtain best performance

To study resilience to residual CD and first-order PMD or DGD for example, [63,64], only the eight even channels are used. Figure 6.21 (top) shows the OSNR penalty for each format versus residual CD for the central channel. As can be seen, NRZ is the most tolerant format to CD accumulation. For a 1 dB OSNR penalty, an acceptance window of 90 ps/nm, 75 ps/nm, and 60 ps/nm is observed for NRZ, CSRZ, and 33% RZ respectively. Clearly, the wider the pulse spectrum, the less resilient is the format to residual CD. Periodic π-phase alternation also increases CD resilience of CSRZ, by reducing intersymbol interference (ISI). Note that slight shifts observed on the CD curves against the 0 ps/nm point are due to the residual chirp of the emitter. Figure 6.21 (middle) shows the OSNR penalty for each format as a function of the DGD. With 33% RZ, the accepted DGD (defined as the level of DGD that leads to 1 dB OSNR penalty) is maximal and equal to nearly 13 ps (it is 10.5 ps and 6.5 ps with the CSRZ and NRZ formats respectively). Not surprisingly, the larger the pulse duty cycle, the lower is the modulation format robustness to DGD. In particular, when considering the NRZ format, the presence of PMD leads to a leaking of the "marks" energy into adjacent "spaces," which enhances the BER. This explains the inferior resilience of the NRZ format in comparison with the 33% RZ format towards PMD. Indeed, the 33% RZ format under the influence of 12 ps of DGD yields an eye diagram which is very close to what is obtained with the NRZ format with 0 ps of DGD. Finally, note that the periodic π-phase shifts of the CSRZ format do not affect the resilience of this format to DGD, unlike the case of CD.

The modulation format tolerance to *output filter detuning* is at least as important as the resilience to optical filter bandwidth variations. Figure 6.21 (bottom) illustrates the observed

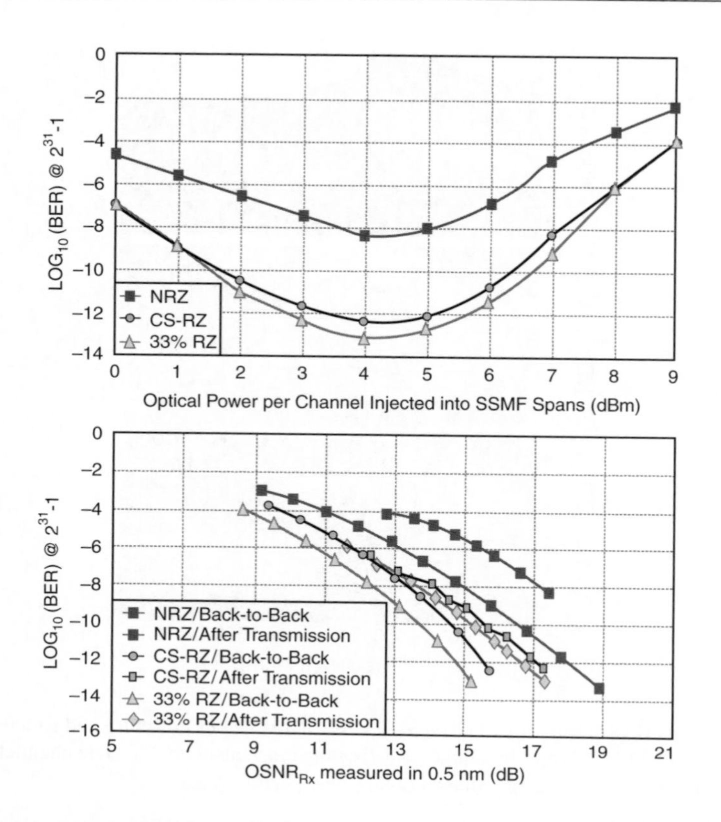

Figure 6.20 BER versus power per channel injected into SSMF spans (top), BER versus the receiver OSNR in back-to-back and after transmission (bottom), for the central channel at 1550.12 nm. OSRN is measured at 0.5 nm

dependence of the OSNR penalty upon optical filter detuning from the channel carrier wavelength. The bandwidth is fixed at its optimal value for each of the formats. As can be seen in Figure 6.21, the most resistant format to output optical filter detuning is 33% RZ: the acceptance window (defined here as the filter detuning that introduces a 2 dB penalty) is equal to 0.2 nm for 33% RZ, and it is equal to 0.15 nm for both CSRZ and NRZ formats. The larger the spectral width (and the corresponding optical filter bandwidth), the higher is the modulation format tolerance to optical filter detuning. Whenever a relatively large filter detuning is applied, penalties increase owing to eye diagram distortions and crosstalk from neighboring channels.

The measurements compare well with other results available in the literature [67] in the particular case of a 100 GHz channel spacing, whenever a rectangular optical filter is employed. For the NRZ and CSRZ formats, Ref. [67] quotes an optimal optical filter bandwidth of about 90–100 GHz, which is close to what is obtained here.

Again, the experimental results are complemented by an extensive numerical investigation of the filtering impact of an OADM cascade on the performance of our three ASK modulation formats [68–70]. Simulation configuration is explained in Appendix 6.6. Two types of multiplexer (MUX)/demultiplexer (DEMUX) were simulated, as discussed in Appendix 6.6.

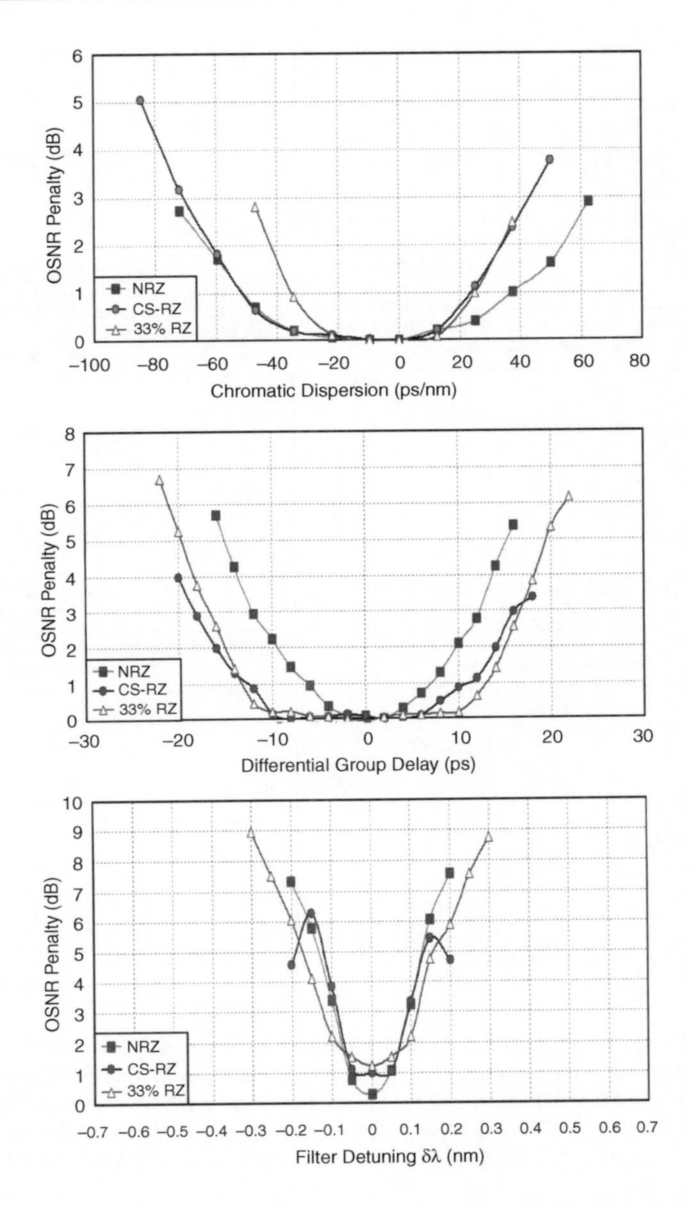

Figure 6.21 OSNR penalty in 0.5 nm at fixed BER $= 10^{-9}$ versus the CD (top), the DGD (middle), and detuning (bottom) for the central channel at 1550.12 nm

The first type has the same characteristics as the square flat-top optical filter already used in the receiver (see Figure 6.55), and the second 100 GHz flat-top MUX/DEMUX under study is an ideal one. The simulation results are detailed in Figure 6.22, where the BER versus transmission distance is plotted for the NRZ, CSRZ, and 33% RZ modulation formats in various configurations (as discussed in the figure legend).

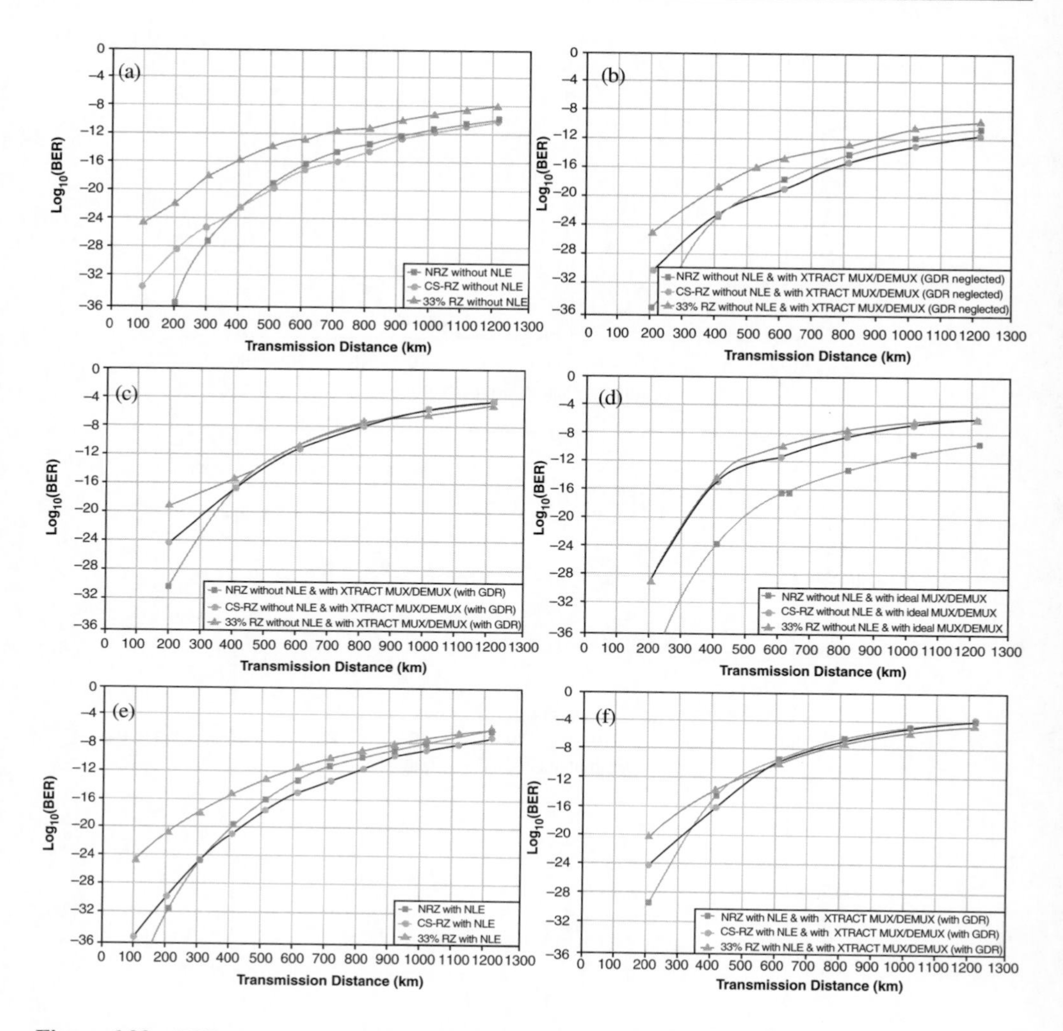

Figure 6.22 BER versus transmission distance for the NRZ, CSRZ, and 33% RZ modulation formats in various configurations: (a) without nonlinear element (NLE) and any MUX/DEMUX; (b) without NLE but with the XTRACT MUX/DEMUX (without GDR); (c) without NLE but with the XTRACT MUX/DEMUX (with GDR); (d) without NLE but with the ideal 100 GHz flat-top MUX/DEMUX; (e) with (NLE) but without any MUX/DEMUX; (f) with NLE and with the XTRACT MUX/DEMUX (with GDR)

When fiber nonlinearity and OADM are disregarded, the 100 GHz channel spacing configuration is very detrimental for the 33% RZ modulation format, as shown in Figure 6.22a. The superior resilience of the 33% RZ format to the accumulation of ASE noise, as observed experimentally with the spectral granularity of 200 GHz on the back-to-back BER measurements of Figure 6.20 (bottom), vanishes when the channel spacing is reduced down to 100 GHz. At the opposite end, Figure 6.22a shows that the CSRZ format is the most resistant to the accumulation of ASE noise: the reduction of the channel spacing to 100 GHz is not sufficient to remove completely its gain in terms of back-to-back sensitivity observed in Figure 6.20 (bottom) when compared with the NRZ format.

Inserting now the *square* MUX/DEMUX every 200 km while neglecting the GDR influence does not degrade the transmission quality at 1200 km, as shown in Figure 6.22b, proving that the bandwidth of this MUX/DEMUX has been properly chosen. When including the GDR, as shown in Figure 6.22c, one obtains a dramatic change of the transmission quality: as can be seen, the 33% RZ, NRZ, and CSRZ formats loose 4.5 decades, 6 decades and 7 decades on the BER respectively. Therefore, the previously observed BER differences among the different formats are virtually gone when GDR is taken into account. Let us consider next replacing the "realistic" MUX/DEMUX by the ideal (without GDR) 100 GHz flat-top MUX/DEMUX. The results in Figure 6.22d show that the reduced bandwidth (with respect to the MUX/DEMUX of Figure 6.22c) of this ideal filter significantly affects the transmission quality of the CSRZ and 33% RZ formats, proving that the square flat-top amplitude transfer function is more adapted to this system). However, as expected, the NRZ format is less affected than the CSRZ and 33% RZ by the strong optical filtering action of the ideal MUX/DEMUX.

The simulation results of Figure 6.22e show the case where only the nonlinear propagation effects (NLE) are taken into account. In this case, by comparing Figure 6.22e with Figure 6.22a one notices a general performance degradation. Moreover, nonlinearity leads to an advantage of 1 decade on the BER for the CSRZ modulation format at 1200 km when compared with 33% RZ and NRZ. This confirms the results obtained above, when using the spectral granularity of 200 GHz.

Finally, the results of Figure 6.22f show the performance comparison in the realistic case where one XTRACT MUX/DEMUX is inserted after every two SSMF spans, while its GDR as well as the fiber nonlinearity are not neglected. When comparing Figure 6.22f with the corresponding results of Figure 6.22c, where the NLE were not taken into account, one can see that the BER is slightly degraded for all modulation formats. In conclusion, the results of Figure 6.22 show that the fine control of the GDR of the OADM is at least as important in determining the overall system performance as the clever design of the dispersion map, or the precise optimization of the span input power.

In conclusion, the studies presented in this section have shown that, although the 33% RZ format is the most robust to OSNR degradation, the CSRZ format is the most tolerant to intrachannel nonlinearities at 40 Gbit/s. Both NRZ and CSRZ formats exhibit the best tolerance to residual CD, whereas RZ performs better than NRZ and (slightly better than) CSRZ as far as DGD is concerned. Finally, the NRZ format is the least penalized by filtering and the 33% RZ format is the most resistant to filter detuning. As far as the impact of an OADM cascade on the transmission quality is concerned, it appears that the GDR of the OADM has to be precisely controlled in order to limit the overall performance degradation. As expected, modulation formats with large spectral occupancies are most impacted by OADM cascades. Overall, it appears that using the CSRZ format provides the best balance when trying to meet all different key requirements of future all-optical networks. Nonetheless, owing to its relatively poor resilience to PMD, the use of the CSRZ format in 40 Gbit/s transmission systems is likely to be limited on the existing, high-PMD long-haul fiber infrastructure. This motivates the current interest in exploring the more complex (and costly) modulation formats such as differential PSK (DPSK), as described in the following sections.

6.3.3 A Power-Tolerant Modulation Format: APRZ-OOK

To compensate for fiber loss, optical amplifiers are inserted periodically in a transmission link. At each amplifier stage, however, the OSNR is further degraded, as mentioned in Chapter 5. The longer a transmission link, the larger is the amount of amplifiers. Eventually, the OSNR may fall to such low levels to make error-free transmission impossible. One obvious way to counter this is to increase the launch power into the fiber, but generally high launch power leads to nonlinear effects that distort the signal. A desirable characteristic of a modulation format, therefore, is to be tolerant to nonlinear effects. An OOK modulation format recently introduced is the APRZ-OOK (or simply APRZ) [71,72], which was shown to achieve nonlinear tolerance levels similar to advanced DPSK schemes [62], while keeping transmitter and receiver complexity comparatively low.

An APRZ signal is an RZ signal in which the phase of neighboring bits alternates between two values. If the proper phase-alternation amplitude is chosen, then destructive interference is induced in the different IFWM contributions accumulating in each bit slot, and increased nonlinear tolerance is achieved. The exact shape of the phase variation can be different in different implementations; what is important is the phase-alternation amplitude. Square-APRZ, in which the pulses experience a constant phase, is in a way ideal, since it is not accompanied by any chirp.

It is noteworthy that in the case of $\Delta\phi = \pi$, square-APRZ reduces to a generalized form of CSRZ [72], in which the duty cycle is a free parameter.

The most intuitive way to implement an APRZ transmitter consists of using a standard RZ pulse generator (also referred to as *pulse carver*), followed by a phase modulator (PM) driven by a clock signal at half the bit rate. The pulse shape and duty cycle are determined by the RZ pulse generator. Ideally, one could drive the PM with any periodic signal and, therefore, obtain any arbitrary phase-modulation shape, including a square-APRZ. In practice, the bandwidth of the electronics is limited and the clock signals available at high speed are normally sinusoidal. The PM implementation, therefore, normally gives rise to a signal like that shown in Figure 6.23 (sin-APRZ). This implementation is relatively easy to build, although one needs to design the delay line in Figure 6.24 carefully in order to align the phase modulation signal and the pulses properly (misalignment $\Delta\tau$ is shown in the figure).

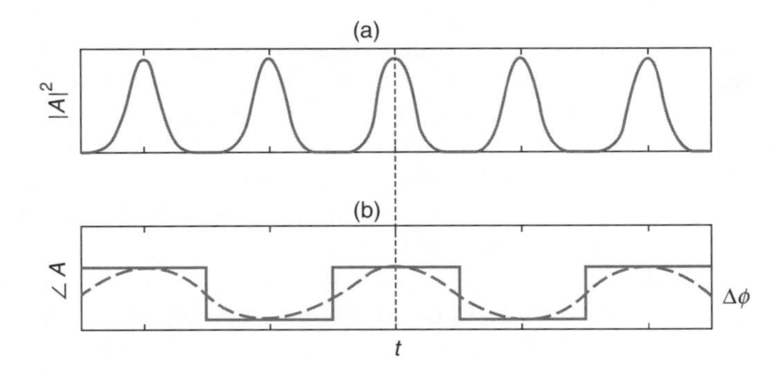

Figure 6.23 Intensity (a) and phase of APRZ signals (b). In (b) the phase variation is shown for square-APRZ (solid line) and for sin-APRZ (dashed line)

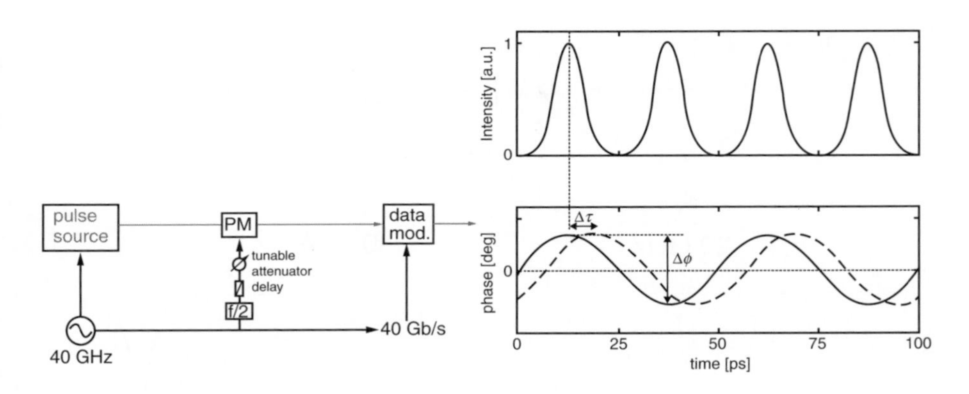

Figure 6.24 (a) Scheme of a 40 Gbit/s APRZ transmitter implemented with an RZ pulse generator and a PM. (b) Intensity and phase of APRZ signals: the phase variation is shown for the case of alignment between data pulses and modulating wave (solid line) and in the case of $\Delta\tau$, $\Delta\varphi$ misalignment (dashed line)

A possible way to implement square-APRZ is by means of optical time-division multiplexing (OTDM) [73]. In this APRZ implementation, a train of narrow pulses at half the bit rate is generated. The signal is then divided by a 3 dB power splitter and one of the two copies of the signal is passed through a phase shifter (PS) and through a 1-bit-slot delay, before being recombined with the other copy, as shown in Figure 6.25. This assures that the phase of the signal is constant in each bit slot, which is ideal in the sense that it is not accompanied by any chirp, but the need for narrow pulses and for proper phase balancing of the two interferometer arms make this a relatively challenging implementation. The pulse shape and duty cycle depend on the technique used to generate RZ pulses.

The PM implementation is relatively easy, but it has the disadvantage that it requires three modulators: one for the generation of the RZ pulses, one for the data modulation, and one for the phase modulation. A more efficient implementation is that proposed in [74], in which only two modulators are needed: one to generate phase-alternated pulses and one for data

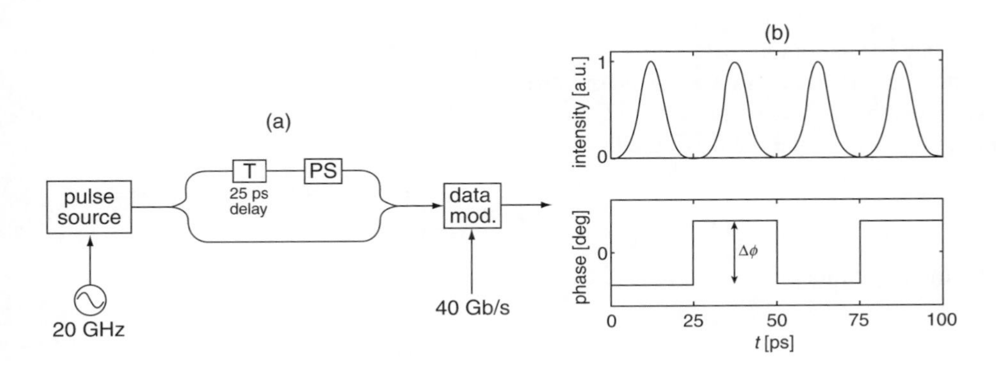

Figure 6.25 (a) OTDM implementation of a square-APRZ transmitter. (b) Intensity and phase of the resulting signal: phase-alternation amplitude $\Delta\varphi$ is adjusted by tuning the phase shifter (PS)

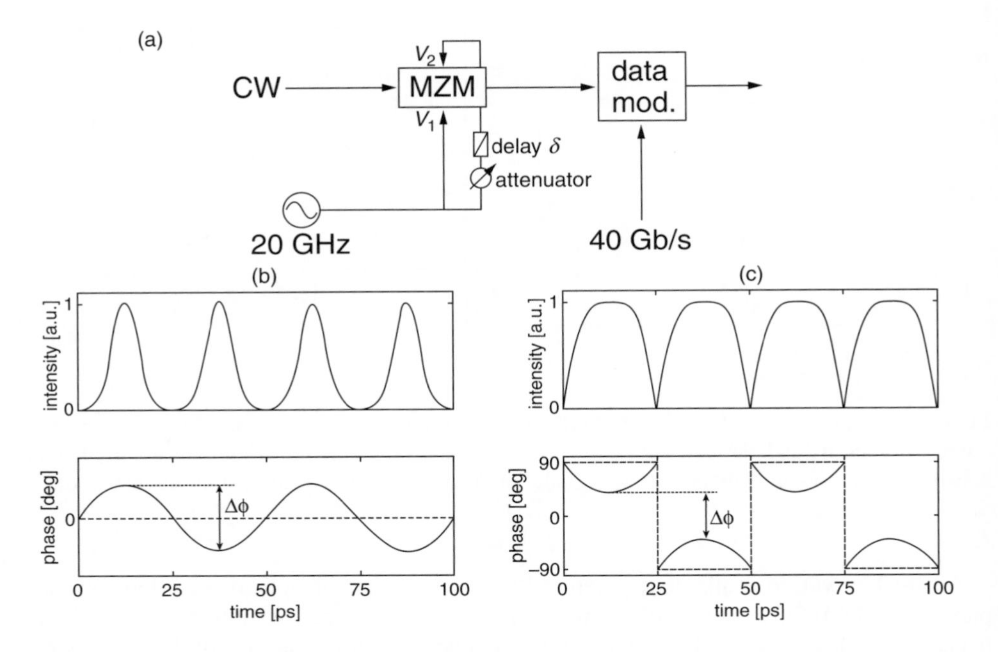

Figure 6.26 (a) Single-MZM APRZ transmitter. By properly unbalancing the amplitude or the phase of V_1 and V_2, pulses with different duty cycles and phase alternation $\Delta\varphi$ can be obtained: e.g. (b) 33% APRZ for maximum-transmission bias and $V_1 = V_2 = 0.5\sin(\delta/2)$, and in particular 33% RZ if $\delta = \pi$, $\Delta\varphi = 0$ and $V_1 = V_2 = V_\pi$ (dashed-line phase profile in the figure); and (c) 67% APRZ for zero-transmission bias, $\delta = \pi$, $\Delta\varphi = \pi$ and $V_1 + V_2 = 2V_\pi$, and in particular standard 67% CSRZ if $V_1 = V_2$ V_π (dashed-line phase profile in the figure)

modulation. This transmitter, shown Figure 6.26, is essentially a modification of the standard CSRZ transmitter: the pulses are generated by a dual-drive MZM, driven by two sinusoidal signals at half the bit rate. Unlike in the CSRZ transmitter, however, the two driving signals are unbalanced in amplitude and/or phase. Depending on the specific configuration of this phase and amplitude unbalance, different signals can be obtained [75].

In one configuration, 33% duty cycle sine-APRZ can be obtained if the MZM is biased at the maximum transmission point and driven by two phase-delayed 20 GHz clock signals. It can be shown that pulses with phase-alternation amplitude $\Delta\phi$ are obtained (Figure 6.26) if the phase delay between the driving clocks is

$$\delta = 2\arctan\left(\frac{\pi}{\Delta\phi}\right) \tag{6.11}$$

and their amplitude is

$$V_1 = V_2 = \frac{V_\pi}{\sin(\delta/2)} \tag{6.12}$$

where V_π is the voltage needed to induce a π phase shift on the signal propagating on one arm of the MZM. A standard 33% RZ is obtained for $\delta = \pi$ and $V_1 = V_2 = V_\pi$.

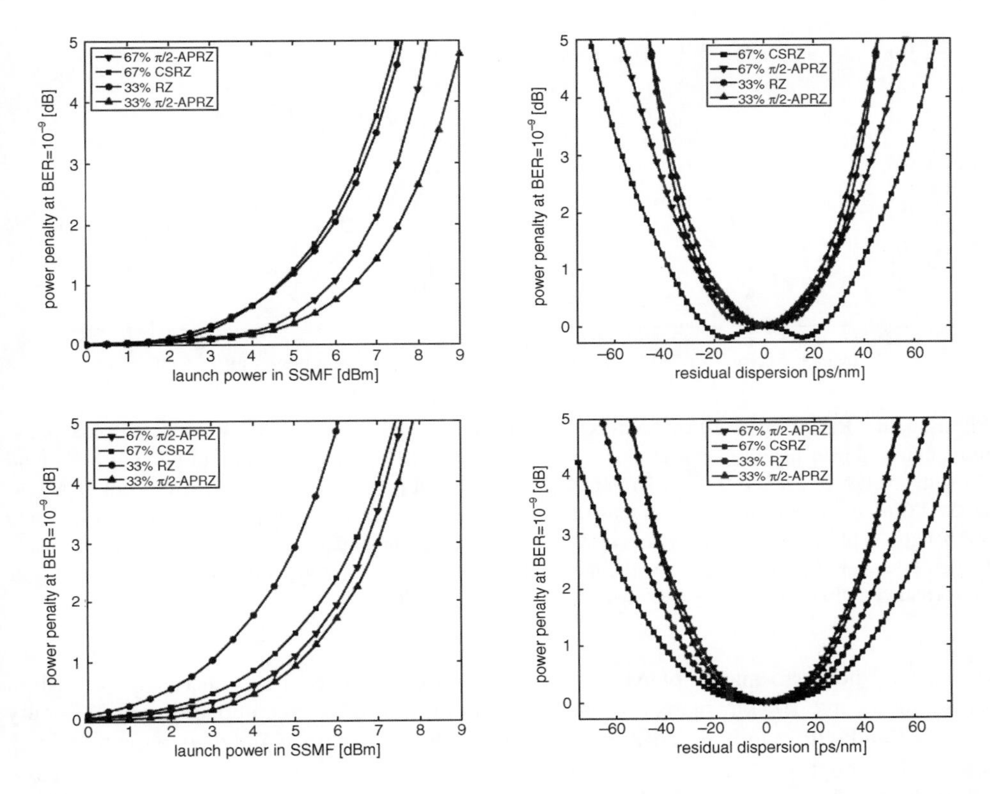

Figure 6.27 (Top-left) Power tolerance on a 5×100 km transmission link for the different modulation formats, without DWDM filtering: (top right) dispersion tolerance for the different modulation formats; (bottom) effect of typical DWDM filtering on power tolerance (bottom left) and dispersion tolerance (bottom right) for the different modulation formats [76]

APRZ pulses can also be obtained by biasing the MZM in Figure 6.26 at the zero-transmission point and keeping the drive signals on the two arms in phase opposition ($\delta = \pi$). When the drive-voltage amplitudes on the two arms, V_1 and V_2, are equal, a CSRZ signal is obtained, in which the pulses acquire a flat alternate phase with $\Delta\phi = \pi$. When $V_1 \neq V_2$, though, the shape of the phase variation is a sinusoidal function with a π discontinuity at the bit transition; see Figure 6.26.The phase difference between the centers of neighboring pulses will be

$$\Delta\phi = \pi - \pi \frac{|V_1 - V_2|}{2V_\pi} \tag{6.13}$$

If $V_1 + V_2 = 2V_\pi$, then the modulator is driven between two transmission maxima and the duty cycle is 67%.[9]

The improved power tolerance of APRZ was verified in a number of numerical and experimental studies. For instance, Figure 6.27 shows the results of a numerical simulation

[9] Lower drive voltages will give narrower pulses at the cost of higher insertion loss.

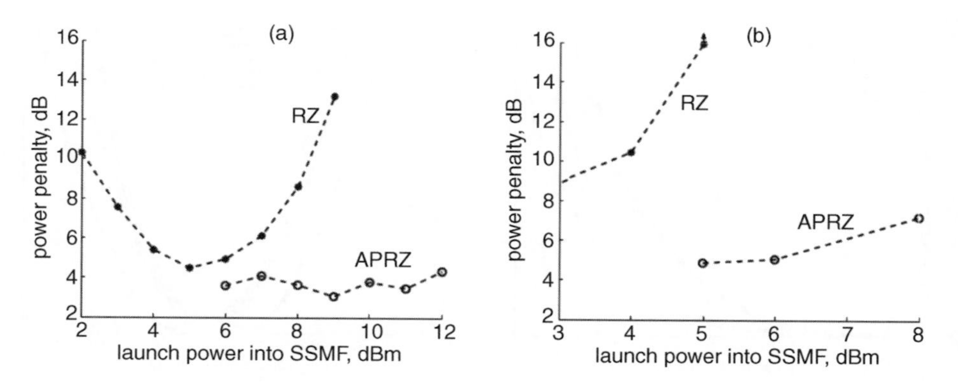

Figure 6.28 Results of a 820 km field-experiment comparing RZ and $\pi/2$-APRZ with sinusoidal phase modulation: power penalty versus launch power over the 820 km link (a) with no MUX/DEMUX and (b) with 75 GHz MUX and 50 GHz DEMUX [77]. At launch-power levels below 6 dBm, without MUX/DEMUX (a), no significant change in sensitivity was observed when applying phase modulation, and only RZ penalty is plotted. Here, the system is ASE-noise-limited and performance improves with increasing launch power. For higher power levels, however, nonlinear effects start to become important and the performance enhancement of APRZ over RZ becomes visible

comparing the performance of APRZ with RZ and CSRZ over a typical transmission link consisting of five 100 km spans [76]. APRZ with optimized $\Delta\phi$ is compared with 33% duty cycle RZ and with standard CSRZ, with respect to power tolerance and dispersion tolerance, with and without MUX/DEMUX for DWDM. The figure shows that, in the unfiltered case, 67% $\pi/2$-APRZ possesses considerably improved nonlinear tolerance with respect to CSRZ, and higher dispersion tolerance than 33% $\pi/2$-APRZ. In the case of strong filtering, APRZ with $\pi/2$ phase shift still gives maximum nonlinear tolerance, although the margin on CSRZ is reduced. Similar results are obtained experimentally, as shown in Figure 6.28 [77]. Initially, transmission with no MUX/DEMUX filtering is studied. It is interesting to note that nonlinear tolerance is degraded considerably by the presence of the MUX/DEMUX. This is because the MUX placed at the transmitter acts as a narrow-band filter, broadening the pulses. Long pulses are less tolerant to nonlinear impairments and, indeed, nonlinear effects. Applying phase modulation (APRZ) reduces the impact of nonlinear effects significantly and an increased error-free power range is obtained. This increased power tolerance can be translated into transmission over longer distances.

6.3.4 DPSK Modulation Formats

In PSK, information is carried by the signal phase. PSK has an inherently better tolerance to noise at the receiver (*receiver sensitivity*) than ASK does [31] and is broadly used in, for example, radio communications.

Receivers able to detect the phase of an optical signal (*coherent detectors* [1]) are not trivial to implement, so pure PSK is not implemented in currently installed systems (although rapid progress in digital signal processing (DSP) in recent years has set in motion an intense research activity in coherent PSK [78,79]). A way around this problem in commercial systems is to use

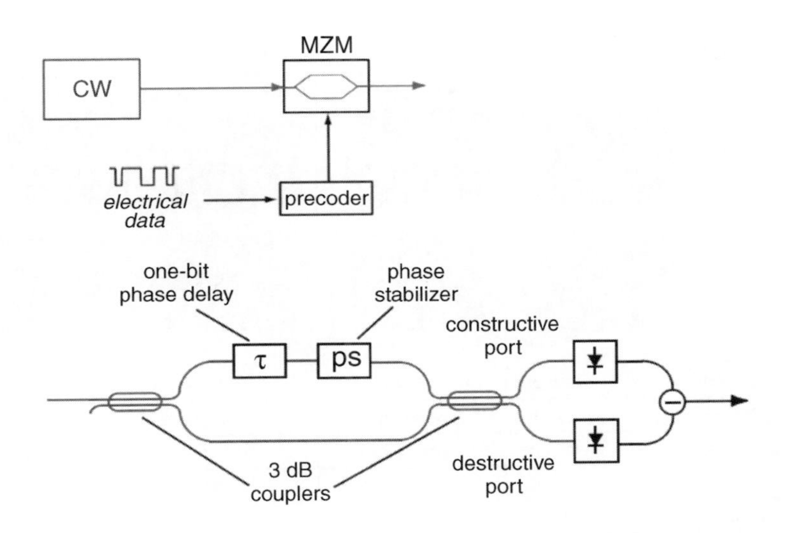

Figure 6.29 The diagram illustrates a DPSK transmitter in MZM configuration (top) and receiver (bottom). Because of the differential detection at the receiver, data needs to be pre-coded before being modulated onto the MZM at the transmitter

DPSK (Figure 6.29), in which the phase of a bit slot is encoded relative to the previous bit slot, so that detection can be done by comparing the phase of consecutive bits using an interferometer and simple intensity detection [80]. A differential pre-coder is needed at the transmitter side, while at the receiver side a differential receiver has to be implemented and combined with a decoder. Moreover, in order to take advantage of the better noise tolerance, balanced detectors need to be used [42].

As in the case of ASK, even DPSK can be implemented as RZ or NRZ. The resulting modulation formats are then referred to as NRZ-DPSK and RZ-DPSK [81]. In fact, a PSK format has also been proposed with CSRZ pulses (CSRZ-DPSK [82]). DPSK generally refers to binary DPSK. Multilevel DPSK is now starting to receive considerable attention, especially in the four-level form of differential quadrature PSK (DQPSK) [83,84], but more recently also in the eight-level form [85]. The primary advantage of multilevel schemes is that a lower symbol rate is needed to achieve a specific bit rate, therefore increasing the bit duration, which, as will be seen, is very beneficial in terms of tolerance to filtering and dispersion (see Section 6.3.5).

As mentioned above, DPSK shows increased tolerance to noise, specifically 3 dB improved receiver sensitivity with respect to ASK, if balanced detection is used, which in principle allows one to double the noise-limited transmission distance. In principle, nonlinear effects are also lower in DPSK with respect to ASK, since a pulse is present in each bit slot of the DPSK sequence, so that the pulse peak power is half that of the corresponding ASK sequence for a given average power. Moreover, as the pulse sequence is not pseudo-random as in ASK, interchannel cross-phase modulation (XPM), which is not sensitive to the bit phase, has a minor impact (because neighboring bits are equally affected). In ASK transmission mode, interchannel XPM generated at each collision between pulses belonging to adjacent channels is responsible for a detrimental timing jitter. When considering intrachannel pulse-to-pulse

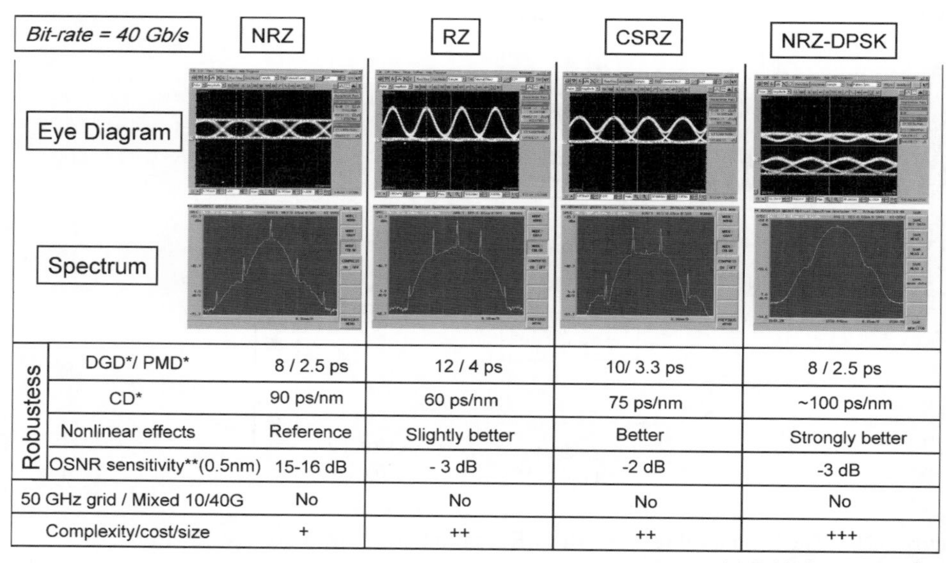

Bit-rate = 40 Gb/s		NRZ	RZ	CSRZ	NRZ-DPSK
Eye Diagram					
Spectrum					
Robustess	DGD*/ PMD*	8 / 2.5 ps	12 / 4 ps	10/ 3.3 ps	8 / 2.5 ps
	CD*	90 ps/nm	60 ps/nm	75 ps/nm	~100 ps/nm
	Nonlinear effects	Reference	Slightly better	Better	Strongly better
	OSNR sensitivity**(0.5nm)	15-16 dB	- 3 dB	-2 dB	-3 dB
50 GHz grid / Mixed 10/40G		No	No	No	No
Complexity/cost/size		+	++	++	+++

* 1-dB OSNR penalty @ 10^{-9}

** OSNR sensitivity in 0.5 nm @ 10^{-9}

Figure 6.30 Resilience to various transmission impairments at 40 Gbit/s per channel for binary RZ-DPSK, compared with different types of basic OOK format (NRZ, RZ, and CSRZ). The ability of a format to resist, for example, ASE noise is measured through different symbols (++ : for a high robustness; + : for a relatively high robustness; −: for a relatively poor robustness; −−: for a very poor robustness)

nonlinear effects – namely, intrachannel XPM (IXPM) and IFWM, discussed in Chapter 5 – the DPSK is also more resilient than ASK, because of the symmetry of the generating pulses (power in different bits is independent of the data sequence). Moreover the IFWM-induced nonlinear phase noise leads to a correlation between the nonlinear phase-shifts that are experienced by any two adjacent bits [86,87]. Clearly, this does not affect the information which is contained, for the DPSK format, in the relative phase difference between these bits.

DPSK, however, is affected by a nonlinear effect not present in ASK: the nonlinear phase noise generated by the Gordon–Mollenauer effect, through which amplitude variation due to noise is translated to phase variations. If it is true in transmission lines using high local dispersion fiber (like SSMF) that pulses broaden dramatically at 40 Gbit/s therefore their several pulses with random relative phase will rapidly overlap so that at specific locations many broadened pulses will sum in phase, giving rise to high power peaks. Overall, however, nonlinear tolerance of DPSK is generally higher than standard ASK formats (e.g., see Ref. [62]). Figure 6.30 tabulates all the major characteristics of NRZ, RZ, CSRZ, and DPSK for 40 Gbit/s transmission systems.

6.3.5 Spectrally Efficient Modulation Formats

The modulation formats described above work well on the standard 100 GHz ITU DWDM channel grid. In order to be able to transmit with tighter channel grids, more spectrally efficient

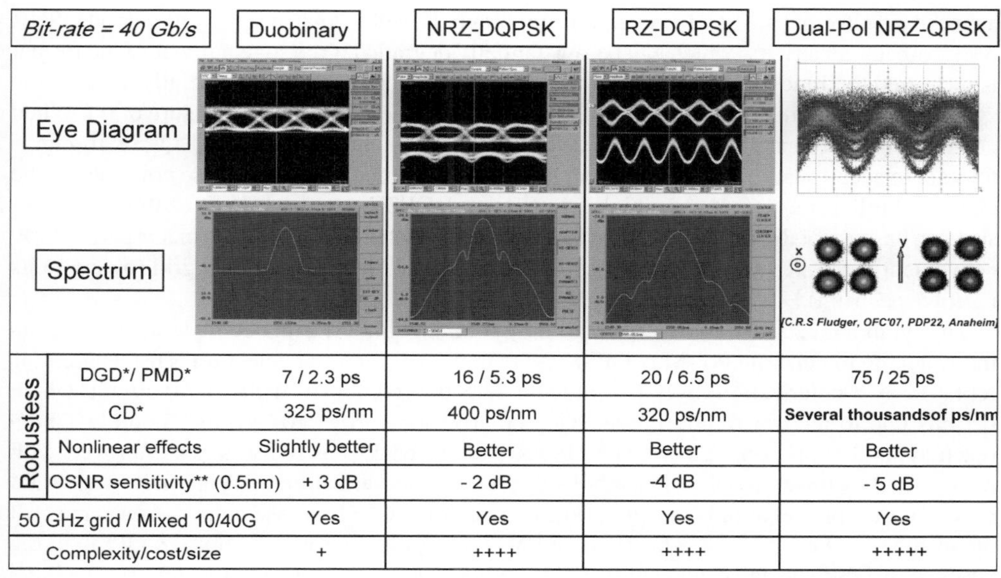

Bit-rate = 40 Gb/s		Duobinary	NRZ-DQPSK	RZ-DQPSK	Dual-Pol NRZ-QPSK
Robustess	DGD*/ PMD*	7 / 2.3 ps	16 / 5.3 ps	20 / 6.5 ps	75 / 25 ps
	CD*	325 ps/nm	400 ps/nm	320 ps/nm	Several thousandsof ps/nm
	Nonlinear effects	Slightly better	Better	Better	Better
	OSNR sensitivity** (0.5nm)	+ 3 dB	- 2 dB	-4 dB	- 5 dB
50 GHz grid / Mixed 10/40G		Yes	Yes	Yes	Yes
Complexity/cost/size		+	++++	++++	+++++

* 1-dB OSNR penalty @ 10^{-9}

** OSNR sensitivity in 0.5 nm @ 10^{-9}

Figure 6.31 Resilience at 40 Gbit/s per channel of different types of modulation format to various transmission impairments. The ability of a format to resist, for example, ASE noise is measured through different symbols (++ : for a high robustness; + : for a relatively high robustness; −: for a relatively poor robustness; −−: for a very poor robustness)

formats are needed. Three such formats are optical duobinary, DQPSK, and Pol-Mux quadrature PSK (QPSK), which have a narrower spectrum, compatible with the 50 GHz ITU grid. Figure 6.31 summarizes the transmission characteristics of these formats, which should be compared with OOK and binary DPSK, as shown in Figure 6.28.

Duobinary (see Section 6.3.1) and DQPSK (see Section 6.2.5) require to be differentially pre-coded (like the DPSK format) and, in the particular case of the duobinary format, the electrical pre-coded binary sequence must be filtered with a fifth-order Bessel filter with a 3 dB bandwidth equal to 11.2 GHz. The main advantages of the duobinary format reside in its ease of implementation and its high tolerance to residual CD (nearly three times the tolerance of the NRZ-OOK format). However, when compared with the NRZ-OOK format, the duobinary presents a degraded OSNR back-to-back sensitivity (nearly 2 dB) as well as a worse PMD tolerance. Duobinary would be particularly convenient for short-reach applications (metro) and for high spectral efficiency transmissions (like transmitting together 10 Gbit/s and 40 Gbit/s channels on a 50 GHz ITU grid). The RZ-DQPSK format has some interesting features for DWDM 40 Gbit/s transmission, even if its implementation complexity and cost are relatively high. First, it has high tolerance to PMD [88] and residual CD, respectively by factors of 2 and 4 better than RZ-OOK. This can be explained by the fact that its symbol rate is half the bit rate (20 Gsymbol/s for the symbol rate, while the bit rate is equal to 40 Gbit/s). Decreasing the symbol rate while maintaining the bit rate is very convenient to increase the robustness of a modulation format to ISI. However, it is important as well to keep a sufficiently good OSNR back-to-back sensitivity, which can be substantially degraded if the state number of the format is increased too much: the DQPSK format represents a good trade-off, especially because the symbols are optimally distributed in the signal constellation diagram (see Figure 6.31). When considering nonlinear effects, as all the formats with constant envelope, the DQPSK format is tolerant to accumulation of nonlinear impairments even if the nonlinear phase noise due to the Gordon–Mollenauer effect has a higher impact on DQPSK than on DPSK, as the number of states in the case of the DQPSK is equal to 4. Note as well that the DQPSK format is particularly convenient for mixed 10/40 Gbit/s WDM transmission with a 50 GHz ITU grid (owing to its narrow spectrum).

The symbol rate can be further decreased by combining multilevel coding with polarization multiplexing to generate Pol-Mux QPSK. Note that no differential detection is used here. It is replaced by a polarization diversity coherent detection combined with a DSP circuit, which permits one to recover the bit phase. Each polarization carries 10 Gsymbol/s and 20 Gbit/s (each symbol supporting 2 bits). The OSNR back-to-back sensitivity is not improved with respect to the conventional 20 Gsymbol/s QPSK (each polarization carrying half of the total power of the channel at the opposite of the conventional QPSK whose polarization carries all the channel power), but the robustness to PMD (nearly 14 ps for Pol-Mux RZ-QPSK in the absence of DSP specially adapted to PMD compensation) is equivalent to that of 10 Gbit/s WDM transmission systems today deployed in the field, while its resilience to CD (still without dedicated signal processing) is 10 times higher than that of the 40 Gbit/s NRZ format. The main advantage of Pol-Mux RZ-QPSK is to permit a 40 Gbit/s WDM transport while keeping the present fiber infrastructure even when substantially affected by PMD. In spite of its considerably higher cost and complexity, probably substantially higher than that of classical 40 Gbit/s NRZ-OOK, incumbent carriers are particularly interested in this technology because the extra cost generated by complex transponders can be compensated by the cost savings realized by not replacing the optical cable affected by PMD. Note at last that DQPSK and Pol-Mux QPSK modulation formats have similar maximum reach ($\sim$1500 km) and they are particularly convenient for crossing transmission lines implementing an OADM cascade.

6.4 Electronic Equalization of Optical Transmission Impairments

6.4.1 Electronic Equalization Concepts

Electronic equalization is used in all types of communication systems for wireline as well as wireless applications. Descriptions of the concepts can be found in standard textbooks [89] about the basics on detection in the presence of ISI and noise. However, the requirements of the signal processing complexity and high speed have prevented application of those techniques for long time in optical transport systems.

Electronic equalization provides an alternative approach to optical distortion mitigation techniques; hence, in order to be competitive, one must provide either improved performance or cost efficiency. In contrast to optical dispersion compensation, electronic signal processing in today's systems suffers from the fact that, after direct detection in optical receivers, all signal distortions become nonlinear in nature. The benefit, on the other hand, comes from the fact that the electronic equalizer does not require any knowledge of the nature of the distortion

(CD, SPM, PMD, etc.); hence, it seems attractive to high bit-rate systems that suffer from other distortions as well. In its simplest approach it may be deployed as a linear filter with adjustable parameters that are tuned on the basis of the channel characteristics. Additionally, decision FB (also known as nonlinear equalization) can be used, exploiting the estimated received bit information for cancellation of ISI. Finally, maximum likelihood sequence detection allows optimization of the bit decisions over a large number of received symbols, selecting the most likely sequence out of all possible ones.

In order to achieve optimum performance the equalizer's channel model must be adapted continuously. This is done by derivation of a signal performance measure (FB signal), which is then used to tune the adjustable parameters of the equalizer within a control loop circuit. It is important to notice that FB signal generation in high-speed/optical applications is a very important part of the equalizer design, since it is dealing with electronic circuits close to the limits of high-speed application-specific integrated circuit technology.

In the following these concepts are briefly outlined and particular aspects of their application in optical communications are discussed. Then, performance characteristics are reviewed for a number of modulation formats currently deployed in optical transmission systems.

6.4.1.1 Feed-forward Equalization: Linear Transversal Filter

As a first example of a typical equalizer structure, a transversal filter setup is illustrated in Figure 6.32. The received time-domain electrical signal $r(t)$ from the photodetector is fed into a linear filter. At the filter output the signal is sampled and a hard decision is applied. The sampling instant and the decision threshold must be controlled for optimum performance.

One important design consideration comes from the fact that the continuous adaptation shown in Figure 6.32 requires sampling and processing of the received signal, requiring at least a full-speed analogue-to-digital conversion and implementation of a mean-square-error optimization [90], applying a stochastic gradient-search algorithm. In order to relax the requirements regarding complex high-speed circuitry, other types of FB signal have been proposed, which are summarized in Section 6.4.3.

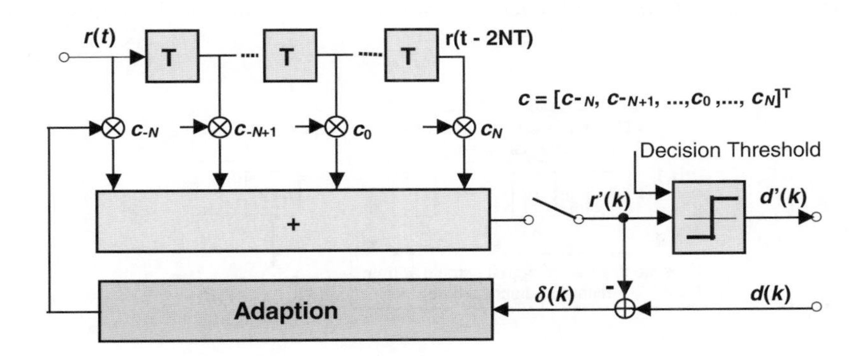

Figure 6.32 Block diagram of FF equalizer (FFE)

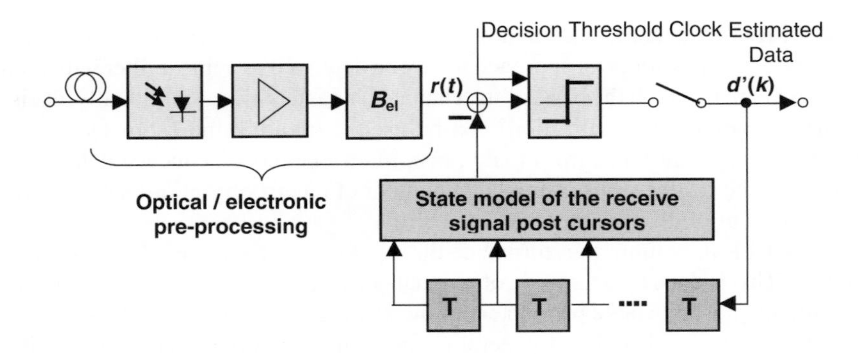

Figure 6.33 Block diagram of decision FB equalizer

6.4.1.2 Decision-Feedback Equalization: Nonlinear ISI Cancellation

After direct detection, the optical channel model has a low-pass characteristic. Linear preprocessing of the received signal by an FF equalizer (FFE) is known to introduce noise enhancement. Therefore, it is well known that a second equalization step should be added. After data decision, estimates of the information bits are available. Those can be applied to an FB filter to remove post cursor distortions from the received signal. Both sets of coefficients must be adapted simultaneously. A potential control concept based on conditional error counts has been proposed in [91] (Figure 6.33).

6.4.1.3 Maximum Likelihood Principle: Sequence Detection

In contrast to the concepts described in preceding sections, a different methodology is used when the data decision is based on the information of an entire sequence of bits.

The most general approach to describe the received signal in a discrete-time representation is shown in Figure 6.34. Under the assumption that the channel has limited memory of length L bit durations, the channel is completely characterized by 2^L possible states. The data samples $y(k)$ are generated from the transmitted binary symbol $a(k)$ by applying the discrete-state channel

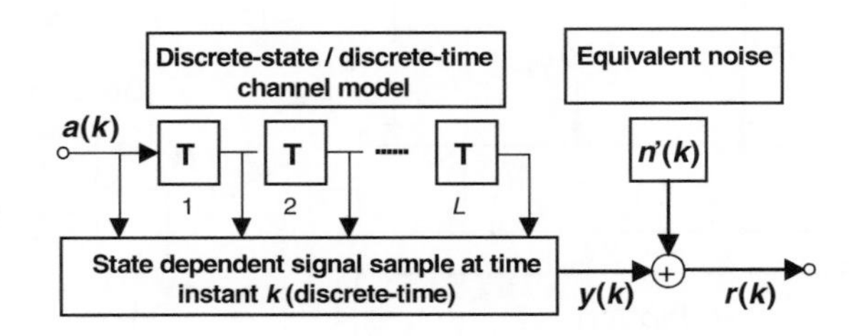

Figure 6.34 State-based digital channel model

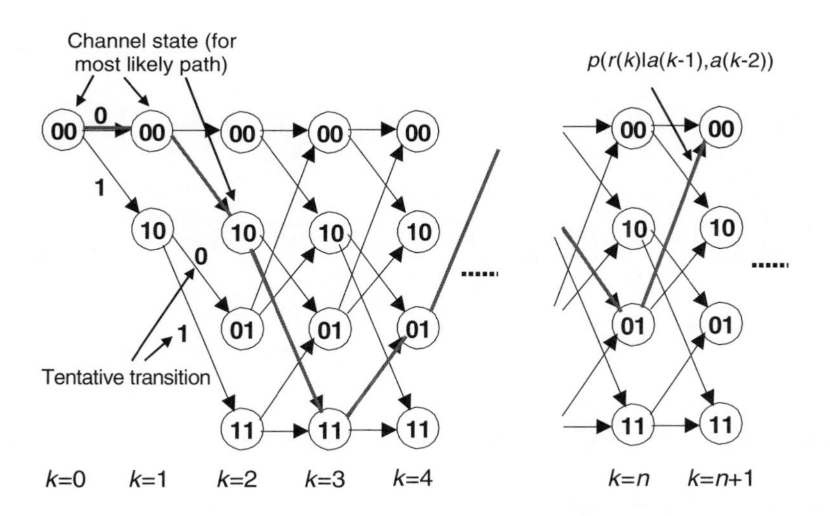

Figure 6.35 Example of trellis diagram for binary symbols and memory length $L=2$

model. Together with noise samples $n'(k)$ they form the received sequence $\{r(k)\}$. The channel model includes all electronic and optical signal modification by filtering, propagation, and signal processing (e.g., analog-to-digital conversion).

At the receiver the optimum estimate of the transmitted data sequence $\{a(k)\}$ is calculated from the observation of the received signal samples $\{r(k)\}$ according to the maximum a posteriori principle. Assuming equal probable data symbols, this leads to the maximum likelihood criterion: $\{\hat{a}(k)\} = \arg\max_{\{a(k)\}} p(r(k)|\{a(k)\})$.

In each clock cycle the channel output $y(k)$ is updated dependent on the channel state and the actual data symbol $a(k)$. In turn, for every received value $r(k)$ there is a probability that, given the actual channel state together with the noise value $n'(k)$, a specific symbol (for binary sequences one out of $\{0, 1\}$) has been sent: $p(r(k)|a(k-1), a(k-2))$. In a second step, these probabilities are used to calculate the data sequence which has highest probability of having been sent. This is illustrated in the trellis diagram (Figure 6.35) by the path which connects the most likely state transitions.

Overall, the path metric is accumulated and few potentially best paths are stored. After a sufficiently long processing window, parts of the path that are unlikely to be required further on are dropped and the data sequence is decided finally. The required calculations are efficiently implemented using the Viterbi algorithm (VA). For operation, first the transition probabilities (branch metric entries) must be estimated and stored in a lookup table. This can be accomplished by histogram approximation (see Figure 6.36). For time-variant channels (e.g., caused by polarization variations) a compromise must be found between fast tracking and accurate calculation of the entries.

Figure 6.37 shows an example for an eye diagram with potential probability density functions for each pattern composed of 3 bits.

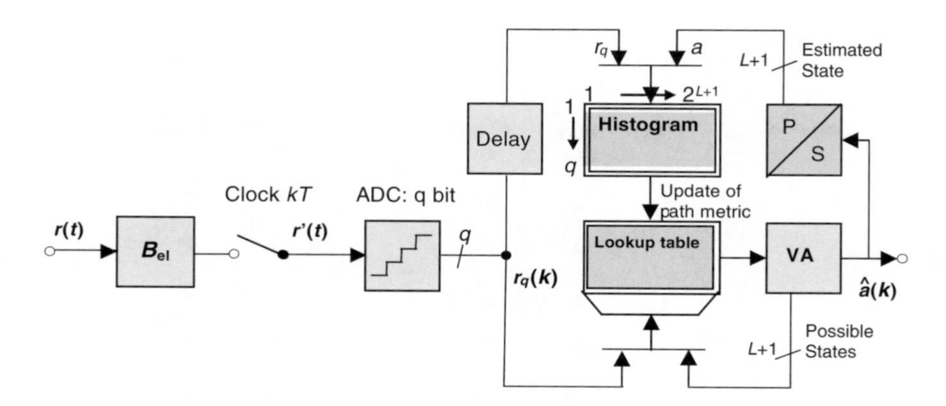

Figure 6.36 MLSE implementation using histogram-based branch metric estimation

6.4.2 Static Performance Characterization

In this section the improvement in dispersion tolerance by electronic equalization is evaluated. The characterization is done by calculation of the required OSNR penalty. In particular, we compare a receiver with optimized decision threshold (automatic threshold control) and an maximum likelihood sequence estimation (MLSE) receiver with four states and 4-bit analog-to-digital converter. Figures 6.38 and 6.39 show simulation results for the required OSNR for different modulation formats with respect to CD and DGD respectively. Although results are given for 10 Gbit/s bit rate, an estimation for higher bit rates can be achieved by linear downscaling of the DGD values and quadratic downscaling in the case of CD.

6.4.3 Dynamic Adaptation of FFE- and DFE-Structures

For all equalization techniques described in the preceding sections, assumptions on the channel model must be made in order to derive control concepts for the optimum compensation

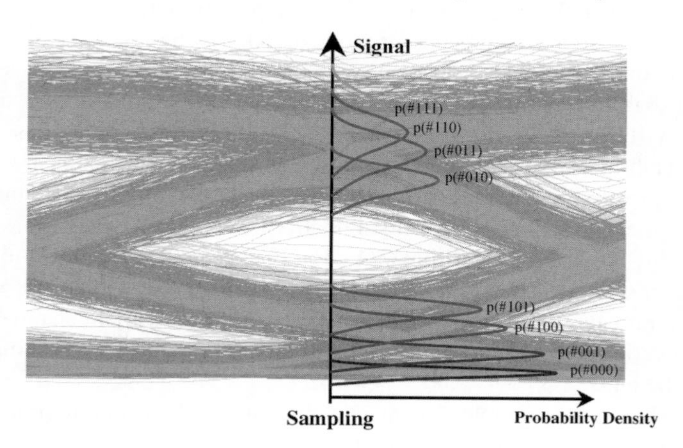

Figure 6.37 Eye diagram and branch metric histogram

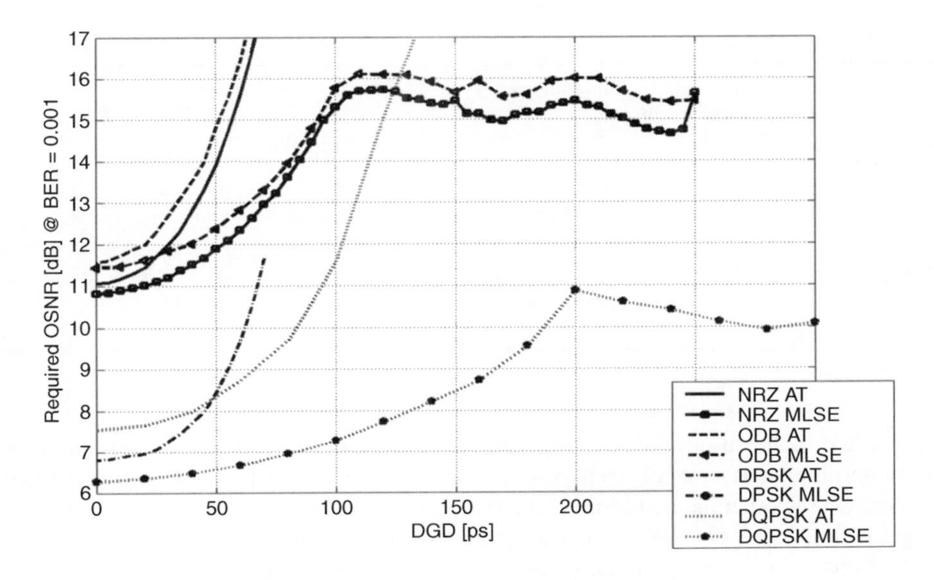

Figure 6.38 Q-factor penalty versus DGD (first-order DGD)

performance. Most common models describe the static behavior, in order to assess the potential benefit of electronic equalization. The results are valid as long as the channel dynamic is slower than the adaptation time of the equalizer. In applications, where fast channel variation must be expected (e.g., for PMD compensation and fast polarization fluctuations), the adaptation becomes a critical part of the equalizer concept. In particular, two modes of operation must be

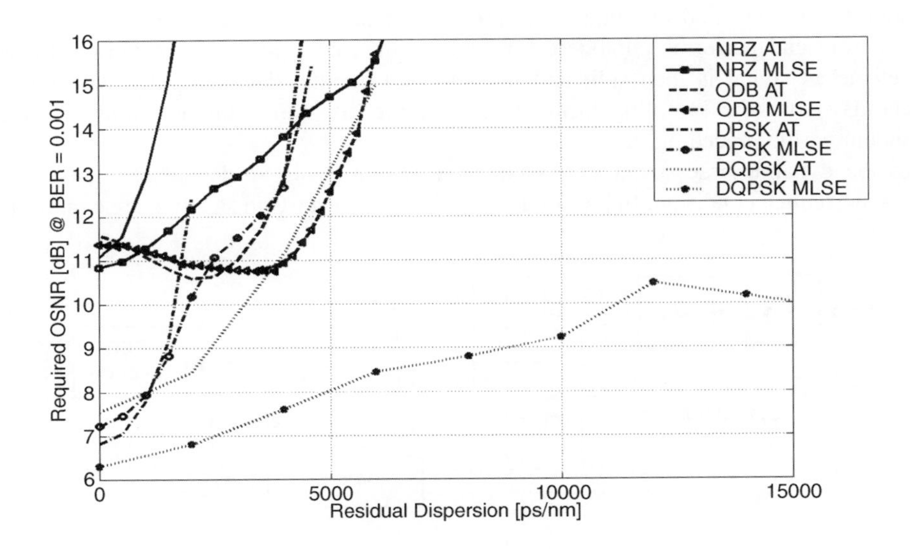

Figure 6.39 Q-factor penalty versus residual CD

Table 6.2 Overview on alternatives for control signal generation

	RF spectrum power/shape	Eye opening monitor	Q factor estimation	Conditional pre-FEC BER
Complexity	Medium	Medium	High (SiGe)	Low (CMOS)
Speed	Fast	Medium	Medium	Slow at low BER
Benefit	No clock sync needed	Correlation with signal distortion	Good correlation to BER	Exact measure as long as error correction works
Drawback	RF design	Needs clock sync, limited correlation to BER	Needs clock sync	Needs clock and frame sync

distinguished: acquisition and tracking. During the acquisition phase, several attempts to achieve optimum performance are allowed, which makes it highly probable to find the global optimum. On the other hand, during tracking of dynamic changes the algorithm has to follow changes without the option to change to a different set of equalizer coefficients. This fact makes it very difficult to prove operation under all circumstances; hence, this topic is still under investigation. The different types of control signal do not provide a complete assessment of the best control concept (Table 6.2):

(1) RF spectrum properties; for example, spectral hole-burning from DGD [92];
(2) eye opening monitor – allows detecting signal distortions [93];
(3) Q-factor estimation – additional estimation of the noise power;
(4) conditional pre-FEC bit error ratio from error counters in FEC decoder [91,94].

As an example we investigate in more detail the one-tap DFE shown in Figure 6.40. The thresholds are controlled by conditional pre-FEC error counters as described in [91]. The optical front end performs photodetection, automatic gain control, and electrical filtering. The electrical signal is then split and fed into two parallel decision gates with different thresholds. After the sampling instance, one of the alternative bit decisions is selected dependent on the preceding bit.

For the dynamic assessment we assume a variation of first-order DGD channel under worst-case launch condition. In the simulation, the system starts in steady-state condition for

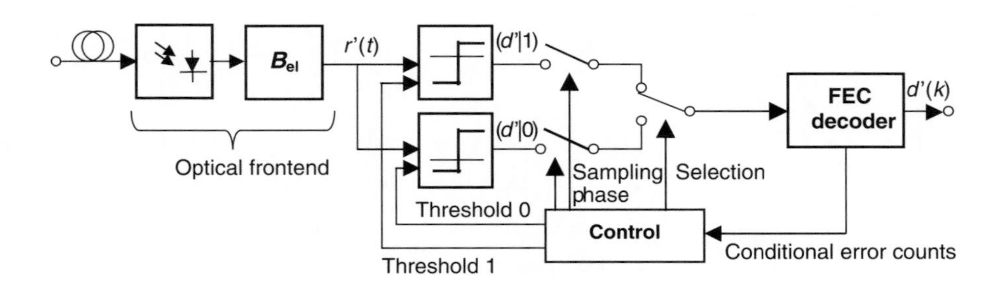

Figure 6.40 Block diagram of DFE control by pre-FEC conditional BER estimation

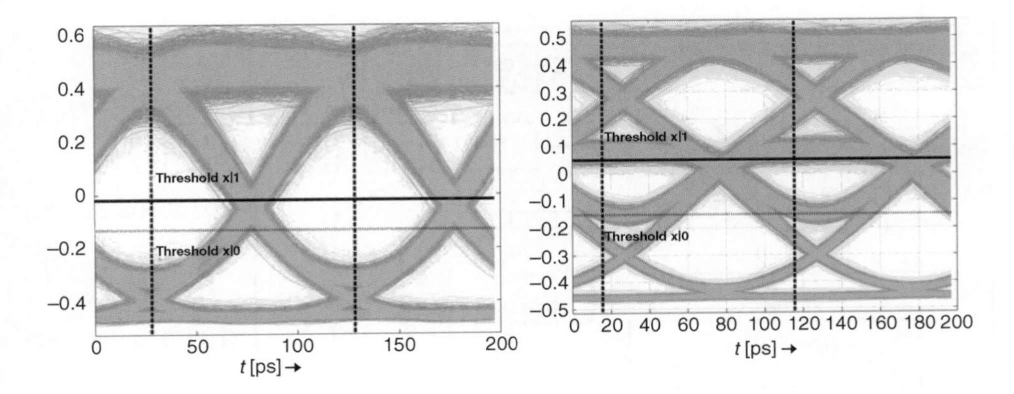

Figure 6.41 Eye diagram for two values of DGD: (a) 0 ps; (b) 80 ps

0 ps DGD. Owing to a residual CD of 800 ps/nm the eye diagram is a little bit closed already (see Figure 6.41a). However, the two thresholds are close to each other, which is an indication that there is only small dependence on the preceding bit. The situation changes when the signal gets degraded by DGD, as is shown in Figure 6.41b. Now the thresholds get more separated.

Figure 6.42a shows the dynamic DGD variation between 0 and 80 ps. The adaptation of the two thresholds is displayed in Figure 6.42b, where the fluctuations indicate that the optimization follows the time-varying distortion in each time step.

6.4.4 General Remarks

The application of electronic equalization is one of the options available for improving transmission in transparent optical networks. It adaptively accounts for signal impairments at the terminals and does not require fine tuning of inline network components. Together with

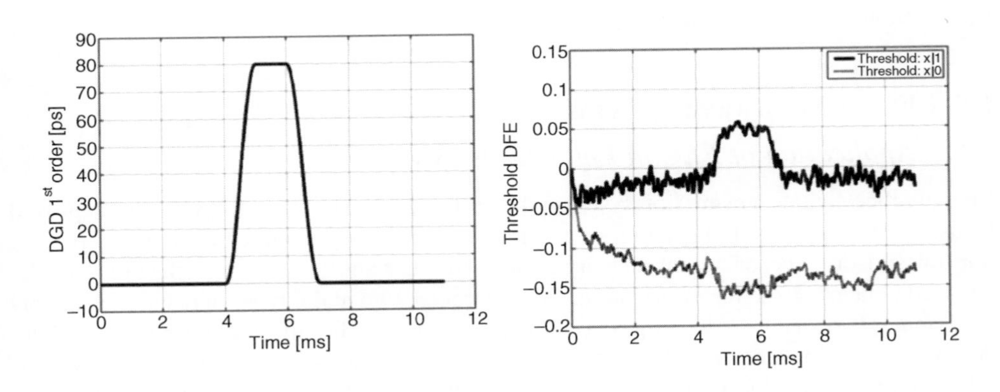

Figure 6.42 DFE tracking behavior for DGD change: (a) DGD time variation; (b) threshold tracking

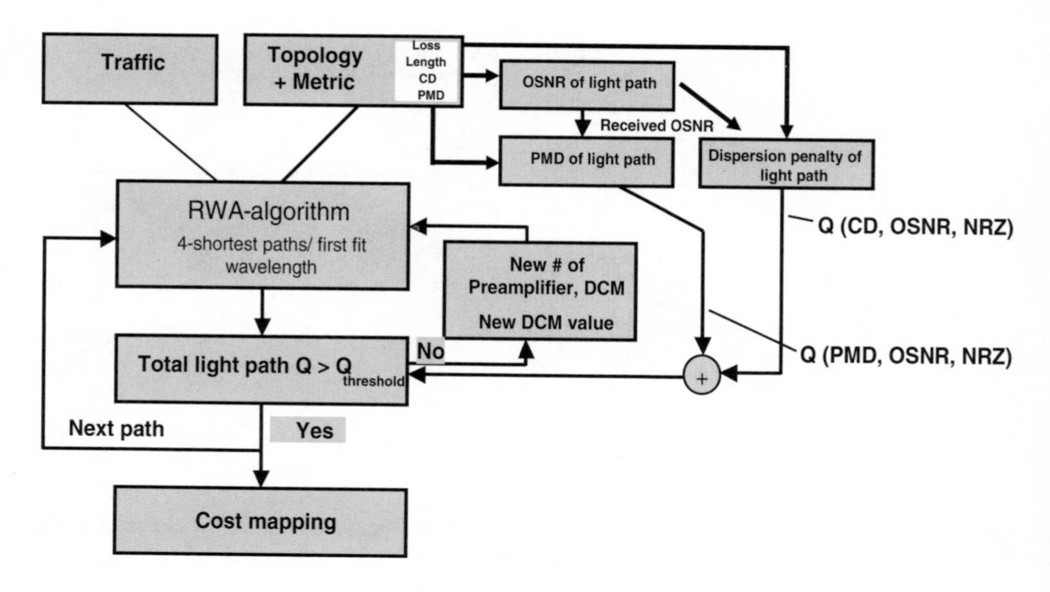

Figure 6.43 *Q* light path setup procedure

advanced modulation formats it is the main contributor towards robust optical transmission. To best utilize this potential, it is required to include the expected transmission characteristic into the path setup procedure. It can be seen from the flow chart in Figure 6.43 that the information about the *Q*-factor (which is directly related to BER) degradation is used to select appropriate light paths to design a network for a given topology and traffic data. The procedure calculates the *Q*-factor penalty for the actual link under the condition of the received OSNR (from the ASE accumulation) and the residual CD along the desired light path. If the *Q*-factor criterion holds, then the light path is accepted; otherwise it is discarded. This procedure has been applied in [95,97] to a number of various transmitter/receiver combinations and can lead to interesting cost savings, as, for example, the number of DCFs can be lowered when electronic equalization is applied at the end terminals [97].

6.5 FEC in Lightwave Systems

6.5.1 Application of FEC in Lightwave Systems

Like in other digital transmission systems, the performance of optical links can be improved by use of FEC. The resulting gain in system margin can be used to increase amplifier spacing, transmission distance, or system capacity. In order to detect and correct errors, the FEC encoder adds redundancy bits to the information. At the receiver, two major types of decoding, namely hard and soft decisions decoding, can be used to recover the information bits. With hard decisions decoding, the receiver first decides on the channel symbols and then passes the information to the FEC decoder (Figure 6.44). With soft decision decoding, the receiver, in principle, would pass on to the decoder the analog signal. Soft decision decoding, thus,

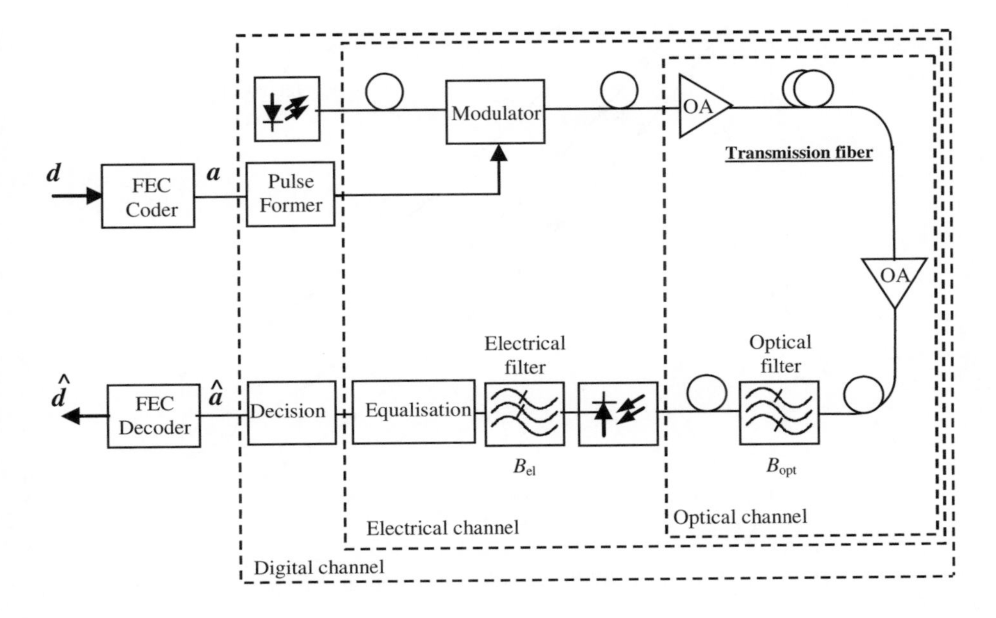

Figure 6.44 Hard decision FEC operates on the digital channel; that is, on binary sequences

is always better performing than hard decisions decoding, at the expense of higher signal processing requirements, like gate count and power consumption.

The benefit of FEC varies between different applications. For interfaces without optical amplifiers, better receiver sensitivity is achieved, which can be used to increase power budget (for interface examples see ITU-T G.959). However, coding gain on the electronic side does only improve optical power budget by half of the coding gain (in decibels), due to the fact that half of the optical signal power (a drop by 3 dB) lowers the signal-to-noise ratio in the electronic domain by 6 dB.

On the contrary, ultra-long-haul systems are OSNR limited. Here, the full FEC gain can be utilized to improved OSNR performance. Additionally, at higher target BER, smaller penalties arise from signal degrading effects like residual CD and receiver noise. An example is shown in Figure 6.45, where the received optical power is chosen 3 dB above the sensitivity for BER of 10^{-12}. It can be seen that the degradation in BER is much less severe at higher BER values, which can be understood as an additional coding gain for distorted signals.

6.5.2 Standards for FEC in Lightwave Systems

Historically, FEC in lightwave systems was first proposed in SDH/SONET standards at 10 Gbit/s. Here, the redundancy bits required by the FEC code are inserted in unused bytes of the SDH/SONET frame; that is, there is no increase in the line rate. Therefore, this approach is called *in-band FEC*. Because of the low number of available unused bytes in the frame, the performance of the in-band FEC is very limited. The code chosen was a 3-bit error-correcting

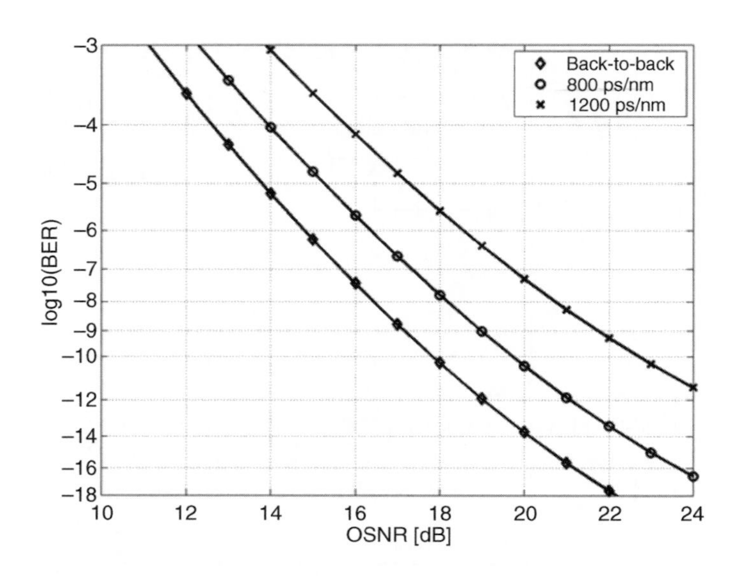

Figure 6.45 BER versus OSNR for a 10 Gbit/s receiver at an optical input power 3 dB above sensitivity for a BER of 10^{-12}

BCH code which was standardized in G.707 [98]. Owing to the limited error correction capability, this code has not gained much attraction.

To achieve better coding performance, additional overhead bits were required, which led to the definition of *out-band FEC*. This approach is mainly used in DWDM systems and is of special interest for submarine and (ultra)-long-haul links. The new frame format for transporting the FEC redundancy bits consists of some overhead bytes (for operations and maintenance purposes), the FEC redundancy bits, and the "protected" data. The increased bandwidth, however, leads to a redefinition of the requirements for most of the optical components, an increase in power consumption, and an increase in the complexity of CMOS circuitry.

6.5.3 FEC Performance Characterization

For the characterization of FEC codes, several assumptions regarding the transmission channel have to be made. The channel is characterized by a set of input symbols, output symbols, and transition probabilities. The coding scheme has to consider these properties of the transmission channel. In a simple case, the transition properties are time invariant and independent from symbol to symbol. This is the so-called discrete memoryless channel (DMC).

The additive white Gaussian noise (AWGN) channel is a widely used model to simulate these channel properties. In that model, the transmitted binary signal $s(t)$ is deteriorated by white Gaussian noise $n(t)$ which is added to this transmitted signal; see Figure 6.46. The power spectral density of that noise has the known "flat" (frequency-independent) characteristics for white noise. The probability density for the received signal magnitude $y(t)$ at any time is Gaussian distributed around the input signal $s(t)$. The impulse response $h(t)$ represents the ideal

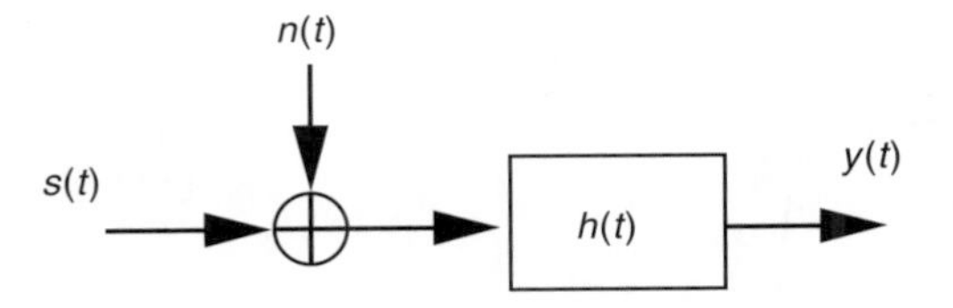

Figure 6.46 AWGN channel

case ISI-free transmission channel. The advantage of that model is that the addition of different noise sources can be modeled easily.

The most commonly encountered case of DMC is the binary symmetric channel. This model is more abstract than the AWGN model. It emphasizes the quality of the received signal much more than the channel properties. It does not refer to the physical properties of the channel, but considers the effects caused by the channel imperfections. An overview on this is given in Figure 6.47. There, the transition probability P_e from state "0" (input of the transmission channel) to state "1" (output of the transmission channel) is the same compared with the transition from "1" to "0," which is characteristic for a BSC.

It must be noted, however, that the BSC/AWGN channel is only an approximation of the optical transmission channel. Depending on modulation format and transmission impairments, the actual channel may behave very differently.

Under this assumption, the performance of a code is described by the BER as a function of E_b/N_0, where E_b is the energy per bit and N_0 is the noise power spectral density. Examples are shown in Figure 6.48.

The implementation of out-band FEC used most is the one specified in G.975 [99], which is based on a Reed–Solomon (255, 239) code with a 7% increase in the data rate. This implementation (referred to as first-generation FEC) achieves a net coding gain of around 5.6 dB at a BER_{out} of 10^{-12} and 6.5 dB at a BER_{out} of 10^{-16} (see Figure 6.48). Also, for the optical transport network defined in ITU-T G.709 [100] the Reed–Solomon (255, 239) is recommended for the optical channel transport unit. Several other types of FEC codes are also suggested in the appendix of G.975; for example, low-density parity check (LDPC) or concatenated codes.

Interleaving/deinterleaving and iterative decoding techniques are used together with the concatenated codes to obtain improved error correction performance. Examples for achievable

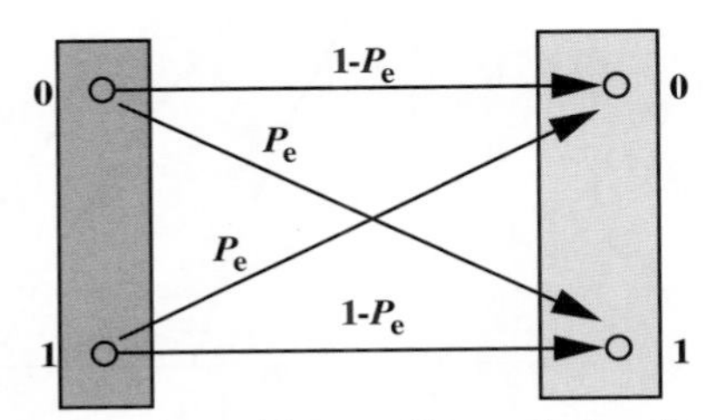

Figure 6.47 BSC

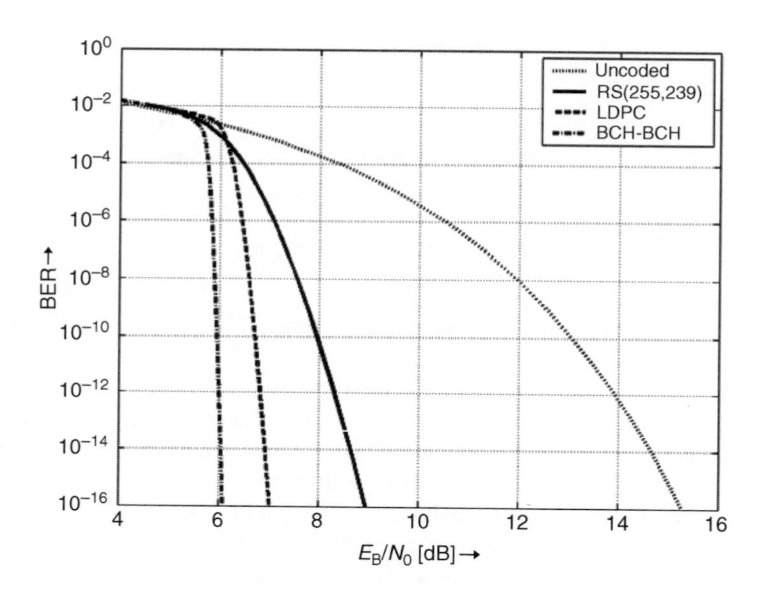

Figure 6.48 Achievable output BER for different FEC codes

net coding gain values are given in Figure 6.49. In [101] a turbo product code with soft-input soft-output decoding and an increased overhead of 21% has been proposed, showing a net coding gain of 10.1 dB (at a BER_{out} of 10^{-13}). However, the larger overhead requires designing optical transmitters and receivers for the higher data rate.

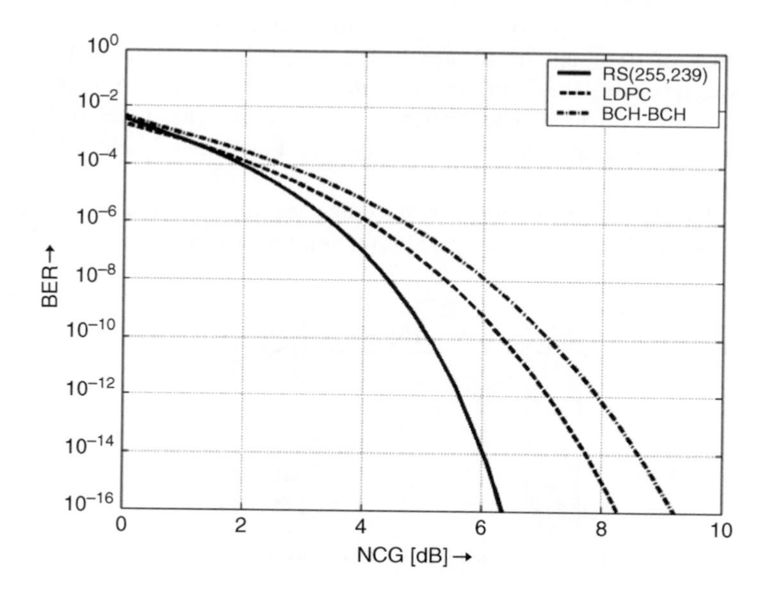

Figure 6.49 Net coding gain for different codes

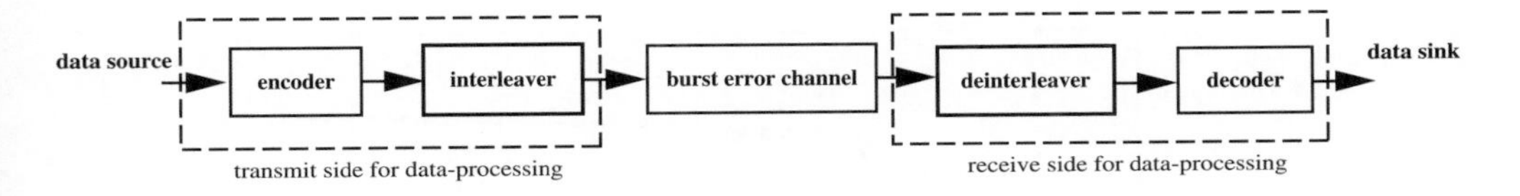

Figure 6.50 Application of an interleaver/deinterleaver

Many performance calculations for FEC codes are made with the assumption of a memoryless channel; that is, the probability of error does not vary with time. This approach cannot be used for burst errors. Therefore, FEC codes are not only characterized by net coding gain and overhead, but also by their ability to correct burst errors. Whereas, for example, BCH codes, concatenated codes, and turbo codes are designed to correct random errors (single errors), Reed–Solomon codes are particularly well suited to correcting burst errors. This is because Reed–Solomon codes handle groups of bits in a symbol, rather than individual bits. The aforementioned Reed–Solomon (255, 239), for example, can correct up to 8(bits) $\times$ 8(symbols) $\times$ 16(interleaving) = 1024 bits successfully, assuming no more than eight symbol errors occur per block (255 symbols).

As shown above, the usage of an additional interleaver/deinterleaver combination together with the encoder/decoder can improve the burst error correction capability. This interleaving rearranges (permutes) the ordering of a sequence of symbols in a deterministic manner. The encoded data stream is interleaved before entering the channel in which burst errors can occur. The deinterleaver on the receive side inverses the functionality of the interleaver and provides the data stream for the decoder. By this technique, error patterns (burst errors) are more uniformly distributed at the decoder input. This is shown in the block diagram in Figure 6.50.

Depending on the required BER_{out}, there is a certain BER_{in} for each FEC code. Improved FEC codes allow for an increase of BER_{in} compared with the BER_{in} of approximately 10^{-5} of the standard Reed–Solomon (255, 239). For example, the LDPC code presented in Figure 6.51 is able to correct a BER_{in} of 1×10^{-3} to a BER_{out} of 1×10^{-16}. This new operating point must be considered when doing the margin allocation for the worst case.

6.5.4 FEC Application in System Design

In optical communication systems, a lighpath is assumed to be performing well when specific criteria are met. The required optical path performance is specified in Q value, as follows:

- In the case of not requiring FEC in the transparent network, $Q = 7$ linear or 16.9 dB at the end of life is required, which corresponds to a BER of $\sim 10^{-12}$. This corresponds to the ITU-T requirement for standardized optical interface parameters.
- In the case of using FEC in the transparent network, $Q = 8$ linear or 18 dB at the end of life is required after FEC, which corresponds to a BER of $\sim 10^{-15}$.

In summary, the target optical path performance at initial installation is (i) a $Q = 18.9$ dB in the case of not requiring FEC in the network and (ii) Q = 20 dB after FEC in the case of using

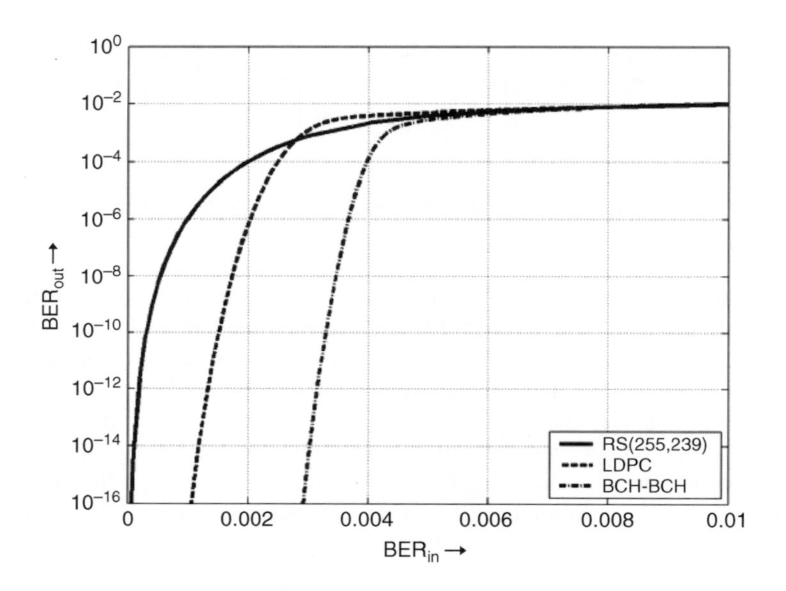

Figure 6.51 Output versus input BER for different FEC codes

FEC in the network, where in both cases a 2 dB margin is built in to cover ageing and extra splice/connector losses. Therefore, the specified end of life performance can be met.

To get a Q factor of 20 dB after an LDPC error correcting code we must have an input Q factor of 10 dB (see Figure 6.52).

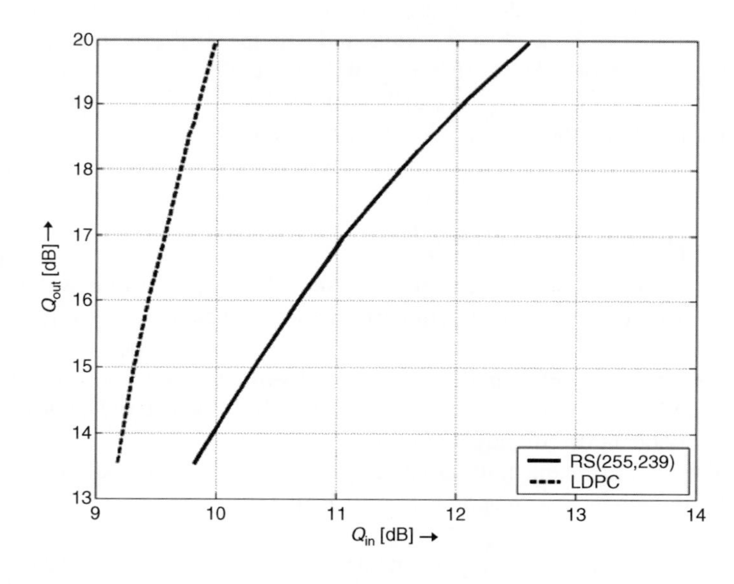

Figure 6.52 Output versus input Q for enhanced FEC compared with standard Reed–Solomon

Moreover, FEC makes it possible to extract the number of bit errors as well as error statistics from the decoder. This information can be used for detailed performance monitoring. As these error counts are a good measure of the signal quality, they can also be used as an FB signal for various control loops, as mentioned in Section 6.4.3. It can be expected that this application of FEC will become even more important in future transparent optical networks. It should be mentioned in this context that in all those cases post-FEC performance must be verified, since there is evidence that the coding gain will decrease in the case of non-AWGN channel responses.

6.6 Appendix: Experimental Configuration and Measurement Procedure for Evaluation and Comparison for Different Modulation Formats for 40 Gbit/s Transmission (Section 6.3.2)

The actual deployment conditions of a 40 Gbit/s transmission system are generally quite different from quasi-ideal laboratory conditions. Therefore, it is important to investigate the role of key practical constraints that may lead to serious obstacles to the successful introduction of 40 Gbit/s WDM EDFA-based transmission systems.

Figure 6.53 illustrates the experimental setup for benchmarking three modulation formats for 40 Gbit/s WDM systems, namely NRZ, RZ, and CSRZ. The transmitter is composed of 16 DFB laser sources, ranging from 1544.53 to 1556.56 nm on a 100 GHz ITU grid. Odd and even channels are separately multiplexed and modulated using independent sets of two in-series $LiNbO_3$ modulators, equipped with automatic bias control (ABC) loop circuits. The task of these circuits is the stabilization of the correct working point of $LiNbO_3$ modulators. This is achieved by means of continuously and automatically changing the modulator bias voltage, in order to keep track of the natural drift of the modulator transmission transfer function.

The first modulators (the pulse carvers) are driven at 20 GHz with a $2V_\pi$ clock and polarized at the null (maximum) transmission point when the CSRZ (33% RZ) format was generated. Each of the second set of modulators is driven by uncorrelated 40 Gbit/s $2^{31} - 1$ PRBSs, obtained by electrically interleaving four delayed copies of 10 Gbit/s $2^{31} - 1$ PRBSs. Switching off the RZ drivers, while polarizing the pulse carvers to their maximum transmission point, permits us to generate the NRZ format. Odd and even wavelengths are recombined through a polarization maintaining 3 dB coupler, so that co-polarized channels could be preserved. Figure 6.18 shows a temporal and spectral characterization of the transmitter for the three modulation formats under test. The extinction ratios of the NRZ, CSRZ, and 33% RZ formats, measured by means of a Tektronix CSA8200 oscilloscope and an 80C10 optical sampling module equipped with a 65 GHz photodiode, are 12.6 dB, 14.1 dB, and 15.3 dB respectively, as indicated on the scope screens shown in Figure 6.18, Section 6.3.2.

First, resilience of modulation formats to intrachannel nonlinear effects is evaluated on a straight transmission line constituted by four distinct 100 km spans of SSMF *to emulate a relatively moderate-length transmission line*. It is a common opinion that ASK formats will be the formats of choice for the deployment of 40 Gbit/s systems in metropolitan and long-haul applications. The cumulated dispersion and slope of SSMF are compensated by DCMs tailored for compensating 100 km SSMF spans (DCM-100). Note that the commercial

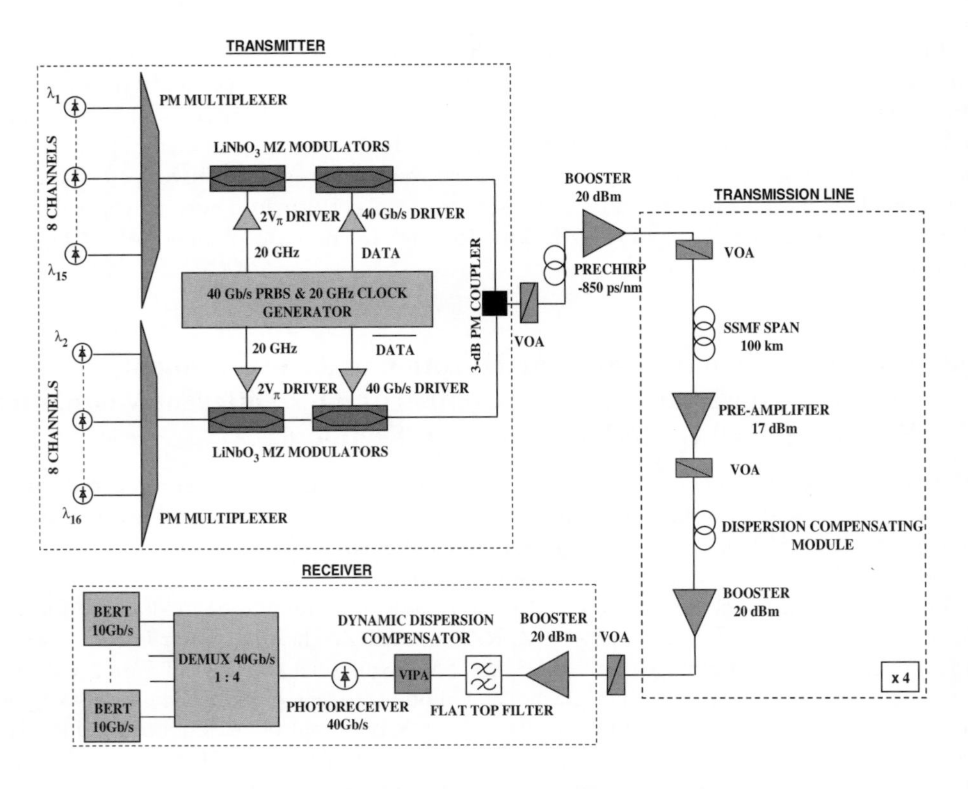

Figure 6.53 Schematic of the experimental setup

DCMs have a tolerance of ±2% [102] of their cumulated dispersion. This translates into a cumulated dispersion in the range [−1665, −1735] ps/nm for the DCM-100 that was used in our experiment. To reproduce practical field conditions, the residual dispersion per span was not precisely tuned (by adding, for example, small pieces of SSMF which would allow for reaching a certain target for the span compensation). The resulting dispersion map for the channel at 1550.12 nm is shown in Figure 6.54, where one can observe a compensation ratio per span which varies between 97.9% and 98.5%. Fiber span (~21 dB) and DCM (~10 dB) losses were compensated by double-stage EDFAs with a global noise figure of 5.5 dB. The optical power injected into the DCM was fixed to −2 dBm per channel. Including a −850 ps/nm pre-chirp at 1550.12 nm reduced the impact of nonlinearities, in particular IXPM and IFWM. This leads to an initial pulse broadening and a symmetric dispersion map [103,104], whereby the first (second) half of span propagation was in the negative (positive) cumulated dispersion regime. The measured channel at the receiver is selected with an XTRACT™ wavelength/bandwidth tunable square flat-top optical filter. The amplitude and group delay transfer function of our receiver optical filter are shown in Figure 6.55 (diamonds). At the transmission end, the residual dispersion is adjusted by a virtually imaged phased-array (VIPA) dispersion compensator (with a nominal dispersion of

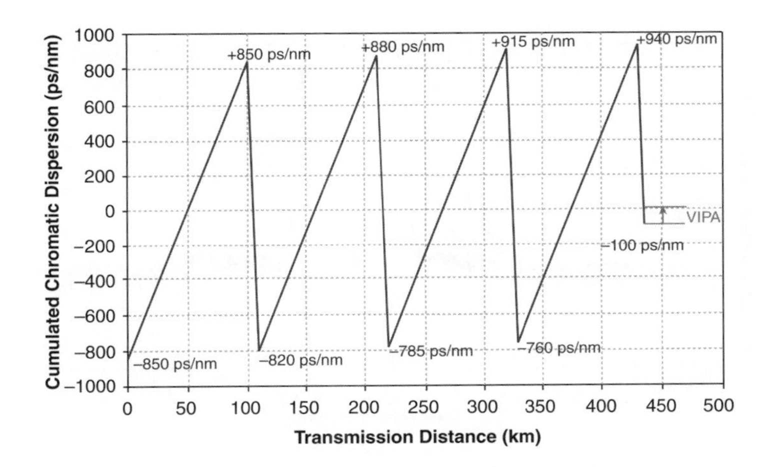

Figure 6.54 Dispersion map used in the experiment for the channel at 1550.12 nm

about + 100 ps/nm) in order to optimize the BER of 10 Gbit/s tributaries after 1 : 4 electrical demultiplexing.

In order to evaluate the robustness *to intrachannel nonlinearities* [61–65], only the eight even channels were used, with a spectral granularity of 200 GHz. In this configuration, interchannel nonlinear effects can be neglected, while preserving the gain flatness of the EDFA as well as the proper operation of the ABC loop circuits (which require around 15 dBm of optical power for their stable operation). At the receiver, the 20 dB bandwidth of the XTRACT square flat-top optical filter was optimized for each format: it was fixed to nearly 0.7 nm for the NRZ and CSRZ format, and to 0.9 nm for the 33% RZ format. The electrical 3 dB bandwidth of the receiver is fixed by its hardware: it consists of a 40 GHz XPRV2021 u^2t photoreceiver [66] connected to an electronic decision circuit and a 1 : 4 electrical demultiplexer. The dispersion map is kept unchanged throughout the measurements, while post-compensation at the receiver side is optimized for each format, by means of fine tuning, about its nominal value of + 100 ps/nm, the extra dispersion that is introduced by the VIPA compensator.

It should be pointed out that no DRA was used in the experiment presented here, and for each format the optical filter bandwidth at the receiver was optimized.

In order to evaluate residual CD and first-order PMD or DGD, only the eight even channels are used and previous tuning of the optical filter is keep fixed. CD increments are equal to + 12.5 ps/nm in the range [−100, + 100] ps/nm. The DGD is produced by means of a first-order PMD emulator. A polarization controller placed at its input permitted one to ensure that the power splitting ratio between the two axes of the emulator was equal to 0.5 (corresponding to the worst case). The received OSNR for zero CD or DGD is fixed in order to have a 10^{-9} BER independently of the modulation format. When varying the residual CD or DGD, OSNR penalties are measured by increasing the received OSNR up to a level where the received BER returned towards 10^{-9} (the BER value obtained at null CD or DGD).

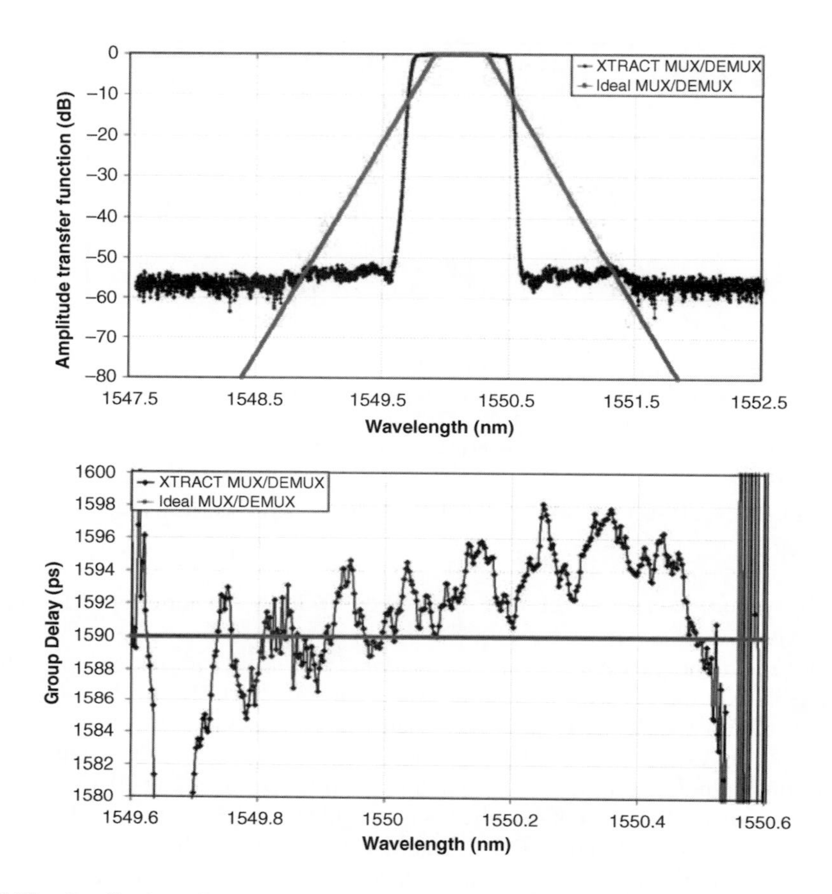

Figure 6.55 Amplitude and group delay transfer function of the XTRACT square flat-top optical filter used in the receiver (diamonds) and of an ideal 100 GHz flat-top demultiplexer (squares)

Note that in these experiments a fine optimization of the output optical filter bandwidth is carried out, which is important in order to ensure a fair comparison among the different formats when considering residual CD or DGD robustness.

6.6.1 Simulation Setup

In order to guide the experiments, the experimental setup was reproduced by numerically solving the nonlinear Schrödinger equation with a split-step Fourier algorithm by the means of a commercial system simulation software package (VPI Transmission Maker).

Initially, for the contour plots of Figure 6.19, only five channels with a channel separation of 200 GHz (to neglect the impact of interchannel nonlinearities) were simulated. For simplicity, a flat-top optical filter was used, with the same bandwidth value of 100 GHz for all formats (in contrast to the experiments where the filter bandwidth was optimized for each modulation format). The dispersion map used in the simulations was identical to the actual map of the experiments (Figure 6.54).

For the investigation of the filtering impact of an OADM cascade the experimental configuration was simulated with VPI Transmission Maker. Five channels separated by a spacing of 100 GHz were used, and a 2048-long PRBS was sent over the simulated link, consisting of twelve 100 km long SSMF spans (17 ps/(nm km), 0.2 dB/km), each of them followed by 16.66 km of DCF (−100 ps/(nm km), 0.6 dB/km) leading to a 98% compensation ratio (in order to match closely the experimental dispersion map). The bandwidth of the square flat-top optical filter located at the receiver was not changed with respect to the optimum value that was found above. In contrast with the previous experimental work, the length of the transmission line was extended up to 1200 km in order to cascade a sufficient number of OADMs. The span loss was compensated for by a double-stage EDFA with 5.5 dB noise figure. The pre-chirp was still fixed at −850 ps/nm in order to obtain a symmetric dispersion map. The post-chirp was optimized in order to obtain the best possible BER. The transmitter OSNR (measured in 0.5 nm) was kept unchanged for all formats (25 dB). When nonlinear effects are taken into account, the optimal span input power is 0 dBm per channel. The channel input power in the DCMs was fixed to −2 dBm. The OADMs were periodically inserted every two spans (corresponding to a total of five OADMs) and were obtained by the concatenation of an optical 100 GHz DEMUX and MUX. Two types of MUX/DEMUX were simulated. The first type has the same characteristics as our XTRACT square flat-top optical filter, already used in the receiver. Its amplitude transfer function and group delay response are represented by diamonds in Figure 6.55. As seen in this figure, the 20 dB bandwidth of this filter is close to 100 GHz, whereas its peak-to-peak GDR is around 4 ps. The second 100 GHz flat-top MUX/DEMUX under study is an ideal one (squares in Figure 6.55). The GDR of this ideal filter is null, whereas its amplitude transfer function is defined by the 1 dB and 20 dB bandwidths, respectively equal to 50 GHz and 145 GHz. A vanishing insertion loss was assumed for the MUX/DEMUX at the maximum transmission point. The simulation results are detailed in Figure 6.22, where the BER versus transmission distance is plotted for the NRZ, CSRZ, and 33% RZ modulation formats in various configurations.

Acknowledgments

A. Schinabeck, A. Zottmann and K. Sticht from Alcatel-Lucent Deutschland AG for contributions to Sections 6.4 and 6.5.

References

[1] Agrawal, G.P. (1989) *Nonlinear Fiber Optics, Quantum Electronics – Principles and Applications*, Academic Press, San Diego, CA.

[2] Kashyap, R. (1999) *Fiber Bragg Gratings*, Academic Press.

[3] Chantry, G.W. (1982) The use of Fabry–Perot interferometers, etalons and resonators at infrared and longer wavelengths – an overview. *Journal of Physics E: Scientific Instruments*, **15**, 3–8.

[4] Kashyap, R. (1999) *Fiber Bragg Gratings*, Academic Press.

[5] Eggleton, B.J., Ahuja, A., Westbrook, P.S. *et al.* (2000) Integrated tunable fiber gratings for dispersion management in high-bit rate systems. *IEEE Journal of Lightwave Technology*, **18**, 1418.

[6] Shu, X., Sugden, K., and Byron, K. (2003) Bragg-grating-based all-fiber distributed Gires–Tournois etalons. *Optics Letters*, **28**, 881–883.

[7] Shu, X., Bennion, I., Mitchell, J., and Sugden, K. (2004) Tailored Gires–Tournois etalons as tunable dispersion slope compensators. *Optics Letters*, **29**, 1013–1015.
[8] Madsen, C.K., Chandrasekhar, S., Laskowski, E.J. *et al.* (2001) Compact integrated tunable chromatic dispersion compensator with a 4000 ps/nm tuning range. Optical Fiber Communication Conference and Exhibit, PD9-1- PD9-3 vol. 4, OFC 2001.
[9] Klekamp, A., Junginger, B., and Bülow, H. (2006) Experimental PMD mitigation for 43 Gb/s DPSK by distributed polarisation scrambling over 4 spans SMF fibre. ECOC 2006, Cannes, post-deadline paper Th4.1.4.
[10] Bülow, H. (2006) Outage vs PMD tolerance trade-off for fast polarization scrambling. Proc. ECOC Sep 2006, Cannes, Th2.5.2.pdf.
[11] Liu, X., Giles, C.R., Wei, X. *et al.* (2004) Experimental demonstration of broadband PMD mitigation through distributed fast polarization scrambling and FEC. Proc. ECOC 2004, Stockholm, post-deadline paper Th4.1.4.
[12] Liu, X., Xie, C., and van Wijngaarden, A.J. (2004) Multichannel PMD mitigation and outage reduction through FEC with sub-burst-error-correction period PMD scrambling. *IEEE Photonics Technology Letters*, **16** (9), 2183–2185.
[13] Liu, X., Giles, C.R., Wei, X. *et al.* (2005) Improved PMD tolerance in systems using enhanced forward error correction through distributed fast polarization scrambling. Proc. ECOC 2005, Glasgow, We1.3.6.
[14] Garthe, D., Milner, G., and Cai, Y. (1998) System performance of broadband dispersion compensating gratings. *Electronics Letters*, **34** (6), 582–883.
[15] Ennser, K., Ibsen, M., Durkin, M. *et al.* (1998) Influence of non-ideal chirped fiber grating characteristics on dispersion cancellation. *Photonics Technology Letters*, **10** (10), 1476–1748.
[16] Evangelides, S.G. Jr, Bergano, N.S., and Davidson, C.R. (1999) Intersymbol interference induced by delay ripple in Bragg gratings. Proc. ECOC-1999, Nice, France, paper FA2-1.
[17] Nielsen, T.N., Eggleton, B.J., and Strasser, T.A. (1999) Penalties associated with group delay imperfections for NRZ, RZ and duo-binary encoded optical signals. Proc. ECOC-1999, Nice, France, 388–389.
[18] Scheerer, C., Glingener, C., Fisher, G. *et al.* (1999) Influence of filter group delay ripples on system performance. Proc. ECOC-1999, Nice, France, 410–411.
[19] Seerer, C., Glingener, C., and Fisher, G. (1999) System impact of ripples in grating group delay. Proc. ICTON'99, paper We.B.5.
[20] Eggleton, B.J., Ahuja, A., Westbrook, P.S. *et al.* (2000) Integrated tunable fiber gratings for dispersion management in high-bit rate systems. *IEEE Journal of Lightwave Technology*, **18** (10), 1418–1432.
[21] Jamal, S. and Cartledge, J.C. (2002) Variation in the performance of multispan 10-Gb/s system due to the group delay ripple of dispersion compensating fiber Bragg gratings. *IEEE Journal of Lightwave Technology*, **20** (1), 28–35.
[22] Yan, L.-S., Luo, T., Yu, Q. *et al.* (2002) System impact of group-delay ripple in single and cascaded chirped FBGs. Proc. OFC-2002, Anaheim, California, paper ThGG63.
[23] Yoshimi, H., Takushima, Y., and Kikushi, K. (2002) A simple method for estimating eye-opening penalty caused by group-delay ripple of optical filters. Proc. ECOC-2002, paper 10.4.4.
[24] Rislet, M.H., Clausen, C.B., and Tkach, R.W. (2003) Performance characterization of components with group delay fluctuations. *IEEE Photonics Technology Letters*, **15** (8), 1076–1078.
[25] Litchnitser, N., Li, Y., Sumetsky, M. *et al.* (2003) Tunable dispersion compensation devices: group delay ripple and system performance. Proc. OFC-2003, Vol. 1, paper TuD2.
[26] Fan, X., Labrake, D., and Brennan, J. (2003) Chirped fiber grating characterization with phase ripple. Proc. OFC-2003, Vol. 2, pp. 638–640.
[27] Bisseur, H., Bastide, C., and Hugbart, A. (2006) Performance characterization of components with phase ripple for different 40 Gb/s formats. Proc. OFC-2006, Anaheim, California, paper OFN4.
[28] Mårtensson, J., Djupsjöbacka, A., Li, J. *et al.* (2006) Field transmission using fiber Bragg grating-based dispersion compensating modules with continuous response over the full C-band. Proc. ECOC-2006, Paper TH3.3.6.
[29] Shannon, C.E. (1948) A mathematical theory of communication. *Bell System Technical Journal*, **27** (3), 379–423.
[30] MacKay, D.J.C. (2003) *Information Theory, Inference, and Learning Algorithms*, Cambridge University Press, Cambridge.
[31] Haykin, S. (1988) *Digital Communications*, John Wiley and Sons, Inc., New York, NY.
[32] Ohm, M. and Speidel, J. (2003) Quaternary optical ASK-DPSK and receivers with direct detection. *IEEE Photonics Technology Letters*, **15** (1), 159–161.

[33] Ip, E. and Kahn, J.M. (2005) Carrier synchronization for 3-and 4-bit-per-symbol optical transmission. *IEEE Journal of Lightwave Technology*, **23** (12), 4110.
[34] Betti, S., De Marchis, G., and Iannone, E. (1992) Polarization modulated direct detection optical transmission systems. *IEEE Journal of Lightwave Technology*, **10** (12), 1985–1997.
[35] Chi, N., Yu, S., Xu, Lin., and Jappesen, P. (2005) Generation of a polarisation shift keying signal and its application in optical labelling. European Conference on Optical Communications (ECOC), Glasgow, UK.
[36] Kaminow, I.P. and Koch, T.L. (1997) *Optical Fiber Telecommunications IIIA*, Academic Press.
[37] Xie, C., Kang, I., Gnauck, A.H. *et al.* (2004) Suppression of intrachannel nonlinear effects with alternate-polarization formats. *IEEE Journal of Lightwave Technology*, **22** (3), 806–812.
[38] Appathurai, S., Mikhailov, V., Killey, R.I., and Bayvel, P. (2004) Effective suppression of intra-channel nonlinear distortion in 40 Gb/s transmission over standard singlemode fibre using alternate-phase RZ and alternate polarisation. *Electronics Letters*, **40** (14), 897–898.
[39] Walklin, S. and Conradi, J. (1997) A 10 Gb/s 4-ary ASK lightwave system. International Conference on Integrated Optics and Optical Fiber Communications/European Conference on Optical Communications (IOOC/ECOC), Edinburgh, UK, p. 255.
[40] Nakamura, T., Kani, J.-I., Teshima, M., and Iwatsuki, K. (2004) A quaternary amplitude shift keying modulator for suppressing initial amplitude distortion. *IEEE Journal of Lightwave Technology*, **22** (3), 733–738.
[41] Walklin, S. and Conradi, J. (1999) Multilevel signaling for increasing the reach of 10 Gb/s lightwave systems. *IEEE Journal of Lightwave Technology*, **17** (11), 2235–2248.
[42] Winzer, P.J. and Essiambre, R.J. (2006) Advanced optical modulation formats. *Proceedings of the IEEE*, **94** (5), 952–985.
[43] Breuer, D. and Petermann, K. (1997) Comparison of NRZ- and RZ-modulation format for 40-Gb/s TDM standard-fiber systems. *IEEE Photonics Technology Letters*, **9** (3), 398–400.
[44] Hayee, M.I. and Willner, A.E. (1999) NRZ versus RZ in 10–40-Gb/s dispersion-managed WDM transmission systems. *IEEE Photonics Technology Letters*, **11** (8), 991–993.
[45] Bosco, G., Carena, A., Curri, V. *et al.* (2002) On the use of NRZ, RZ, and CSRZ modulation at 40 Gb/s with narrow DWDM channel spacing. *IEEE Journal of Lightwave Technology*, **20** (9), 1694–1704.
[46] Kanaev, A., Luther, G.G., and Kovanis, V. (2002) Nonlinear dynamics of modulation formats for 40 Gb/s transmission on standard single-mode fiber and dispersion-managed fiber. The 15th Annual Meeting of the IEEE Lasers and Electro-Optics Society (LEOS), vol. 1, pp. 315–316.
[47] Ennser, K., Laming, R.I., and Zervas, M.N. (1998) Analysis of 40 Gb/s TDM-transmission over embedded standard fiber employing chirped fiber grating dispersion compensators. *IEEE Journal of Lightwave Technology*, **16** (5), 807–811.
[48] Hodzic, A., Konrad, B., and Petermann, K. (2002) Prechirp in NRZ-based 40-Gb/s single-channel and WDM transmission systems. *IEEE Photonics Technology Letters*, **14** (2), 152–154.
[49] Killey, R.I., Mikhailov, V., Appathurai, S., and Bayvel, P. (2002) Investigation of nonlinear distortion in 40-Gb/s transmission with higher order mode fiber dispersion compensators. *IEEE Journal of Lightwave Technology*, **20**, 12.
[50] Hodzic, A. (2004) Investigation of high bit rate optical transmission systems employing a channel data rate of 40 Gb/s, Ph.D. thesis, Technical University of Berlin.
[51] Mu, R.M., Yu, T., Grigoryan, V.S., and Menyuk, C.R. (2002) Dynamics of the chirped return-to-zero modulation format. *IEEE Journal of Lightwave Technology*, **20** (1), 47–57.
[52] Miyamoto, Y., Hirano, A., Yonenaga, K. *et al.* (1999) 320 Gb/s (8 × 40 Gb/s) WDM transmission over 367 km with 120 km repeater spacing using carrier-suppressed return-to-zero format. *Electronics Letters*, **35** (23), 2041–2042.
[53] Forzati, M., Berntson, A., Mårtensson, J., and Davies, R.J. (2006) Performance analysis of single-MZM APRZ transmitters. *IEEE Journal of Lightwave Technology*, **24** (5), 2006–2014.
[54] Morita, I., Suzuki, M., Edagawa, N. *et al.* (1997) Performance improvement by initial phase modulation in 20 Gb/s soliton-based RZ transmission with periodic dispersion compensation. *Electronics Letters*, **33** (12), 1021–1022.
[55] Bergano, N.S., Davidson, C.R., Mills, M.A. *et al.* (1997) Long-haul WDM transmission using optimum channel modulation: A 160 Gb/s (32 × 5 Gb/s) 9,300 km demonstration. Optical Fiber Communication Conference (OFC), vol. PD16, p. 1.

[56] Ohhira, R., Ogasahara, D., and Ono, T. (2001) Novel RZ signal format with alternate-chirp for suppression of nonlinear degradation in 40 Gb/s based WDM. Optical Fiber Communication Conference (OFC), vol. 3, pp. WM2–1–WM2–3.
[57] Liu, X., Wei, X., Gnauck, A.H. *et al.* (2002) Suppression of intrachannel four-wave-mixing-induced ghost pulses in high-speed transmissions by phase inversion between adjacent marker blocks. *Optics Letters*, **27** (13), 1177–1179.
[58] Fukuchi, K., Ono, T., and Yano, Y. (1997) 10 Gb/s–120 km standard fiber transmission employing a novel optical phase-encoded intensity modulation for signal spectrum compression. Optical Fiber Communication Conference (OFC), pp. 270–271.
[59] May, G., Solheim, A., Conradi, J. *et al.* (1994) Extended 10 Gb/s fiber transmission distance at 1538 nm using a duobinary receiver. *IEEE Photonics Technology Letters*, **6** (5), 648–650.
[60] Penninckx, D., Chbat, M., Pierre, L., and Thiery, J.-P. (1997) The phase-shaped binary transmission (PSBT): a new technique to transmit far beyond the chromatic dispersion limit. *IEEE Photonics Technology Letters*, **9** (2), 259–261.
[61] Bissessur, H., Bresson, C., Hébert, J. *et al.* (2003) Transmission of 40 × 43 Gb/s over 2540 km of SMF with terrestrial 100-km spans using NRZ format. Proceedings OFC 2003, CA, USA, paper FN3.
[62] Gnauck, A.H., Liu, X., Wei, X. *et al.* (2004) Comparison of modulation format for 42.7 Gb/s single-channel transmission through 1980 km of SSMF. *IEEE Photonics Technology Letters*, **16** (3), 909–911.
[63] Idler, W., Klekamp, A., and Dischler, R. (2003) System performance and tolerances of 43 DPSK modulation formats. Proceedings ECOC 2003, Rimini, Italy, paper Th2.6.3.
[64] Klekamp, A., Dischler, R., and Idler, W. (2005) Impairments of bit-to-bit alternate-polarization on non-linear threshold, CD and DGD tolerance of 43 Gb/s ASK and DPSK formats. Proceedings OFC 2005, CA, USA, paper OFN3.
[65] Dischler, R., Klekamp, A., Lazaro, J., and Idler, W. (2004) Experimental comparison of non linear threshold and optimum pre dispersion of 43 Gb/s ASK and DPSK formats. Proceedings OFC 2004, CA, USA, paper TuF4.
[66] XPRV2021 data sheet; DSCR404 data sheet, www.u2t.de, www.chipsat.com.
[67] Castañón, G., Vassilieva, O., Choudhary, S., and Hoshida, T. (2001) Requirements of filter characteristics for 40 Gb/s-based DWDM systems. Proceedings ECOC 2001, Amsterdam, Nederlands, paper Mo.F.3.5.
[68] Gnauck, A.H., Winzer, P., and Chandrasekhar, S. (2005) Hybrid 10/40 G transmission on a 50-GHz ITU grid through 2800 km of SSMF and seven optical add-drops. *IEEE Photonics Technology Letters*, **17** (10), 2203–2205.
[69] Gnauck, A.H., Winzer, P.J., Chandrasekhar, S., and Dorrer, C. (2004) Spectrally efficient (0.8 b/s/Hz) 1-Tbps (25 × 42.7 Gb/s) RZ-DQPSK transmission over 28 100-km SSMF spans with 7 optical add/drops. Proceedings ECOC 2004, Stockholm, Sweden, Postdeadline paper Th4.4.1.
[70] Agarwal, A., Banerjee, S., Grosz, D.F. *et al.* (2003) Ultralong-haul transmission of 40-Gb/s RZ-DPSK in a 10/40 G hybrid system over 2500 km of NZ-DSF. *IEEE Photonics Technology Letters*, **15** (12), 1779–1781.
[71] Johannisson, P., Anderson, D., Marklund, M. *et al.* (2002) Suppression of nonlinear effects by phase alternation in strongly dispersion-managed optical transmission. *Optics Letters*, **27** (12), 1073–1075.
[72] Forzati, M., Martensson, J., Berntson, A. *et al.* (2002) Reduction of intrachannel four-wave mixing using the alternate-phase RZ modulation format. *IEEE Photonics Technology Letters*, **14** (9), 1285–1287.
[73] Randel, S., Konrad, B., Hodzic, A., and Petermann, K. (2002) Influence of bitwise phase changes on the performance of 160 Gb/s transmission systems. European Conference on Optical Communications (ECOC), Copenhagen, p. P3.31.
[74] Gill, D.M., Liu, X., Wei, X. *et al.* (2003) $\pi/2$ alternate-phase on–off keyed 40-Gb/s transmission on standard single-mode fiber. *IEEE Photonics Technology Letters*, **15** (12), 1776–1778.
[75] Winzer, P.J., Dorrer, C., Essiambre, R.-J., and Kang, I. (2004) Chirped return-to-zero modulation by imbalanced pulse carver driving signals. *IEEE Photonics Technology Letters*, **16** (5), 1379–1381.
[76] Forzati, M., Berntson, A., Mårtensson, J., and Davies, R.J. (2006) Performance analysis of single-MZM APRZ transmitters. *IEEE Journal of Lightwave Technology*, **24** (5), 2006–2014.
[77] Forzati, M., Berntson, A., Mårtensson, J. *et al.* (2006) 40-Gb/s field experiment over an 820-km transmission link designed for 10 Gb/s, using the APRZ modulation format. *Electronics Letters*, **42**, 991–992.

[78] Ly-Gagnon, D., Tsukamoto, S., Katoh, K., and Kikuchi, K. (2006) Coherent detection of optical quadrature phase-shift keying signals with carrier phase estimation. *IEEE Journal of Lightwave Technology*, **24**, 12–21.

[79] Tsukamoto, S., Ishikawa, Y., and Kikuchi, K. (2006) Optical homodyne receiver comprising phase and polarization diversities with digital signal processing. European Conference on Optical Communications (ECOC), Cannes, France, paper Th4.1.4.

[80] Brady, D. and Verdù, S. (1989) Performance analysis of an asymptotically quantum-limited optical DPSK receiver. *IEEE Transactions on Communications*, **37** (1), 46–51.

[81] Gnauck, A.H., Raybon, G., Chandrasekhar, S. *et al.* (2002) 2.5 Tbps (64×42.7 Gb/s) transmission over 40×100 km NZDSF using RZ-DPSK format and all-Raman-amplified spans. Optical Fiber Communication Conference and Exhibit (OFC).

[82] Zhu, B., Nelson, L.E., Stulz, S. *et al.* (2003) 6.4-Tbps (160×42.7 Gb/s) transmission with 0.8 bit/s/Hz spectral efficiency over 32×100 km of fiber using CSRZ-DPSK format. Optical Fiber Communications Conference (OFC).

[83] Wree, C., Leibrich, J., and Rosenkranz, W. (2002) RZ-DQPSK format with high spectral efficiency and high robustness towards fiber nonlinearities. European Conference on Optical Communication (ECOC), vol. 4, p. 9.6.6.

[84] Griffin, R.A., Carter, A.C., Components, M.O., and Towcester, U.K. (2002) Optical differential quadrature phase-shift key (oDQPSK) for high capacity optical transmission. Optical Fiber Communication Conference and Exhibit (OFC), p. WX6.

[85] Ohm, M. (2004) Optical 8-DPSK and receiver with direct detection and multilevel electrical signals. IEEE/LEOS Workshop on Advanced Modulation Formats, pp. 45–46.

[86] Hodzic, A., Winter, M., Konrad, B. *et al.* (2003) Optimized filtering for 40-Gb/s/ch-based DWDM transmission systems over standard single-mode fiber. *IEEE Photonics Technology Letters*, **15** (7), 1002–1004.

[87] Bosco, G., Carena, A., Curri, V. *et al.* (2002) On the use of NRZ, RZ, and CSRZ modulation at 40 Gb/s with narrow DWDM channel spacing. *IEEE Journal of Lightwave Technology*, **20** (9), 1694–1704.

[88] Charlet, G., Tran, P., Mardoyan, H. *et al.* (2005) 151×43 Gb/s transmission over 4880 km based on RZ-DQPSK. Proc. ECOC 2005, Glasgow, Scotland, paper Th4.1.4.

[89] Proakis, J.G. (1995) *Digital Communications*, McGraw Hill.

[90] Koc, U.V. *et al.* (2002) Adaptive electronic equalization using higher-order statistics for PMD compensation in long-haul fiber-optic systems. Proc. of ECOC, Copenhagen (Denmark).

[91] Haunstein, H. *et al.* (2003) Control of combined electrical feed-forward and decision feedback equalization by conditional error counts from FEC in the presence of PMD. Proceedings OFC 2003, Anaheim, Paper ThG5.

[92] Heismann, F. *et al.* (1998) Automatic compensation of first order polarization mode dispersion in a 10 Gb/s transmission system. Proc. of ECOC, Madrid (Spain).

[93] Buchali, F., Lanne, S., Thiery, J.-P. *et al.* (2001) Fast eye monitor for 10 Gbit/s and its application for optical PMD compensation. Optical Fiber Communication Conference and Exhibit, OFC 2001.

[94] Haunstein, H. *et al.* (2001) Adaptation of electronic PMD equaliser based on BER estimation derived from FEC decoder. Proc. of ECOC, Amsterdam (The Netherlands).

[95] Zhou, Y.R., Lord, A., Schinabeck, A., and Haunstein, H. (2005) Potential benefit of using electronic equalisation in all-optical networks. The IEE Photonics Professional Network Seminar on Optical Fibre Communications and Electronic Signal Processing, London.

[96] Zhou, Y.R., Lord, A., Santoni, S. *et al.* (2006) PMD rules for physical constraint-based routing in all optical networks. Optical Fiber Communications Conference (OFC), Anaheim, JThB17.

[97] Schinabeck, A., Haunstein, H., Zhou, Y.R., and Lord, A. (2006) All optical network dimensioning considering transmission impairments and mitigation using electronic equalization. 7. ITG Fachtagung Photonische Netze, Leipzig, Germany, April.

[98] ITU-T (2000) Network node interface for the synchronous digital hierarchy (SDH). ITU-T G.707 Recommendation.

[99] ITU-T (2000) Forward error correction. ITU-T G.975 Recommendation.

[100] ITU-T (2003) Network node interface for the optical transport network (OTN). ITU-T G.709 Recommendation.

[101] Masashi, A., Fujita, H., Mizuochi, T. *et al.* (2002) Third generation FEC employing turbo product code for long-haul DWDM transmission systems. Proceedings OFC 2002, Anaheim, Paper WP2.

[102] Specialty photonics products overview, Rightwave EWBDK-C data sheet, www.ofs.dk.

[103] Pincemin, E., Grot, D., Borsier, C. *et al.* (2004) Impact of the fiber type and dispersion management on the performance of an NRZ 16 × 40 Gb/s DWDM transmission system. *IEEE Photonics Technology Letters*, **16** (10), 2362–2364.

[104] Mecozzi, A., Clausen, C.B., Shtaif, M. *et al.* (2001) Cancellation of timing and amplitude jitter in symmetric links using highly dispersed pulses. *IEEE Photonics Technology Letters*, **13** (5), 445–447.

Dictionary of Optical Networking

Didier Colle, Chris Matrakidis and Josep Solé-Pareta

Term	**Definition**
Accounted failure	A failure for which particular "healing" measures are provided to overcome the failure. The set of accounted failure is typically very limited (e.g., a network is often designed to survive only single link or node failures).
Accounting management	A set of functions that (i) enables network service use to be measured and the costs for such use to be determined and (ii) includes all the resources consumed, the facilities used to collect accounting data, the facilities used to set billing parameters for the services used by customers, maintenance of the databases used for billing purposes, and the preparation of resource usage and billing reports.
Administrative domain	An administrative domain is a bounded entity within which all encompassed constituent elements are under common ownership, operation and management [Eurescom].
All-optical transport network	An all-optical transport network is a transport network whose functional resources (conveying user information between locations) are fully optical (no electronics).
Alternative path	See "Protection path."
Application bearer = bearer services	Bearer service is a telecommunications term referring to a service that allows transmission of information signals between network interfaces. Telecommunication bearer services reside in the lowest, physical layer. The choice of bearer service for a given offering of telecommunications services is not always self-evident [Ericsson].
Application category	An application category is a set of applications characterized by a set of common basic performance requirements (real-time,

Core and Metro Networks Edited by Alexandros Stavdas

interactivity, reliability, etc.) and related performance constraints [Eurescom].

Assigned resource — The set of resources that is exclusive used/possessed by a path.

Autodiscovery — Autodiscovery is the process, run at node start-up time, of discovering the characteristics of each level of the node's processes, automatically informing the network operator, and hence reducing the operational tasks. Concept: assign a unique Internet protocol (IP) address to the node; this will then be automatically configured and completely integrated to the network.

Automatic bandwidth control — The aim of automatic bandwidth control is to adapt the current available bandwidth on a link of a connection to fit the required amount of input traffic.

Availability — Symbol: A

The probability that an item is in a state to perform a required function under given conditions, assuming that the required external resources are provided [IEC].

Availability can be specified in different ways:

- Instantaneous availability $A(t)$: is the availability at a certain instant t in the time.
- Mean availability $A(t_1, t_2)$: is the availability averaged over a time interval $[t_1, t_2]$.
- Asymptotic availability: the limit, if this exists, of the instantaneous availability when the time tends to infinity. *Note*: under certain conditions, for instance constant failure rate and constant repair rate, the asymptotic availability may be expressed by the ratio of the mean up-time to the sum of the mean up-time and mean down-time.

Availability can be specified on a component, system, network, or service level.

Backhauling — The phenomenon that traffic crosses a particular segment of the network (e.g. a link) more than once. This may occur when traffic is recovered after a failure or when the traffic grooming is optimized.

Backup path — See "Protection path."

Bandwidth on demand — Deliver bandwidth on demand calls for automated processes that efficiently deliver on customer demands and allow for minimizing operational costs.

Broadband — Describes a class of telecommunications access services, such as ADSL, HFC cable, and WiFi, offering a data rate greater than narrowband services. These services are usually "always-on" and do not tie up a telephone line exclusively for data. Broadband is defined in this report to mean any Internet connection with an access data rate (both directions) greater than 2 Mbit/s. However, in many definitions the broadband term is identified with digital signal of the H6 ISDN channel – 386 kbit/s.

	Broadband is a descriptive term for evolving digital technologies offering consumers a single switched facility offering integrated access to voice, high-speed data services, video-demand services, and interactive information delivery services. Broadband also is used to define an analog transmission technique for data or video that provides multiple channels [FCC].
Burstification	The process of grouping a number of (possibly quite short) packets (e.g., IP packets) into a single burst in order to transport them efficiently over, for example, optical packets. This implies that a single optical packet may contain (pieces of) more than one (e.g., IP) packet.
Call	An association between endpoints that supports an instance of a service [G.8080]. A call can consist of several connections crossing different domains.
Capital expenditure (CAPEX)	Expenditure for the acquisition of long-term assets. Investments made in the long-term assets of the company.
Churn	This word has two meanings: • Transfer of a customer's telecommunications service from one provider to another. • The term used to describe turnover rate of subscribers to a service or product [FCC].
Class of service	The class of service (CoS) is a broad term describing a set of characteristics available with a specific service. Both the IETF and ITU-T define the CoS term. It is defined by the IETF as "The definitions of the semantics and parameters of a specific type of QoS" [RFC2386]. The ITU-T definition of the CoS term can be found in [E.493, E.720, E.721, E.771]. Services belonging to the same class are described by the same set of parameters, which can have qualitative or quantitative values. Usually, the set of parameters within the class is defined without assignment of concrete values, but these values can be bounded [CommMag1].
Classification	The process of sorting packets based on the packet headers according to defined rules.
Client interface	The client interface describes the interface of any server layer to its client layer(s) (e.g., to upper service layers like STM-N, ATM, Ethernet). Data, management, and control plane requirements need to be specified.
Colored interface	Interface with a defined wavelength for wavelength division multiplexing (WDM) (usually long reach).
Configuration management	The collection of management processes responsible for discovering and configuring network devices and connections.
Congestion	The condition that occurs in a network when the number of requests for service exceeds the capacity of a selected path in such a way that it results in a irreversible performance degradation (substantial increase of connection blocked, packet lost or

delay). Congestion is caused by the saturation of all or part of the network links.

Connection — A connection is a concatenation of link connections and subnetwork connections (as described in [G.805]) that allows the transport of user information between the ingress and egress points of a subnetwork. [G.8080].

One or more connections, crossing different domains, form a call.

Connectivity — The connectivity number K is defined as the minimum number of items whose removal from the network results in a disconnected network. A network is said to be k-connected if $K \geq k$.

In case of link connectivity, the minimum number of items is the minimum number of links.

In case of node connectivity, the minimum number of items is the minimum number of nodes.

Content service provider — Content service providers collect, organize, and present information. There are content providers as, for instance, CNN, who specialize on certain topics. Other content providers help people to find information quicker (e.g., Yahoo).

Contention — The situation occurring when two or more entities simultaneously want to access the same resource.

Context — Context defines the geographical environment and the requirement of the network at the highest (IP) level. Traffic requirements are between a given set of network sites at IP level (routers, content servers) and they can be defined separately for each class of application/service.

Control plane — The control plane performs the call control and connection control functions including routing. Through signaling, the control plane sets up and releases connections, and may restore a connection in case of a failure [G.8080].

Customer premise equipment — Service provider equipment that is located on the customer's premises (physical location) rather than on the provider's premises or in between.

Data communications network — Data communications network (DCN): the DCN is a network that supports Layer 1 (physical), Layer 2 (data-link), and Layer 3 (network) functionality. A DCN can be designed to support transport of distributed management communications related to the telecommunications management network (TMN), distributed signaling communications related to the automatic switched transport network (ASTN), and other operations communications (e.g., orderwire/voice communications, software downloads, etc.).

See also [G.7713]

Data plane — IETF term for the ITU transport plane.

Dedicated protection — In dedicated protection, one protection entity protects exactly one working entity. This protection entity may be used to carry extra traffic in case the entity is not in use for protection purposes (1 : 1) or the traffic on the working entity can simply be bridged continuously on the protection entity (1 + 1).

Defect A decrease of the ability of a network element to perform a required function.

Degree of survivability The extent to which a network is able to recover (or thus the restorability) from single and multiple network failures, taking into account the probability of each failure to occur.

Distributed computing environment (also known as GRID) A distributed computing environment (DCE) is an industry-standard software technology for setting up and managing computing and data exchange in a system of distributed computers. DCE is typically used in a larger network of computing systems that include different-size servers scattered geographically. DCE uses the client/server model. Using DCE, application users can use applications and data at remote servers. Application programmers need not be aware of where their programs will run or where the data will be located [Eurescom].

Distributor A person/company that distributes signals from a carrier and provides that transmission either directly to individual subscribers for private home viewing or to other program distribution companies for transmission [FCC].

Domain The term domain is used to indicate an administrative or management domain. The context will explicitly indicate what is meant.

Dual homing Two internetwork domain connections for survivability reasons. Both connections are established between the same networks (in contrast to IP dual/multi-homing where a network may have connections to different networks).

Dynamic routing Dynamic routing algorithms are able to autonomously adjust to traffic and/or network topology changes.

Effective bandwidth (equivalent bandwidth) Effective bandwidth is used to calculate the quantity of resources that are necessary to have available in order a particular traffic flow be transmitted with a given quality of service (QoS). Effective bandwidth depends, of course, of the traffic characteristics, but also on the surrounding environment; namely, the other traffic flows and the capacity of the link.

Ethernet LAN service Ethernet local area network (LAN) service enables the connection of multi-LAN sites to multi-LAN sites.

Ethernet line service Service provision of an Ethernet line service means providing a point-to-point connection over a wide area network (WAN).

External network to network interface A bidirectional signaling interface between control plane entities belonging to different domains [G.8080].

Failure The termination of the ability of a network element to perform a required function. In contrast to a fault, a failure refers to a single point in time.

Fault — The inability itself of a network element to perform a required function. In contrast to a failure, a fault refers to a certain time interval (that starts at the time of the failure).
Synonym: outage.

Fault management — The set of functions that (i) detect, isolate, and correct malfunctions in a telecommunications network, (ii) compensate for environmental changes, and (iii) include maintaining and examining error logs, accepting and acting on error detection notifications, tracing and identifying faults, carrying out sequences of diagnostics tests, correcting faults, reporting error conditions, and localizing and tracing faults by examining and manipulating database information.

FCAPS management — Functional areas identified in [M.3010] that are performed in a TMN: fault, configuration, accounting, performance, and security management

Flow — A flow is a stream of packets that is transmitted between a source and a destination. Flows generally follow the same route through a network, although that route may change at any time to bypass downed links and other problems. Flows may be implicit or explicit. An implicit flow is one in which the router detects a flow by inspecting header information in packets and then manages the flow as necessary. An explicit flow is a flow that is predefined; in other words, an end device tells the network that a flow is about to begin and the network sets itself up to handle the flow. In both cases, the network manages the flow in order to allocate resources (e.g., a circuit, bandwidth, buffers, etc.) for the flow (for more details see [Linctionary1]).

Forwarding layer — IETF term for the ITU transport plane.

Full mesh — Full mesh is a term describing a network in which devices are organized in a mesh topology, with each network node having either a physical circuit or a virtual circuit connecting it to every other network node. A full mesh provides a great deal of redundancy, but because it can be prohibitively expensive to implement, it usually is reserved for network backbones [Eurescom].

Gateway — A network node (hardware and software) that permits devices on one network (traditionally used in LAN environments) to communicate with or gain access to the facilities of another possibly dissimilar network. It can translate the protocols up to the application layer (layer 7 of the ISO/OSI reference stack).

Generic routing encapsulation — Generic routing encapsulation is a tunneling protocol that can encapsulate a wide variety of protocol packet types inside IP tunnels, creating a virtual point-to-point link to routers at remote points over an IP internetwork [Eurescom].

Grade of service — As meant by ITU, grade of service (GoS) is a term used in the context of the phase of connection setup. It appears in recommendations related to telephone and ISDN networks [E.493, E.720, E.721,

E.771]. GoS parameters applicable for optical switched connections are, for example: connection setup delay, probability of end-to-end blocking, delay in authentication.

The notion of the GoS is sometimes used to categorize services with respect to high-level requirements. Survivability issues or probability of a physical damage of a connection due to natural disasters (such as earthquakes, volcano eruptions, etc.) may be taken into account. However, when we introduce the notion of quality of protection (QoP) these issues are suited better to QoP rather than to GoS.

Grade of survivability — The number of simultaneous failures that a network can "survive" in the worst case.

Alternative: see also *degree of survivability*.

Gray interface — Interface with an arbitrary wavelength (usually short reach).

Hybrid node — A hybrid node consists of a mixed structure combining different technologies, protocols, services (e.g., a transparent and an opaque part, a circuit and a packet switching part, etc.).

Initial network — A set of network requirements, like topology, technologies, number of layers, number of nodes and its location, and so on, taken into account as a starting point for any migration scenarios, for obtaining of cost model and other activities [Eurescom].

Interface — An interface is a common boundary between two communicating entities; for example, two subsystems or two devices. An interface is used to specify once the interconnection between the two sides of it. The specification includes the type, quantity, and function of the interconnecting means and the type, form, and sequencing order of the signals to be interchanged via those means. One or more protocols may be implemented across an interface [Eurescom].

Internal network-to-network Interface — A bidirectional signaling interface between control plane entities belonging to one or more domains having a trusted relationship [G.8080].

Internet SCSI — Internet SCSI extends SCSI functionality to operate over TCP/IP. It enables any machine on an IP network (client) to contact any other remote machine (a dedicated host-server) and perform block I/O transfer on it just as it would do on its own local hard disk [FreeDictionary].

Internet service provider — An Internet service provider (ISP) is an entity that provides individuals and other companies access to the Internet and other related services such as website building and virtual hosting. An ISP has the equipment and the telecommunication line access required to have POP on the Internet for the geographic area served. An ISP is also sometimes referred to as an Internet access provider [Eurescom].

Interworking function — Interworking function is the function connecting two networks of different signaling and/or transport technology [Eurescom].

IP-based service An IP-based service is defined as a service provided by the service plane to an end user (e.g., a host (end system) or a network element) and which utilizes the IP transfer capability and associated control and management functions, for delivery of the user information specified by the service level agreement (SLA) [Eurescom].

IrDI A physical interface that represents the boundary between two administrative domains [G.872].

Layer The term layer will be reserved for protocol layers (e.g., in OSI reference model: IP is on layer 3) or management layers (adaptation layer, OCh layer, etc.).

Layer network A transport network can be decomposed into a number of independent transport layer networks with a client/server association between adjacent layer networks. Each layer network can be separately partitioned in a way which reflects the internal structure of that layer network or the way that it will be managed.

A layer network is a "topological component" that includes both transport entities and transport processing functions that describe the generation, transport, and termination of a particular characteristic information.

The layer networks which have been identified in the transport network functional model should not be confused with the layers of the OSI Model [X.200]. An OSI layer offers a specific service using one protocol among different protocols. On the contrary, each layer network offers the same service using a specific protocol (the characteristic information) [G.805].

Light path A light path is an end-to-end connection in an optical transport network. It can be transparent or opaque, depending on the network.

Link connection A "transport entity" that transfers information between "ports" across a link [G.805].

Link-state information Set of parameters, which define link performances. These parameters are based on available resources, such as maximum reservable bandwidth, unreservable bandwidth, available wavelength, and so on.

Link-state routing Routing technique based on using the link state information to take routing decisions. Link-state routing is a better technique for larger networks. Routers use it to build a topological database that describes routes on the entire internetwork. This information is used to build routing tables with more accurate routing information. Link-state routing also responds faster to changes in the network. Link-state routing is now the preferred routing method for most organizations and ISPs [Linktionary2].

Local area network A LAN is a network that spans a small area, typically an office, building, or even university campus.

Maintainability (performance) The ability of an item, under given conditions of use, to be retained in, or restored to, a state in which it can perform a required function, when maintenance is performed under given conditions and using stated procedures [IEC].

Management domain A set of managed objects, to which a common systems management policy applies [X.701].

Management layer An architectural concept that reflects particular aspects of management and implies a clustering of management information supporting that aspect [M.3010].

Management plane It performs management functions for the transport plane, the control plane, and the system as a whole. It also provides coordination between all the planes. The TMN architecture is described in [M.3010]; additional details of the management plane are provided by the M series Recommendations [G.8080].

Mean nodal degree The average of the nodal degrees for all the nodes present in the network.

Mean operating time between failures Symbol: MTBF

The expectation of the operating time between failures [IEC].

Mean time to recover See "Mean time to restore."

Mean time to repair The expected total time to replace or repair a failed component/unit/item in the network. During the mean time to repair, recovery techniques may help in reducing the mean time to restore.

Mean time to restore The expected total elapsed time from a customer-reported fault to service restoration [WirelessReview]. More generally, the total time elapsed since a fault has been detected and/or reported (thus, soon after a failure occurrence).

Synonyms: mean time to recover.

Metropolitan-area network A metropolitan-area network (MAN) is a network that spans a metropolitan area. Generally, a MAN spans a larger geographic area than a LAN, but a smaller geographic area than a WAN [Eurescom].

Multilayer traffic engineering See "Traffic engineering."

Multi-exchange unit The cross-connect unit supports grooming on timeslot level for optimal use of bandwidth and hardware capacity. It also enables ring structures for protection, a cost-efficient alternative for high-availability networks [Ericsson].

Narrow-band Definitions vary according to reference, but in this book narrow-band means a less than 64 kbit/s data steam.

Network architecture 1. The design principles, physical configuration, functional organization, operational procedures, and data formats used as the bases for the design, construction, modification, and operation of a communications network.

2. The structure of an existing communications network, including the physical configuration, facilities, operational structure, operational procedures, and the data formats in use [FreeDictionary] [FCC].

Network integrity — The ability of a network to provide the desired QoS to the services, not only in normal (i.e., failure-free) network conditions, but also when network congestion or network failures occur [Wu].

Network operator — A network operator is an entity that is responsible for the development, provisioning, and maintenance of telecommunications services and for operating the corresponding networks.

Network performance — Network performance (NP) is defined in [E.800] as "the ability of a network or network portion to provide the functions related to communications between users." NP is defined and measured with the parameters of the network components involved in providing a particular service. Some general NP parameters are defined by ITU-T in [E.800], while NP parameters for IP networks can be found in [I.380] Recommendation.

Network performance level — A network performance level is a set of target values (or range of values) and related guarantees for a set of network performance parameters to be provided at the edges of the provider network.

Network performance parameter — A network performance parameter is a variable that is used to assess network performance.

Network provider — Authority that owns and/or manages a provider network.

Network scenario — Technological solution applied for the network involving multilayer protocol stack specification (IP over SDH over WDM, for instance).

Network state information — Set of parameters obtained by means of flooding mechanisms, which are used in each node to distribute both link and node state information.

Network survivability — 1. The ability of a network to maintain or restore an acceptable level of performance during network failures by applying various restoration techniques. The acceptable level of service can be defined by the SLA.

2. The mitigation or prevention of service outages from network failures by applying preventative techniques.

Alternative: a subset of network integrity. It is the ability of a network to recover the traffic in the event of a failure, causing little or no consequences for the users [G.841].

Network topology — A network may be represented as a collection of nodes, some of which are connected by links. A given node may have links to many others. Network topology is determined only by the configuration of connections between nodes; it is therefore a part of graph theory. Distances between nodes, physical interconnections, transmission rates, and/or signal types are not

a matter of network topology, although they may be affected by it in an actual physical network [FreeDictionary].
In our investigations as network topology we mean also topology with techniques implemented in it.

Network-attached storage — Network-attached storage systems are computing-storage devices that can be accessed over a computer network, rather than directly being connected to the computer (via a computer bus). This enables multiple users to share the same storage space at once, and often minimizes overhead by centrally managing hard disks [FreeDictionary].

Next shortest (separated) path — This definition consists of two conditions: (i) the path is the second in a list of paths between two nodes, ordered by increasing lengths; (ii) it is physically distinct (except for the two common terminations) from the first shortest path.

Nodal degree — The number of links incident to a node.

Node — Network site of traffic generation/flexibility/termination.

Node functionality — A set of functions implemented in a single node unit.

Node integration — Combination of multiple network functionality options in one node unit.

Node state information — Set of parameters which define a node characteristics. These parameters are: utilization, packet delay, packet loss, and so on [RFC3272].

Opaque (logical)

- Contents of data are opaque if they will be not processed (e.g., in a network node) even when they are terminated (example multiprotocol label switching (MPLS): on a label-switched path the data are opaque to intermediate nodes, but are simply forwarded by means of label swapping (header processing)).
- Specific contents may be broadcasted/flooded but used only by specific usergroups or applications for which the information is dedicated (example: opaque link state advertisements (OSPF; [RFC2370])).
- Generally: contents/information that is hidden to some receivers (or parts of data/information that are hidden).

Opaque (node) — If there is any E/O or O/E conversion on the optical layer of a node, then this node is opaque.

Opaque network — Optical-transmission-based network making use of 3R signal regeneration at (every) intermediate node along a connection.

Opaque WIXC-E — O/E/O at ingress and egress with electrical core.

Opaque WIXC-O — O/E/O at ingress and/or egress with optical core.

Operating expenses (OPEX) — OPEX are the ongoing costs to keep the network operational. Those are the costs that arise in the normal course of running a business.

Operational support system — An operations support system generally refers to a system (or systems) that performs management (e.g. FCAPS), inventory, engineering, planning, and repair functions for telecommunications networks.

Operations, administration, maintenance, and provisioning — This comprises all the activities of network management systems and personnel in the network operating and maintenance centers.

Optical add/drop multiplexer — Within an optical add/drop multiplexer (OADM) it is possible to bypass traffic or to terminate arbitrary wavelengths. There are two different kinds of OADMs: in the static case, the dropped and added wavelengths are fixed, while in a flexible OADM the wavelengths can be chosen by reconfigurable transmitters and receivers and a special coupling scheme.

Optical cross-connect — An optical cross-connect is a more advanced OADM which allows interaction between fibers with multiple inputs and outputs. Switching between ingress and egress is controlled by the control plane or management plane. It is also possible to drop and add wavelengths; it has the same effect as another input or output.

Optical transparency — A transparent optical channel is practically defined by the absence of O/E (or E/O) conversion within the optical channel.

An optical network is transparent when any end-to-end optical channel belonging to the network is transparent.

Optical transport network — The optical transport network is defined by a layered structure comprising the optical channel, optical multiplex section, and optical transmission section layer networks. Motivation for this three-layer structure is reported in [G.872].

Outage — See “Fault.”

Path reversibility — Path reversibility is the ability to return to the working path from the protection path after a failure is solved.

Peer to peer — “Peer-to-peer” applications are applications that exchange and use resources (CPU cycles, storage capacity, contents, etc.) distributed in hosts with access to telecommunication networks.

An application is peer to peer if the answer to both of these questions is yes:

1. Does it treat variable connectivity and temporary network addresses as the norm?
2. Does it give the nodes at the edges of the network significant autonomy?

Per hop behavior — The externally observable behavior of a packet at a DiffServ-router.

Performance management — (i) A set of functions that evaluate and report the behavior of telecommunications equipment and the effectiveness of the network or network element and (ii) a set of various subfunctions, such as gathering statistical information, maintaining and examining historical logs, determining system performance under natural and artificial conditions, and altering system modes of operation.

Permanent connection A permanent connection is a connection type that is provisioned by the management system [G.8080].

Point of presence The access network for dial-up access connects the subscriber to the Internet operator's connection point. It usually is a demarcation or interface point between communications entities. A point of presence is also a location where a long-distance carrier could terminate services and provide connections into a local telephone network.

Presently it is a term often used by ISPs with relation to Internet exchange points and collocation centers. It is common, that medium or large ISPs have many points of presence [Ericsson].

Predicted information Predicted information would be obtained from the "history" of the more recent network states by applying certain heuristics.

Prediction-based routing The routing inaccuracy and signaling overhead problems can be tackled by applying a new prediction-based routing approach. This routing algorithm would use both the "usual" (and perhaps inaccurate) network state information and the "predicted" routing information to compute paths.

Primary failure See "Root failure."

Primary path See "Working path."

Protection A recovery mechanism that makes use of pre-assigned capacity between nodes to protect working paths. The protection path has the exclusive use/possession over its capacity and thus intermediate nodes on this path are not involved in the signaling process at the time of a failure. Both dedicated protection and shared protection exist.

Protection path A protection path is a path that a recovery mechanism can use to recover traffic on a working path affected by a failure.

Synonyms: alternative path, backup path, secondary path, recovery path.

Provider-provisioned VPN See "Virtual private LAN services."

Quality of protection The term QoP is used to describe all aspects of the service related to protection and restoration. QoP parameters may encompass: availability of protection, criteria of service availability (determine when to start protection/restoration process), recovery time, recovery coverage, redundancy, quality of recovery path/link. It should be noted that QoP term is also widely used for security issues and then has completely different meaning.

Quality of service "QoS is the complete set of service requirements to be met by each layer of the communication stack to satisfy the overall user requirements." It is necessary that all these requirements are adjusted to each other to cooperate in the most efficient way and to provide the optimal overall service. The service parameters required by the users are often fixed in a so-called SLA and are the basis for the decision on the service requirements of the underlying communication stack.

Recovery — Recovery of a network is a general term that can be used to group all the different possibilities to recover the network after a failure. Recovery mechanisms can be categorized in many ways:

- Path computation: preplanned versus dynamic and distributed versus centralized recovery path computation
- Path establishment/resource assignment: protection (pre-established recovery path) versus restoration (dynamic establishment of recovery path) and distributed versus centralized path establishment.
- Resource usage: shared versus dedicated recovery.
- Recovery extent: global versus local versus segment recovery.
- Topology: ring-based versus mesh-based.
- Path reversibility: revertive versus nonrevertive mode.

Recovery domain — A set of nodes and spans over which one or more recovery schemes are provided. A recovery domain served by one single recovery scheme is referred to as a "single recovery domain," while a recovery domain served by multiple recovery schemes is referred to as a "multirecovery domain" [RFC4427].

Recovery extent — Refers to the segment that is delimited by the recovery head end and the recovery tail end. A distinction can be made between end-to-end (global), segment and local (span) recovery [RFC4427]: i.e., on an end-to-end, on subnetwork connection, or on an link-per-link/node-per-node basis respectively.

Synonym: recovery scope.

Recovery head end — The network node upstream from the failure where the affected traffic is redirected onto the recovery path.

Recovery path — See "Protection path."

Recovery scope — See "Recovery extent."

Recovery tail end — The network node downstream from the failure where the recovery path again joins the working path and thus from where the traffic can continue along its original route.

Reference point — An architectural component, which is formed by the binding between inputs and outputs of transport processing functions and/or transport entities [G.805].

Regeneration — The transmission of information over an optical network is hindered by the accumulation of impairments that need to be mitigated to maintain signal quality. It is recognized from a modeling viewpoint that these compensations need to be described in terms of processes. In particular, the description of processes involved in so-called 1R, 2R, and 3R regeneration are of interest. A transport function must be described in terms of the processes associated with the relevant adaptation and termination functions in each layer and a simple statement of 1R, 2R, or 3R regeneration is insufficient. However, because

1R, 2R, and 3R regeneration are commonly used terms, the following classification is provided as an aid to understanding them.

These forms of regeneration are composed of a combination of the following processes:

1. Equal amplification of all frequencies within the amplification bandwidth. There is no restriction upon client layers.
2. Amplification with different gain for frequencies within the amplification bandwidth. This could be applied to both single-channel and multichannel systems.
3. Dispersion compensation (phase distortion). This analog process can be applied in either single-channel or multichannel systems.
4. Noise suppression.
5. Digital reshaping (Schmitt trigger function) with no clock recovery. This is applicable to individual channels and can be used for different bit rates but is not transparent to line coding.
6. Complete regeneration of the pulse shape, including clock recovery and retiming within required jitter limits.

1R regeneration is described as any combination of processes (1) to (3). 2R regeneration is considered to be 1R regeneration together with processes (4) and (5), whilst 3R regeneration is considered to be 2R regeneration together with process (6).

An informal description of 1R regeneration is that 1R regeneration is based on analog techniques; 2R involves digital processing of the signal levels, while 3R regeneration also involves digital processing of the signal timing information [G.872].

Reliability
"$R(t_1, t_2)$"
The probability that an item can perform a required function under given conditions for a given time interval. *Notes*: (1) It is generally assumed that the item is in a state to perform this required function at the beginning of the time interval. (2) The term "reliability" is also used to denote the reliability performance quantified by this probability [IEC].

Reserved resource
The set of resources that can be shared by a set of paths (for example, backup paths). One of these paths obtains the exclusive use/possession over these resources when these resources are assigned to this path.

Restorability
The restorability under failure scenario f is defined as

$$R(f) = 1 - N_{\mathrm{nr}}(f)/N_{\mathrm{aff}}(f)$$

where $N_{\mathrm{nr}}(f)$ and $N_{\mathrm{aff}}(f)$ represent respectively the number of nonrestorable and affected service paths [ONM].

Restoration
A recovery mechanism which makes use of any capacity that is not pre-assigned, but possibly reserved to some backup paths.

Because the (possibly reserved) spare capacity still needs to be assigned exclusively to the appropriate backup paths at the time of a failure, the intermediate nodes on these backup paths have to be involved in the signaling process at the time of the failure. Some restoration mechanisms can be very flexible and rely on dynamic routing when the routes of the backup paths are not preplanned.

Risk domain — A group of arbitrarily connected nodes and links that together can provide certain like-capabilities (such as a chain of dedicated/shared protected links and nodes, or a ring forming nodes and links, or a protected Forwarding Adjacency).

Root failure — The basic, original failure occurring in the network (e.g., a cable cut). Synonym: primary failure.

Route management — Route management provides means to optimize the network resources utilization while providing the required QoS to the users by using a set of different routing strategies in a multilayer architecture.

Routing — The process of determining and prescribing the path or method to be used for establishing connections or forwarding messages.

The routing decision can be made in a centralized way (using a common algorithm for the whole network), therefore facilitating the achievement of a consistent view of the network or in a distributed way (based on the particular network view of each node).

The routing process should be handled differently depending on the scope of the path decision. Hence, intradomain routing (inside a single administrative domain) and interdomain routing (among different administrative domains) tackle the routing problem differently.

Routing inaccuracy problem — This problem consists of the negative effects on the global network performance produced by path selections made using inaccurate network state information.

It has been widely demonstrated that global network performance is hugely sensitive to routing paths set up with inaccurate network state information. Triggering policies used to update the routing information constitute one of the causes of this inaccuracy problem. These triggering policies indeed are used to reduce the signaling overhead.

Saturation — The condition in which a link of a network has reached its maximum traffic-handling capacity.

Secondary failure — A failure or symptom that is caused by another failure (the root failure).

Secondary path — See "Protection path."

Security management — The set of functions (i) that protects telecommunications networks and systems from unauthorized access by persons, acts, or influences and (ii) that includes many subfunctions, such as creating, deleting, and controlling security services and mechanisms; distributing security-relevant information; reporting security-relevant events; controlling the distribution of cryptographic keying material; and authorizing subscriber access, rights, and privileges.

Service access point — A service access point is a physical point at which a circuit may be accessed. In an Open Systems Interconnection (OSI) layer, a service access point is a point at which a designated service may be obtained [Eurescom].

Service level agreement — In compliance with the ITU definition, SLA is "a negotiated agreement between a customer and the service provider on levels of service characteristics and the associated set of metrics. The content of SLA varies depending on the service offering and includes the attributes required for the negotiated agreement" [Y.1241]. The IETF defines SLA in a similar way as "a service contract between a customer and a service provider that specifies the forwarding service a customer should receive" [RFC2475]. The SLA may be in the form of a document containing names of the parties signing the contract. It should encompass all parameters defining the service and/or class of the service provided, the SLA expiry date, specification of the way of measuring service quality and other parameters used to assess whether the service complies with the SLA. It may also include an agreement on form and frequency of delivering the report on service usage. It includes also billing options and penalties for breaking the contract by the service provider as well as by the customer. A technical part of SLA is the service level specification (SLS).

Service level specification — The notion of SLS was introduced to separate the technical part of the contract from the SLA. It is defined as "a set of parameters and their values which together define the service offered to a traffic" [RFC3260]. In other words, it specifies a set of values of network parameters related to a particular service. The SLS consists of a selected set of QoS, GoS, and QoP parameters that are key for service definition. This set of parameters differentiates a particular service from other services and constitutes a basis for the price differentiation of service. The SLS parameters may be categorized as quantitative (such as bandwidth, delay, BER) or qualitative (such as provision (or not) of protection path, confidentiality, etc.).

Service provider — Authority that offers application service. Service providers exploit network services in order to provide added value application services to users. A service provider is an entity that provides services to its service subscribers on a contractual basis and who is responsible for the services offered. The same entity may act as both a network operator and a service provider [Eurescom].

Shared protection — In shared protection, M protection entities protect (and thus are shared between) N working entities (M: N), where N is strictly larger than M.

Shared risk group — Represents the risk domains' capabilities and other parameters, which assist in computing diverse paths through the domain and in assessing the risk associated with the risk domain.

Shared risk link groups address only risks associated with the links (physical and logical) and locations within the risk domain, whereas shared risk groups contains nodes and other topological information in addition to links.

Shared risk link group — The set of links sharing a common physical resource. (IETF ipo-framework [RFC3717].)

Signaling plane — Part of the control plane responsible for call/connection control. The routing part is excluded from the signaling plane.

Single-layer traffic engineering — Single-layer traffic engineering is the sum of all functions in a network layer that are applied to provide QoS.

SLS in optical domain — The definition of SLS and SLA for the optical domain is a new issue. Specifies SLS parameters for the optical domain. See "Service level specification" and [CommMag2].

Soft-permanent connection — A soft-permanent connection is a user-to-user connection whereby the user-to-network portion of the end-to-end connection is established by the network management system as a permanent connection. The network portion of the end-to-end connection is established as a switched connection using the control plane. In the network portion of the connection, the management plane initiates the request and the control plane establishes the connection [G.8080].

Spare resource — The set of all the resources available in the network (i.e., that are not currently in use).

Statistical multiplexing — The fact that asynchronous multiplexing can yield the same QoS with a smaller total bit-rate than the sum of the individual peak rates, due to the stochastic properties of traffic, is called (statistical) multiplexing gain.

Storage area network — A storage area network (SAN) is a network which is designed to attach computer storage devices such as disk array controllers and tape silos to servers. SANs are increasingly considered the best way of storing large volumes of data [FreeDictionary].

Switched connection — A switched connection is a connection established between connection end-points as a result of an end-user's request, using a signaling/control plane. It involves the dynamic exchange of signaling information between signaling elements within the control plane(s) [G.8080].

Terminal (equipment) — A device used at the network edge, where all ingress and egress traffic is terminated.

Traffic engineering — Traffic engineering is concerned with performance (e.g., QoS) and resource optimization (e.g., for costs reduction) of networks in response to dynamic traffic demands and other stimuli, like node and link failures.

In a multilayer network (e.g., IP over ASON/GMPLS) traffic engineering is referred as MTE and it comprises traffic management and capacity management.

Traffic management ensures that traffic performances are maximized both in normal and fault conditions at minimum cost.

Traffic management means moving the traffic where the bandwidth is.

According to [G.805] functional modeling, traffic management is an intra-layer process.

(Traffic management is based on routing methods, QoS resource management methods, routing tables information management.)

Capacity management ensures that the circuit connections are provisioned to meet performance objectives at minimum cost.

Capacity management is moving the bandwidth where the traffic is (or is expected to be).

According to [G.805] functional modeling, capacity management is an inter-layer process.

(Capacity management is based on dynamic setup and tear down of switched connections.)

Traffic modeling — This consists of capturing one or more traffic characteristics (such as connection interarrival time, connection holding time, packet interarrival time, packet size, etc.) so that they can be used in an analytical or empirical model (traffic model) in order to artificially generate that traffic.

Traffic profile — A repetitive count of the amount of traffic offered to/carried by a network, which is done each period of time (e.g., each second or minute) during a certain time (e.g., an hour, a day, a week, a month, or a year), and that is one of the possible metrics to characterize the traffic.

Traffic shaping — The process of delaying packets within a traffic stream to make the stream conformant to some defined traffic profile.

Trail — A "transport entity" which consists of an associated pair of "unidirectional trails" capable of simultaneously transferring information in opposite directions between their respective inputs and outputs [G.805].

Trail termination — A transport processing function which inserts/extracts trail overhead information for the purpose of validating and supervising the connection.

Translucent network — Infrastructure where a signal is made to propagate in the optical domain as long as possible with respect to optical transmission impairments and signal adaptation or wavelength conversion requirements.

Transparent LAN — A service enabling transparent interconnection of remote LAN networks. Flexible, high-speed service that hides the complexity of WAN technology from enterprise users. It solves customers' current problems and establishes the network service provider as a forward-thinking solution provider and business partner; provides native LAN speed interconnection of corporate sites within a local metropolitan area.

Transparent LAN service — A transparent LAN service is a service provided by a provider to interconnect LANs transparently. The WAN lying between those LANs is not considered.

Transparent network — Infrastructure where the transmission of the optical signal is independent of the specific characteristics (digital or analog type, modulation scheme, signal format, bit rate) of the actual data to be transported through the optical layer.

Transparent wavelength interchange cross-connect — Wavelength interchanging is done only in the optical domain; there is no conversion to the electric domain; that is, the node is transparent from the point of view of a single wavelength.

Transparent wavelength-selective cross connect — Switching in a transparent wavelength-selective cross-connect (WSXC) is done in the optical domain; no O/E conversion is needed.

Transponder — A full duplex device, which receives and amplifies the incoming signal and transmits an outgoing signal.

Transport interface — The transport interface connects the node to the core/metro network. It describes the physical requirements like wavelength agility, tolerance to optical power fluctuations, and so on.

Transport network — Transport network: the functional resources of the network which convey user information between locations [G.805].

Transport plane — The transport plane provides bidirectional or unidirectional transfer of user information, from one location to another. It can also provide transfer of some control and network management information. The transport plane is layered; it is equivalent to the transport network defined in [G.805] ([G.8080]).

Type of service — This is sometimes wrongly used to mean CoS; type of serviceis a 1-byte field in the IP header.

User-to-network interface — A bidirectional signaling interface between service requester and service provider control plane entities [G.8080].

Virtual leased line — Virtual leased line services can be broadly classified into two types:

- a service that is primarily focused on transporting IP, known as IP VLL;
- a service that can be used to transport any Layer 2 technology, including Ethernet, Frame Relay, ATM, PPP.

The primary purpose of this service type is to transport IP in a point-to-point manner. Connectivity between the edge device and provider router is, therefore, always an IP connection. This IP trunk may emulate a voice trunk or may simply transport data between a backup site and a data center. In each case, the QoS requirements are distinct. In the former, tight QoS guarantees are needed, while in the latter, loose QoS guarantees are required [Eurescom].

Virtual private LAN services — Virtual private LAN services (VPLS) is an emerging standard whose objective is to connect two or more sites over a managed IP/MPLS networks. In difference to TLAN, VPLS enables a smooth handoff between a user's Ethernet traffic and the IP core (either public or private).

Virtual private network — A group of sites that, as the result of a set of administrative policies, are able to communicate with each other over a shared backbone network.

Virtual private network — Network that shares resources with other virtual private networks (VPNs) but provides privacy, confidentiality, integrity, and possibly also separation of capacity. Several means exist to implement VPNs; for example, circuit switching, connection-oriented packet switching, or a shared connectionless IP network infrastructure overlaid by technologies such as network address translation, firewalls, tunneling protocols, security gateways and network-based routing solutions.

A VPN can be contrasted with a system of owned or leased lines that can only be used by one company. The idea of the VPN is to give the company the same capabilities at much lower cost by using the shared public infrastructure rather than a private one.

A specific site may be a member of one or more VPNs; that is, specific sites can have additional capabilities that allow them access to sites outside the group, and/or to be accessed from sites outside the group [Eurescom].

Virtual wavelength path — A path across an optical transport network where wavelength conversion occurs between the end points.

Wavelength banding — Or "hierarchical WDM," a grouping of all wavelengths belonging to a dense WDM comb into sub-bands that have different routing and adaptive compensation in the network.

Wavelength conversion — An optical network functionality changing the specific wavelength assigned to the optical channel. Wavelength conversion is performed by the OMS/OCh adaptation function. At the OCh layer the wavelength is undefined. The OMS/OCh_A source assigns a specific wavelength to the optical channel [G.872].

Wavelength converter — This functionality allows converting a wavelength to another by using a fully optical or O/E/O conversion technique and is optional for transparent nodes to reduce blocking.

Wavelength interchange cross-connect — Wavelength interchange cross-connect (WIXC) has the same properties as a WSXC and additionally the possibility to convert the wavelength.

Wavelength path	A path across an optical transport network where the same wavelength is employed
Wavelength routing	Data routing from source to destination based on the wavelength of the signal. When wavelength conversion is not available, only "wavelength paths" are established; otherwise "virtual wavelength paths" may be used.
Wavelength-selective cross-connect	With a WSXC each wavelength on each fiber can be switched through a non-blocking fabric to each output fiber. The wavelength remains untouched.
Wavelength-routed network	A network making use of wavelength-routing.
Wide area network	A WAN is a network that spans a large area (regional or multiregional). It may encircle more than 100 km and use narrowband links with WAN protocols (like frame relay, ISDN, PPP).
Wideband	Describes a class of telecommunications access services, such as ISDN, offering a data rate greater than narrowband services but smaller than broadband services. In this book wideband is defined as a data stream service with throughput higher than 64 kbit/s but less than 2 Mbit/s.
Working path	A path that carries traffic that can be recovered by a network recovery mechanism. Synonyms: primary path.
WSXC/WIXC hybrid or L-WIXC	The number of wavelengths to be converted is limited. It is only possible to convert a subset of all wavelengths of the input or output of a node.
WSXC/WIXC-LS	L-WIXC with converter sharing on per-link basis.
WSXC/WIXC-NS	L-WIXC with converter sharing on per-node basis.

References

[CommMag1] Gozdecki J., Jajszczyk A., Stankiewicz R. Quality of Service Terminology in IP Networks, *IEEE Communications Magazine*, vol. 41, no. 3, March 2003, pp. 153–159.

[CommMag2] W. Fawaz, B. Daheb, O. Audouin, B. Berde, M. Vigoureux, M. Du-Pond, G. Pujolle, "Service Level Agreement and Provisioning in Optical Networks", *IEEE Communications Magazine*, vol. 42, no. 1, January 2004, pp. 36–43.

[E.493] ITU-T Recommendation E.493, "GRADE OF SERVICE (GOS) MONITORING", February 1996.

[E.720] ITU-T Recommendation E.720, "ISDN GRADE OF SERVICE CONCEPT", 1993.

[E.721] ITU-T Recommendation E.721, "Network grade of service parameters and target values for circuit-switched services in the evolving ISDN", May 1999.

[E.771] ITU-T Recommendation E.771, "Network grade of service parameters and target values for circuit-switched public land mobile services", October 1996.

[E.800] ITU-T Recommendation E.800: Terms and definitions related to quality of service and network performance including dependability.

[Ericsson] http://www.ericsson.com/

[Eurescom] http://www.eurescom.de/
[FCC] http://www.fcc.gov/
[FreeDirectionary] http://www.thefreedictionary.com/
[G.7713] ITU-T Recommendation G.7713 /Y.1704: Distributed Call and Connection Management (DCM), May 2006.
[G.805] ITU-T Recommendation G.805: Generic functional architecture of transport networks, March 2000.
[G.808.1] ITU-T Recommendation G808.1: Generic protection switching - Linear trail and sub-network protection, March 2006.
[G.8080] ITU-T Recommendation G.8080/Y.1304: Architecture for the automatically switched optical network (ASON), June 2006.
[G.841] ITU-T Recommendation G.841, "Types and characteristics of SDH network protection architectures", October 1998.
[G.872] ITU-T Recommendation G.872: Architecture of optical transport networks, November 2001.
[I.380] ITU-T Recommendation I.380/Y.1540: Internet protocol data communication service - IP packet transfer and availability performance parameters, November 2007.
[IEC] IEC International Standard 50 (191), International Electrotechnical Vocabulary, Chapter 191: Dependability and Quality of Service, First Edition, 1990-12.
[ITG] M. Barry, S. Bodamer, J. Späth, M. Jäger, R. Hülsermann, "A Classification Model for Network Survivability Mechanisms", to appear in Proceedings of the 5th ITG Workshop on Photonic Networks ", Leipzig, May 2004.
[Linktionary1] http://www.linktionary.com/f/flow.html
[Linktionary2] http://www.linktionary.com/l/link_state_routing.html
[M.3010] ITU-T Recommendation M.3010: Principles for a telecommunications management network, February 2000.
[ONM] J. Doucette, M. Clouqueur, W. D. Grover, "On the availability and capacity requirements of shared backup path-protected networks,"*Optical Networks Magazine*, November/December 2003, pp. 29–44.
[RFC2370] R. Coltun, "The OSPF Opaque LSA Option", *IETF RFC 2370*, July 1998.
[RFC2386] E. Crawley, R. Nair, B. Rajagoplan, H. Sandick, "A Framework for QoS-based Routing in the Internet", *IETF RFC 2386*, August 1998.
[RFC2475] S. Blake *et al.*, "An Architecture for Differentiated Services", *IETF RFC 2475*, December 1998.
[RFC3260] D. Grossman, "New Terminology and Clarifications for Diffserv", *IETF RFC 3260*, April 2002.
[RFC3272] D. Awduche *et al.*, "Overview and Principles of Internet Traffic Engineering", *IETF RFC 3272*, May 2002.
[RFC3717] B. Rajagopalan, J. Luciani, D. Awduche, "IP over Optical Networks: A Framework", *IETF RFC 3717*, March 2004.
[RFC44271] E. Mannie, D. Papadimitriou (Eds.), "Recovery (Protection and Restoration) Terminology for Generalized Multi-Protocol Label Switching (GMPLS), "*IETF RFC 4427*, March 2006.
[WirelessReview] http://www.wirelessreview.com/ar/wireless_network_fault_management
[Wu] T.-H. Wu, N. Yoshikai, "*ATM transport and network integrity*", Academic Press, 1997.
[Y.1241] ITU-T Recommendation Y.1241, "Support of IP-based Services Using IP Transfer Capabilities", March 2001.
[X.200] ITU-T Recommendation X.200: Information technology - Open Systems Interconnection - Basic Reference Model: The basic model, July 1994.
[X.701] ITU-T Recommendation X.701: Information technology - Open Systems Interconnection - Systems management overview, August 1997.

Acronyms

Abbreviation	Description
2R	reamplification and reshaping
3R	reamplification, reshaping, and retiming
4R	reamplification, reshaping, retiming, and whole optical spectrum recovery
ABC	automatic bandwidth control
ABR	area border router
AC	application category
AC	attachment circuit
ADC	analog-to-digital converter
ADM	add/drop multiplexer
AGC	automatic gain control
AIS	alarm indication system (or signal?)
AOLS	all-optical label swapper
AOM	acusto-optic modulator
AOWC	all-optical wavelength converter
APS	automatic protection switching
ARP	address resolution protocol
AS	autonomous system
ASE	amplified spontaneous emission
ASIC	application specific IC
ASK	amplitude shift keying
ASON	automatically switched optical network (ITU-T G.8080)
ASP	access service provider
ASP	application service provider
ASTN	automatic switched transport network
ATM	asynchronous transfer mode
AWG	arrayed waveguide grating
B&S	broadcast & select
BDI	backward defect indication

Core and Metro Networks Edited by Alexandros Stavdas

BER	bit error rate
BERT	bit error rate tester
BFD	bidirectional forwarding (failure) detection
BGP	border gateway protocol
BiCMOS	bipolar complementary metal oxide semiconductor
BMT	burst mode transceiver
BoD	bandwidth on demand
BR	border router
CAPEX	capital expenditures
CC	control channel
CC	cross-connect
CCID	control channel ID
CD	chromatic dispersion
CDC	chromatic dispersion compensator
CDR	clock and data recovery
CE	customer (client) edge (device)
CLI	command line interface
CMEMS	compliant MEMS
CMISE	common management information service element
COPS	common open policy service/server
CORBA	common object request broker architecture
CoS	class of service
CP	control plane
CPE	customer premise equipment
CPM	channel power monitor
CPS	common pool survivability
CR-LDP	constraint-based routing – label distribution protocol
CS	carrier suppressed
CSP	content service provider
CSPF	constraint shortest path first (= constraint OSPF)
CW	continuous wave
DB	database
dB	decibel
DC	direct current
DCC	decision circuit
DCC	data communication channel
DCE	distributed computing environment
DCF	dispersion compensating fiber
DCM	dispersion compensating module
DCN	data communication network
DDR	double data rate
DEMUX	demultiplexer
DFB	distributed feedback laser
DFE	decision feedback equalizer
DGD	differential group delay
DLL	delay-locked loop

DOCS	dispersion overcompensation scheme
DOP	degree of polarization
DP	dedicated protection
DP	data plane
DPSK	differential phase shift keying
DQPSK	differential quadrature phase shift keying
DRA	distributed Raman amplifier
DRAM	dynamic RAM
DSF	dispersion-shifted fiber
DSL	digital subscriber line
DUCS	dispersion undercompensation scheme
DWDM	dense wavelength division multiplexing
DXC	digital cross-connect (see EXC)
E/O	electrical-to-optical conversion
E2E	end-to-end
EAM	electro-absorption modulator
EAS	electrical amplitude sampling
ECL	emitter coupled logic
EDC	electronic dispersion compensator
EDE	electronic dispersion equalization
EDFA	erbium-doped fiber amplifier
EDWA	erbium-doped waveguide amplifier
EFEC	enhanced forward error correction
eLSP	EXP inferred label-switched path
EML	element management layer
EMS	element management system
E-NNI	external NNI
EPL	Ethernet private line
EPS	equipment protection switching
ERO	explicit routing object
ESA	electrical spectrum analyzer
ESCON	enterprise system connection
Eth	Ethernet
EVC	Ethernet virtual circuit (MEF)
EVC	Ethernet virtual connection
EVPL	Ethernet virtual private line
EVPLAN	Ethernet virtual private LAN
EVPN	Ethernet virtual private network
EXC	electrical cross-connect (see DXC)
EXOR	EXclusive OR gate
FA	forwarding adjacency
FAS	frame alignment signal
FBG	fiber Bragg grating
FCAPS	fault, configuration, accounting, performance and security management
FCS	frame check sequence
FDI	forward defect indication

FDL	fiber delay line
FE	fast Ethernet
FEC	forward error correction
FEC	forwarding equivalent class
FF	flipflop
FFE	feed forward equalizer
FIFO	first in first out (in registers and buffers)
FOCS	full optical compensation scheme
FPGA	field programmable gate array
FR	frame relay
FSK	frequency shift keying
FSM	finite state machine
FSR	free spectral range
FT	transit frequency
FTP	file transfer protocol
FWM	four-wave mixing
FXC	fiber cross-connect
GaAs	gallium arsenide
GDR	group delay ripple
GE	gain equalizer
GFP	generic framing procedure
GFP-F	frame-mapped GFP
GFP-T	transparent GFP
GigE	gigabit Ethernet (GEth G(b)E)
GMPLS	generalized multiprotocol label switching
GoS	grade of service
GPRS	general packet radio service
GPS	global positioning system
GRE	generic routing encapsulation
GTE	gain tilt equalizer
GTM	gain tilt monitor
GVD	group velocity dispersion
HBT	hetero bipolar transistor
H-LSP	hierarchical-LSP
HICUM	high current bipolar compact transistor model
HO	higher order
HOM fiber	higher order mode fiber
HW	hardware
I/O	input/output
IANA	internet assigned numbers authority
IC	integrated circuit
ID	identifier (address)
IETF	Internet Engineering Task Force
IF	interface
IL	insertion loss
ILM	integrated laser-modulator

IM	intensity modulation
I-NNI	internal NNI
InP	indium phosphide
IP	Internet protocol
IPCC	IP control channel
IPG	Inter-packet gap
IPv4	Internet protocol version 4
IPv6	Internet protocol version 6
IrDI	interdomain interface
IRTF	Internet Research Task Force
ISCSI	Internet SCSI
ISDN	integrated services digital network
IS-IS	intermediate system to intermediate system (protocol)
ISO	International Standard Organization
ISP	Internet service provider
ITU	International Telecommunication Union
ITU-T	ITU – Telecommunication sector
L2	layer 2
L2SC	layer 2 switching capable
L2VPN	layer 2 virtual private network
L3	layer 3 (network layer)
LAN	local area network
LC	line card
LCAS	link capacity adjustment scheme
LDP	label distribution protocol
LH	long haul
LMP	link management protocol
LO	lower order
LOF	loss of frame
LOP	loss of pointer
LOS	loss of signal
LPDP	local policy decision point
LQ	liquid crystals
LRA	lumped Raman amplifier
LRD	long-range dependent
LSA	link state advertisement
LSP	label switched path
LSR	label switching router
MAC	medium/media access control
MAN	metropolitan area network
MCU	main control unit
MDT	mean down time
MEF	Metro Ethernet Forum
MEMS	micro-electro-mechanical system
MFAS	multiframe alignment signal
MIB	management information base

MLS	multilayer switch
MLSD	maximum likelihood sequence detection
MLSE	maximum likelihood sequence estimation
MM	multimode
MMI	multimode interference
MPLS	multiprotocol label switching
MRN	multiregion network
MS	multiplex section
MSA	multisource agreement
MSB	most significant bit
MSE	mean square error
MSOH	MS overhead
MSP	MS protection
MS-SPRing	multiplex section shared protection ring
MTBF	mean time between failure(s)
MTE	multilayer traffic engineering
MTTR	mean time to restore
MUT	mean up-time
MUX	multiplexer
MWSS	multiwavelength selective switch
MXU	multi-exchange unit
MZI	Mach–Zehnder interferometer
NAS	network-attached storage
NAT	network address translation
NCI	NE control interface
NDSF	non-dispersion-shifted fiber
NE	network element
NLE	nonlinear element
NM	network management
NMF	Network Management Forum, now renamed to TMF, Telemanagement Forum
NMI	NM interface
NMS	network management system
NNI	network(-to-)network interface (network node interface)
NP	network performance
NPL	network performance level
NPP	network performance parameter
NRZ	nonreturn to zero (signal format)
NS	(fiber) normalized section
NSAP	network service access point
NSP	native service processing
NZDSF	nonzero dispersion shifted fiber
O/E	optical to electrical conversion
O/E/O	optical–electrical–optical
O/O/O	all-optical (device or functionality)
OA	optical amplifier
OADM	optical add drop multiplexer

OAM	operation (administration) and maintenance
OAM&P	operations, administration, maintenance and provisioning
OAS	optical amplitude sampling
OBS	optical burst switching
OCC	optical cross-connect
OCh	optical channel
OCS	optical circuit switching
OCU	optical channel unit
ODU	optical data unit
ODXC	optical digital cross-connect
OFA	optical fiber amplifier
OH	overhead
OIF	optical internetworking forum
OLO	other local operator
OLR	optical line regenerator
OLT	optical line terminal
ONE	optical network element
OPA	optical performance analyzer
OPADM	optical packet add drop multiplexer
OPEX	operational expenditures
OPEX	operating expenses
OPM	optical performance monitoring
OPR	optical packet router
OPS	optical packet switching
OPU	optical payload unit
OQM	optical Q-meter
ORION	overspill routing in optical networks
OS	operating system
OSA	optical spectrum analyzer
OSAAR	optical signal to added ASE ration
OSC	optical supervisory channel
OSI	Open System Interconnection
O-SLS	SLS in optical domain
OSNR	optical signal-to-noise ratio
OSPF	open short path first (protocol)
OSQA	optical signal quality analyzer
OTDM	optical time-domain multiplexing
OTH	optical transport hierarchy
OTM	optical transport module
OTN	optical transport network
OTT	optical termination terminal
OTU	optical transport unit
OXC	optical cross-connect
P2P	peer to peer
PBM	policy based management
PCB	printed circuit board
PCC	path computation client

PCE	path computation element
PCF	photonic crystal fiber
PD	photodiode
PDB	per domain behavior
PDF	probability density function
PDFFA	praseodymium-doped fluoride fiber amplifier
PDG	polarization-dependent gain
PDH	plesiochronous digital hierarchy
PDL	polarization-dependent loss
PDP	policy decision point
PE	provider edge (device)
PEP	policy enforcement point
PHB	per hop behavior
PLL	phase-locked loop
PM-AM	phase-modulation to amplitude-modulation
PMD	polarization-mode dispersion
PMDC	polarization-mode dispersion compensator
POM	path overhead monitoring
PON	passive optical network
POP	point of presence
POS	packet over SONET
POTS	plain old telephony service
PPVPN	provider-provisioned virtual private network
PRBS	pseudorandom bit sequence
PSK	phase-shift keying
PSN	packet-switched network
PSP	principle state of polarization
PXC	photonic cross-connect
PW	pseudo-wire
Q-factor	quality factor of the optical signal converted into the electrical domain
QoP	quality of protection
QoR	quality of resilience
QoS	quality of service
RAM	random access memory
RC	route controller
RCA	root cause analyzer
RF	radio frequency
RFA	Raman fiber amplifier
RFC	request for comments (IETF)
ROADM	reconfigurable optical add/drop multiplexer
RPR	resilient packet ring
RRO	record route object
RSOH	regeneration section overhead
RSVP	resource reservation protocol
RSVP-TE	RSVP with traffic engineering extensions

RWA	routing and wavelength assignment
RX	receiver
RZ	return to zero (signal format)
SAN	storage area network
SAP	service access point
SAR	segmentation and reassembly
SBS	stimulated Brillouin scattering
SCN	signaling control network
SD	signal degrade
SDH	synchronous digital hierarchy
SF	signal fail
SFD	start frame delimiter
SFP	small form plugable
SiGe	silicon–germanium technology
SLA	service level agreement
SLM	single longitudinal mode (laser)
SLS	service level specification
S-LSP	LSP segment
SME	small and medium-sized enterprise
SMF	single mode fiber
SNCP	subnetwork connection protection
SNMP	simple network management protocol
SNR	signal-to-noise ratio
SOA	semiconductor optical amplifier
SOH	section overhead
SOHO	small office, home office
SOI	silicon on insulator
SONET	synchronous optical network
SOP	state of polarization
SOS	start-of-slot
SP	shared protection
SP	service provider
SPC	soft permanent connection
SPM	self-phase modulation
SPOF	single point of failure
SRD	short-range dependent
SRG	shared risk group
SRLG	shared risk link group
SRS	stimulated Raman scattering
SSB	single side band
SSM	space switching matrix
SSMF	standard single mode fiber
STE	single-layer traffic engineering
STM	synchronous transport module
STP	spanning tree protocol
SW	software

TCM	tandem connection monitoring
TCP	transmission control protocol
TDC	tunable dispersion compensator
TDCM	tunable dispersion compensating module
TDM	time-division multiplexing
TE	traffic engineering
TFF	thin-film filter
TIM	trace identifier mismatch
TLAN	transparent LAN
TLS	transparent LAN service
TLV	type-length-value format
TMF	Telemanagement Forum, formerly NMF, Network Management Forum
TNA	transport network assigned address
TNE	transport network element
TNMS	transport network management system
TOC	table of contents
ToS	type of service
TP	termination point
TSI	time slot interchanging
TWC	tunable wavelength converter
TX	transmitter
UDP	user datagram protocol
ULH	ultra-long-haul
UMTS	universal mobile telecommunications system
UNI	user-to-network interface
UNI-C	UNI signaling controller – client side (edge node)
UNI-N	UNI signaling controller – network side (edge core node)
UTC	coordinated universal time
VAS	value added services
VC	virtual channel (ATM)
VC	virtual concatenation (SONET/SDH)
VC	virtual circuit
VC	virtual container (SONET/SDH)
VCAT	virtual concatenation
VCG	virtual concatenation group
VE	Viterbi equalizer
VFI	VPN forwarding instance
VIPA	virtually imaged phase array
VLAN	virtual local area network
VLH	very long-haul
VLL	virtual leased line
VOA	variable optical attenuator
VPLS	virtual private LAN service
VPN	virtual private network
VSB	vestigial side band

WAN	wide area network
WB	wavelength blocker
WDM	wavelength division multiplexing
WIXC	wavelength interchange cross-connect
WLAN	wireless local area network
WRN	wavelength-routed network
WSS	wavelength selective switch
WSXC	wavelength selective cross connect
WXC	waveband cross-connect
XC	cross-connect
XGM	cross-gain modulation
XML	extensible markup language
XPM	cross-phase modulation
XRO	exclude route object

Index

Core and Metro Networks Edited by Alexandros Stavdas